*A. Castigliano*

# THE THEORY OF EQUILIBRIUM OF ELASTIC SYSTEMS AND ITS APPLICATIONS

*by*

CARLO ALBERTO PIO CASTIGLIANO

Translated from
*Théorie de l'Équilibre des Systèmes Élastiques et ses Applications*
by
EWART S. ANDREWS

With a New Introduction and
Biographical Portrait Section by

GUNHARD Æ. ORAVAS
*Chairman, Department of Civil Engineering and Engineering Mechanics*
*McMaster University*
*Hamilton, Ontario*

DOVER PUBLICATIONS, INC., NEW YORK

This Dover edition, first published in 1966, is an unabridged and corrected republication of the work originally published by Scott, Greenwood & Son, London, in 1919 under the title *Elastic Stresses in Structures*. This edition also contains a new Introduction and a biographical portrait section prepared especially for this edition by Gunhard Æ. Oravas.

*Library of Congress Catalog Card Number: 64-18843*

Manufactured in the United States of America
Dover Publications, Inc.
180 Varick Street
New York, N. Y. 10014

# INTRODUCTION TO THE DOVER EDITION

Castigliano's magisterial monograph on minimal principles of structural mechanics, *Théorie de l'équilibre des systèmes élastiques et ses applications*, has suffered only slightly from obsolescence with the passage of time as far as structural mechanics is concerned, while its educational value and its depth of study are still without parallel on this subject. Originally published by Editore Augusto Federico Negro of Turin in 1879, this work of great merit was translated into German by Emil Hauff in 1886 and into English by Ewart Sigmund Andrews in 1919. The latter translation, the better of the two, is the basis of the present edition.

This monograph represents in fact the lifework of a very ingenious engineering scientist; or perhaps it is more accurate to term it a half-lifework, since premature death took its author at the very pinnacle of his intellectual powers, thus inflicting a severe loss upon the engineering profession and the progress of structural mechanics.

In historical perspective, Castigliano has represented the forgotten giant among engineering scientists. He was no more than the inanimate name of a theorem in structural mechanics until Professor Stephen P. Timoshenko, in his *History of Strength of Materials* (1953), pointed out Castigliano's great services to the science of structural mechanics.

In this long-overdue reprint of Castigliano's major treatise, which may be considered to constitute a summary of the major part of his original researches, it is valuable to append a detailed description of his life and work and a brief historical review of energetical elastomechanics, a knowledge of which is essential for a proper assessment of Castigliano's contribution in the field of structural mechanics.

Castigliano's life constitutes a classic example of the insatiable desire for learning of an underprivileged but proud youth in the face of severe penury.

Carlo Alberto Pio Castigliano was born on November 8, 1847, in Asti, Province of Asti, Italy, the only child of Giovanni Michele Castigliano and Orsola Maria Cerrato, both of whom were of the humblest station in life. When Alberto was a few years old, his father died. Later on his mother married Luigi Mozzi, but his stepfather too was from a very humble background and Alberto's fortunes did not take a turn for the better except, perhaps, for the fact that his stepfather encouraged and supported, through great sacrifices, Alberto's aspiration to procure an education. When his stepfather, who was advanced in years, suffered ill health for quite some time, Castigliano's mother supported the family (by selling fruit) with a self-sacrificing courage which Alberto never permitted himself to forget.

When Alberto graduated from elementary school, where he had been an outstanding student, he was fortunate to be able to procure the financial support of wealthy citizens in order to attend the Istituto Industriale, Sezione di Meccanica e Costruzioni (later renamed Istituto Tecnico e per Geometri Statale, "Giovanni A. Giobert") in Asti in 1864–1865. He graduated with honors. Subsequently he attended the fourth class in Sezione di Meccanica e Costruzioni at the Reale Istituto Industriale e Professionale in Turin from December, 1865, until his graduation in July, 1866. Being financially unable to continue his education for an engineering degree, he brilliantly completed a special three-month course sponsored by the government at the Reale Museo Industriale in Turin, earning the *diploma di professore di meccanica* on October 26, 1866, thus allowing the fulfillment of his initial desire to become a teacher in the newly established chain of technical schools.

At the age of nineteen Castigliano won, in a public contest sponsored by ministerial decree, an instructor's position in the Department of Applied Mechanics, Mechanical Engineering and Construction at the Reale Istituto Tecnico in Terni. His mother had again been widowed, and Castigliano supported her from his very meagre salary. He was relieved from his post in response to his request in 1870, and subsequently, by the decree of the Royal Prefect in Turin in November of that year, he was transferred to instruct at the all-male Reale Collegio Convitto Nazionale in Turin. At this time he passed the entrance examinations of the Faculty of Pure Mathematics at the Reale Università degli Studi in Turin, becoming a pupil of the famous Italian mathematician and jurist, Angelo Genocchi (1817–1889), one of the founders of the Italian analytical school of mathematics.

With a brilliant performance unparalleled in the annals of the university, Castigliano was able to pass in a single year all required examinations in pure mathematics, normally a three-year course of study. This was made possible by a special authorization from the university. He obtained his *diploma di licenza in matematiche pure* on November 11, 1871. During this year Castigliano earned his sustenance by giving private lessons and by translating foreign scientific books, mainly German and French, into Italian. The superhuman effort required for this feat apparently affected Castigliano's health, and paved the way for his premature demise.

Upon graduation from the university, Castigliano entered into studies of engineering at the Reale Scuola d'Applicazione degli Ingegneri in Turin, and graduated with *laurea di ingegneri civile* in two years from the normally five-year course on September 30, 1873, after earning enduring respect and admiration as a brilliant scholar from his schoolmates and the professorial staff. One of Castigliano's closest friends and competitors for scholastic honors at the Reale Scuola d'Applicazione degli Ingegneri was Valentino Cerruti (1850–1909), who later had an important influence on Castigliano's effort to perfect and generalize his principle of the work of deformation in structural mechanics. Cerruti also gained renown in his own right, and had his name immortalized in 1882 by his solution of the classic problem of a tangential force impressed upon the boundary of a semi-infinite elastic continuum, now known as "the Cerruti Problem."

Both Castigliano and Cerruti had chosen the analysis of statically indeterminate frameworks as the topic of their laureate theses. Castigliano's thesis was entitled *Intorno ai sistemi elastici* (About elastic systems) and Cerruti's was named *Sui sistemi elastici articolati* (On articulated elastic systems). Cerruti employed the classical geometric method of displacement in the solution of his articulated framed structure, while Castigliano undertook to develop the abstract method of work in the solution of statically indeterminate linear elastic systems. It is of interest here to relate how Castigliano came to choose this difficult principle, which had resisted the efforts of a number of outstanding scientists to discover its general proof, as the subject of his *dissertazione di laurea*.

Count Luigi Federigo Menabrea, Marquis di Valdora (1809–1896), a famous general, military engineer, and statesman of the Italian Risorgimento, had published two treatises, in 1858 and 1867, in which he announced his general Principle of Elasticity based on an extremal consideration of the quantity of the internal work of deformation, but had failed in his attempts to give a logical general proof for this principle and consequently his work drew criticism from his countrymen; especially severe critiques originated from General Emilio Francesco Sabbia (1838–1914), professor at the Accademia Militaire di Torino, and Angelo Genocchi. The validity of the Menabrea Theorem became a popular subject of discussion in Italian scientific and engineering circles, but since no valid general proof to this theorem had been given very few engineers and scientists took it seriously.

Castigliano had witnessed the presentation, at the Società degli Ingegneri e degli Industriali di Torino, of a memoir by one of his professors, Giovanni Sacheri, entitled "Estensione del principio di elasticità ad un qualsiasi sistema articolato complesso e non equilibrato, sua applicazione al calculo di stabilità delle Centine poligonali della tettoia principale nella stazione di Arezzo" (published in *Atti della Società degli Ingegneri e degli Industriali di Torino*, Anno VI, 1872, pp. 68–116), dealing with the extension of the Menabrea Theorem and its application to the stress analysis of the great arched truss roof over the railroad station in Arezzo.

Early in 1873 Castigliano, having already considered other methods for his researches on equilibrium of elastic systems which would have led to certain and exact results, finally decided to accept the challenge posed by the Menabrea Theorem and consider the solution of indeterminate elastic structural frameworks by means of work consideration. What Castigliano had in mind was to find out, once and for all, whether the minimal criterion of analysis of indeterminate trusses propounded by Menabrea was a tenable proposition by considering a number of revealing examples, and, if it proved to be valid, to attempt to find a general and rigorous proof for it. Castigliano was successful in demonstrating the validity of such a principle, which he called *principio del minimo lavoro* (principle of least work), a rather unfortunate terminology, first coined by Menabrea, for the method he actually employed.

In his disseration Castigliano considered only elastic systems which were internally redundant yet externally statically determinate, and gave a proof by reducing the minimum internal work principle into an algebraic form. In his proof he extremized the quantity of internal work of deformation of the articulated elastic system augmented by the constraint conditions in the form of equilibrium equations multiplied by unknown Lagrangian multipliers, and showed that the resulting equations were identical to the equations given by the classical geometric method. Castigliano demonstrated that the extremum of this augmented function yields the compatibility conditions, and that the Lagrangian multipliers are identical with the displacement of the joints of the articulated structure. His approach to the elasticity of solids was based on the molecular hypothesis; he did not arrive at his theorem in the pristine form given in his thesis through the generality of Legendre and Green, for he made no reference to the concept of internal elastic potential and Euler-Legendre Transformation, but rather used the Principle of Conservation of Work due to General Lazare Nicolas Marguerite Carnot (1753–1823), General Jean Victor Poncelet (1788–1867), and Émile Paul Benoît Clapeyron (1799–1864). Castigliano also proved Clapeyron's Three Moment Equation by means of the least work theorem.

Both Castigliano and Cerruti defended their theses successfully, and subsequently graduated in late September of 1873. Their laureate theses were published by the Reale Scuola d'Applicazione degli Ingegneri through Editore Vincenzo Bona, Turin, in November of the same year. Castigliano's success in resolving some of the difficulties in the proof of the minimal theorem for statically indeterminate structures earned him considerable reputation and attention among the profession as well as with the scholars of the Reale Accademia di Torino and the Reale Accademia dei Lincei di Roma, the oldest and most revered learned society in Italy.

After Castigliano's graduation, Professor Prospero Richelmy, rector of the Reale Scuola d'Applicazione degli Ingegneri, recommended him for the position of resident engineer of maintenance with the Alba Division of the Strade Ferrate Alta Italia (Northern Italian Railroad), a post for which Cerruti also competed. Cerruti was rejected on the basis of his medical examination, but Castigliano was accepted on November 19, 1873, and remained with the railroad for the remainder of his life. After a transfer and promotion to sectional engineer of the Turin Division in April, 1874, he was promoted to the posts of chief sectional engineer in February, 1875, and chief inspector of maintenance in 1883. In July, 1884, he was to attain the position of chief engineer of the Principal Engineering Office of Maintenance and Construction of the railroad in Milan, one of the most prominent civil engineering posts in Italy.

After Castigliano entered the engineering profession, his researches consisted of extending and refining the principal idea of his thesis, which he called *teorema del minimo lavoro di deformazione* (theorem of the least work of deformation). Castigliano's further studies on the subject of his thesis were presented by Professor Giovanni Curioni to the Reale Accademia delle Scienze di Torino in the form of two memoirs entitled "Intorno all'equilibrio dei sistemi elastici" (About the equilibrium of elastic systems) and "Nuova teoria intorno all'equilibrio dei sistemi elastici" (New theory about the equilibrium of elastic systems), subsequently published in *Atti della Reale Accademia delle Scienze di Torino*, Volumes X and XI, respectively, Paravia, Turin, 1875.

In the first memoir Castigliano extended his considerations and proved that the derivative of the quantity of internal work, with respect to the applied force, yields the displacement component in the direction of the force. By means of this theorem he considered externally redundant structures with immobile supports.

In the second memoir Castigliano proposed another principle, consisting of two parts, which he called *teorema delle derivate del lavoro di deformazione* (theorem of the derivatives of the work of deformation) and gave a number of important applications for his theorem. "Parte Seconda" had already appeared in the first memoir. "Parte Prima" stated that the derivative of the internal work of deformation, with respect to the displacement component in the direction of the force, yields the force.

Castigliano expressed the further opinion that Menabrea's Theorem or Principle of Elasticity was nothing more than a corollary and special case of the "Parte Seconda" of his theorem of the derivative of the work of deformation, and he maintained this opinion for the rest of his life. Since Castigliano considered only linearly elastic materials, and his derivations were of a formal mathematical nature, he failed to observe that the practically important "Parte Seconda" did not deal conceptually with internal work, but rather with a mathematical quantity possessing the dimensions of work but otherwise having no physical meaning. This is also the case with the Menabrea Theorem. ("Parte Prima," in contrast, deals with the derivative of the true internal work.) Thus Castigliano overlooked the fact that, even though both his and Menabrea's quantities possess dimensions of work for linear elastic materials, and in spite of the fact that in this particular case they possess an identical mathematical form, they are not in fact the same conceptual quantities.

Castigliano also proved, by means of the "Parte Seconda" of his theorem of differential coefficients of internal work, that the derivative of internal work, with respect to a force of a set of equal and opposite collinear forces, yields their relative displacement, and that the derivative of internal work, with respect to a couple, gives the component of rotation in the direction of the couple as a vector. Moreover, he proved the reciprocal law of displacement in an elementary manner from "Parte Seconda" of his theorem, basing it on the principle of superposition of Euler.

Shortly after the publication of Castigliano's memoirs, Menabrea published in *Atti della Reale Accademia dei Lincei* (Series II, Volume 2, 1875) a memoir, "Sulla determinazione delle tensioni e delle pressioni ne' sistemi elastici" (On the determination of tensions and pressures in elastic systems), in which he made a new attempt to prove his theorem. But the methods he used, as well as his algebraic formulas and symbols, were similar to those employed by Castigliano,* and, inadvertently, Menabrea committed the "sin" of misspelling Castigliano's name as "Castiglione."

In a letter dated March 11, 1875, to the civil engineer and mathematician Luigi Cremona (1830–1903), president of the Reale Accademia dei Lincei, Castigliano accused Menabrea of not properly recognizing his own (Castigliano's) priority in the proof of the theorem, and of reproducing his own (Castigliano's) algebraic formulas as his (Menabrea's) incontestable proof.†

* On the other hand, it might also be said that Castigliano, in his thesis, had used many symbols originated by Menabrea.

† Castigliano's attitude toward Menabrea had perhaps been influenced by his master, Angelo Genocchi, who intensely disliked Menabrea and had, in 1871–1872, carried on a polemic against him. The attack seems to have had its roots in the accusation by Felice Chiò (1813–1870), professor of mathematics at the Reale Università degli Studi in Turin and a colleague of Genocchi, that his 1843 memoir, "Recherche sur la série de Lagrange," pointing out fundamental errors in Lagrange's work related to the nonconvergency of his series, had been a victim of academic passion and discrimination at the hands of Menabrea. Chiò accused Menabrea, who, together with Carlo Ignazio Giulio (1803–1859), later rector of Università di Torino, was one of the referees to the memoir's acceptability, of catering to his former master, astronomer and mathematician, Baron Giovanni Antonio Amadeo Plana (1781–1864), president of the Accademia delle Scienze of Turin. Being a relative of Lagrange, Plana evidently resented Chiò's criticism of the work of his wife's uncle who was, and still is, considered to be the greatest Italian mathematician, and who founded the society that later became the Accademia delle Scienze di Torino. Chiò was forced to submit his memoir elsewhere, and it was eventually accepted by Augustin Louis Cauchy (1789–1857) and published by the Académie des Sciences in Paris. (*Continued on p. xii.*)

This letter stunned the Italian professional society since, despite the similarity of the formulas involved, Menabrea's proof, from a conceptual point of view, was notably different in its origin, substance, and method of procedure.

In answer to Castigliano's accusation, Menabrea pointed out in his letter of March 27, 1875, to Cremona that he had had no intention of opening a polemic with the publication of his memoir, and expressed the opinion that perhaps Castigliano did not fully understand the full scope and purpose of the memoir. He further stated that since Castigliano's 1873 thesis still seemed to have left many mathematicians in doubt as to certain aspects of the general validity of Menabrea's Theorem, he himself had tried to find different demonstrations for it which might be useful in dispelling their incredulity.

It is pertinent to point out that Castigliano's first memoir was presented to the Reale Accademia delle Scienze di Torino on January 24, 1875, and Menabrea's memoir was submitted to the Reale Accademia dei Lincei on January 3, 1875. Thus Menabrea could only have been aware of Castigliano's laureate thesis of 1873, which was rather restricted in scope concerning Menabrea's Theorem. Also significant is the fact that since 1869 Menabrea had been in possession of a correct general proof of his theorem, given to him by the French mathematician Joseph Louis François Bertrand (1822–1900). In answer to a critical note by Emilio Sabbia, Menabrea had published Bertrand's proof in an article entitled "Sul principio di elasticità—dilucidazioni" (On the principle of elasticity—explanation) in *Atti della Reale Accademia delle Scienze*, Turin, Volume V, May 1, 1870. Menabrea attempted to develop a different proof of his own, but again failed to use an acceptable procedure, even though he paid some attention to Bertrand's and Castigliano's methods. Castigliano, who certainly must have seen Bertrand's proof for Menabrea's Theorem, apparently did not recognize it as such, which plainly indicates that his way of thinking about this theorem was entirely different from that

(*Continued from p. xi.*)

In this case Chiò's criticism and Genocchi's resentment of Menabrea, a biographer of Lagrange, seem to have been justified. But Genocchi also criticized Menabrea's proof of his Principle of Elasticity as absurd, appropriated from Antoine Augustin Cournot (1801–1877), and not more general than Cournot's contribution; in this assessment Genocchi was in error. But the fact that he and his circle considered Menabrea capable of professional misconduct certainly must have influenced Castigliano's interpretation of Menabrea's intentions in writing his memoir.

of Menabrea and Bertrand. Under these circumstances, an honest difference of opinion was inevitable.

Cremona, president of the Reale Accademia dei Lincei and a noted civil engineer and mathematician, published his famous "Salomonical decision" on the dispute in the *Atti della Reale Accademia dei Lincei*, Series 2, Volume II, 1874–1875, pp. lix–lxvi. It appears that he did not go into the background of the theorem in any great depth, but rather made his assessment on the basis of the two letters and Castigliano's dissertation forwarded to him by Castigliano. Cremona stated that even though Castigliano had the honor of having produced a beautiful work, which was not completely free of *all* objections, his claims were insufficiently founded. In Cremona's opinion no one had the right to take from Menabrea the merit of having made a general theorem popular for common usage. Cremona expressed his conviction that this theorem certainly would receive more extensive applications in the future.

In Menabrea's opinion, priority for the proof of the Principle of Elasticity was a question of purely personal judgment, and science could have no special interest in it. In retrospect it appears that Menabrea was somewhat unjustly accused by Castigliano, even though at that time Castigliano's attitude was not unreasonable. The moderate tone of Menabrea's letter indicates that even he seems to have recognized this. He was certainly convinced of the professional honesty of Castigliano's position, and recognized that the root of the matter was in the widely divergent conceptual views held by the principals in this confrontation, a situation which, as he ruefully admitted, he had faced several times before with different adversaries. He retained no rancor against Castigliano for this painful incident, but rather recognized and even expressed esteem for the young man's genius by contributing, some years later, to a fund for the erection of a monument on the tomb of Castigliano.*

Since Castigliano's two memoirs of 1875 had not attracted the attention he felt they deserved, he finally composed the present treatise, *Théorie de l'équilibre des systèmes élastiques et ses applications*, which he wrote in French in order to capture a wider circle of readers, and to which he appended an extensive section on practical applications. In this book he considered structures subjected to initial as well as thermal stresses, and the theory of elasticity in cases where the elastic solid is considered to be a limiting case of a molecular framed structure of a

* See the Historical Review section (p. xx).

high degree of redundancy [in this last consideration Castigliano was influenced by the molecular view of solids of Adéhmar Jean Claude Barré, Count de Saint-Venant (1797–1886), whom he admired above all other contemporary engineering scientists]. Despite the elapsed time, Castigliano's last major work is still one of the most penetrating expositions of structural mechanics based on the principle of work.

Curiously, the success of this book depended more on the extensive number of problems it treated than on the rigorous proofs of its theorems. The greatest merit of the treatise, as well as Castigliano's most essential contribution to the progress of structural mechanics, lies in the wide influence it exerted on engineers, thus making the statically indeterminate elastic structure a practical proposition in engineering design.

Castigliano carried out many scientific studies of construction which were always tempered by a high degree of practical sense. He devoted as much of his effort to engineering design, construction and cost estimates as to the theory of structural mechanics—a prevalent attitude among Italian engineers of the period. This acute awareness of the practical always dominated his work as an engineer, despite the fact that he had at his command rather exceptional powers in abstract mathematics. In 1880 he carried out experimental strain measurements on the metal truss of the Bridge of Fella a Pontedimuro using a strain gauge of his own design. A model of a new slide rule with a high reading accuracy invented by Castigliano was exhibited at the National Exhibition in Milan in 1881.

Castigliano was a gentle, compassionate, and serene man who rarely lost his modest attitude. Only once in his life, in the dispute with General Menabrea, did he feel compelled to come forward in defense of his work and professional status.

Castigliano's education extended to languages, arts, and the science of economics. As a member of the administrative council of the Pension Fund of the Strade Ferrate Alta Italia, he carried out a mathematical analysis of the financial condition of the fund and, predicting that the existing plan was rapidly approaching its failure, developed an alternate scheme which was instituted at a critical moment. The five thousand employee-subscribers of the fund conferred on Castigliano a gold medal in token of their gratitude. The unusual extent of his interests was also demonstrated by his mastery of stenography, enabling him to write with greater efficiency.

In September, 1878, Castigliano married Felicità Pennachio of Milan, who bore him a son, Luigi (who was to follow in his father's footsteps by becoming a noted civil engineer), and three daughters, two of whom died at very early ages. While returning from a visit to his vacationing family near Merate in August, 1884, Castigliano was caught in a violent thunderstorm and, failing to take the prompt precautionary measures necessary under such circumstances, contracted pulmonary pneumonia. Exhausted from overwork and early deprivation, he was physically incapable of offering the necessary resistance to the grave malady, and not even the efforts of an eminent physician could render effective help to Castigliano's condition. After a long and painful illness, Castigliano died on October 25, 1884, and was buried in the Cimitero Maggiore of Milan, attended by a large group of noted engineers and professional men.

The writer is greatly indebted to his colleague, Professor James N. Siddall, for his help and advice throughout the writing of this article. Thanks are also due to Col. Giovanni B. Lenuzza, Scuole di Applicazione d'Arman di Torino, for locating copies of original memoirs on energy principles in elastomechanics and other pertinent information over the span of many years; to Professor Dr. Luigi Stabilini, Politecnico di Milano, for locating the rare autograph of Castigliano—thus ending a two-year search; to Professor Stefano Bruera, Istituto Technico di Stato "G. A. Giobert" di Asti, for providing the photographs of Castigliano's bust in the hall of the institute; and to Dr. Arturo Gaj, President of the Ordine degli Ingegneri della Provincia di Asti, for obtaining the photograph of the portrait of Castigliano.

GUNHARD-ÆSTIUS ORAVAS

*McMaster University*
*Hamilton, Ontario, Canada*
*October, 1965*

## Castigliano's Honors and Awards

Castigliano received the following distinctions and honors during his professional career:

1880 August 30. He became a member of the commission of general examinations for civil engineering students at the Reale Istituto Tecnico Superiore of Milan.

1881 November 20. He was elected to membership in the Collegio degli Ingegneri ed Architetti (Society of Engineers and Architects) of Milan.

1881 He was appointed to the jury of the Exposition of Italian Industry in Milan, for which he received an official commemorative medal.

1882 March 12. He was elected corresponding member of the Reale Áccademia delle Scienze di Torino, in the section of physical sciences and mathematics.

1882 July 21. He was named Cavalier of the Crown of Italy No. 26631 by royal decree.

1882 September 2. He was appointed by the ministry to the Examination Commission for competitive examinations in the Department of Mechanical Sciences at the Reale Scuola d'Applicazione in Rome.

1883 November 27. He was appointed a member of the commission for the habilitation examination in the Department of Projective Geometry at the Reale Istituto Tecnico Superiore in Milan.

1884 March 9. He was elected Vice-President of the Collegio degli Ingegneri ed Architetti of Milan.

1884 July 15. He was elected to membership on the administrative council of the Pension Fund of the Strade Ferrate Alta Italia.

1884 August. He became a member of the Examination Commission for the promotion of new engineers and chief engineers for the departments and sections of the Strade Ferrate Alta Italia.

1884 September. He was given a gold medal by the five thousand employee-subscribers of the Pension Fund of the Strade Ferrate Alta Italia for his study to improve its system.

## PUBLISHED WORKS OF CASTIGLIANO

1872 "Modo di eseguire alcune delle operazioni che devono precedere l'esecuzione degli sterri" (A method of performing some of the operations which should precede the execution of excavations). *Atti della Società degli Ingegneri e degli Architetti di Torino,* Volume V, 1872, Folio II, Editore Favale, Turin.

1873 *Intorno ai sistemi elastici* (About elastic systems). Dissertazione presentata da Castigliano Alberto alla Commissione Esaminatrice della Reale Scuola d'Applicazione degli Ingegneri in Torino per ottenere la Laurea di Ingegnere Civile, 1873, Vincenzo Bona, Tipografo di S. M., Turin, 52 pp.

1874 "Intorno alla resistenza dei tubi alle pressioni continue e ai colpi d'ariete" (About the resistance of pipes subject to continuous pressure and to water hammer). *Atti della Reale Accademia delle Scienze di Torino,* Volume IX, 1873/1874, pp. 222–252, Stamperia Reale di G. B. Paravia e C., Turin (presented on January 25, 1874).

1875 "Intorno all'equilibrio dei sistemi elastici" (About the equilibrium of elastic systems). *Atti della Reale Accademia delle Scienze di Torino,* Volume X, 1874/1875, pp. 380–422, Stamperia Reale di G. B. Paravia e C., Turin (presented on January 10, 1875).

"Nuova teoria intorno all'equilibrio dei sistemi elastici" (New theory about the equilibrium of elastic systems). *Atti della Reale Accademia delle Scienze di Torino,* Volume XI, 1875/1876, pp. 127–286, Stamperia Reale di G. B. Paravia e C., Turin (presented on November 28, 1875).

1876 "Formole razionali ed esempi numerici per il calcolo pratico degli archi metallici e delle volte a botte murali" (Rational formulas and numerical examples for the practical calculation of metal arches and masonry barrel vaults). *L'Ingegneria Civile e le Arti Industriali,* Volume II, 1876; Sept., No. 9, pp. 129–135; Oct., No. 10, pp. 145–153, Tipografia e Litografia dell' Editore Camilla e Bertolero, Turin.

1878 "Applicazioni pratiche della teoria sui sistemi elastici" (Practical applications of the theory of elastic systems). *Studi dell'Ufficio d'Arte delle Strade Ferrate dell'Alta Italia,* 1878, Stabilimento G. Civelli, Milan, 146 pp.

1879 *Théorie de l'équilibre des systèmes élastiques et ses applications* (Theory of equilibrium of elastic systems and its applications). Auguste Frédéric Negro, Éditeur, Turin, 1879, 480 pp. (This work was translated into German by Emil Hauff and published under the title *Theorie des Gleichgewichtes elastischer Systeme und deren Anwendung*, Verlag von Carl Gerold's Sohn, Vienna, 1886.)

1881 "Descrizione ed uso di un micrometro moltiplicatore per misurare gli allungamenti e gli accorciamenti prodotti dai carichi nelle costruzioni metalliche" (Description and use of a multiplicative micrometer for the measurement of extension and contraction produced by loads in metal structures). *Memoria dell'Ufficio d'Arte delle Strade Ferrate dell'Alta Italia*, 1881, Stabilimento G. Civelli, Milan, 16 pp. Also in *Il Politecnico*, Volume XXIX, 1881, July, No. 7, pp. 398–409, Premiata Tipografia e Litografia degli Ingegneri dell'Editore B. Saldini, Milan.

"Descrizione ed uso di un nuovo aritmografo" (Description and use of a new arithmograph). *Il Politecnico*, Volume XXIX, 1881, October/November, Nos. 10/11, pp. 564–576, Premiata Tipo-Litografia degli Ingegneri dell'Editore B. Saldini, Milan.

"Descrizione ed uso di un nuovo aritmografo modificato" (Description and use of a new modified arithmograph). *L'Ingegneria Civile e le Arti Industriali*, Volume VII, 1881, May, No. 5, pp. 71–75, Tipografia e Litografia dell'Editore Camilla e Bertolero, Turin.

1882 "Alcune considerazioni sopra i ponti in ferro" (Some considerations on iron bridges). *Atti del Collegio degli Ingegneri ed Architetti in Milano*, Anno XV, 1882, Fasc. 1, January/March, Folio I, pp. 44–51. Also in *Il Politecnico*, Volume XXX, 1882, June/July, Nos. 6/7, pp. 353–360, Premiata Tipo-Litografia degli Ingegneri dell'Editore B. Saldini, Milan.

"Esame di alcuni errori che si trovano in libri assai reputati" (Examination of some errors encountered in highly esteemed books). *Il Politecnico*, Volume XXX, 1882, January/February, Nos. 1/2, pp. 66–82, Premiata Tipo-Litografia degli Ingegneri dell'Editore B. Saldini, Milan.

"Intorno ad una proprietà dei sistemi elastici" (About a property of elastic systems). *Atti della Reale Accademia delle Scienze di Torino*, Volume XVII, 1881/1882, pp. 705–713, Ermanno Loescher, Librario della R. Accademia delle Scienze, Turin (presented on June 11, 1882).

*Studio intorno alle condizioni della Cassa Pensioni delle Strade Ferrate dell'Alta Italia* (Study about the condition of the Pension Fund of Strade Ferrate Alta Italia). Stabilimento G. Civelli, Milan, 1882, 50 pp.

*Studio intorno alle condizioni della Cassa Pensioni delle Strade Ferrate dell'Alta Italia* (*Appunti al lavoro statistico del Cav. Fochesato*) (Study about the condition of the Pension Fund of Strade Ferrate Alta Italia—Notes to the statistical work of Cav. Fochesato). Stabilimento G. Civelli, Milan, 1882, 16 pp.

1883 *Complemento allo studio intorno alle condizioni della Cassa Pensioni* (Complement to the study about the condition of the Pension Fund). Stabilimento G. Civelli, Milan, 1883.

1884 "Intorno ai muri di sostegno delle acque" (About retaining walls for water). *Il Politecnico*, Volume XXXII, 1884, January/February, Nos. 1/2, pp. 67–81, Premiata Tipo-Litografia degli Ingegneri dell'Editore B. Saldini, Milan.

*Teoria delle molle* (*a balestra ed a spirale piana*) (Theory of leaf and plane coil springs). Editore Augusto Federico Negro, Turin, 1884, 100 pp. (posthumous). (This book was translated into German by R. Trotz and published under the title *A. Castigliano: Theorie der Biegungs- und Torsionsfedern*, Verlag von Carl Gerold's Sohn, Vienna, 1888.)

*Manuale practico per gli ingegneri* (Practical manual for engineers). Editore Augusto Federico Negro, Turin. Volume I—Parte Prima, *Raccolta delle Tavole Numeriche che Spesso Occorrono nei Calcoli* (Collection of numerical tables which frequently occur in computation), 1882, 96 pp. Volume II—Parte Seconda; Sezione Prima: *Geometria e Trigonometria* (Geometry and Trigonometry); Sezione Seconda: *Misura dell Volte, Ossia, Loro Superficie e volume* (Measurement of vaults, or their area and volume); Sezione Terza: *Topografia e Livellazione Barometrica* (Topography and barometric leveling); Sezione Quarta: *Meccanica* (Mechanics) 1882, 122 pp. Volume III—Parte Terza, *Resistenza dei Materiali* (Strength of Materials), 1884, 120 pp. Volume IV—Parte Quarta: This volume was to treat "Stabilità delle Costruzioni" (Stability of Constructions), but at the time of the author's death only its first chapter, "Opere in Muratura" (Works in Masonry) had been completed. The remainder of the volume, which was intended for the treatment of the metal structures, was virtually untouched by the author. Consequently this volume was never published.

## Historical Review of Extremum Principles in Elastomechanics

There is a touch of irony in Castigliano's dispute with Menabrea over the priority for the first valid proof of the Menabrea Theorem, since it had been already proved by Cotterill nearly a decade before Castigliano's work. A brief review of the history of this theorem is given below for orientation:

The Italian universal genius, Lionardo da Vinci (1452–1519), seems to have been the first scientist to propose the intuitive concept of Least Action as a General Law of Nature: "Ogni azione fatta dalla natura é fatta nel più breve modo".

The origin of the minimum methods in solid mechanics dates back to a memoir of Christian Huygens (1629–1695) in 1662, who extremized the potential energy of a rigid beam failing by fracture.

In 1738 Daniel Bernoulli (1700–1782) became the first scientist to extend Huygen's extremal principle to flexible elastic beams by formulating an extremal problem involving a quantity proportional to elastic potential, whose extremum yields the Bernoulli Equation of Flexure.

In 1748 Leonhard Euler (1707–1783), after years of effort to extend Daniel Bernoulli's elastic potential concept, was successful in establishing a Potential Energy Principle as an extended Principle of Least Action for a loaded flexible elastic beam. In the development of this principle the work performed by a stress couple is evaluated for the first time.

In 1750 Jacopo Francesco, Count Riccati (1676–1754), speculated that, since kinetic energy is able to deform a solid and, consequently, can be stored in the solid as latent internal potential energy representable in the form of a mutually equal and opposite interacting force system acting in the deformed solid, it subsequently would be available for at least partial restoration of the initial configuration of the deformed solid. He could not follow it up with a mathematical formulation.

In 1774 Euler solved the first statically indeterminate problem, dealing with an infinitely rigid table on a linearly elastic foundation, in which Euler's Hypothesis of Elastic Foundation, postulating a linear displacement-reactive foundation force relation, made its first appearance. This work of Euler was responsible for all the subsequent research which finally led to the Complementary Potential Energy Principle.

Subsequent to 1818 Antoine Alfred Vène (1789–1863) had searched for an extremal solution of the indeterminate problem of a rigid body supported at more than three nonyielding points. Vène considered a rigid body to be a limiting case of an elastic body. In 1821 he proposed the extremal condition that the reactions should possess the maximum possible uniformity among themselves; in 1825 he advanced the criterion that the product of all reactions be the maximum possible; and finally in 1836 he proposed the principle that the sum of the squares of reactions should assume its minimal value, a result already given by Pagani and Cournot. Vène made the first attempt to prove the minimal nature of this extremum.

In 1821 and 1823 Henri Louis Marie Claude Navier (1785–1836) became the first scientist to find an expression for the general internal elastic potential of an elastic solid as a stored quantity of virtual work in his pioneering memoir on the molecular theory of elasticity.

In 1823 and 1834 Gaspare Michele Mario Pagani della Torre (1796–1855) investigated Euler's problem of a rigid body on elastic supports, and proved that Euler's solution results from an *a priori* minimal condition, which is identical to the principle later given by Cournot and Vène.

In 1827–1828 Antoine Augustin Cournot (1801–1877) published two memoirs in which he considered the same problem as Vène. Cournot based his arguments on the Virtual Work Principle and he derived the same principle as Pagani had obtained by an *a priori* minimal method equivalent to Euler's solution. Cournot's method served as the basis for Dorna's work on the subject. In his solution Cournot permitted the rigid body to become, temporarily, elastic around the support points in order to use Euler's Hypothesis. His solution was based on his misunderstanding of another problem solved correctly by Poisson in 1811.

Riccati's concept was first realized by Émile Paul Benoît Clapeyron (1799–1864) who in 1827 established his Conservation of Work Principle for elastic solids based on the Virtual Work Principle. It was first published in application to elastostatics only in 1852 in Lamé's book on elasticity, since Clapeyron himself used it mainly in dynamical investigations. Rev. Henry Gwynne Moseley (1801 1872) formulated his erratic Principle of Least Resistance in 1834 In 1838 George Green (1793–1841) propounded his epoch-making potential theory of elastic continuum in which he postulated an *a priori* existence of Elastic Potential as a function of strains.

Gabriel Lamé (1795–1870) used the Clapeyron Principle, in 1852, as an integral form of the Virtual Work Principle for the calculation of the displacement of a point in an articulated elastic framework.

Alessandro Dorna (1825–1886) resumed Cournot's investigations of a redundant rigid body, in 1857, in which he replaced the elastic regions of the body around the points of support by hypothetical elastic bars. His arguments, based on order of magnitude considerations of virtual displacements initiated by Cournot, were fallacious.

In 1857 Menabrea became the first scientist to abandon completely the rigid body concept and to formulate correctly this problem for an internally redundant elastic framework with frictionless joints. He arrived at the correct solution by a faulty order of magnitude argument following Cournot and Dorna. Menabrea's solution was first published in 1858.

In 1858 Clapeyron published a brief note on his Conservation of Work Principle.

In 1864 James Clerk-Maxwell (1831–1879) established the kinematic relationship existing between the point displacement and elongation of members in a truss by the Clapeyron Conservation of Work Principle. He was able to solve for statically indeterminate quantities as well as for displacements by means of his ingenious method of superposition of auxiliary problems in truss analysis. A special case of the Reciprocal Displacement Law was present in his work, which gives the first complete solution for articulated elastic frameworks.

In 1865 the remarkable English engineer James Henry Cotterill (1836–1922) published a set of four epoch-making memoirs containing not only the first valid proof of the Menabrea Theorem but also the first statement of the reciprocal Theorems of the Derivative of Internal Work of Deformation, which antedate Castigliano's independent derivation of the same theorems (in 1875) by ten years. Cotterill gave two different formal proofs to his theorems, both of which were based on Clapeyron Conservation of Work Principle. In the first proof he appealed to Euler's theorem of homogeneous functions of second degree, and in the second proof he imposed the Virtual Work Principle upon the *total* formal variation of the Clapeyron Principle.

Cotterill was led to these results by attempting to extend the vague Principle of Least Resistance of Moseley to the study of elastic bodies. In the case of linear elastic members, Cotterill restricted his study to discrete applied force systems by postulating stresses to be linear functions of the discrete applied forces, which make the Internal Work of Deformation, or the Elastic Potential, a homogeneous quadratic function of discrete forces.

Cotterill demonstrated that the Euler Extremal Equation obtained from the stationary value of the Internal Work of Deformation or the Elastic Potential which is expressed as a functional of stresses, appears as a compatibility condition in the form of stress–strain relation if the constraint condition imposed upon the variation of stresses represents the equilibrium equation in the interior of the body. Cotterill employed Euler–Lagrange Multipliers, which he demonstrated to represent displacements in this constrained variational problem.

Cotterill proposed his principle, which he considered to be an extended Least Action Principle, as a fundamental principle of elasticity. He was unaware of Euler's similar work on beams in 1748. He speculated that his principle may actually represent a restricted case of a more general principle. He was probably led to this conclusion by the fact that in one of his formal proofs both displacements and forces were subjected to simultaneous formal variations. Such a general Canonic Potential Energy Principle, vaguely anticipated by Cotterill, was ultimately established by Max Born and Gustav Prange.

Cotterill studied his extended Least Action Principle from the point of view of variational calculus. He extremized the Internal Deformation Work for a three-dimensional elastic continuum as a functional of stresses subject to equilibrium conditions in the interior of the body, which represent a constraint condition on the admissible variations of stresses, and subject to prescribed boundary values. He also, briefly, considered the problem of constrained variation of the Internal Work of Deformation, which satisfies all extremal conditions in the interior of the body, i.e. both the equilibrium and compatibility conditions, but without having *all* the boundary values prescribed.

In the process of establishing his theorems for slender members, Cotterill converted the variational problem into an ordinary extremum problem by introducing a discrete boundary force system.

At the time Cotterill was not aware of Menabrea's work. During the publication of his papers Cotterill discovered the brief Appendix by William Thomson, later Lord Kelvin (1824–1907), published in the latter's memoir of 1863. In this work Thomson demonstrates that, if the elastic solid is subjected to prescribed kinematic boundary conditions, the Elastic Potential assumes a *minimum* value at stable equilibrium configurations and yields a solution which is unique; but that for prescribed stress boundary conditions the solution, in general, is not unique and there may exist configurations of unstable equilibrium. The last conclusion Thomson credited to George Gabriel Stokes (1819–1903).

Thomson was influenced by the research of Gustav Robert Kirchhoff (1824–1887) who, in 1858, proved that the stable equilibrium condition of elastic bars is governed by the criterion

$$\Phi - \mathcal{U} = \text{Maximum}$$

where $\Phi$ denotes the work function of applied forces and $\mathcal{U}$ refers to the Elastic Potential, which is equivalent to

$$\mu - \Phi = \text{Minimum.}$$

Cotterill's work was totally overlooked by the profession, and it remained for Castigliano to rediscover Cotterill's theorems, independently, ten years later.

In 1869 Joseph Louis François Bertrand (1822–1900) sent a proof of the Menabrea Theorem in his letter to Menabrea, in response to Menabrea's submission of a copy of his memoir of 1867 to the French Academy of Sciences. Bertrand's proof is based upon the Virtual Work Principle in which the varied state of stress resultants in an elastic framework is postulated to be an equilibrium force system whose virtual work must vanish. He then imposed on the Virtual Work Principle the so-called Bertrand Hypothesis, which replaces virtual displacements by *actual* displacements of the equilibrated system, barring gross deformations of the system, resulting in the Menabrea Theorem. Bertrand's proof used the fundamental conceptual argument of the Complementary Strain Energy principle, which clearly demonstrates the abstract mathematical nature of this theorem.

A general form of Bertrand's proof was independently rediscovered by Wilhelm Fränkel (1841–1895) in 1882 with respect to three-dimensional elastic continua and elastic frameworks.

In 1875 Valentino Cerruti (1850–1909) published his famous critical note on the works of Menabrea and Castigliano, which were purely algebraic and inextensible to continuum problems. Cerruti believed that the Menabrea Theorem must be based on Green's elastic potential concept, but his own attempt to establish such an analytical foundation was faulty. However, it induced Castigliano to undertake further research on the Menabrea Theorem.

In his monograph of 1879, Castigliano, in contrast to Cotterill's continuum concept, regarded continuous elastic bodies to be limits of "molecular frameworks." He implied the extensibility of the elastic framework theory to the study of continuous bodies, but failed to elaborate as to how such a transfer could be accomplished.

Castigliano postulated a mathematical model of the continuous elastic body that had the form of a polyhedron with plane faces which remain plane after deformation. This hypothesis, which is analogous to Poisson's concept of molecules of solids, permitted him to apply Euler's Kinematic Formula to the description of displacements of any point on the boundary and to replace the boundary stresses by their statically equipollent resultant force system acting at a convenient reference point in each polyhedral face. In the last substitution Castigliano relied heavily on the Saint-Venant Principle insofar as he assumed that the substitute resultant force system does not alter the actual quantity of Internal Work of Deformation of the solid body. He assumed, moreover, that the fundamental equations of elasticity are known *a priori*, which permitted him to evaluate the Internal Work of Deformation in terms of the boundary stresses. Finally, Castigliano used the Clapeyron Principle in order to express the Internal Work of Deformation in terms of a boundary force system, from which he derived the two theorems of Cotterill by pursuing the same procedure he used for articulated frameworks. Castigliano, therefore, used only the second form of Cotterill's theory. By virtue of his assumptions Castigliano was successful in reducing a variational problem to an ordinary extremal problem of infinitesimal analysis. Because of these contrived procedures employed by Castigliano the whole conceptual foundation of Castigliano's theory remained obscure to his followers, and became a direct cause of a bitter and prolonged controversy between the German engineers, Otto Christian Mohr (1835–1918) and Heinrich Franz Bernhard Müller-Breslau (1851–1925).

In an article published in 1883, Müller-Breslau demonstrated that for many engineering problems Castigliano's Theorems yield the same results as the Maxwell–Mohr Reciprocal Work Principle, which Mohr had based directly on the Virtual Work Principle. Müller-Breslau's attention was called to Castigliano's monograph by a note, appended by Fränkel to his paper of 1882, in which Fränkel recognizes Castigliano's priority in the proof of his theorem, based on information given to him by Emil Winkler (1835–1888). In response to Müller-Breslau's article Mohr pointed out the apparently dissident definitions of the Internal Work of Deformation used in Castigliano's Theorems, and he also discussed with some precision the nebulous nature of the variational procedure employed by Castigliano. This dispute grew into a long and vitriolic controversy, especially after Reinhold Krohn (1852–1932), Georg Christoph Mehrtens (1843–1917), and the acerbic mathematician Julius Ludwig Weingarten (1836–1910) joined Mohr's side of the argument against Müller-Breslau, Fränkel, Julius Jakob von Weyrauch (1845–1917) and August Hertwig (1872–1955).

Actually none of the participants in this bitter controversy, except Hertwig, fully comprehended the axiomatico-conceptual foundation of the theorems under discussion which, however, were clearly outlined by a close friend of Castigliano.

Castigliano, himself, was not interested in the philosophico-conceptual foundation of the Menabrea Theorem, since his aims were solely directed towards the practical needs of his engineering activity. For instance, he never attempted to prove the minimal nature of the extremum in the Theorem of Least Work (Menabrea Theorem) by considering varied comparison states of the work of deformation.

Castigliano's colleague and close friend, Francesco Crotti (1835–1892), a former pupil of the mathematician Gaspare Angelo Mainardi (1800–1879) at the University of Padua, was keenly interested in the logico-conceptual foundation of the Menabrea Theorem as a compartment of general rational mechanics and he evidently had resolved the entire matter by 1884.

In 1877 and 1878 Crotti gave a demonstration for the general analytic foundation in its infinitesimal form and the proof for Cotterill–Castigliano Theorems based on Green's influence-coefficient method from which he derived Clerk-Maxwell's reciprocal displacement theorem. He expressly pointed out that Castigliano's Theorem is not equivalent to the Virtual Work Principle nor to the Clapeyron Principle and that the two Castigliano Theorems, in fact, are reciprocal, or complementary to each other. Crotti, who was a keen proponent of the consistent axiomatic foundation of rational mechanics and Green's continuum theory of solid bodies, is the first scientist to recognize in the Menabrea Theorem the canonization of the fundamental energetical theory of solid mechanics corresponding to the canonical treatment of analytical mechanics initiated by Simeon Denis Poisson (1780–1840) in 1809, adopted by Joseph Louis Lagrange (1736–1813) in 1811, and systematically investigated by William Rowan Hamilton (1805–1865) and Carl Gustav Jacob Jacobi (1804–1851) from 1834 on.

It appears that by 1884 Crotti had resolved all the axiomatico-conceptual difficulties in connection with Menabrea's and Castigliano's Theorems, as his eulogy of Castigliano indicates. Crotti published that part of his researches which pertains to structural mechanics in his remarkable monograph on the application of the theory of elasticity in 1888. Crotti demonstrated that Castigliano's Theorems do not constitute a general physical principle, but represent a special mathematical rule whose validity is restricted solely to linearly elastic solids.

Crotti demonstrated that the general Menabrea Theorem actually deals with a hypothetical canonical function of "work": $\Lambda = \sum F_i u_i - L$ which is procured from the quantity of physical work $L$ by means of the Euler–Legendre Canonical Transformation, in analogy to the canonical theory of analytical mechanics, where the new independent variable $F_i$ is given by Green's Theorem $\partial L(u_i)/\partial u_i = F_i$. Crotti showed that only for linearly elastic systems will the two reciprocal or complementary work quantities $L$ and $\Lambda$ become mathematically—but *not* conceptually—identical, i.e., $L = \Lambda$. He also proved the positive definite nature of both $L$ and $\Lambda$ and the reciprocal relations $\partial F_i/\partial u_j = \partial F_j/\partial u_i$, and $\partial u_i/\partial F_j = \partial u_j = \partial F_i$. Crotti established the general principle $\partial\Lambda/\partial F_i = u_i$, which embraces Castigliano's Theorem as a mere special case for linearly elastic systems. This relation represents the Crotti Principle of the Derivative of Complementary Strain Energy, and, as Crotti pointed out, it is valid for nonlinear as well as for linear elastic systems. For statically indeterminate structures he also established the Menabrea–Crotti Principle, $\delta\Lambda = \sum (\partial\Lambda/\delta F_i)\delta F_i = 0$ or for independent $F_i$ $\partial\Lambda/\partial F_i = 0$, which is valid for nonlinear as well as linear elastic redundant systems.

Crotti was the first scientist since Lord Kelvin to prove that for stable equilibrium of linear elastic bodies the work of deformation $L$ assumes its *minimum* value. Crotti, in contrast to Castigliano, also expressed the Elastic Potential in terms of stresses $T_{rs}$ and strains $t_{rs}$ and used the multiconstant elastic continuum theory of Green and Augustine Louis Cauchy (1789–1857). Castigliano employed the rariconstant theory of elasticity of Navier and Cauchy in which Poisson's ratio is $\nu = \frac{1}{4}$ for all materials and consequently $E = (\frac{5}{2})\mu$.

In 1883 Crotti developed an original formalized symbolic method for the calculation of planar elastic systems which permitted him to give simplified and concise formal solutions to a host of important structural problems independent of particular applied loads. Crotti, like Josiah Willard Gibbs (1839–1903), paid relentless attention to the logic and axiomatic precision of his physical arguments as well as to the importance of consistent mathematical notation. Crotti's monograph is probably the only major work published before 1900 in which a consistent tensorial double-suffix notation is employed for the tensor components of stresses $T_{rs}$ and strains $t_{rs}$.

In 1884–1885 Jacob Johann von Weyrauch (1845–1917) had all the essential ingredients of the complementary work quantity present in his books on linear elasticity, but he failed to isolate the complementary work as a fundamental conceptual quantity.

In 1886 Silvio Canevazzi (1852–1918) published a monograph on trusses in which he followed up Cerruti's suggestion by treating the Menabrea Theorem from the point of view of elastic potential theory.

In 1889 Friedrich Engesser (1848–1931) studied at some length the Crotti Complementary Strain Energy Principle, including thermoelastic effects, but he did not refer to the original source of his study and, moreover, he overlooked the importance of the unifying canonical feature of its foundation within the framework of general mechanics.

Luigi Donati (1846–1932), under the influence of Canevazzi's monograph, wrote three memoirs (in 1888, 1889 and 1894) in which he successfully differentiated between the Menabrea Theorem and the Potential Energy Principle as well as between the two alternate forms of the Menabrea Theorems. Donati is the first scientist who drew attention to the fact that the Castigliano Theorems actually deal with functional derivatives, first investigated by Vito Volterra (1860–1940) in 1887 in connection with the latter's pathbreaking theory of functionals.

In 1889 Canevazzi published an extensive memoir on the Menabrea and Castigliano Theorems which are employed in the solution of a number of practical problems.

In 1890 Carlos Mario Viola (1855–1925) studied the theorems of Menabrea and Castigliano in their application to elastic continuum and elastic frameworks from the point of view of variational methods, in which he pointed out that the quantity extremized in these theorems have no mechanical significance, and that the absurdity present in the Theorem of Least Work has resulted from interpreting this quantity in general to be the work of deformation.

In 1894 Francisco Siacci (1839–1907) studied the Menabrea Theorem in connection with an initial state of strain and demonstrated that this theorem yields stress resultants caused by applied forces which can be superimposed on stress resultants caused by initial strain.

In 1905 Carl Johann Kriemler (1865–1936) considered the initial and final equilibrium configurations of an elastic system as special cases of its state of motion. His analysis was based on the Dynamic Virtual Work Principle in which his concept of Surplus Work, defined as the difference between the work performed by the applied forces and the internal stresses, becomes identical to Crotti's Complementary Strain Energy for equilibrium problems.

In 1905 Pierre Cérésole (1879–1945), inspired by Born's lecture on variational treatment of elastica, rediscovered Crotti's method of Euler–Legendre Canonical Transformation connecting the Strain Energy with the Complementary Strain Energy as a unifying counterpart to the canonical theory of analytical mechanics, which was disseminated by Arnold Sommerfeld (1868–1951) to August Hertwig.

In 1906 Hertwig demonstrated the application of Crotti's Complementary Strain Energy Principle, including thermal effects, to flexural problems of beams. Hertwig seemed to be unaware of Crotti's work since he accredited the Complementary Strain Energy Principle to Engesser.

In 1906 Max Born (1882–      ), the most outstanding pupil of the greatest contemporary mathematician, David Hilbert (1862–1943), published his pathbreaking prize-winning habilitation dissertation (presented early in 1905) on the problem of elastica, which he investigates by variational methods based on energetical considerations. He broke new ground in elastomechanics by expounding two new important formulations of variational methods in the energy principles of elasticity. In the appendix of this memoir, Born demonstrates that the *minimum* variational problem

$$\int_x \int_y F\, dxdy = \int_A F\,(u', u^{\cdot}, u; x, y) dA = \text{min.}$$

can be converted into a corresponding maximo-minimum variational problem

$$\int_A [F(p_1, p_2, u) + (u' - p_1)F_{p1} + (u^{\cdot} - p_2)F_{p2}] dA = \text{max.} - \text{min.},$$

where $p_1 = u' = \partial u/\partial x$, $p_2 = u^{\cdot} = \partial u/\partial y$, $q_1 = F_{p1} = \partial F/\partial p_1$, $q_2 = F_{p2} = \partial F/\partial p_2$, in which the maximum of the integral is determined for an arbitrary function $u$, and then the function $u$ is evaluated in such a manner that the integral assumes, under the maximal values $q_1$ and $q_2$, its minimum value. Born, moreover, demonstrated that if a new canonic functional

$$H(q_1, q_2, u) = [F(u', u^{\cdot}, u) - u'q_1 - u^{\cdot}q_2]$$

is introduced by means of the Euler–Legendre Canonical Transformation, then the maximo-minimum variational problem changes into a corresponding canonic variational problem.

$$\int_A \Pi(q_1, q_2, u)dA = \text{max.} - \text{min.},$$

where the original functional appears in its canonical form

$$F(u', u^{\cdot}, u) = \Pi(q_1, q_2, u) = [H(q_1, q_2, u) + u'q_1 + u^{\cdot}q_2],$$

whose Born Canonic Extremal Equations $(\partial H/\partial q_1) + u' = 0$, $(\partial H/\partial q_2) + u^{\cdot} = 0$, $(\partial q_1/\partial x) + (\partial q_2/\partial y) - (\partial H/\partial u) = 0$ are fully equivalent to the Euler Extremal Equations of the original problem. The remarkable aspect of Born's Canonic Variational Principle is the fact that the dependent variables $u$, $q_1$ and $q_2$, which actually are functionally dependent on each other, can be varied independent of each other as if they were *pseudo-independent* variables in this canonic variational problem, since due to the Euler–Legendre Transformation the coefficients of the variations $\delta q_1$ and $\delta q_2$ vanish identically.

Born's Canonic Variational Principle came at the tail end of his dissertation, and he never returned to the application of his principle to the general elastic continuum, due to his interest in modern physics. Born had even chosen the subject of his habilitation dissertation against his own better judgment under the pressure exerted by the great mathematician Felix Klein (1849–1925), a king-maker of professors in Germany at the time. The extension of his principle to three dimensions and to more than two dependent variables is a rather formal and trivial matter. It remained for Born's close friend Hellinger to apply his Canonic Variational Principle as a Canonic Potential Energy Principle to the elastic continuum eight years later.

In 1912 Gustavo Colonnetti (1886– ) wrote an extensive historically-oriented study on energy principles in elastomechanics in which he reviewed the contributions of Menabrea, Bertrand, Castigliano, Cerutti, Canevazzi and Donati.

In 1914 Oskar Domke (1874–1955) published an investigation on energetical methods in elastomechanics based on the variational calculus point of view. He demonstrated that the Virtual Work Principle is the foundation of both the Potential Energy and Complementary Potential Energy Principles.

Domke obtained the stress-strain relations by extremizing the Complementary Strain Energy subject to the constraint of equilibrium, and the Beltrami-Michell compatibility condition $\nabla^4 F = 0$ if variations of stresses are carried out in terms of the Airy stress function $F$ for a plane stress problem.

In 1914 Ernst Hellinger (1883–1950) applied Born's Canonic Variational Principle to elastic continuum by extending the parameter $u$ to three displacement components $u_x$, $u_y$, $u_z$ and the two canonical variables $q_1$, $q_2$ to six stress components $\sigma_{xx} = \partial\phi/\partial u'_x$, $\sigma_{xy} = \partial\phi/\partial u^{\cdot}_x$, $\ldots$, $\sigma_{zz} = \partial\phi/\partial u^*_z$, where the Elastic Potential $\phi = F$, $u^*_z =$

$\partial u_z/\partial z$, and the surface integral is extended into a volume integral. Then the Born Canonic Potential Energy Principle assumes the form

$$\int_v \Pi(\sigma_{\alpha\beta}, u_\alpha)dv = \int_v \Big[H(\sigma_{xx}, \sigma_{xy}, \ldots, \sigma_{zz}, u_x\, u_y, u_z) + \sum_\alpha \sum_\beta (\partial u_\alpha/\partial\beta)\sigma_{\alpha\beta}\Big]dv = \text{max.}-\text{min.} \qquad (\alpha, \beta = x, y, z),$$

where the Canonical Elastic Potential $\Pi(u_\alpha, \sigma_{\alpha\beta})$ and the Complementary Elastic Potential $H = \phi - \sum_\alpha \sum_\beta (\partial u_\alpha/\partial\beta)\sigma_{\alpha\beta}$. The corresponding Born Canonic Extremal Equations

$$\sum_\beta (\partial\sigma_{\alpha\beta}/\partial\beta) - (\partial H/\partial u_\alpha) = 0 \qquad (\alpha = x, y, z)$$

$$(\partial H/\partial\sigma_{\alpha\beta}) + (\partial u_\alpha/\partial\beta) = 0 \qquad (\alpha, \beta = x, y, z)$$

represent equilibrium equations in the interior of the elastic body and the compatibility conditions for the stress and strain. If the Potential Energy density is independent of displacements $u_\alpha$, then the Born Canonic Potential Energy Principle reduces to the Menabrea–Crotti Principle $\int_v H(\sigma_{\alpha\beta})dv = \text{max.}$ subject to the constraint condition in the form of equilibrium equation $\sum_\beta (\partial\sigma_{\alpha\beta}/\partial\beta) = 0$ $(\alpha = x, y, z)$, whose six Euler Extremal Equations

$$\partial(\partial H/\partial\sigma_{\alpha\alpha})/\partial\beta = \partial(\partial H/\partial\sigma_{\alpha\beta})/\partial\alpha \qquad (\alpha, \beta = x, y, z)$$

represent kinematic compatibility conditions of the elastic continuum.

It seems that Hellinger considered his contribution a straightforward application of the Born Canonic Variational Principle to elastic continuum to which he devoted only one page of text without advancing beyond the variational theorem proved by Born.

The next advance in Born's Canonic Potential Energy Principle was achieved by Prange through the introduction of the Canonic Potential Energy of the boundary conditions.

In 1916 Georg Heinrich Friedrich Wilhelm Prange (1885–1941) gave a general fundamental clarification of all extremal principles of solid mechanics, excluding thermodynamic considerations, in his celebrated habilitation dissertation. Prange demonstrated that all known variational principles in solid mechanics are mere subsidiary cases of a general Canonic Variational Principle, or *Canonic Potential Energy Principle*, whose Born Canonic Extremal Equations represent equilibrium equations in the interior, boundary conditions, and compatibility conditions in the form of stress-strain relations. He showed that the Canonic Potential Energy Principle reduces to the Potential Energy Principle if the kinematic compatibility condition is assumed to be satisfied *a priori* for all variations of the displacement, and to the Menabrea–Crotti Principle if the equilibrium equations are assumed to be satisfied *a priori* for all variations of stresses. He also considered mixed boundary conditions and extremum extremorum problems.

In 1919 Prange published an extensive study on beams, in which the Born Canonic Potential Energy Principle plays a central role. Prange demonstrated the analogy existing between the Extremal Principles in solid mechanics and general mechanics: the *Principle of Least Internal Work of Deformation* corresponds to the *Principle of Least Action*, the *Equations of Equilibrium* and *Compatibility Equations* correspond to the *Differential Equations of Motion*, the *Cotterill–Castigliano Theorems* are the counterpart to the *Derivatives of Hamilton's Varying Action*.

Prange's work completed the classical mechanistic theory of variational principles in elasticity and structural mechanics. It is still the most comprehensive and definitive work on the subject from the point of view of the unified theory of classical mechanics.

In 1921 Domke studied thermoelastic problems in structural mechanics in connection with the Complementary Potential Energy Principle in which he drew upon some thermodynamic considerations. Castigliano in 1879, Joseph Melan (1853–1941) in 1883, Müller-Breslau in 1884, Hertwig in 1906, Weyrauch, Weingarten and Engesser in 1907, and Kriemler in 1905 and 1911 considered temperature effects on deformation work without general thermodynamic considerations.

In 1929 Kurt Otto Friedrichs (1901– ) published an important memoir in which he clarifies the relationship existing between the general Menabrea Theorem, the Minimum Potential Energy Principle and the Direct Minimal Method of Erich Immanuel Trefftz (1888–1937). Friedrichs was unaware of Crotti's and Born's tracts, and of Cérésole's rediscovery of Crotti's concept of the Euler–Legendre Canonical Transformation as the vital connecting link between the Potential Energy Principle and the Menabrea–Crotti Principle. Friedrichs was the first scientist since Born to discover that the canonical transformation changes a *minimum* problem of Potential Energy into a *maximum* problem of Complementary Potential Energy. Concerning the reciprocity-transformation of the variational problem, Friedrichs' theory is more detailed than Born's concise appendix.

In 1938 Arno Schleusner (1882– ) again demonstrated that the extremum in the Complementary Potential Energy Principle describes a *maximum*, and not a minimum, condition for Complementary Potential Energy. He did not refer to Born's or Friedrichs' work on the same subject. In his memoir the differences existing between the various extremal methods used in structural mechanics in their relation to the Virtual Work Principle are discussed with considerable precision.

Georg Karl Wilhelm Hamel (1877–1954) gave occasional lectures on the Born–Prange Canonic Potential Energy Principle, and on variational principles in general, at the Technological University of Berlin. In 1949 he published this principle briefly in his highly individualistic and the most impressive contemporary pedagogical tract "Theoretische Mechanik," which served to disseminate this principle among his contemporaries.*

However, practically all the works mentioned in this brief panoramic history, except the work of Castigliano, have been more or less overlooked by the profession. This is one of the primary reasons why Castigliano's great monograph occupies a position of preeminence in this field of research (especially on the strength of its admirable section on applications), and why this reprinting of it is most welcome.

G.-Æ. O.

---

* After nearly a half-century of obscurity, the Born–Prange Canonic Potential Energy Principle has been enjoying a well-deserved contemporary revival in the researches of Eric Reissner, B. Fraeys de Vuebeke, L. I. Balabukh, Hai-Chang Hu, K. Washizu, I. I. Gol'denblat, I. N. Slezinger and D. Rüdiger. Despite the fact that Eric Reissner established the Born–Prange Canonic Potential Energy Principle empirically, thereby missing its important unifying canonical foundation altogether (later corrected by de Veubeke), thus being unable to deduce this canonic principle from the Potential Energy Principle as it's primitive form, his subsequent contributions on this variational principle are of solid substance of original research. Reissner is the first engineering scientist to apply this principle to finite elasticity and the first investigator, after Born, to study thoroughly the extremal properties of this Canonic Potential Energy Principle in a work distinguished by classic simplicity and clarity that makes him, in addition to Born and Prange, an important contributor to this principle.

ENGINEERING SCIENTISTS WHOSE RESEARCHES LED TO POTENTIAL AND COMPLEMENTARY POTENTIAL ENERGY PRINCIPLES IN ELASTOMECHANICS

*Christian Huygens*

### Christian Huygens (1629–1695)

Dutch philosopher, astronomer, mathematician and physicist. Pupil of van Schooten at the University of Leyden. He was the only truly rigorous mathematician of the seventeenth century and a physicist's physicist. In originality, rigorous logic and profound conceptual intuition, he was distinctly superior to Newton, and the only scientist comparable to Euler. Like Hooke, he was a master technologist and invented improved telescopes and pendulum clocks of great accuracy, which made possible progress in astronomical measurements. First to establish: the kinematic law of curvilinear motion as a vectorial composition of tangential and centrifugal motion (later adopted by Hooke, then by Newton); the vectorial axiom of motion (axiom of centrifugal motion) and the indirect energetical conservation principle which recognized kinetic energy as a fundamental quantity of motion (also by Hooke); the unity between celestial and terrestrial mechanics after Kepler and Descartes; extremal principle to the limiting strength of beams in flexure; and a nontrivial solution of rigid-body mechanics by explaining rotational motion of pendulums, the moment of inertia concept, and the inertia force concept. Replaced Fermat's long-distance Least Action Principle in optics by his contact extremal principle in Hooke's undulatory theory of light. His optical work represents the first book on geometric calculus of variations (used by Johann Bernoulli), continued by Hamilton and Jacobi. Suggested the application of his conservation principle to fluids, which was carried out by Daniel Bernoulli in the latter's tract on hydrodynamics. His book on mathematical theory of probability is a great classic. He founded "local" geometric study of curves. Leibniz was one of his pupils. Member of the French Academy of Sciences and a Fellow of the Royal Society.

### Daniel Bernoulli (1700–1782)

Swiss mathematician, physicist, botanist, anatomist and physician. Studied mathematics with his friend Euler under his father Johann, and medicine in Basel, Heidelberg, Strassburg and Venice (under Michelotti). Professor of physics at the Russian Academy of Sciences, of anatomy and botany, and finally physics and natural philosophy, at Basel. First to apply momentum principle to continuum, to linearize and solve Jacob Bernoulli's equation of flexure, to investigate beam-column problems, conservation of energy principle, vibration of string problem (solved by his novel semi-direct product solution in the form of Fourier series), for which he was the first to locate simple modes and frequencies and to employ Bessel series. He was also the first to use and extend Jacob Bernoulli's Lost Force Principle (D'Alembert Principle), later generalized by Euler and D'Alembert. Also studied ocean currents and tides. He was the greatest proponent of his time of Huygens' Conservation of Energy Principle, extending it to the problem of flexible elastic beams as a Least Elastic Strain Energy Principle. Wrote the first treatise on hydrodynamics in 1738, based on Huygens' principle. Introduced piezometric opening in conduit walls for pressure measurement. Member of the French Academy of Sciences.

### Leonhard Euler (1707–1783)

Swiss mathematician, physicist and astronomer. Pupil of Johann Bernoulli. Ranks as the greatest mathematical genius of all time. In fecundity, intuition, originality, inventiveness, and powers of memory, he stands head and shoulders above all other scientists. At first inspired by Daniel Bernoulli's researches, but soon left his friend far behind. Euler's contributions are so extensive that they cannot be accounted here in any brief summary. It suffices to mention only a few of his relatively unknown contributions: potential energy principle for flexible elastic members, differential equations of equilibrium for curved elastic members, elastic modulus E as a material constant, stress principle (action-reaction for contact forces), indirect vector analysis, principle of superposition, free-body and section principle, hypothesis of elastic foundation, Euler-multiplier method for constrained extremum (Lagrangian multiplier), potential function and bipotential equation in hydrodynamics (so-called Laplace's equation), moving-triad method in differential geometry, etc. About three-quarters of classical mechanics, including the D'Alembert Principle, are Euler's contributions. About the same proportion applies to applied mathematics. Most of Gauss's work was an extension of Euler's researches.

### Col. Antoine Alfred Vène (1789–1863)

French military engineer. Graduate of the École Polytechnique. Took part, with distinction, in Napoleonic campaigns in Africa, Spain and Russia. Decorated with the Cross of St. Louis, Legion of Honor, and the Cross of Charles III of Spain. Chief of the military installations in Paris. Member of the Society of Encouragement of National Industry. Research on variational methods, limits of roots of polynomial equations, infinitesimal methods, and on statically indeterminate systems by extremal methods.

*Daniel Bernoulli*

*Leonhard Euler*

*Gaspare Michele Mario Pagani della Torre*

*Antoine Augustin Cournot*

*Émile Paul Benoît Clapeyron*

### Gaspare Michele Mario Pagani della Torre (1796–1855)

Italian mathematician. Studied mathematics and mechanics at Turin. Taught in Genf and Brussels, finally becoming professor of mathematics at the University of Louvain. Member of the Academy of Sciences of Brussels. First scientist to give an extremal formulation to Euler's Hypothesis of Elastic Foundation, which he used in the study of simple statically indeterminate elastic systems and equilibrium of columns.

### Antoine Augustin Cournot (1801–1877)

French mathematician, philosopher of mathematics and science, historian and political economist. Graduate of the École Normale. Private tutor, professor of mathematics on the faculty of science at Lyon, Rector of the Academy at Grenoble, Inspector General of Studies, Rector of the Academy of Sciences at Dijon. Became an important dignitary, despite of which he lived a modest and unpretentious life. Research on theory of wealth, doctrine of chance, probability, symbolic logic, political economy, history of science and mathematics, theory of functions, differential equations, and astronomy and mechanics in general. First to investigate the statically indeterminate problem based on the Virtual Work Principle, on which all later works in this subject are based.

### Émile Paul Benoît Clapeyron (1799–1864)

French civil engineer. Studied at the École Polytechnique and École des Mines. Professor of applied mathematics at the new Institute of Engineers of Ways of Communication in St. Petersburg (together with his friend Gabriel Lamé). Lieutenant-Colonel in the corps of engineers in the Russian Army, in charge of engineering-design. Wrote in 1827, together with Lamé, a classic memoir on theory of elasticity and its applications, in which all known results were considered, Navier's molecular theory was compared with Cauchy's phenomenological theory, and important applied problems were solved. At the same time, he wrote a memoir (never published) on his classic Conservation of Work Principle, which was based upon the Virtual Work Principle and represented an extension of General Lazare Carnot's Conservation of Work Principle to elastic bodies. It served as a basis for the researches of Clerk-Maxwell, Cotterill, Castigliano, Crotti, and Müller-Breslau. Developed, with Lamé, membrane theory of shells and arches. Upon his return to France in 1831, he became the most important French railroad engineer. Developed in 1848 the "three moment equation" for the design of continuous girder-bridges. Designed locomotives and was the chief promoter of Sadi Carnot's theory of thermodynamics. Became a distinguished lecturer on steam engines at the École des Ponts et Chaussées and a member of the French Academy of Sciences.

### Alessandro Dorna (1825–1886)

Italian civil engineer. Professor of rational mechanics at the Military Academy in Turin. Professor of astronomy at the University of Turin and director of the observatory. Professor of geodesy at the Higher Military School in Turin. Editor of the meteorological and astronomical bulletin of the observatory. Research on rational mechanics, astronomy, instrumentation, hydraulics and geodesy. First to reassume Cournot's investigations on the statically indeterminate problem. Member of the Academy of Sciences in Turin.

### Count Luigi Federigo Menabrea (1809–1896)

Italian nobleman. Pupil of Amadeo Plana. Professor of mathematics at the University of Turin, and of mathematics and structural mechanics at the Royal Military Academy in Turin. Became deputy of parliament and fought with General Garibaldi in the Austrian campaign at the rank of colonel. Promoted general in charge of the military engineering corps. Became minister of the Marine, and in 1866 signed the peace treaty for Italy with Austria (at Vienna). Became Minister of Public Works, president of the Council of Ministers and Head of the Government. He was an intimate friend of King Vittorio Emanuel II of Sardinia, later King of Italy, and represented the king at the coronation of King Oscar II of Sweden. Ambassador to England and France. Member of the French Academy of Sciences and the Royal Academy of Sciences in Rome and Turin. He was a promoter of Babbage's calculating machine on which he wrote a long memoir. First formulation of the Menabrea Theorem. Became Marquis (Marchese) di Valdora.

### James Henry Cotterill (1836–1922)

English engineer and applied mathematician. Apprentice to Fairbairn in Manchester; later studied at St. John's College in Cambridge. A scholar at St. John's, then lecturer at the Royal School of Naval Architecture and Marine Engineering in South Kensington, where, in 1870, he became vice-principal. Upon the merger of this school with the Royal Naval College in Greenwich, he occupied the chair of applied mechanics until his retirement. His researches encompassed Dynamic Principle of Least Action, propeller theory, superheated steam, homogeneous functions of second degree, energy distribution in liquids in steady state of motion and reaction of elastic fluids issuing from orifices. He was a pedagogue of great reputation, and published several books on applied mechanics and heat engines.

### Joseph Louis François Bertrand (1822–1900)

French child prodigy. Educated by his uncle, Jean Marie Constant Duhamel. Received his doctoral degree at the age of

*Alessandro Dorna*

*Count Luigi Federigo Menabrea, Marquis di Valdora*

*James Henry Cotterill*

*Joseph Louis François Bertrand*

*Wilhelm Fränkel*

seventeen. Professor of mathematics at the École Polytechnique and of mathematical physics at the Collège de France in Paris. Perpetual secretary of the French Academy of Sciences. Fundamental contributions in all facets of mathematics and physics. First to give the conceptual reasoning leading to the Principle of Complementary Strain Energy.

### Wilhelm Fränkel (1811–1895)

A native of Russia. Studied at the Politechnicum in Dresden. After graduation, worked as an engineer for the Sachsian Railroad. Became docent and professor at the newly established Polytechnical Institute of Dresden. Son-in-law of the mathematician, Oskar Schlömilch. Invented an extensometer and deflectometer permitting experimental verification of the theoretical analysis of bridges. Wrote a number of treatises on bridge and railroad engineering. Published numerous important theoretical papers on structural mechanics, especially on influence lines. Discovered, independently, the general form of Bertrand's proof of the Complementary Strain Energy Principle in its application to elastic continuum, as well as to elastic frameworks.

### Valentino Cerruti (1850–1909)

Italian engineer. Graduate of the Politechnicum in Turin and a close friend of Castigliano. Professor of mathematical physics and rational mechanics, later of mathematics, at the University of Rome. Founder of the library at the Politechnical Institute. Secretary-general of public instruction, rector of the university, and minister of public instruction. Member of Società Italiana dei Quaranta, Istituto Lombardo, Academy of Leopold-Caroline in Halle, and secretary of the Academia dei Lincei in Rome. Wrote an extensive number of works on elasticity, rational mechanics, and mathematics. Emphasized the central importance of potential theory concept in Menabrea and Castigliano's theory.

### Francesco Crotti (1839–1896)

Italian engineer. Graduate of the department of mathematics of the University of Padua. Pupil of Gaspare Mainardi and Delfino Codazzi. Railroad engineer in Bergamo and Milan. A colleague and close friend of Castigliano. Contributed in theoretical and practical aspects of engineering-design, elasticity, economics, geodesy, and mathematics. Established the conceptual foundation of the Complementary Energy Principle for nonlinear elasticity and demonstrated its relation to canonical theory of mechanics. He relinquished research on energy principles in elastomechanics in view of the tragic death of his friend Castigliano.

### Luigi Donati (1846–1932)

Italian mathematical physicist, graduate of the University of Pisa, and pupil of Enrico Betti and Ulisso Dini. Docent at the Politechnical Institute in Bologna, professor of engineering physics at the University of Bologna. Succeeded Silvio Canevazzi as the director of this school. Research in electrodynamics, electromagnetism, X-rays, electrotechnology, and thermodynamics. His three early researches are the epochal memoirs on Menabrea's theorem and elasticity. Member and president of the Royal Academy of Sciences in Bologna.

### Max Born (1882– )

German physicist and applied mathematician. One of the best pupils of Hilbert and probably the greatest contemporary physicist and pedagogue. Professor of physics at Berlin, Frankfurt am Main, Göttingen and Edinburgh. Founder of modern wave mechanics. Performed profound researches in quantum mechanics, crystal lattices and relativity. Both of his outstanding memoirs on elastomechanics were awarded prizes.

### Ernst Hellinger (1883–1950)

German mathematician. Pupil of Hilbert at the University of Göttingen. Professor of mathematics at the University of Frankfurt am Main. Wrote numerous works on differential and integral equations and other topics in abstract mathematics. Published a famous encyclopedic article on continuum mechanics containing the formulation of the Canonic Elastic Potential concept.

### George Heinrich Friedrich Wilhelm Prange (1885–1941)

German applied mathematician. Pupil of Klein and Hilbert at the University of Göttingen. Professor of mathematics at the Technological University in Hanover. His doctoral dissertation extended the principle of varying action to multidimensional variational problems leading to the definition of functional-fields which satisfy Hilbert's theorem. His work on general integration methods in analytical mechanics has a far-reaching significance. His habilitation dissertation contains the first fundamental conceptual clarification of all classic variational principles in elastomechanics and the establishment of the Born–Prange Canonic Potential Energy Principle. Wrote an extensive memoir on the variational principles in their application to the analysis of elastic beams. Investigated fundamental switching processes and electromagnetic fields in electrotechnology, and the three-body problem in celestial mechanics. His extensive researches were few in number, but profound in scientific thought, despite the fact that he was afflicted with a grave malady throughout his entire productive life.

*Valentino Cerruti*

*Francesco Crotti*

*Luigi Donati*

*Max Born*

*Ernst Hellinger*

*George Heinrich Friedrich Wilhelm Prange*

# THE THEORY OF EQUILIBRIUM OF ELASTIC SYSTEMS AND ITS APPLICATIONS

Carlo Alberto Castigliano

# AUTHOR'S PREFACE

THIS book contains the theory of elastic stresses in structures explained in accordance with a new method, based upon theorems which are either quite new or little known; incorporated in this theory will be found the mathematical theory of elastic solids, considered particularly from the standpoint of the strength of materials.

We believe that the time has arrived for introducing into our courses of instruction this scientific method of treating the strength of materials and thus abandoning the older methods which Lamé has fairly characterised as "neither scientific nor empirical, serving only to hinder the approach of true science".

We will now give some historical notes of the discovery of the theorems of which we make repeated use in this book.

These theorems are the following three :—

1. That of the differential coefficients of internal work, 1st part.

2. That of the differential coefficients of internal work, 2nd part.

3. That of the least internal work.

The first has already been employed by the celebrated English astronomer Green, but only for a particular problem; it has not been enunciated and proved in a general way such as we give here.

The second is the converse of the first, and we believe that it was first enunciated and proved in our thesis for the diploma of engineering at Turin; we have given it at greater length in our memoir entitled "Nuova teoria intorno all' equibrio dei sistemi elastici," published in the Transactions of the Academy of Science of Turin in 1875.

The third theorem can be regarded as a corollary of the second; but as in some other problems of maxima and minima, it has been partly known several years before the discovery of the principal theorem.

In the year 1818 Captain Vène, of the French Engineer Corps, enunciated a principle which was absolutely incorrect under the conditions to which he wished to apply it, but which, by one of those peculiar combination of circumstances of which science presents several examples, was destined to lead later to the discovery of the theorem of least work.

After this first step, the distinguished scientists, MM. A. Cournot, Pagani, Mossotti, A. Dorna, and General L. F. Ménabréa investigated the question. The last-mentioned gave the name "principle of elasticity" to the theorem of least work, and made it the subject of his researches, in a first memoir presented in 1857 to the Academy of Science of Turin, later in a second presented in 1858 to the Academy of Science of Paris, and again in a third submitted in 1868 to the Turin Academy. Since, however, the proofs given by M. Ménabréa were not exact, the "principle of elasticity" was not accepted by the greater number of the authorities, and some of them published memoranda to show the fallacy of it. It was not until 1873 that we gave, in our above-mentioned thesis, the first rigorous proof, in a form which appeared to us clear and exact, of the theorem of least work. Afterwards, in our thesis of 1875, we demonstrated that the theorem of least work is only a corollary of that of the differential coefficients of internal work.

We can thus state that the present book, comprising the complete theory of elastic stresses in structures, of which the mathematical theory of the elasticity of solid bodies comprises only a chapter, is wholly based on the theorems of the differential coefficients of internal work. As our object is not only to expound a theory, but further to show its advantages of brevity and simplicity in practical applications, we have solved, according to the new method, the greater number of the general problems dealt with in courses on the strength of materials, and have added several numerical examples for the calculation of the stresses in the more important types of structure. Each of these examples is, so to speak, a particular application of the theory to one of these structures, but in order to give our book a more practical value we have put each example in the form which would arise in an actual design, in order to justify the dimensions of the principal members.

Moreover, since in our view these examples ought to serve as a model for similar calculations, we have always examined several cases of loading and taken account of the effects due to temperature variations; that is to say, that for each assumed load we have determined the stresses occurring at different temperatures. By taking account of these circumstances and by following the new method of calculation, which permits of the solution of all questions on the stresses in elastic structures without the introduction of any arbitrary assumption, we have the advantage of being able to adopt higher working stresses; for one of the causes which often compels, in practice, the adoption of small values for these stresses, is the imperfection of the principles upon which the calculation of the stresses is based; and another reason for the adoption of low working stresses is that we cannot take account of all the circumstances liable to cause these stresses to increase. With regard to the calculations, we may say that they are but a little longer than in methods ordinarily followed, and moreover, they can nearly always be materially shortened by neglecting some terms having small influence on the result.

In conclusion, we have heartily to thank M. Louis Reeb, Permanent Way Superintendent of the Northern Italian Railway, who has kindly undertaken the proof revision.

## TRANSLATOR'S PREFACE

CASTIGLIANO's work is referred to as a classic in the leading text-books dealing with the advanced Theory of Structures, but few students in modern times have had an opportunity of studying it, since, as far as we have been able to ascertain, there is no previous translation in English, and the copy in French is out of print and very scarce.

With the development in practical design of methods of calculation involving considerations of internal work arising in "Higher Structures," the work of Castigliano becomes of fresh interest to engineers. Although the book is now practically forty years old, it is surprising that it is by no means out of date; the reason for this is that though practice may vary, principles are almost invariable. Castigliano's work gives us the most complete analysis of the theory of elasticity applied to the determination of stresses in structures that we have yet met, and we believe that it deserves to receive a close study by all engineers and students who wish to follow the logical development of structural theory and its application to practical design.

In view of the very large number of tabulated numerical values and of the resulting labour involved in translating these values from metric into British units, it has been decided to preserve the metric units; it is thought that this does not detract from the value of the book in demonstrating methods of design for British designers. In order, however, to bring the notation into more modern and standardised form, the mnemonic notation prepared by the Concrete Institute has been followed as far as possible.

The author's thanks are due to Mr. C. Paice of the Examining Staff of H.M. Patent Office for assistance in the troublesome task of reading the proofs, and to Mr. C. Wyndham Hulme, Librarian of H.M. Patent Office, for courtesy in giving special access to the copy of M. Castigliano's book in the Patent Office Library.

EWART S. ANDREWS.

Rolls Chambers, 89 Chancery Lane,
London, W.C. 2, *July*, 1918.

# CONTENTS

# PLATES

*(Following Index)*

# NOTATION

The following is the general notation adopted throughout the book; wherever additional notation is necessitated, its explanation is given in the text:—

| | |
|---|---|
| A | = Area of cross section. |
| B | = Bending moment. |
| C | = Constant. |
| E | = Elastic modulus (Young's modulus). |
| F | = Force (general). |
| F | = Coefficient of shear stress in beam formulæ. |
| G | = Glide, shear or rigidity modulus. |
| H | = Horizontal thrust in arches. |
| I | = Moment of inertia (about a line). |
| $I_p$ | = Polar moment of inertia. |
| M | = Moment (general). |
| M | = Modulus of section. |
| $M_1$ | = First moment of a force. |
| | = First moment about N.A. of area above N.A. in shear in beams. |
| N.A. | = Neutral axis. |
| N | = Node. |
| P | = Pressure or thrust. |
| R | = Radius of an arch. |
| R | = Resultant force. |
| R | = Reaction. |
| S | = Shearing force. |
| T | = Tensile force. |
| $T_M$ | = Twisting moment. |
| V | = Volume. |
| W | = Weight. |
| $W_i$ | = Internal work. |
| X, Y, Z | = Components of a force parallel to the corresponding axes. |
| $a$ | = An arm or distance (general). |
| $a$ | = Length of prismatic body of cylinder. |
| $a$ | = Breadth of a rectangle. |
| $a$ | = Semi-axis of an ellipse. |
| $b$ | = Semi-axis of an ellipse. |
| $b$ | = Depth of a rectangle. |
| $c$ | = Compressive stress. |
| $d$ | = Diagonal length. |
| $d$ | = Depth of a beam (general). |
| $d_n$ | = Deflection. |
| $e$ | = Eccentricity of load. |

| | |
|---|---|
| $f$ | = Force per unit area. |
| $g$ | = Gyration radius. |
| $l$ | = Length (general). |
| | = Span of a beam or arch. |
| $n$ | = A number. |
| | = Neutral axis depth from compressed edge of beam. |
| $r$ | = Rise of an arch. |
| $r$ | = Radius or radial length. |
| $s$ | = Shear stress. |
| $t$ | = Tensile stress. |
| $t^\circ$ | = Temperature. |
| $u$, $v$, $w$ | = Displacements (p. 36). |
| $w$ | = Weight or load per unit length. |
| $x$ | = Extension. |
| $x, y, z$ | = co-ordinates of a point. |
| $\alpha$, $\beta$, $\gamma$ | = Angles. |
| $\epsilon$ | $= \frac{EA}{l}$. |
| $\epsilon$ | = A coefficient (p. 40). |
| $\xi$ | = Increment of $x$. |
| $\eta$ | = Increment of $y$. |
| $\zeta$ | = Increment of $z$. |
| $\lambda$ | = Increment of $l$. |
| $\rho$ | = Increment of $r$. |
| $\kappa$ | = Compressive strain. |
| $\sigma$ | = Shear strain. |
| $\tau$ | = Tensile strain. |
| $\nu$ | = Dilatation or volume strain. |
| $\Pi$ | = Poisson's ratio. [Lateral contraction to longitudinal extension.] |
| $\pi$ | = Circle function 3·14159 . . . |
| $\theta$ | = Angle of torsion. |
| $\Theta$, $\phi$ | = Angles generally. |

# PART I.

## THEORY.

# CHAPTER I.

## FRAMED STRUCTURES.

**1. Definition.**—We will use the term *Rigid Structure* to denote any system which cannot be subjected to any other deformations than those due to its elasticity.

**2. Experimental Law.**—*A rod or homogeneous bar, which is straight and of constant cross-section, will, when pulled or compressed in the direction of its axis, elongate or contract by an amount proportionate to the force to which it is subjected.*

*For specimens of the same material, but of different lengths and cross-sections, the elongations and contractions are directly proportional to the forces and lengths of the bars and inversely proportional to the areas of cross-section.*

If then we call T the tension in a homogeneous bar, $l$ its length, A the area of its cross-section, $x$ its extension, and E a numerical coefficient depending only upon the material of which the bar is composed, we may write

$$x = \frac{\mathrm{T}l}{\mathrm{EA}} \quad . \quad . \quad . \quad . \quad . \qquad (1)$$

from which we obtain

$$\mathrm{T} = \frac{\mathrm{EA}}{l} \, . \, x$$

and writing for brevity

$$\frac{\mathrm{EA}}{l} = \epsilon \quad . \quad . \quad . \quad . \quad . \qquad (2)$$

the formula becomes

$$\mathrm{T} = \epsilon x \quad . \quad . \quad . \quad . \quad . \quad . \qquad (3)$$

**3. Internal Work or Strain Energy of a Rod Stretched or Compressed along its Axis.**—If a rod of original length $l$ is stretched to $l + x$ or compressed to $(l - x)$ by an axial force increasing slowly and uniformly from zero to T, the work done by the force, which is equal to the internal work or strain energy of the rod, will be

$$\int_0^x \mathrm{T}dx = \epsilon \int_0^x x dx = \frac{\epsilon x^2}{2} = \frac{1}{2}\frac{\mathrm{T}^2}{\epsilon} \quad . \quad . \quad . \qquad (4)$$

If, instead of increasing the force on the rod slowly and uniformly from zero to T, a force equal to T is suddenly applied, the elongation will be greater than $\frac{T}{\epsilon}$; in fact, if we call $x'$ the elongation which would have taken place, it is clear that the strain energy will be $\frac{\epsilon x'^2}{2}$, while the work done by the force will be $Tx'$, or since these two must be equal, we shall have the equation

$$Tx' = \frac{\epsilon x'^2}{2}$$

from which we deduce that

$$x' = \frac{2T}{\epsilon}.$$

The force T will thus produce twice the elongation which it would have produced if slowly and uniformly increased from O to T, i.e. twice that which it can maintain; from this it follows that after extension the bar will return to its initial length and will then continue to vibrate, the amplitude of the vibration being $x'$.

4. **Properties of Framed Structures.**—We will call a framed structure one which is composed of bars pin-jointed at their ends and in which each bar is secured only at the ends so that it could easily turn around one end if the other end were free.

These structures have the property that the bars of which they are composed are subjected to axial forces only and that the bending moments are necessarily zero at the nodes.

Let us imagine that there are $n$ nodes or points $N_1$, $N_2$, . . . $N_n$ which have to be connected by bars so that a framed structure of invariable form results. Choose three nodes, e.g. $N_1$, $N_2$, $N_3$, and connect them by three bars forming a triangle and then take another node $N_4$ which is not in the plane of the other three and connect it to them by three more bars; we have thus constructed with six bars a firm frame of four nodes. Now take a fifth node $N_5$ and connect it by means of three bars to three of the four other nodes which are not in the same plane with it, thus giving a firm frame having five nodes and composed of nine bars. Continuing in this way with each additional node and connecting it by three bars to three nodes which are not in the same plane with it, we shall have a firm frame in which the number of nodes will be $n$ and the number of bars $(3n - 6)$.

It will be seen that it is not possible to obtain a firm frame with a smaller number of bars; but that with this number we shall usually be able to build it up in a number of different ways.

If we wish to find the maximum number of bars which we

can employ in forming a framed structure with $n$ nodes, it is first necessary to connect the two nodes $N_1$, $N_2$, and then the node $N_3$ to these two, then the node $N_4$ to all three, and so on, connecting in turn each node to all the preceding ones. Then the number of bars will be

$$1 + 2 + 3 + 4 + \ldots + (n - 1) = \frac{n(n - 1)}{2}.$$

We usually have to consider framed structures possessing more than $(3n - 6)$ bars ; then from among these we will select $(3n - 6)$ so arranged as to be sufficient to form a firm frame, even if the others were absent, and we will call these *essential bars* and the others *redundant bars*.

In general there will be more than one such firm frame ; i.e. there will be several ways of choosing $(3n - 6)$ bars which will give a firm frame. Our reasoning will apply to all the different possible firm frames.

We note that $(3n - 6)$ and $\frac{n(n - 1)}{2}$ are equal for $n = 3$ and $n = 4$ ; i.e. framed structures made with three or four nodes cannot have any redundant bars ; for values of $n$ greater than 4, $\frac{n(n - 1)}{2}$ will always be greater than $(3n - 6)$.

From the manner in which we can build up a firm frame with $n$ nodes and $(3n - 6)$ bars, it is quite clear that in such a structure no stresses can be induced until external forces are applied. Suppose that no external force is applied and that other bars have to be introduced : when these bars are of the exact lengths between the nodes that they have to connect, it is clear that in this structure formed of more than $(3n - 6)$ bars no stresses can occur in the members until external forces are applied. But if the lengths of the additional bars are not exactly equal to the distances between the nodes which they have to connect, it is evident that to get them into the structure they will have to be increased or decreased in length, while the nodes which they have to connect will be brought closer together or driven further apart, and this cannot be done without inducing stresses ; in such a structure, therefore, the bars are stressed even before the loads are applied.

**5. Strain of Framed Structures.**—Now let us assume that we have a framed structure with $n$ nodes $N_1$, $N_2$, . . . $N_n$, and that a force is applied at each node.

Since the stresses in the bars depend upon their strains and therefore upon the relative displacements between the nodes, we will consider these with reference to three axes, taking for origin the node $N_1$, for the axis of $x$ the line $N_1N_2$, and for the plane of $xy$ the plane of the three nodes $N_1$, $N_2$, and $N_3$, which must not be in a straight line.

Now let $N_p$, $N_q$ be any two nodes connected by a bar and let their co-ordinates before strain be $x_p$, $y_p$, $z_p$ and $x_q$, $y_q$, $z_q$; $l_{pq}$ the length of the bar connecting them; and $\alpha_{pq}$, $\beta_{pq}$, $\gamma_{pq}$ the angles which this bar makes with the axes.

After strain let the co-ordinates of $N_p$ be $x'_p = x_p + \xi_p$, $y'_p = y_p + \eta_p$, $z'_p = z_p + \zeta_p$, and let the length of the bar $l'_{pq} = l_{pq} + \lambda_{pq}$ and $\alpha'_{pq}$, $\beta'_{pq}$, $\gamma'_{pq}$ the angles which the bar makes with the axes, and let $T_{pq}$ be the tension in the bar.

Then from § 2 we have

$$T_{pq} = \epsilon_{pq}\lambda_{pq} \left(\epsilon_{pq} \text{ being constant and equal to } \frac{EA_{pq}}{l_{pq}}\right).$$

Now we have

$$l_{pq} = \sqrt{(x_q - x_p)^2 + (y_q - y_p)^2 + (z_q - z_p)^2},$$

$$l_{pq} + \lambda_{pq} = \sqrt{(x_q + \xi_q - x_p - \xi_p)^2 + (y_q + \eta_q - y_p - \eta_p)^2 + (z_q + \zeta_q - z_p - \zeta_p)^2}.$$

We may write the last relation as follows :—

$$l_{pq} + \lambda_{pq} = \sqrt{\left\{\begin{array}{c} l^2_{pq} + 2(x_q - x_p)(\xi_q - \xi_p) + 2(y_q - y_p)(\eta_q - \eta_p) \\ + 2(z_q - z_p)(\zeta_q - \zeta_p) + (\xi_q - \xi_p)^2 + (\eta_q - \eta_p)^2 + (\zeta_q - \zeta_p)^2 \end{array}\right\}}.$$

If the strain of the structure is very small, as we will always assume to be the case, the differences $(\xi_q - \xi_p)$, $(\eta_q - \eta_p)$ and $(\zeta_q - \zeta_p)$ will be very small in comparison with $(x_q - x_p)$, $(y_q - y_p)$, and $(z_q - z_p)$, and we can expand the square root in a converging series, thus obtaining

$$\lambda_{pq} = (\xi_q - \xi_p) \cdot \frac{x_q - x_p}{l_{pq}} + (\eta_q - \eta_p) \cdot \frac{y_q - y_p}{l_{pq}} + (\zeta_q - \zeta_p) \cdot \frac{z_q - z_p}{l_{pq}} + \theta_{pq}$$

where $\theta_{pq}$ represents the sum of all the terms of the series which contain powers of $(\xi_q - \xi_p)$, etc., above the first. The ratio $\frac{\theta_{pq}}{l_{pq}}$ has a limiting value of zero when these differences tend towards zero.

Now since

$$\left.\begin{aligned} \cos \alpha_{pq} &= \frac{x_q - x_p}{l_{pq}} \\ \cos \beta_{pq} &= \frac{y_q - y_p}{l_{pq}} \\ \cos \gamma_{pq} &= \frac{z_q - z_p}{l_{pq}} \end{aligned}\right\} \quad . \quad . \quad . \quad . \quad (5)$$

the above formula may be written

$$\lambda_{pq} = (\xi_q - \xi_p) \cos \alpha_{pq} + (\eta_q - \eta_p) \cos \beta_{pq} + (\zeta_q - \zeta_p) \cos \gamma_{pq} + \theta_{pq},$$

and writing for brevity,

$$\lambda'_{pq} = (\xi_q - \xi_p) \cos \alpha_{pq} + (\eta_q - \eta_p) \cos \beta_{pq} + (\zeta_q - \zeta_p) \cos \gamma_{pq},$$

we have

$$\lambda_{pq} = \lambda'_{pq} + \theta_{pq}.$$

After strain we have

$$\cos \alpha'_{pq} = \frac{x'_q - x'_p}{l_{pq}} = \frac{x_q + \xi_q - x_p - \xi_p}{l_{pq}}$$

$$\cos \beta'_{pq} = \frac{y'_q - y'_p}{l_{pq}} = \frac{y_q + \eta_q - y_p - \eta_p}{l_{pq}}$$

$$\cos \gamma'_{pq} = \frac{z'_q - z'_p}{l_{pq}} = \frac{z_q + \zeta_q - z_p - \zeta_p}{l_p}.$$

We can expand these expressions in converging series in increasing powers of $(\xi_q - \xi_p)$, $(\eta_q - \eta_p)$, $(\zeta_q - \zeta_p)$, giving

$$\left.\begin{aligned}\cos \alpha'_{pq} &= \frac{x_q - x_p}{l_{pq}} + \omega_{pq} = \cos \alpha_{pq} + \omega_{pq}\\ \cos \beta'_{pq} &= \frac{y_q - y_p}{l_{pq}} + \omega'_{pq} = \cos \beta_{pq} + \omega'_{pq}\\ \cos \gamma'_{pq} &= \frac{z_q - z_p}{l_{pq}} + \omega''_{pq} = \cos \gamma_{pq} + \omega''_{pq}\end{aligned}\right\} \quad . \quad . \quad (6)$$

where $\omega_{pq}$, $\omega'_{pq}$, $\omega''_{pq}$ represent quantities of the same order of magnitude as the differences $(\xi_q - \xi_p)$, etc., and which therefore also tend towards zero when these quantities do so.

**6. Equations for the Equilibrium of Framed Structures.**—Since after strain the structure is in equilibrium, it is clear that the stresses in all the bars meeting at any node must equilibrate the external force applied at this node.

If now we call $X_p$, $Y_p$, $Z_p$ the components parallel to the axes of the external force applied at the node $N_p$ we shall have for the equilibrium of this node the three equations

$$\left.\begin{aligned}X_p + \Sigma T_{pq} \cos \alpha'_{pq} &= 0\\ Y_p + \Sigma T_{pq} \cos \beta'_{pq} &= 0\\ Z_p + \Sigma T_{pq} \cos \gamma'_{pq} &= 0\end{aligned}\right\} \quad . \quad . \quad . \quad . \quad (7)$$

which from the previously found values of $\cos \alpha'_{pq}$, etc., become

$$\left.\begin{aligned}X_p + \Sigma\left(T_{pq} \cdot \frac{x'_q - x'_p}{l_{pq}}\right) &= 0\\ Y_p + \Sigma\left(T_{pq} \cdot \frac{y'_q - y'_p}{l_{pq}}\right) &= 0\\ Z_p + \Sigma\left(T_{pq} \cdot \frac{z'_q - z'_p}{l_{pq}}\right) &= 0\end{aligned}\right\} \quad . \quad . \quad . \quad (8)$$

Multiplying the second of these equations by $z'_p$ and subtracting it from the third multipled by $y'_p$, we obtain the first of the following equations, and the other two are similarly obtained :—

$$\left.\begin{aligned}Z_p y'_p - Y_p z'_p + \Sigma\left(T_{pq} \cdot \frac{y'_p z'_q - y'_q z'_p}{l_{pq}}\right) &= 0\\ X_p z'_p - Z_p x'_p + \Sigma\left(T_{pq} \cdot \frac{x'_q z'_p - x'_p z'_q}{l_{pq}}\right) &= 0\\ Y_p x'_p - X_p y'_p + \Sigma\left(T_{pq} \cdot \frac{x'_p y'_q - x'_q y'_p}{l_{pq}}\right) &= 0\end{aligned}\right\} \quad . \quad . \quad (9)$$

We thus have for each node of the structure two sets of equations similar to (8) and (9). Now if we add together all the equations deduced from each of these six equations by putting successively $p = 1, p = 2$, etc. . . . $p = n$, we obtain the following six equations which are independent of the stresses in the bars:—

$$\left.\begin{aligned} \Sigma X_p = 0, \ \Sigma(Z_p y'_p - Y_p z'_p) = 0 \\ \Sigma Y_p = 0, \ \Sigma(X_p z'_p - Z_p x'_p) = 0 \\ \Sigma Z_p = 0, \ \Sigma(Y_p x'_p - X_p y'_p) = 0 \end{aligned}\right\} \quad . \quad . \quad (10)$$

It is easily seen that these six equations are the six fundamental equations for stable structures; it follows from this that the structure which is considered stable after strain is in equilibrium under the action of the external forces.

Since for each node of the structure there are three equations similar to equations (8), it will be seen that for the whole structure we shall have $3n$ equations between the stresses in the bars: and since we can by combining them deduce the six equations (10) which are independent of the stresses, it follows that the useful equations for the determination of the stresses reduce to $3n - 6$.

But it should not be thought that the six equations (10) can be substituted for any six equations similar to (8); in fact the latter are such that if the stresses in all the bars were given, we could find from them the external forces. If we take only $(3n - 6)$ of the equations similar to (8) they would determine only $(3n - 6)$ of the external forces in terms of the stresses in the bars, and we should have to use the six equations (10) to determine the remaining six external forces.

It will be clear that these equations would give results of the form $\frac{0}{0}$ if we wished to employ them to determine the six external forces applied at only two nodes; we could not then substitute equations (10) for the two groups of equations similar to (8) corresponding to the two nodes only.

On the other hand, if we continue for the sake of simplicity to take $N_1$ as the origin and make the axis of $x$ pass through $N_2$ and the plane of $xy$ through the node $N_3$, we see that the six equations (10) are sufficient to determine the three forces $X_1, Y_1, Z_1$ applied at the first node, the two forces $Y_2, Z_2$ applied at the second, and the force $Z_3$ applied at the third; they can then be substituted in the three equations of equilibrium with regard to the node $N_1$, in the two equations with regard to the node $N_2$ containing the components parallel to the axes of $y$ and $z$, and finally in the equation with regard to the node $N_3$ containing the components parallel to the axis of $z$. We will therefore in future write only these six equations.

It should be noted that the remaining $(3n - 6)$ equations between the stresses in the bars and the external forces are exactly those which we should obtain if the node $N_1$ were fixed, the node $N_2$ could only slide along the axis of $x$, and the node $N_3$ were forced to remain in the plane of $xy$.

This must be the case because with those restrictions while it is true that we prevent general movement of the structure, the relative movements can take place freely. In this case the equations (10) will still serve to determine the forces $X_1$, $Y_1$, $Z_1$, $Y_2$, $Z_2$, $Z_3$, which will express the reaction forces caused by this method of fixing.

**7. Simplification of the Equations (7).**—The $(3n - 6)$ equations (7) contain the final stresses in the bars and the angles which they make with the axes after deformation.

We have seen from § 5 that

$$T_{pq} = \epsilon_{pq}\lambda_{pq} = \epsilon_{pq}(\lambda'_{pq} + \theta_{pq})$$

and

$$\cos \alpha'_{pq} = \cos \alpha_{pq} + \omega_{pq}$$

where $\lambda'_{pq}$ is a function of the first degree of the displacements of the nodes $N_p$, $N_q$ given by the formula

$$\lambda'_{pq} = (\xi_q - \xi_p) \cos \alpha_{pq} + (\eta_q - \eta_p) \cos \beta_{pq} + (\zeta_q - \zeta_p) \cos \gamma_{pq} ;$$

the quantity $\theta_{pq}$ contains only powers of $(\xi_q - \xi_p)$, etc., above the first, and the quantity $\omega_{pq}$ contains the first and higher powers of these differences.

Regarding these quantities as small quantities of the first order, we see that $\lambda'_{pq}$ and $\omega_{pq}$ are of the same order, while $\theta_{pq}$ is of the second order.

Now we may write for the quantity $T_{pq} \cos \alpha'_{pq}$ which occurs in equations (7),

$$\begin{aligned} T_{pq} \cos \alpha'_{pq} &= \epsilon_{pq}(\lambda'_{pq} + \theta_{pq})(\cos \alpha_{pq} + \omega_{pq}) \\ &= \epsilon_{pq}(\lambda'_{pq} \cos \alpha_{pq} + \lambda'_{pq}\omega_{pq} + \theta_{pq} \cos \alpha_{pq} + \theta_{pq}\omega_{pq}) ; \end{aligned}$$

the first term $\lambda'_{pq} \cos \alpha_{pq}$ inside the brackets is a small quantity of the first order, the next two are of the second order, and the last is of the third order. If, therefore, we restrict our degree of accuracy to quantities of the first order as has always been done up till now, which degree is moreover nearly always sufficient in questions upon elastic structures, we can neglect the last three terms, thus having

$$T_{pq} \cos \alpha'_{pq} = \epsilon_{pq}\lambda'_{pq} \cos \alpha_{pq},$$

then calling $T_{pq}$ the final stress in the bar $N_pN_q$ which is obtained by neglecting the quantity $\theta_{pq}$ of the second order in comparison with the first order quantity $\lambda_{pq}$, i.e. putting

$$T_{pq} = \epsilon_{pq}\lambda_{pq}$$

we have
$$T_{pq} \cos a'_{pq} = T_{pq} \cos a_{pq}.$$

We see, therefore, that, to the degree of accuracy which we have in view, the equations (7) may be written

$$\left.\begin{aligned} X_p + \Sigma(T_{pq} \cos a_{pq}) &= 0 \\ Y_p + \Sigma(T_{pq} \cos \beta_{pq}) &= 0 \\ Z_p + \Sigma(T_{pq} \cos \lambda_{pq}) &= 0 \end{aligned}\right\} \quad . \quad . \quad . \quad (11)$$

Although these equations are of the same form as equations (7), we see that they are much simpler because they contain the angles which the bars make with the axes before strain instead of after strain; and instead of the exact stresses $T'_{pq}$ they contain the approximate stresses $T_{pq}$ which are linear functions of the displacements $\xi_p$, $\eta_p$, $\zeta_p$.

In future, therefore, we will always use equations (11).

**8. Determination of the Stresses in the Bars.**—The $(3n - 6)$ equations (11) between the stresses in the bars and the external forces enable these stresses to be determined whatever their number may be.

We will examine separately two cases.

(1) The number of bars will be taken as exactly equal to $(3n - 6)$, i.e. equal to the number required for a perfect or firm frame. In this case if we cut through one of the bars, the structure ceases to be rigid and some movement will tend to take place; to prevent this movement we must apply to the severed bar a force which can be determined by the ordinary laws of statics, and which is equal to the stress to which the bar was subjected before being severed. We see, therefore, that when a framed structure is composed of only $(3n - 6)$ bars the stresses in the bars can be found by means of the ordinary laws of statics. This follows also from the analysis, because we have then $(3n - 6)$ equations similar to (11) between the $(3n - 6)$ unknown tensions.

(2) Now suppose that the number of bars is greater than $(3n - 6)$. In this case some bars could be omitted without causing collapse of the structure. It is now no longer possible to determine the stresses by the ordinary laws of statics; but, if we take elasticity into account, we see that all the stresses can be found, because, as the structure becomes strained, all the bars will elongate or contract, and we know already that these elongations and contractions determine the tension or compression in each bar.

Now let us see in what manner we may proceed to determine the unknown stresses.

Taking the node $N_1$ as origin and making the axis of $x$ pass through $N_2$ and the plane of $x$, $y$ through $N_3$, the following are the $(3n - 6)$ equations which express the elastic equilibrium of the structure, assuming the six fundamental equations of equilibrium to be satisfied already:—

$$\left.\begin{array}{l}
X_2 + \Sigma T_{2q} \cos \alpha_{2q} = 0 \\
X_3 + \Sigma T_{3q} \cos \alpha_{3q} = 0 ; \ Y_3 + \Sigma T_{3q} \cos \beta_{3q} = 0 \\
X_4 + \Sigma T_{4q} \cos \alpha_{4q} = 0 ; \ Y_4 + \Sigma T_{4q} \cos \beta_{4q} = 0 ; \ Z_4 + \Sigma T_{4q} \cos \gamma_{4q} = 0 \\
\cdot \quad \cdot \quad \cdot \quad \cdot \quad \cdot \quad \cdot \quad \cdot \quad \cdot \quad \cdot \quad \cdot \quad \cdot \quad \cdot \\
X_p + \Sigma T_{pq} \cos \alpha_{pq} = 0 ; \ Y_p + \Sigma T_{pq} \cos \beta_{pq} = 0 ; \ Z_p + \Sigma T_{pq} \cos \gamma_{pq} = 0 \\
\cdot \quad \cdot \quad \cdot \quad \cdot \quad \cdot \quad \cdot \quad \cdot \quad \cdot \quad \cdot \quad \cdot \quad \cdot \quad \cdot
\end{array}\right\} \quad (12)$$

If now we call $\xi_2$ the displacement of the node $N_2$ on the axis of $x$; $\xi_3$, $\eta_3$ the displacements of $N_3$ parallel to $x$ and $y$; and $\xi_4\eta_4\zeta_4 \ldots \xi_p\eta_p\zeta_p$ the displacements of $N_4 \ldots N_p$ parallel to these three axes, we shall have the following formulæ for the stresses in all the bars:—

$$\left.\begin{array}{l}
T_{1q} = \epsilon_{1q}[\xi_q \cos \alpha_{1q} + \eta_q \cos \beta_{1q} + \zeta_q \cos \gamma_{1q}], \\
T_{2q} = \epsilon_{2q}[(\xi_q - \xi_2) \cos \alpha_{2q} + \eta_q \cos \beta_{2q} + \zeta_q \cos \gamma_{2q}], \\
T_{3q} = \epsilon_{3q}[(\xi_q - \xi_3) \cos \alpha_{3q} + (\eta_q - \eta_3) \cos \beta_{3q} + \zeta_q \cos \gamma_{3q}], \\
\cdot \quad \cdot \quad \cdot \quad \cdot \quad \cdot \quad \cdot \quad \cdot \quad \cdot \quad \cdot \quad \cdot \quad \cdot \quad \cdot \\
T_{pq} = \epsilon_{pq}[(\xi_q - \xi_p) \cos \alpha_{pq} + (\eta_q - \eta_p) \cos \beta_{pq} + (\zeta_q - \zeta_p) \cos \gamma_{pq}] \\
\cdot \quad \cdot \quad \cdot \quad \cdot \quad \cdot \quad \cdot \quad \cdot \quad \cdot \quad \cdot \quad \cdot \quad \cdot \quad \cdot
\end{array}\right\} \quad (13)$$

If therefore we know the $(3n - 6)$ displacements

$$\begin{array}{c}
\xi_2 \\
\xi_3\eta_3 \\
\xi_4\eta_4\zeta_4 \\
\cdot \quad \cdot \quad \cdot \\
\xi_p\eta_p\zeta_p \\
\cdot \quad \cdot \quad \cdot
\end{array}$$

we shall obtain the stresses in all the bars.

If, therefore, in equations (12) we substitute for these stresses their expressions (13), the only unknowns will be the $(3n - 6)$ displacements of the nodes, i.e. a number of unknowns exactly equal to the number of equations.

As these equations are of the first degree, their solution will not present any difficulty.

Having found these displacements, we substitute in equations (13) to obtain the stresses in all the bars of the structure.

**9. Notes on Absolutely Rigid Structures.**—The values of the displacements which we obtain from the above formulæ can be expressed in the form of two determinants of order $(3n - 6)$. The determinant forming the denominator is the same for all the displacements, and each of its terms contains as a factor one of the coefficients $\epsilon$ so that it is a homogeneous function of the $(3n-6)$th degree with respect to these coefficients. The determinants which form the numerators can be all deduced from that of the denominator by substituting for the elements of one column the known terms of equations (12), i.e. the forces $X_2$, $X_3$, $Y_3$; . . . $X_p$, $Y_p$, $Z_p$ . . .; the numerators, therefore, are also determinants of the $(3n-6)$th order, but with respect to the coefficients $\epsilon$ they are homogeneous functions of the order $(3n-7)$.

If then we substitute in equations (13) these expressions for the displacements, we see that each numerator will be multiplied by one of the quantities $\epsilon$ and that it becomes then a homogeneous function of the $(3n-6)$th order of these quantities. Thus each of the stresses will be expressed in terms of two homogeneous functions of the $(3n-6)$th order of the coefficients $\epsilon$ and therefore will not depend upon the absolute values of these quantities, but only upon their relative values.

It follows, therefore, that if we increase, for example, all the sections of the bars in the same proportion, which will not alter the relative values of the coefficients $\epsilon$, the stresses in the bars of the structure will remain exactly the same.

We can easily see the reason for this property if we note that when all the coefficients $\epsilon$ are increased in the same proportion, the elongations or contractions of the bars diminish in the same proportion, and that, therefore, the ratios between these elongations and contractions will remain the same.

But in absolutely rigid structures, as there are no elongations and contractions of the bars there will be no fixed ratio between these quantities and therefore none between the coefficients $\epsilon$; it follows, therefore, that the stresses in the bars are indeterminate.

These considerations show the error which MM. Vène, Cournot, and other authorities have made in believing possible the determination of the stresses in absolutely rigid structures, considered by them as the limits of elastic structures in which the elasticity diminishes indefinitely. In fact, even admitting the soundness of this idea, it does not follow that the stresses can be found in rigid structures; because, for that to be possible, it would be necessary that when the elasticity diminishes indefinitely, i.e. when the coefficients of $\epsilon$ increase indefinitely, the ratios between these quantities should have finite, determinate limits, which is not the case.

Moreover, the determination of the stresses in absolutely rigid structures is of no importance since such structures do not exist.

It should be noted that the above reasoning applies only to

structures having $n$ nodes and more than $(3n - 6)$ bars, because in those which have only $(3n - 6)$ bars the stresses are determined, as we have seen, by the ordinary laws of statics.

**10. Expression of the Internal Work as a Function of the External Forces.**—We have derived the formula

$$\frac{T_{pq}}{\epsilon_{pq}} = (\xi_q - \xi_p)\cos\alpha_{pq} + (\eta_q - \eta_p)\cos\beta_{pq} + (\zeta_q - \zeta_p)\cos\gamma_{pq}:$$

If we note that

$$\cos\alpha_{pq} = -\cos\alpha_{qp};\ \cos\beta_{pq} = -\cos\beta_{qp};\ \cos\gamma_{pq} = -\cos\gamma_{qp};$$

this formula becomes

$$\frac{T_{pq}}{\epsilon_{pq}} + (\xi_p\cos\alpha_{pq} + \eta_p\cos\beta_{pq} + \zeta_p\cos\gamma_{pq}) + (\xi_q\cos\alpha_{qp} + \eta_q\cos\beta_{qp} + \zeta_q\cos\gamma_{qp}) = 0.$$

Multiply all these terms by $\frac{T_{pq}}{2}$ and then add together the similar equations for all the bars of the structure. If we note that any displacement, e.g. $\xi_p$, occurs in all the equations involving the bars meeting at the node $N_p$, we can write the sum in the following way:—

$$\begin{aligned}\tfrac{1}{2}\Sigma\left(\frac{T^2_{pq}}{\epsilon_{pq}}\right) &+ \tfrac{1}{2}\xi_2\Sigma T_{2q}\cos\alpha_{2q}\\ &+ \tfrac{1}{2}\xi_3\Sigma T_{3q}\cos\alpha_{3q} + \tfrac{1}{2}\eta_3\Sigma T_{3q}\cos\beta_{3q}\\ &+ \tfrac{1}{2}\xi_4\Sigma T_{4q}\cos\alpha_{4q} + \tfrac{1}{2}\eta_4\Sigma T_{4q}\cos\beta_{4q} + \tfrac{1}{2}\zeta_4\Sigma T_{4q}\cos\gamma_{4q}\\ &\quad\cdot\quad\cdot\quad\cdot\quad\cdot\quad\cdot\quad\cdot\quad\cdot\quad\cdot\\ &+ \tfrac{1}{2}\xi_p\Sigma T_{pq}\cos\alpha_{pq} + \tfrac{1}{2}\eta_p\Sigma T_{pq}\cos\beta_{pq} + \tfrac{1}{2}\zeta_p\Sigma T_{pq}\cos\gamma_{pq}\\ &\quad\cdot\quad\cdot\quad\cdot\quad\cdot\quad\cdot\quad\cdot\quad\cdot\quad\cdot\end{aligned}$$

Substituting for $\Sigma T_{2q}\cos\alpha_{2q}$, etc., their values $-X_2$, etc., given by equations (12) this formula becomes

$$\tfrac{1}{2}\Sigma\left(\frac{T_{pq}^{\ 2}}{\epsilon_{pq}}\right) = \tfrac{1}{2}\left(\begin{aligned}X_2\xi_2 + X_3\xi_3 + Y_3\eta_3 + X_4\xi_4 + Y_4\eta_4 + Z_4\zeta_4 +\\ \ldots + X_p\xi_p + Y_p\eta_p + Z_p\zeta_p + \ldots\end{aligned}\right)$$

or $$\tfrac{1}{2}\Sigma\left(\frac{T_{pq}^{\ 2}}{\epsilon_{pq}}\right) = \tfrac{1}{2}\Sigma(X_p\xi_p + Y_p\eta_p + Z_p\zeta_p).$$

Now the first term of this equation expresses the internal work of the whole structure (§ 3) so that we have:

Internal work of structure

$$= W_i = \tfrac{1}{2}\Sigma(X_p\xi_n + Y_p\eta_p + Z_n\zeta_p) \qquad . \quad . \quad (14)$$

*This depends only upon the final values of the exterior forces and not upon the manner in which these forces have increased from zero to their final values.*

If we call

$F_p$ = the force of which $X_p$, $Y_p$, $Z_p$ are the components parallel to the axes,

$\rho_p$ = the length of the short straight line which joins the final position of this force to its initial position, i.e.

the length of which $\xi_p$, $\eta_p$, $\zeta_p$ are the projections on the axes,

$a, b, c =$ the angles which $F_p$ makes with the axes,

$\alpha, \beta, \gamma =$ " " $\rho_p$ " " "

we then have

$$X_p = F_p \cos a,\ Y_p = F_p \cos b,\ Z_p = F_p \cos c,$$
$$\xi_p = \rho_p \cos \alpha,\ \eta_p = \rho_p \cos \beta,\ \zeta_p = \rho_p \cos \gamma,$$

and therefore

$$X_p\xi_p + Y_p\eta_p + Z_p\zeta_p = F_p\rho_p (\cos a \cos \alpha + \cos b \cos \beta + \cos c \cos \gamma).$$

Now if $\theta$ is the angle between $F_p$ and $\rho_p$ we have

$$\cos \theta = \cos a \cos \alpha + \cos b \cos \beta + \cos c \cos \gamma,$$

$\therefore$ the right-hand side of the above equation becomes $F_p\rho_p \cos \theta$. But $\rho_p \cos \theta$ represents the projection in the direction of the force $F_p$ of the short displacement $\rho_p$, and if we call this $r_p$ we have

$$X_p\xi_p + Y_p\eta_p + Z_p\zeta_p = F_p r_p$$

and formula (14) becomes

$$W_i = \tfrac{1}{2}\Sigma F_p r_p. \quad . \quad . \quad . \quad . \quad (15)$$

If $F_p$ is the resultant of several forces $Q_p$, $S_p$, etc., applied at the node $N_p$ in any direction, and if we call $q_p$, $s_p$ the projections of the displacement of $N_p$ in these directions, we know that by the general principle of work done

$$F_p r_p = Q_p q_p + S_p s_p + \ldots .$$

Now the terms of the right-hand side are exactly similar to that on the left-hand side, i.e. each is a product of a force and the projection in its direction of the displacement of its point of application.

The internal work of the whole structure will therefore be still expressed by a formula similar to (15), but into which each of the forces applied will enter. We can still continue to use formula (15), but it is convenient for $F_p$ to represent one of the forces applied at the node $N_p$ instead of their resultant. We thus obtain the following rule :—

*The internal work of a framed structure is equal to half the sum of the products obtained by multiplying each external force applied to the structure by the displacement of its point of application in the direction of the force.*

It should be noted that in the product $F_p\rho_p \cos \theta$ we suppose that the force $F_p$ and the effective displacement $\rho_p$ of its point of application are positive; because the positions of the lines represented by $F_p$ and by $\rho_p$ around the node $N_p$ are determined by the angles $a, b, c, \alpha, \beta, \gamma$.

Now we know that the value of $\cos \theta$ will be positive or negative according to whether $\theta$ is less or greater than a right angle, i.e. according to whether the projection of $\rho_p$ upon the direction of $F_p$ falls on the same or on the opposite side of the point as the line of action of the force.

*Therefore the projection of the displacement of a point upon the line of action of a force at that point will be taken as positive or negative according as it falls on the same or on the opposite side of the point as the direction of the force.*

For brevity we will call in future the projection $r_p$ of the displacement in the direction of a force $F_p$ the *relative displacement of the point of action of the force.*

**11. Theorem of the Differential Coefficients of the Internal Work.**—*Part* 1.—*If the internal work of a framed structure is expressed as a function of the relative displacements of the external forces applied at its nodes, the resulting expression is such that its differential coefficients with regard to these displacements give the values of the corresponding forces.*

*Part* 2.—*If, on the contrary, the internal work of a framed structure is expressed as a function of the external forces, the resulting expression is such that its differential coefficients give the relative displacements of their points of application.*

In fact if we give to the external forces $F_p$ infinitely small increments $dF_p$, the frame will be subjected to infinitely small strains in which the relative displacement of the point of application of the force $F_p$ will increase from the value $r_p$ by an amount $dr_p$ and the work done by this force during the small displacement will be $F_p dr_p$.

It follows that the total increase of internal work due to the increments $dF_p$ of the external forces will be given by

$$dW_i = \Sigma F_p dr_p.$$

Now if we call $W_i$ the total internal work of the structure due to the forces $F_p$, expressed as a function of the relative displacements of the points of application of the external forces, it is clear that the increase of $W_i$ due to the increments $dr_p$ of the relative displacements of the nodes will be given by

$$\Sigma \frac{dW_i}{dr_p} dr_p.$$

Since it follows that this result must be identically the same as the previous one whatever be the values of $dr_p$, it follows that for each force we must have

$$F_p = \frac{dW_i}{dr}$$

which proves the first part of the theorem.

For the second part we will note that increase of the internal work of the structure, due to the increments $dF_p$ of the external forces, must be also represented by the differential of formula (15), i.e.

$$dW_i = \tfrac{1}{2}\Sigma F_p dr_p + \tfrac{1}{2}\Sigma r_p dF_p.$$

We therefore have the equation

$$\Sigma F_p dr_p = \tfrac{1}{2}\Sigma F_p dr_p + \tfrac{1}{2}\Sigma r_p dF_p$$

from which we get

$$\Sigma F_p dr_p = \Sigma r_p dF_p;$$

and as the left-hand side represents the internal work of the structure for the increments $dF_p$ of the external forces, the right-hand side must express the same quantity.

Now it is clear that the increment of internal work due to the increments $dF_p$ will be expressed by the formula

$$dW_i = \Sigma \frac{dW_i}{dF_p} \cdot dF_p.$$

As this result must be identical with $\Sigma r_p dR_p$, it follows that for each force we must have

$$r = \frac{dW_i}{dF_p}$$

which proves the second part of the theorem.

**12. Continuation of the Theorem of the Differential Coefficients of the Internal Work.**—To avoid falling into an error in the application of the preceding theorem, it is necessary to regard all the external forces as independent variables, and therefore to represent them by different letters.

Let us examine what we obtain by taking the differential coefficient of the internal work of the structure with regard to one letter P when two equal forces P, parallel and opposite in direction, are applied to the structure.

Suppose at first that the two forces P are represented by different letters P, P′, and that we take the differential coefficient of the internal work of the whole structure with regard to P, treating P′ as a function of P. If we represent by $W_i$ the function of the external forces which gives the strain energy, its differential coefficient with regard to P will be

$$\frac{dW_i}{dP} + \frac{dW_i}{dP'} \cdot \frac{dP'}{dP},$$

but since $P' = P$ and $\frac{dP'}{dP} = 1$, the differential coefficient of $W_i$ with regard to P becomes

$$\frac{dW_i}{dP} + \frac{dW_i}{dP'}.$$

Now $\frac{dW_i}{dP}$, $\frac{dW_i}{dP'}$ express the paths * of the two forces P and P′; and it should be noted that since these forces are in opposite directions, the differential coefficients will be of the same or opposite signs according as the paths A*a*, A′*a*′ take place in the opposite or the same sense (Pl. I, Fig. 1).

Let us now examine two very important special cases :—

CASE 1.—*The two forces act in the line joining their points of application.*

In this case the sum $\frac{dW_i}{dP} + \frac{dW_i}{dP'}$, i.e. the sum of the two paths A*a*, A′*a*′, is the distance by which the points of application A, A′ get further apart, supposing that the forces P, P′ have the directions shown in Fig. 2. In the same case if the sum $\frac{dW_i}{dP} + \frac{dW_i}{dP'}$ comes negative it means that the points A, A′ have come closer together.

When the two forces P, P′ have the directions shown in Fig. 3, i.e. when they tend to make the points A, A′ approach each other, the sum $\frac{dW_i}{dP} + \frac{dW_i}{dP'}$ expresses the contraction of the line AA′ if positive and the elongation if negative.

CASE 2.—*The forces P, P′ are at right angles to the line AA′* (Fig. 4, Pl. I) *joining their points of application.*

In this second case, calling M the moment of the couple formed by the two forces P, P′, we have

$$M = P \,.\, AA' = P' \,.\, AA',$$

and we can regard the internal work $W_i$ as a function of the variables P, P′ which in their turn are regarded as functions of M: thus the differential coefficient of $W_i$ with regard to the latter quantity will be

$$\frac{dW_i}{dM} = \frac{dW_i}{dP} \cdot \frac{dP}{dM} + \frac{dW_i}{dP'} \cdot \frac{dP'}{dM}$$

but since

$$\frac{dP}{dM} = \frac{1}{AA'} = \frac{dP'}{dM},$$

we have

$$\frac{dW_i}{dM} = \frac{1}{AA'}\left(\frac{dW_i}{dP} + \frac{dW_i}{dP'}\right).$$

If the straight line joining the points of application of the forces, which had the position AA′ before deformation, and which after

* Our friend, Engineer Frances. Crotti, who has read some papers before the Milan Society of Engineers and the Lombardy Institute upon our previous works, has proposed to call the relative displacement of the point of application of a force its "path".

deformation occupies such a position that its projection upon the plane of the couple is CC′, it is clear that the angle through which AA′ will have turned around an axis perpendicular to the plane of the couple will be the angle between the lines AA′, CC′. Now if C*b* is drawn parallel to AA′ and C′*b* is drawn perpendicular to AA′, and AA′ is produced to *a*, we see that C′*a* and *ab* will be the projections of the displacements A′C′, AC on the directions of the forces, i.e. the paths of these forces, and will be given by $\frac{dW_i}{dP}$ and $\frac{dW_i}{dP'}$.

Tan $\angle C'Cb = \frac{C'b}{Cb} = \frac{C'b}{AA'}$, neglecting the very small quantities A′*a*, C*c* in comparison with AA′. As the $\angle C'Cb$ is very small, it may be taken as equal to its tangent, and we may say, therefore, that it is expressed by the formula

$$\angle C'Cb = \frac{C'b}{AA'} = \frac{C'a + ab}{AA} = \frac{1}{AA'}\left(\frac{dW_i}{dP} + \frac{dW_i}{dP'}\right) = \frac{dW_i}{dM}.$$

The result proves the following theorem.

*The differential coefficient of the internal work of a framed structure, with reference to the moment of a couple composed of two forces perpendicular to the line joining their points of application, gives the angular rotation of this line about an axis perpendicular to the plane of the couple.*

As the partial differential coefficients $\frac{dW_i}{dP}, \frac{dW_i}{dP'}$ are positive or negative according as the displacements A*c*, *a*C′ fall in the same direction or the opposite direction as the forces, it follows that $\frac{1}{AA}\left(\frac{dW_i}{dP} + \frac{dW_i}{dP'}\right)$ will be positive or negative according as the rotation of AA′ is in the same or the opposite sense as that of the couple.

**13. The Theorem of Least Work.**—*If we find the least value of the function* $\frac{1}{2}\Sigma\frac{T_{pq}^2}{\epsilon_{pq}}$ *which gives the internal work of a framed structure taking into account the* $(3n - 6)$ *equations* (12) *between the stresses in all the bars of the system, we obtain the values of these stresses.*

We will note in the first place that after deformation each bar $N_pN_q$ may be replaced by two forces equal to $T_{pq}$ applied at the nodes, these forces acting towards each other if the bar is in tension and away from each other if it is in compression.

Now from among the bars, whose number is assumed to be greater than $(3n - 6)$, choose $(3n - 6)$ such that they form in themselves a firm frame. Then all the other bars might be omitted without altering the conditions of equilibrium of the structure, provided that at the two nodes connected by each we apply forces equal to the stress in the bar and directed in the manner set out above. The $(3n - 6)$ equations (12) enable us to determine the stresses in the essential bars, as a function of the forces applied at the nodes and of the stresses in the redundant bars.

But as the latter are unknown, it will be necessary to determine them by expressing the geometrical conditions which the structure must satisfy.

These are very simple conditions because, if, in the structure reduced to the essential bars with the redundant bars replaced by the equivalent forces, we consider two nodes connected by one of the bars omitted, it is clear that in the deformation of the structure these nodes will recede from or approach each other by the same amount as in the firm frame, i.e. by an amount equal to the extension or contraction of the omitted bar.

Now let $W_i$ be the internal work of the frame reduced to $(3n - 6)$ essential bars, this work being expressed as a function of the external forces and of the stresses in the omitted bars.

If $N_p$, $N_q$ are two nodes connected by one of the latter bars and $T_{pq}$ the tension in it, it follows from case (1) of § 12 that the differential co-efficient of $W_i$ with regard to $T_{pq}$ expresses the amount by which the nodes approach each other.

But under the action of the tension $T_{pq}$ the bar $N_p$, $N_q$ extends by an amount equal to $\frac{T_{pq}}{\epsilon_{pq}}$, so that we have

$$\frac{dW_i}{dT_{pq}} = -\frac{T_{pq}}{\epsilon_{pq}}$$

i.e.

$$\frac{dW_i}{dT_{pq}} + \frac{T_{pq}}{\epsilon_{pq}} = 0.$$

We shall have a similar equation for each of the omitted bars and therefore as many equations as unknowns.

Now the internal work of the original structure, i.e. that containing not only the essential bars but also the redundant ones, is given by the formula

$$W_i + \tfrac{1}{2}\Sigma\left(\frac{T_{pq}^2}{\epsilon_{pq}}\right)$$

the sign $\Sigma$ containing as many terms as there are redundant bars.

The above equations which express all the geometrical conditions which the structure must satisfy are the equations to zero of the differential co-efficients of the internal work for the whole structure; and therefore the stresses which occur after strain are those which make this work a minimum.

**14. Alternative Proof of the Theorem of Least Work.**—We will give here another proof, given first in 1873, which results in a remarkable property.

To find the minimum value of the function $\frac{1}{2}\Sigma\frac{T_{pq}^2}{\epsilon_{pq}}$, taking into account the equations (12) in which the external forces must be regarded as constants, we have the equation

$$\Sigma\frac{T_{pq}dT_{pq}}{\epsilon_{pq}} = 0 \quad . \quad . \quad . \quad . \quad (16)$$

in which the tensions $T_{pq}$ are given by the $(3n - 6)$ equations (12) and their differentials $dT_{pq}$ by the following $(3n - 6)$ equations obtained by differentiation of (12).

$$\left.\begin{array}{l} \Sigma dT_{2q} \, . \cos \alpha_{2q} = 0 \\ \Sigma dT_{3q} \, . \cos \alpha_{3q} = 0, \ \Sigma dT_{3q} \, . \cos \beta_{3q} = 0 \\ \Sigma dT_{4q} \, . \cos \alpha_{4q} = 0, \ \Sigma dT_{4q} \, . \cos \beta_{4q} = 0, \ \Sigma dT_{4q} \, . \cos \gamma_{4q} = 0 \\ \quad . \quad . \quad . \quad . \quad . \quad . \quad . \quad . \quad . \quad . \end{array}\right\} \quad (17)$$

We have now to eliminate $(3n - 6)$ $dT$'s between these $(3n - 6)$ equations and equation (16); but instead of effecting this elimination directly we can proceed by the method of multipliers. We will multiply each equation (17) by a constant to be determined and add all the products to equation (16); we will then equate to zero the coefficients of all the $dT$'s which will give as many equations as there are tensions. It is true that we have thus introduced $(3n - 6)$ new unknowns, i.e. the constants to be determined; but as we have also the $(3n - 6)$ equations (12) we see that there are as many equations as unknowns, so that these constants can be completely determined.

Now let us call

$$\begin{array}{l} A_2 \\ A_3 B_3 \\ A_4 B_4 C_4 \\ . \ . \ . \end{array}$$

the constants by which we multiply the equations (17), and let us find the equation obtained by equating to zero the coefficient of any one of these differentials, e.g. $dT_{pq}$.

It is clear that this differential can only occur in one term of equation (16) and in one term of each of the six equations (17) relating to the nodes $N_p$, $N_q$: the required equation is therefore

$$\begin{aligned} \frac{T_{pq}}{\epsilon_{pq}} &+ A_p \cos \alpha_{pq} + B_p \cos \beta_{pq} + C_p \cos \gamma_{pq} \\ &+ A_q \cos \alpha_{qp} + B_q \cos \beta_{qp} + C_q \cos \gamma_{qp} = 0 \end{aligned}$$

and noting that $\cos a_{qp} = -\cos a_{pq}$, etc., and multiplying by $\epsilon_{pq}$ we have

$$T_{pq} = \epsilon_{pq}\{(A_q - A_p)\cos a_{pq} + (B_q - B_p)\cos \beta_{pq} + (C_q - C_p)\cos \gamma_{pq})\} = 0 \quad (18)$$

we shall have then as many equations similar to this as there are stresses, i.e. as there are bars in the structure, and all these stresses are expressed as functions of the $(3n - 6)$ constants $A_2$, $A_3$, $B_3$, etc.

If we now compare the equations thus obtained with equations (13), we recognise that they differ from each other only in the substitution of A, B, C for $\xi$, $\eta$, $\zeta$, the subscripts remaining the same.

Now to find the tension which will make $\frac{1}{2}\Sigma\frac{T_{pq}^2}{\epsilon_{pq}}$ a minimum, we must first substitute in equations (12) the values of the tensions given by the equations (13) and then solve the resulting equations which will give the displacements of the nodes.

But we see that these operations are absolutely identical except for the difference of the letters representing the unknowns: *thus we will find for the constants $A_p$, $B_p$, $C_p$ the same values as for the displacements $\xi_p$, $\eta_p$, $\zeta_p$, and therefore the stresses obtained in finding the least internal work of a structure are the same as those which result after deformation.*

From this proof there results the remarkable property to which we wish to direct attention and which consists in the fact that *the values of the constants which are introduced in the calculations to find the least internal work give the displacements of the nodes.*

This proof of the theorem of least work is, we believe, the first exact one that has been given.

**15. Cases in which the Theorem of Least Work is not Applicable.**—We will not stop here to show the manner in which the theorem of least work is applied to the cases in which some of the nodes are fixed or are constrained to remain on frictionless surfaces or lines, because there no difficulties arise.

But one case which requires consideration is that in which some of the nodes are constrained to remain on lines or surfaces which present a frictional resistance to sliding.

We will assume for purposes of simplification that there is only one node of the structure thus constrained and that this node is forced to remain upon a given surface.

In this case it may happen that the friction is sufficient to prevent sliding of the node; and then the node should be regarded as fixed, so that the difficulty disappears.

But it may happen also that sliding occurs. Then if we call P the pressure on the surface, the resistance to sliding will be $\Phi$P, $\Phi$ being the coefficient of friction.

To discover which of these two possibilities will occur, and to find at the same time the direction of the force $\Phi$P, we will first regard as fixed the node under consideration and will find the three components, parallel to the axes, of the reaction at the node to satisfy the condition that the displacements of this node parallel to the axes are zero. We will then reduce these three components to two forces, one, P, normal and the other tangential to the surface, and we will note whether the latter is less or greater than $\Phi$P; if it is less sliding cannot occur, but if it is greater sliding will occur in the direction of the tangential force, assuming that the coefficient of friction is the same in all directions.

In the latter case we will remake the calculation on the assumption that the node is free but is subjected to an unknown force P normal to the surface and to a force $\Phi$P tangential to the surface and parallel to the direction of sliding.

The forces P and $\Phi$P being regarded as external forces, we can express as a function of the single unknown P the stresses in all the bars of the structure and also, therefore, the internal work $W_i$ of the structure.

In accordance with the theorem of least work, to determine the unknown P it will be necessary to make $W_i$ a minimum, i.e. to equate to zero the differential coefficient of $W_i$ with regard to P.

If, for greater clearness, we represent by S the tangential or shear force $\Phi$P, the internal work will be expressed as a function of P and S, and on differentiating and equating to zero we have

$$\frac{dW_i}{dP} + \frac{dW_i}{dS} \cdot \Phi = 0.$$

If, on the contrary, we keep out of consideration the geometrical conditions which the structure must satisfy, we see that we must express the fact that the node resting on the surface must be displaced only in that surface, i.e. that its displacement in the direction of the normal force P must be zero.

In accordance with the theorem of the differential coefficients of the internal work, we must then equate to zero the differential coefficient of $W_i$ with regard to P, S = $\Phi$P being regarded as constant; this gives

$$\frac{dW_i}{dP} = 0.$$

Now this equation cannot be identical with the preceding, because for that we should have to have $\frac{dW_i}{dS} = 0$, which is not possible in view of the fact that $\frac{dW_i}{dS}$ gives the displacement of the node in the surface, and this is by hypothesis not zero since we have assumed sliding to occur.

We will conclude by saying that when the displacement of nodes is hindered by friction, it is better always to work by the theorem of the differential coefficients of internal work, because errors are likely to arise if we apply the theorem of least work

**16. Structures in which Stresses Occur Before the External Loads are Applied.**—We have considered up to the present framed structures in which the bars are not stressed when no force is acting on them; for this it is necessary that all the bars shall be of the exact lengths necessary for the erection of the structure, so that after the $(3n - 6)$ essential bars have been put in place, the redundant bars are of the exact lengths between the nodes which they have to connect.

We have now to study the very important case in which after the $(3n - 6)$ essential bars have been put in place, we have to add other bars which are not of the exact lengths between the nodes which they have to connect, but differ by very small quantities, comparable with changes in length which do not exceed the elastic limit.

In this case it is clear that if, after having erected the $(3n - 6)$ bars, we wish to add one which is not of the exact length required, we can only do this by altering the distance between the two nodes, so that after the insertion of the new bar all the others are extended or compressed, although no external load has been applied.

A change in these stresses in the bars will occur for each new bar which we introduce into the structure.

When the structure is completed, if we apply loads to the nodes there will be a fresh change in the stresses in the bars. We now propose to solve the following problem:—

*Given a framed structure in which the lengths of the redundant bars before insertion are not exactly equal to the distances between the nodes which they have to connect, such distances being measured when the structure is composed of essential bars only and is unloaded; to find the stresses in all the bars after the loads are applied.*

Let A, B (Fig. 5, Pl. I) be two nodes joined by a redundant

bar, and suppose that after the essential bars have been erected the distance between these two nodes is $l$, while the length of the redundant bar is $l - \lambda$, $\lambda$ being very small in comparison with $l$ and comparable to a strain within the elastic limit.

To introduce this bar into the structure it will be necessary to extend it and at the same time to reduce the distance between the two nodes, thus inducing stresses in all the other bars.

If, after the strain caused by the introduction of redundant bars, the loads are applied, fresh strains will occur which will again alter the stresses in the bars.

Now it is clear that the deformation of the structure and the final stresses in the bars would be the same if instead of introducing the redundant bar AB we applied to the nodes A, B axial forces equal to the final tension T in this bar.

The same reasoning holds for all the other redundant bars. But it follows from the theorem of the differential coefficients of internal work that if, after having expressed the internal work $W_i$, for the frame composed of essential bars only, as a function of the loads and of the stresses in the redundant bars, we take the differential coefficient $\frac{dW_i}{dT}$, we obtain the distance by which the nodes A, B are caused to approach. Now this length of approach added to the elongation of the bar, which is $\frac{T}{\epsilon}$, must give the length $\lambda$ by which the length of the bar falls short of the distance between the nodes, i.e.

$$\frac{dW_i}{dT} + \frac{T}{\epsilon} = \lambda \quad . \quad . \quad . \quad . \quad . \quad (19)$$

We shall have as many equations similar to this as there are redundant bars, and therefore by combining them with the $(3n - 6)$ equations (12) we shall have as many equations as there are bars in the structure and shall thus be able to determine the stresses in all the bars.

As a corollary we may observe that all the equations similar to (19) are obtained by equating to zero the differential coefficients of the function

$$W_i + \Sigma\frac{T^2}{2\epsilon} - \Sigma T\lambda$$

in which the $\Sigma$ applies only to the redundant bars. Now since the function $W_i + \Sigma\frac{T^2}{2\epsilon}$ expresses the internal work of the whole structure, it follows that *the stresses in the bars after strain are those which will make a minimum the internal work of the system diminished by the sum of the products* $T\lambda$ *for all the redundant bars, having regard to the* $(3n - 6)$ *equations* (12) *between the stresses in the bars.*

**17. Corollary.**—If the differences $\lambda$ are all zero, i.e. if the redundant bars are all of the exact length required for introduction without inducing initial stress, we see that the term $\Sigma T\lambda$ will be zero, and thus that the function that has to be made a minimum to satisfy the $(3n - 6)$ equations (12) will be simply the internal work of the whole system, i.e. $W_i + \Sigma\frac{T^2}{2\epsilon}$. We thus discover afresh the theorem of least work.

**18. Temperature Stresses.**—The problem that we have considered in § 16 has an immediate application to the variation in the stresses in the bars of a framed structure caused by changes of temperature.

For if we suppose that we have a framed structure subjected to given loads, and we consider it in any temperature condition, it is clear that to obtain the stresses we may proceed as follows :—

We will assume in the first place that all the external loads and redundant bars are absent : we then have only a structure composed of the essential bars in which no stresses occur. The distance between the nodes of this system and the natural length of the redundant bars at the assumed temperature will be fully known, so that we know the small quantity $\lambda$ by which each of the redundant bars falls short of the required length.

To now determine the final stresses we must suppose that to the firm frame we add the redundant bars and apply the given loads.

We see then that this brings us to the problem which we considered in § 16.

**19. The Principle of Superposition.**—We shall see that in problems on elasticity the principle of superposition always holds, but to make the matter more clear we will consider two separate cases.

1. Suppose in the first place that we are dealing with a structure in which there are no initial stresses.

We have seen in § 8 that in this case the stresses in the bars are given by the solution of two series of equations (12) and (13) ; now, as all the equations are linear functions of the stresses and the external forces, it follows that the first will be linear functions of the second.

If then we call T the tension in any one bar and P, Q, R . . . the forces applied at the nodes, we shall have

$$T - aP + \beta Q + \gamma R + \dots$$

$a$, $\beta$, $\gamma$ being constants.

If instead of the forces P, Q, R we have $P+P'$, $Q+Q'$, $R+R'$. etc., acting in the same direction and $T + T'$ is then the tension in the bar previously considered, we shall have

$$T + T' = a(P + P') + \beta(Q + Q') + \gamma(R + R') + \dots,$$

and, by subtraction, $T' = aP' + \beta Q' + \gamma R' + \dots$

i.e. the stress caused by forces $P + P'$, $Q + Q'$, $R + R'$, etc., applied to the structure is the sum of the stresses which would occur if the forces P, Q, R . . . and P', Q', R' . . . were separately applied.

We see then that the principle of superposition holds, and that therefore we can study separately the effects of different forces applied to the structure, for we have only to add the separate stresses to obtain the stresses for the case when the forces are applied simultaneously.

2. Now consider the case in which the bars are not originally of the exact length required for the building up of the structure, i.e. when there are initial stresses.

We have seen that we shall have as many equations similar to (19) as there are redundant bars in the structure.

Now the function $W_i$ which enters into this equation, and which gives the internal work of the essential bars only, is a summation of the form $\frac{1}{2}\Sigma\frac{T^2}{\epsilon}$, the summation being extended to the essential bars only. By means of equations (12) we can express the stresses in these bars as linear functions of the external forces and of the stresses in the redundant bars, so that the above summation may be expressed by a function of the second degree in these quantities, and therefore all the differential coefficients $\frac{dW_i}{dT}$ will be linear functions of the same quantities.

Then on solving the series of equations similar to (19) to determine the stresses in the redundant bars, we shall obtain linear functions of the external forces and of the quantities $\lambda$ for the stresses in all the bars.

If then T is the tension in any particular bar, P, Q, R, etc., the external forces applied at the nodes, and $\lambda_1$, $\lambda_2$, $\lambda_3$, etc., the values of the small quantities $\lambda$: we shall have

$$T = aP + \beta Q + \gamma R + \dots a_1\lambda_1 + a_2\lambda_2 + a_3\lambda_3 + \dots,$$

$a, \beta, \gamma \dots, a_1, a_2, a_3 \dots$ being the constant coefficients.

Now if we call $T_1$ the tension in the bar under consideration after all the redundant bars have been inserted, but before the external loads are applied, it is clear that $T_1$ is obtained by putting $P = Q = R$, etc., $= 0$ in the general expression for T, thus getting

$$T_1 = a_1\lambda_1 + a_2\lambda_2 + a_3\lambda_3 + \ldots$$

If we next call $T_2$ the tension in the bar under consideration after the loads have been applied but when the redundant bars are of the exact initial length required, we get $T_2$ by putting $\lambda_1 = \lambda_2 = \lambda_3$, etc., $= 0$.

$$\therefore T_2 = aP + \beta Q + \gamma R + \ldots$$
$$\therefore T = T_1 + T_2,$$

which shows that the principle of superposition is applicable to this case.

When, therefore, we wish to study the case of a framed structure in which the redundant bars are not of the exact length required, so that initial stresses are thus caused in the structure, we can divide the question into two parts. We will first determine the stresses which would occur in the structure before the application of the external loads, and then determine those which the external loads would produce in the same structure if no initial stresses were induced. We shall thus find two stresses for each bar and their sum will give the effective stress required.

**20. Example.**—Let ABCD (Fig. 6, Pl. I) be a quadrilateral frame composed of 6 bars having at temperature $t$ the exact lengths required for erection without initial stress.

If the bars are not of the same material or if they are not equally heated, it will happen that at some other temperature the distance between the nodes C, D will exceed the length of the bar CD by an amount $\lambda_3$.

We see that this quantity can easily be obtained by simple geometrical calculation. When it has been found we can calculate the stresses corresponding to the new temperature, taking account of the external loads at the same time, because these stresses must render a minimum the function

$$\frac{1}{2}\left(\frac{T_1^2}{\epsilon_1} + \frac{T_2^2}{\epsilon_2} + \frac{T_3^2}{\epsilon_3} + \frac{T_4^2}{\epsilon_4} + \frac{T_5^2}{\epsilon_5} + \frac{T_6^2}{\epsilon_6}\right) - T_3\lambda_3, \quad . \quad (20)$$

having regard to the static equations between the unknowns.

These equations are five; for supposing A fixed and the node B constrained to remain on AB, there will be only one equation for the node B, and two equations for each of the nodes C, D.

To obtain the equations of equilibrium around the node C in

the simplest form, we will first express the condition that the sum of the projections of the forces meeting at this point on a straight line at right angles to the diagonal AC must be zero; and then that the sum of the projections of the same forces upon a straight line at right angles to BC must also be zero.

For equilibrium about the node D, we will treat similarly for projections at right angles to the diagonal DB and to DA. This gives the following equations :—

$$\left.\begin{aligned} T_1 + T_2 \cos \angle ABC + T_6 \cos \angle ADB - P_b &= 0 \\ T_2 \sin \angle ACB - T_3 \sin \angle ACD - P_c \sin \angle ACP_c &= 0 \\ T_5 \sin \angle ACB + T_3 \sin \angle BCD - P_c \sin \angle BCP_c &= 0 \\ T_4 \sin \angle ADB - T_3 \sin \angle BDC - P_d \sin \angle BDP_d &= 0 \\ T_6 \sin \angle ADB + T_3 \sin \angle ADC + P_d \sin \angle ADP_d &= 0 \end{aligned}\right\} \quad (21)$$

The four last equations give $T_2$, $T_4$, $T_5$, and $T_6$ in terms of $T_3$, and on substituting in the first we then get $T_1$ in terms of $T_3$: then formula (20) can be expressed in terms of $T_3$ so that by equating to zero its differential coefficient with regard to $T_3$, we obtain a linear equation for this quantity.

The stress $T_3$ being found in terms of $\lambda_3$, the other stresses can be found by equations (21) in terms of the same quantity.

# CHAPTER II.

## ELASTIC STRUCTURES.

**1. Definitions and Fundamental Principles.**—We have studied up to the present framed structures, i.e. those composed of straight bars hinged to each other at their ends only, so that any bar could turn freely about one of its ends if the other were free.

We now pass to elastic structures, i.e. those composed of solid members rigidly fixed to each other so that none can turn about their ends even if the other ends were free.

We will include in elastic structures those which are composed of solid members rigidly fixed to each other and also of bars hinged to each other and to the solid members.

We will now prove that the *theorem of the differential coefficients of internal work* and the *theorem of least work* will hold equally well for elastic structures as for framed structures.

But we will first explain certain properties of the molecular constitution of such bodies as we are considering.

We will follow the example of Navier, Poisson, Cauchy, Lamé, Barré de Saint-Venant and the majority of other physicists in assuming that solid bodies are composed of very small molecules separated by very small distances.

When we say that a force acts throughout a whole body, as, for instance, the force of gravity, we mean that a portion of the force acts upon each molecule proportionally to its mass.

We often say that over the surface of a body there acts a distributed force of $w$ kg. per sq. metre; we then mean that to the body under consideration we have applied another whose molecules act on a very small layer of the first so that the resultant of all the forces acting through an element $\delta A$ of the surface between them is equal to $w \times \delta A$.

These considerations show that a body subjected to forces acting either over its whole mass or only over the molecules at the surface, may be regarded as a system of points, each of which is subjected to certain forces.

If a body is not subjected to any external forces, we say that it is in its natural condition.

When we have a body in this condition or even subjected to external forces but in equilibrium under their action, and we apply new forces, strain will result and the distance between the molecules will change. Now we will assume the following principle on the ground of common experience.

*When two molecules of a body approach or recede from each other by a very small distance compared with their normal distance apart, there is induced, along the line joining their centres, an elastic force which is proportional to the amount of the approach or retrogression, and which acts towards or away from the molecules according as the distance between them increases or decreases.*

Assuming this law, we shall see that all bodies and all structures composed of bodies connected in any way whatever can be regarded as framed structures and will therefore possess the same properties.

In fact, if, in a body in equilibrium under a certain system of forces, we consider a molecule $N_p$, it is clear that all the forces acting on it must be in equilibrium. Now suppose that the body is referred to three axes at right angles, of which the origin is at the centre of the molecule, the axis of $x$ passes through another molecule, and the plane of $x$, $y$ contains a third molecule; and let us call $x_p, y_p, z_p$ and $x_q, y_q, z_q$ the coordinates of any two molecules $N_p$, $N_q$ before strain;

$(x_p + \xi_p)$, $(y_p + \eta_p)$, $(z_p + \zeta_p)$ and $(x_q + \xi_q)$, $(y_q + \eta_q)$, $(z_q + \zeta_q)$ the corresponding quantities after strain; $a_{pq}, \beta_{pq}, \gamma_{pq}$ the angles which the straight line $N_p$, $N_q$ makes with the axes before strain.

The increase of distance between the two molecules during strain will be

$$\lambda_{pq} = (\xi_q - \xi_p)\cos a_{pq} + (\eta_q - \eta_p)\cos \beta_{pq} + (\zeta_q - \zeta_p)\cos \gamma_{pq}; \quad (1)$$

and the elastic force or stress induced between the two molecules will be

$$T_{pq} = \epsilon_{pq}[(\xi_q - \xi_p)\cos a_{pq} + (\eta_q - \eta_p)\cos \beta_{pq} + (\zeta_q - \zeta_p)\cos \gamma_{pq}], \quad . \quad (2)$$

the coefficient $\epsilon_{pq}$ being different for each pair of molecules.

If we now call $X_p$, $Y_p$, $Z_p$ the components parallel to the axes of the external force applied to the molecule $N_p$, we shall have the three equations of static equilibrium

$$\left.\begin{aligned} X_p + \Sigma T_{pq}\cos a_{pq} &= 0 \\ Y_p + \Sigma T_{pq}\cos \beta_{pq} &= 0 \\ Z_p + \Sigma T_{pq}\cos \gamma_{pq} &= 0 \end{aligned}\right\} \quad . \quad . \quad . \quad (3)$$

It will be seen that equations (1), (2), (3) are the same as those obtained for framed structures, so that we can make the same deductions from them; therefore the principles and theorems that we have established for framed structures will also hold for elastic structures. We will explain these principles and theorems separately, because for elastic structures they can be given in a form that is more convenient in practice.

**2. The Principle of Superposition.**—For all bodies or structures the principle of superposition holds, that the effect which a number of forces produces upon a body or structure is the sum of the effects which the separate forces will produce if they are separately applied.

This very important principle can be expressed in another manner by saying that the *displacements or strains of different points, and also the elastic forces or stresses, are linear functions of the external forces.*

It should always be noted that this principle holds only when the increase or decrease of the distances between two molecules is very small compared with their initial distance apart; for if this were not the case we could not introduce in equations (1), (2), and (3) the angles which the line joining the molecules makes with the axes before strain in place of those which occur after strain.

**3. Property of Internal Work.**—*For all bodies or structures which are subjected to forces increasing by infinitely small amounts from zero to their final value, the internal work remains the same whatever may be the law according to which these forces increase.*

If we multiply formula (2) § 1 by $\frac{1}{2}T_{pq}$ and add together all the similar equations for all the other pairs of molecules, combining all the terms containing the same strain, and having regard to the static equations (3), as we did for framed structures in Chap. I, § 10, we obtain as an expression for the internal work

$$W_i = \tfrac{1}{2}\Sigma(X_p\xi_p + Y_p\eta_p + Z_p\zeta_p); \quad . \quad . \quad . \quad (4)$$

from this follows the property enunciated above, because the strains $\xi_p$, $\eta_p$, $\zeta_p$ depend solely upon the final values of the external forces.

If we call, as in Chap. I, § 10, $F_p$ the resultant force at $N_p$ and $r_p$ the strain relative to the point of application of this force, i.e. the projection upon the direction of $F_p$ of the displacement of its point of application, we have as before

$$F_pr_p = X_p\xi_p + Y_p\eta_p + Z_p\zeta_p,$$

so that the total internal work of the structure will be expressed by the formula

$$W_i = \tfrac{1}{2}\Sigma F_pr_p \quad . \quad . \quad . \quad . \quad (5)$$

**4. Theorems of the Differential Coefficients of Internal Work.** PART 1.—*If we express the internal work of a body or elastic structure as a function of the relative displacements of the points of application of the external forces, we shall obtain an expression whose differential coefficients with regard to these displacements give the values of the corresponding forces.*

PART 2.—*If we express the internal work of a body or elastic structure as a function of the external forces, the differential coefficient of this expression, with regard to one of the forces, gives the relative displacement of its point of application.*

The two parts of this theorem are proved in the same way as for framed structures.

**5. Continuation of the Theorems of the Differential Coefficients of Internal Work.**—*If we express the internal work of a body or elastic structure as a function of the external forces and moments of couples, we obtain an expression the differential coefficient of which with regard to one of the couples gives the relative angular movement of the line joining the points of application of the two forces composing the couple.*

We understand here by "couple" a system of two equal and opposite parallel forces acting at right angles to the line joining their points of application.

We also understand by "relative angular movement" of this line the angular movement of its projection on the plane of the couple.

This being made clear, the proof follows in the same manner as for framed structures.

**6. Extension of the Theorem of the Differential Coefficients of Internal Work.**—Consider a small volume in a body or structure, the molecules of which are displaced relatively to each other by very small distances in the general strain, so that we may neglect the relative displacements of these molecules in comparison with their absolute displacements. Draw through a point in this element of volume three axes at right angles which before strain are parallel to the fixed axes to which the whole structure is referred; and call X, Y, Z the components parallel to these axes of the external force applied at any molecule of the structure, and $\xi$, $\eta$, $\zeta$, the absolute displacements or strains of this molecule. We know that the internal work of the whole structure can be expressed by each of the formulæ

$$W_i = \Sigma(X d\xi + Y d\eta + Z d\zeta) \quad . \quad . \quad . \quad (6)$$

$$W_i = \Sigma(\xi dX + \eta dY + \zeta dZ) \quad . \quad . \quad . \quad (7)$$

the summation extending throughout the body or structure.

Now for all the molecules comprised within the element of volume considered, the displacements $\xi$, $\eta$, $\zeta$ can be expressed as functions of six quantities only, for the movement of this element from its original to its final position can be effected by the combination of a linear displacement and a rotation; or rather by means of three displacements $\xi_0$, $\eta_0$, $\zeta_0$ parallel to the fixed axes to which the whole structure is referred, and which move to its final position the origin of the movable axes to which the element of volume is referred, and by three rotations $\theta_x$, $\theta_y$, $\theta_z$ about the latter axes.

If we call $x$, $y$, $z$ the ordinates with reference to the latter of a molecule in the element under consideration, we can express its absolute displacements by the formulæ

$$\left.\begin{aligned} \xi &= \xi_0 + z\theta_y - y\theta_z \\ \eta &= \eta_0 + x\theta_z - z\theta_x \\ \zeta &= \zeta_0 + y\theta_x - x\theta_y \end{aligned}\right\}, \qquad (8)$$

whose differentials are

$$\left.\begin{aligned} d\xi &= d\xi_0 + zd\theta_y - yd\theta_z \\ d\eta &= d\eta_0 + xd\theta_z - zd\theta_x \\ d\zeta &= d\zeta_0 + yd\theta_x - xd\theta_y \end{aligned}\right\}, \qquad (9)$$

for the co-ordinates $x$, $y$, $z$ of the molecule considered with regard to the movable axes are practically constant. We then have for this molecule

$$\begin{aligned} \mathrm{X}d\xi + \mathrm{Y}d\eta + \mathrm{Z}d\zeta = \mathrm{X}d\xi_0 + \mathrm{Y}d\eta_0 + \mathrm{Z}d\zeta_0 &+ (\mathrm{Z}y - \mathrm{Y}z)d\theta_x \\ &+ (\mathrm{X}z - \mathrm{Z}x)d\theta_y + (\mathrm{Y}x - \mathrm{X}y)d\theta_z \end{aligned}$$

and

$$\begin{aligned} \xi d\mathrm{X} + \eta d\mathrm{Y} + \zeta d\mathrm{Z} = \xi_0 d\mathrm{X} + \eta_0 d\mathrm{Y} + \zeta_0 d\mathrm{Z} &+ (yd\mathrm{Z} - zd\mathrm{Y})\theta_x \\ &+ (zd\mathrm{X} - xd\mathrm{Z})\theta_y + (xd\mathrm{Y} - yd\mathrm{X})\theta_z. \end{aligned}$$

Now for the element of volume under consideration let $F_X$, $F_Y$, $F_Z$ be the sums of the components X, Y, Z of the external forces applied at all its molecules, and $M_X$, $M_Y$, $M_Z$ the sums of the moments $\mathrm{Z}y - \mathrm{Y}z$, $\mathrm{X}z - \mathrm{Z}x$, $\mathrm{Y}x - \mathrm{X}y$ of these forces about the movable axes. As the co-ordinates $x$, $y$, $z$ are those which occur before strain, they should be regarded as constant for each molecule of the element of volume; we shall therefore have

$$yd\mathrm{Z} - zd\mathrm{Y} = d(\mathrm{Z}y - \mathrm{Y}z), \text{ etc.,}$$

and, therefore, the sums of the six quantities $d\mathrm{X}$, $d\mathrm{Y}$, $d\mathrm{Z}$, $yd\mathrm{Z} - zd\mathrm{Y}$, $zd\mathrm{X} - xd\mathrm{Z}$, $xd\mathrm{Y} - yd\mathrm{X}$ will be $dF_X$, $dF_Y$, $dF_Z$, $dM_X$, $dM_Y$, $dM_Z$.

Then the sums of the quantities

$$Xd\xi + Yd\eta + Zd\zeta,$$
$$\xi dX + \eta dY + \zeta dZ,$$

extended to all the molecules of the element considered, will be

$$F_X d\xi_0 + F_Y d\eta_0 + F_Z d\zeta_0 + M_X d\theta_x + M_Y d\theta_y + M_Z d\theta_z \qquad (10)$$
$$\xi_0 dF_X + \eta_0 dF_Y + \zeta_0 dF_Z + \theta_x dM_X + \theta_y dM_Y + \theta_z dM_Z$$

Now the sum (6) which gives the differential of the internal work of the body or system regarded as a function of the relative displacements of the external forces, will contain the differentials $d\xi_0$, $d\eta_0$, $d\zeta_0$, $d\theta_x$, $d\theta_y$, $d\theta_z$, only, because it contains formula (10); and, moreover, the sum (7) which gives the differential of the internal work considered as a function of the external forces, will contain the differentials $dF_X$, $dF_Y$, $dF_Z$, $dM_X$, $dM_Y$, $dM_Z$, only, because it contains formula (11).

From the above the following two theorems result:—

(1) *The resultant forces* $F_X$, $F_Y$, $F_Z$ *and the resultant moments* $M_X$, $M_Y$, $M_Z$ *are the differential coefficients of the internal work of the structure with regard to the displacements* $\xi_0$, $\eta_0$, $\zeta_0$ *and the angular movements* $\theta_x$, $\theta_y$, $\theta_z$.

(2) *The three displacements* $\xi_0$, $\eta_0$, $\zeta_0$ *and the three angular movements* $\theta_x$, $\theta_y$, $\theta_z$ *are the differential coefficients of the internal work of the structure with regard to the resultant forces* $F_X$, $F_Y$, $F_Z$ *and the resultant moments* $M_X$, $M_Y$, $M_Z$.

**7. Corollary.**—Now let us consider at the surface of a body or structure a plane element $\delta A$, which we can still regard as plane after strain of the body or structure.

When we say that a force is applied only at the surface of a body, we mean that it acts upon a very small layer of molecules near to the surface: now we assume the dimensions of the surface $\delta A$ very small in comparison with the body or structure, but very large compared with the very small thickness of this superficial layer.

We can apply to the very small prism, having for base the surface $\delta A$ and for height the thickness of this layer, the two foregoing theorems; but as the forces applied to this prism are those which we have applied to the elemental surface $\delta A$, we see that these theorems apply equally well for an element of surface as for an element of volume.

When an element of surface is under consideration, we will take in general its centroid for the origin of co-ordinates, its normal for the axis of $x$ and its principal axes of inertia for the axes of $y$ and $z$. Then $\theta_X$ will be the rotation of the surface $\delta A$ about its centroid, and we will call that the *angle of torsion*, and $\theta_Y$, $\theta_Z$ will be the rotations of the same surface about its principal axes, and we will call them the *angles of flexure*. In the same way $M_X$ will be the moment of the forces applied to the surface $\delta A$ and which tend to make it turn about the axis of $x$, i.e. about its centroid, and we will call it the *Twisting Moment;* whilst $M_Y$, $M_Z$ will be the moments which tend to turn the surface about its principal axes, and we will call them the *Bending Moments.*

**8. Theorem of Least Work.**—Proceeding with equations (2) and (3) of § 1 of the present chapter in the same way as we have done in §§ 13, 14 of Chapter I, we shall prove the following theorem which is called the theorem of least work:—

*The stresses which occur between the molecular couples of a body or structure after strain are such as to render the internal work a minimum, regard being had to the equations which express equilibrium between these stresses around each molecule.*

This theorem is very important.

We must here explain certain considerations to show the use of the theorem, because it is clear that we cannot, as for framed structures, write the three equations of equilibrium for each node, i.e. for each molecule.

In general, for bodies or structures, it happens that the unknowns, which have to be determined in order to obtain the stress at any point, are either the resultants or resultant moments of the stresses acting across certain interior planes or the reactions at points or surfaces subjected to particular conditions. The internal work of the structure may be expressed as a function of these unknowns; and if between them we have some fundamental equations, we can use them to eliminate an equal number of unknowns from the expression for the internal work. Then the remaining unknowns, which are functions of the stresses, can be used to express the stress at any point of the body or structure, i.e. the stresses of all the molecular couples: now since these stresses must render the internal work a minimum, it follows that the same must occur for the values of the unknowns in terms of which this work has been expressed.

We can thus enunciate the theorem of least work in its most general form as follows:—

*Whatever may be the unknown quantities in terms of which the internal work of a structure is expressed, the values which they must have after strain of the structure are those which render this internal work a minimum, having regard to the fundamental equations which obtain between them.*

# CHAPTER III.

## GENERAL EQUATIONS OF THE ELASTIC EQUILIBRIUM OF A SOLID.

1. **Strain of a Body.**—Consider a body in its natural state and refer it to three rectangular axes of which the origin is in a molecule of the body. The axis of $x$ passes through this and through another molecule, and the plane $xy$ passes through these two and through a third molecule.

Take in this body a parallelopiped element having edges, which we will call $\Delta x$, $\Delta y$, $\Delta z$, parallel to the three axes, and which element, while itself very small, contains a very large number of molecules.

We will call the *Sphere of Action* of a molecule a sphere having its centre in this molecule and containing all the molecules upon which this molecule exercises any action. The radius of this sphere is comparable with molecular distances, and is therefore extremely small.

Now we assume that the edges $\Delta x$, $\Delta y$, $\Delta z$ of the elementary parallelopiped, although very small, are large in comparison with the radii of the spheres of action of the molecules, so that in considering on the surface of the parallelopiped a layer having this radius for thickness, the volume of the layer may be neglected in comparison with that of the parallelopiped.

If any forces are applied to the body under consideration, it is strained or deformed, its molecules being displaced. We will call the displacements parallel to the axis of any given molecule $u$, $v$, $w$; the original co-ordinates of the molecule having been $x$, $y$, $z$. The displacements will be functions of the co-ordinates.

This being given, take in the very small parallelopiped a straight line of length $r$, joining the centres of two molecules, and making angles $\alpha$, $\beta$, $\gamma$ with the axes. If $x$, $y$, $z$ are the original co-ordinates of one of these molecules, then

$$(x + r\cos\alpha),\ (y + r\cos\beta),\ (z + r\cos\gamma)$$

are those of the other.

The displacements of the first molecule parallel to the axes being $u$, $v$, $w$, those of the second will be (neglecting the terms containing powers of $r\cos\alpha$, $r\cos\beta$, $r\cos\gamma$ above the first)

$$\left.\begin{aligned} u' &= u + \frac{du}{dx} r \cos a + \frac{du}{dy} r \cos \beta + \frac{du}{dz} r \cos \gamma \\ v' &= v + \frac{dv}{dx} r \cos a + \frac{dv}{dy} r \cos \beta + \frac{dv}{dz} r \cos \gamma \\ w' &= w + \frac{dw}{dx} r \cos a + \frac{dw}{dy} r \cos \beta + \frac{dw}{dz} r \cos \gamma \end{aligned}\right\} \quad . \quad (1)$$

**2. Unit Tensile Strain.**—Now the total elongation of the straight line of length $r$ is expressed by the formula

$$(u' - u) \cos a + (v' - v) \cos \beta + (w' - w) \cos \gamma \quad . \quad (2)$$

and its *elongation per unit of length*, i.e. what is called the *unit tensile strain of a solid at the point* $(x, y, z)$ *in the direction* $(a, \beta, \gamma)$, is equal to the total elongation divided by the initial length $r$.

Thus, if we call the unit tensile strain thus defined $\tau_r$ and substitute for $u' - u$, $v' - v$, $w' - w$ their values given by formula (2) we get

$$\left.\begin{aligned} \tau_r = {} & \frac{du}{dx} \cos^2 a + \frac{dv}{dy} \cos^2 \beta + \frac{dw}{dz} \cos^2\gamma \\ & + \left(\frac{dv}{dz} + \frac{dw}{dy}\right) \cos \beta \,.\, \cos \gamma + \left(\frac{dw}{dx} + \frac{du}{dz}\right) \cos \gamma \cos a \\ & + \left(\frac{du}{dy} + \frac{dv}{dx}\right) \cos a \,.\, \cos \beta \end{aligned}\right\} \quad (3)$$

Applying this formula successively in three directions parallel to the axes $x$, $y$, and $z$, and calling $\tau_x$, $\tau_y$, $\tau_z$ the three corresponding strains, we get

$$\tau_x = \frac{du}{dx}, \ \tau_y = \frac{du}{dy}, \ \tau_z = \frac{du}{dz} \quad . \quad . \quad . \quad (4)$$

It follows that the three edges of the parallelopiped will have, after strain, the lengths

$$(1 + \tau_x)\Delta x, \ (1 + \tau_y)\Delta y, \ (1 + \tau_z)\Delta z,$$

and that its volume will be, neglecting the products of the strains,

$$(1 + \tau_x + \tau_y + \tau_z)\Delta x \ \Delta y \ \Delta z.$$

The increase of volume will be

$$(\tau_x + \tau_y + \tau_z)\Delta x \ \Delta y \ \Delta z.$$

And the unit volume strain or dilatation, which we will call $v$, that is the change in volume per unit volume, will be,

$$v = \tau_x + \tau_y + \tau_z = \frac{du}{dx} + \frac{dv}{dy} + \frac{dw}{dz} \quad . \quad . \quad (5)$$

**3. Shear.**—Consider in the parallelopiped a third molecule such that the straight line of length $r_1$, joining it with the first shall be, before strain, perpendicular to the straight line of length $r$ which joins the first to the second.

The angle between the two straight lines $r$ and $r_1$ will be an acute angle after strain, and its cosine will be equal to the sine of its complement, or, in fact, to the complement itself, since this will be a very small angle.

This complement is the sum of the two small quantities by which the right angle would have diminished, if each of the two straight lines $r$ and $r_1$ had separately changed its direction, the other remaining fixed.

Now, to determine the quantity by which the right angle would have diminished, if the straight line $r$ had alone changed in direction, it would be necessary to project the three displacements of the extremity of this straight line upon the original direction of the straight line $r_1$; and to divide the sum of the three projections by the final length of the straight line $r$, or by the initial length, neglecting small quantities of the second order in comparison with those of the first. Thus, calling $\alpha'$, $\beta'$, $\gamma'$ the angles which the straight line $r_1$ makes with its axes, we have the following expression to express the above angle :—

$$\begin{aligned}&\left(\frac{du}{dx}\cos\alpha + \frac{du}{dy}\cos\beta + \frac{du}{dz}\cos\gamma\right)\cos\alpha'\\ &+ \left(\frac{dv}{dx}\cos\alpha + \frac{dv}{dy}\cos\beta + \frac{dv}{dz}\cos\gamma\right)\cos\beta'\\ &+ \left(\frac{dw}{dx}\cos\alpha + \frac{dw}{dy}\cos\beta + \frac{dw}{dz}\cos\gamma\right)\cos\gamma'.\end{aligned}$$

The amount by which the right angle would have decreased if the straight line $r_1$ alone had changed its direction is expressed by the preceding formula, changing the angles $\alpha$, $\beta$, $\gamma$ to $\alpha'$, $\beta'$, $\gamma'$, and vice-versa ; which gives

$$\begin{aligned}&\left(\frac{du}{dx}\cos\alpha' + \frac{du}{dy}\cos\beta' + \frac{du}{dz}\cos\gamma'\right)\cos\alpha\\ &+ \left(\frac{dv}{dx}\cos\alpha' + \frac{dv}{dy}\cos\beta' + \frac{dv}{dz}\cos\gamma'\right)\cos\beta\\ &+ \left(\frac{dw}{dx}\cos\alpha' + \frac{dw}{dy}\cos\beta' + \frac{dw}{dz}\cos\gamma'\right)\cos\gamma.\end{aligned}$$

Adding together these two expressions, we obtain for the decrease of the original right angle the following formula :—

$$\begin{aligned}\sigma_{rr_1} = {}& 2\frac{du}{dx}\cos\alpha\cos\alpha' + 2\frac{dv}{dy}\cos\beta\cos\beta' + 2\frac{dw}{dz}\cos\gamma\cos\gamma'\\ &+ \left(\frac{dv}{dz} + \frac{dw}{dy}\right)(\cos\beta\cos\gamma' + \cos\beta'\cos\gamma)\\ &+ \left(\frac{dw}{dx} + \frac{du}{dz}\right)(\cos\gamma\cos\alpha' + \cos\gamma'\cos\alpha)\\ &+ \left(\frac{du}{dy} + \frac{dv}{dx}\right)(\cos\alpha\cos\beta' + \cos\alpha'\cos\beta)\ . \quad . \quad . \quad (6)\end{aligned}$$

This quantity $\sigma_{rr_1}$ is called *Shear Strain.* It may be considered in two different ways; for if we imagine a rectangle of which the straight lines $r$ and $r_1$ shall be two sides, we see that after the deformation this rectangle will be changed into a rhombus, and we can suppose that this deformation has taken place, either by the slide of the two sides $r$ one upon the other, or by the slide of the two sides $r_1$; in both cases this slide or shear strain will be measured by the cosine of the angle between the straight lines $r$ and $r_1$ after the deformation. If we apply formula (6) to the case where the sides $r$ and $r_1$ are parallel to the axes, taking them first parallel to the axes of $y$ and $z$, secondly to the axes of $z$ and $x$, thirdly to the axes of $x$ and $y$, we get

$$\sigma_{yz} = \frac{dv}{dz} + \frac{dw}{dy}, \quad \sigma_{zx} = \frac{dw}{dx} + \frac{du}{dz},$$
$$\sigma_{xy} = \frac{du}{dy} + \frac{dv}{dx} \quad . \quad . \quad . \quad . \quad . \quad . \quad (7)$$

By means of these relations, and those given in the preceding paragraph, we can write the expressions for $\tau_r$ and $\sigma_{rr_1}$ in the following way :—

$$\left.\begin{aligned}\tau_r = \tau_x \cos^2 a + \tau_y \cos^2 \beta + \tau_z \cos^2 \gamma + \sigma_{yz} \cos \beta \cos \gamma \\ + \sigma_{zx} \cos \gamma \cos a + \sigma_{xy} \cos a \cos \beta\end{aligned}\right\} \quad (8)$$

$$\left.\begin{aligned}\sigma_{rr_1} = 2\tau_x \cos a \cos a' + 2\tau_y \cos \beta \cos \beta' + 2\tau_z \cos \gamma \cos \gamma' \\ + \sigma_{yz} (\cos \beta \cos \gamma' + \cos \beta' \cos \gamma) \\ + \sigma_{zx} (\cos \gamma \cos a' + \cos \gamma' \cos a) \\ + \sigma_{xy} (\cos a \cos \beta' + \cos a' \cos \beta)\end{aligned}\right\} \quad (9)$$

These formulæ serve to express the three direct strains $\tau_x$, $\tau_y$, $\tau_z$, and the three shears $\sigma_{zy}$, $\sigma_{zx}$, $\sigma_{xy}$ as functions of the three direct strains $\tau_{x'}$, $\tau_{y'}$, $\tau_{z'}$, and of the three shears $\sigma_{y'z'}$, $\sigma_{z'x'}$, $\sigma_{x'y'}$ referred to three other rectangular axes. We will examine the case, which will prove useful later, where the axis of $x'$ coincides with the axis of $x$, and where, consequently, the axes of $y'$ and $z'$ are in the plane $yz$. We can see by Fig. 7, Plate I, that in this case the axis of $x$ will make angles of 0°, 90°, 90°, with the axes $x'$, $y'$, $z'$; the axis of $y$, angles 90°, $-\beta$, $-(90° + \beta)$, and the axis of $z$, angles of 90°, 90° $-\beta$, $\beta$.

Substituting then in formulæ 8 and 9 after changing $x$, $y$, $z$ in the second member of them into $x'$, $y'$, $z'$, we get

$$\left.\begin{aligned}\tau_x &= \tau_{x'}. \\ \tau_y &= \tau_{y} \cos^2 \beta + \tau_{z'} \sin^2 \beta - \sigma_{y'z'} \cos \beta \sin \beta \\ \tau_z &= \tau_{y'} \sin^2 \beta + \tau_{z'} \cos^2 \beta + \sigma_{y'z'} \cos \beta \sin \beta \\ \sigma_{yz} &= 2(\tau_{y'} - \tau_{z'}) \cos \beta \sin \beta + \sigma_{y'z'} (\cos^2 \beta - \sin^2 \beta) \\ \sigma_{zx} &= \sigma_{z'x'} \cos \beta + \sigma_{x'y'} \sin \beta \\ \sigma_{xy} &= - \sigma_{z'x'} \sin \beta + \sigma_{x'y} \cos \beta.\end{aligned}\right\} \quad (10)$$

**4. Internal Work of a very small Parallelopiped.**—In the elementary parallelopiped of which the edges are $\Delta x$, $\Delta y$, $\Delta z$, consider the small straight line $r$ joining the very close molecules. In the deformation of the body, this straight line increases from the original length $r$ to the value $r(1 + \tau_r)$ and the tension between the two molecules increases proportionately to the strain, in such a way that when the straight line has the length $r + \rho$, $\rho$ being a quantity smaller than $r\tau_r$, the tension between the two molecules will be $\epsilon\rho$, $\epsilon$ being a coefficient, constant for each pair of molecules but different for different pairs.

The internal work or strain energy of the straight line $r$ will be

$$\int_0^{r\tau_r} \epsilon\rho \, d\rho = \tfrac{1}{2}\epsilon r^2 \tau_r^{\,2},$$

that is, substituting for $\tau_r$ its value given by formula (8)

$$\tfrac{1}{2}\epsilon r^2(\tau_x \cos^2 a + \tau_y \cos^2 \beta + \tau_z \cos^2 \gamma + \sigma_{yz} \cos \beta \cos \gamma \\ + \sigma_{zx} \cos \gamma \cos a + \sigma_{xy} \cos a \cos \beta)^2$$

in which we can see, that in developing the square and putting together the terms containing the same products of the cosines $\cos a$, $\cos \beta$, $\cos \gamma$, the distinct terms will be reduced to fifteen.

To find the internal work of the whole parallelopiped, it is necessary to add up expressions analogous to these for all the pairs of molecules that it contains. For complete accuracy it would be necessary to add a part of the internal work of the pairs of molecules, having one of the molecules in the parallelopiped and the other outside. For each of these pairs, we shall have an expression analogous to the preceding one, and, although we need only take a part of the internal work of these pairs, the multiplying fraction can be considered included in the factor $\epsilon$, so that the form of the expression is unchanged.

Now, the sum of the values of the coefficient $\epsilon r^2 \cos^4 a$ for all the pairs of molecules which we must consider to find the internal work of the parallelopiped, is evidently proportional to the volume of that element. The same thing is true for the other coefficients. Thus, if in order to simplify, we put

$$\Delta x \Delta y \Delta z = \Delta V$$

we get

$$\Sigma \epsilon r^2 \cos^4 a = a_1 \Delta V, \quad \Sigma \epsilon r^2 \cos^4 \beta = a_2 \Delta V$$
$$\Sigma \epsilon r^2 \cos^4 \gamma = a_3 \Delta V, \quad \Sigma \epsilon r^2 \cos^2 \beta \cos^2 \gamma = b_1 \Delta V$$
$$\Sigma \epsilon r^2 \cos^2 \gamma \cos^2 a = b_2 \Delta V, \quad \Sigma \epsilon r^2 \cos^2 a \cos^2 \beta = b_3 \Delta V$$
$$\Sigma \epsilon r^2 \cos^2 a \cos \beta \cos \gamma = c_1 \Delta V$$
$$\Sigma \epsilon r^2 \cos a \cos^2 \beta \cos \gamma = c_2 \Delta V$$

$$\Sigma \epsilon r^2 \cos a \cos \beta \cos^2 \gamma = c_3 \Delta V$$
$$\Sigma \epsilon r^2 \cos \beta \cos^2 \gamma = e_1 \Delta V, \ \Sigma \epsilon r^2 \cos^2 a \cos \gamma = e_2 \Delta V$$
$$\Sigma \epsilon r^2 \cos a \cos^2 \beta = e_3 \Delta V, \ \Sigma \epsilon r^2 \cos^2 \beta \cos \gamma = f_1 \Delta V$$
$$\Sigma \epsilon r^2 \cos a \cos^2 \gamma = f_2 \Delta V, \ \Sigma \epsilon r^2 \cos^2 a \cos \beta = f_3 \Delta V$$

$a_1, a_2, a_3, b_1, b_2$ . . . being fifteen numerical coefficients variable from point to point of the solid, if it is not homogeneous, but constant at each point.

With this nomenclature the internal work of the parallelopiped is expressed by the following formula :—

$$\frac{\Delta V}{2}\left\{\begin{array}{l} a_1\tau_x^2 + a_2\tau_y^2 + b_1(2\tau_y\tau_z + \sigma_{zy}^2) + b_2(2\tau_z\tau_x + \sigma_{zx}^2) \\ \quad + b_3(2\tau_x\tau_y + \sigma_{xy}^2) + 2c_1(\tau_x\sigma_{yz} + \sigma_{zx}\sigma_{xy}) \\ \quad + 2c_2(\tau_y\sigma_{zx} + \sigma_{yx}\sigma_{xy}) + 2c_3(\tau_z\sigma_{xy} + \sigma_{yz}\sigma_{zx}) \\ \quad + 2e_1\tau_z\sigma_{yz} + 2e_2\tau_x\sigma_{zx} + 2e_3\tau_y\sigma_{xy} + 2f_1\tau_y\sigma_{yz} \\ \quad + 2f_2\tau_z\sigma_{zx} + 2f_3\tau_x\sigma_{xy} \end{array}\right\} \quad (11)$$

**5. Elastic Forces or Stresses on the Faces of the Parallelopiped.**—Consider the two faces of the parallelopiped that are perpendicular to the axis of $x$, and of which the area is $\Delta x \Delta z$; let A be the face of which the abscissa is $x$, and A′ the parallel face of which the abscissa is $x + \Delta x$. Imagine on one side or the other of the face A two very thin layers, the thickness of which is equal to the radius of the sphere of action of the molecules, and therefore negligible in comparison with the edge $\Delta x$ of the parallelopiped. The molecular actions between the two layers are those applied to the face $\Delta y \Delta z$; now, each of these molecular actions can be resolved into three stresses parallel to the axes, and the sum of these components parallel to each axis referred to the unit of surface we can call

$$t_x, \ s_{xy}, \ s_{xz}$$

the first index showing to which axis the face under consideration is perpendicular, and the second to which axis the component of force applied to this face is parallel. We will call the force $t_x$ which is perpendicular to the face $\Delta y \Delta z$ the *tensile stress ;* and the two other forces $s_{xy}$, $s_{xz}$ parallel to the same face *shear stresses.*

The stresses applied to the face A′ will be

$$t_x + \frac{dt_x}{dx}\Delta x, \ s_{xy} + \frac{ds_{xy}}{dx}\Delta x, \ s_{xz} + \frac{ds_{xz}}{dx}\Delta x;$$

they become reduced to $t_x$, $s_{xy}$, $s_{xz}$ as on face A if we neglect the very small quantities $\frac{dt_x}{dx}\Delta x$, etc., as we must do in order to find an expression for the stresses $t_x$, $s_{xy}$, $s_{xz}$ as a function of the three direct strains and the three shear strains.

If we consider the two faces B, B′, perpendicular to the axis of $y$, the elastic stresses applied upon these faces, referred to unit surface, will be

$$t_y, s_{yx}, s_{yz}:$$

finally, if we consider the two faces C, C′, perpendicular to the axis of $x$, the elastic stresses applied to each and referred to unit surface will be

$$t_z, s_{zx}, s_{zy}.$$

We will now find an expression for the nine stresses considered above, as a function of the three direct strains and the three shear strains.

To do this we will note first that the faces AA′ are displaced with reference to one another by a quantity $\tau_x \Delta x$ in the direction of $x$, and by quantities $\sigma_{xy}\Delta x$, $\sigma_{zx}\Delta x$ in the directions of $y$ and $z$. Thus, *if we take the differential coefficients of the internal work of the parallelopiped* (last formula of No. 4) *with regard to the relative displacements* $\tau_x\Delta x$, $\sigma_{xy}\Delta x$, $\sigma_{zx}\Delta x$, we shall get the absolute values of the forces applied to the face $\Delta y \Delta z$, and dividing these by the product $\Delta y \Delta z$ we shall obtain the values of $t_x$, $\sigma_{xy}$, $\sigma_{yz}$. We see then that these last are the differential coefficients with regard to $\tau_x$, $\sigma_{xy}$, $\sigma_{zx}$ of the internal work of the parallelopiped, divided by its volume $\Delta V = \Delta x \Delta y \Delta z$.

Similarly the stresses

$$t_y, s_{yx}, s_{yz} \text{ and } t_z, s_{zx}, s_{zy}$$

are the differential coefficients of the same expression with regard to the corresponding strains, from which it follows immediately that the nine stresses are reduced to six, for there are only six differential coefficients to take, and we shall have

$$s_{yx} = s_{xy}, \; s_{xz} = s_{zx}, \; s_{zy} = s_{yz}.$$

Henceforth, therefore, we shall always employ the notation $s_{xy}$, $s_{zx}$, $s_{yz}$ equally to represent the shear stresses under consideration on the faces A, C, B, and parallel respectively to the axes of $y$, $x$, and $z$, and the shear stresses on the faces B, A, C and parallel to the axes of $x$, $z$, and $y$.

The following are the expressions for the six stresses obtained in the manner indicated :—

$$\left.\begin{aligned}
t_x &= a_1\tau_x + b_3\tau_y + b_2\tau_z + c_1\sigma_{yz} + e_2\sigma_{zx} + f_3\sigma_{xy}\\
t_y &= b_3\tau_x + a_2\tau_y + b_1\tau_z + f_1\sigma_{yz} + c_2\sigma_{zx} + e_3\sigma_{xy}\\
t_z &= b_2\tau_x + b_1\tau_y + a_3\tau_z + e_1\sigma_{yz} + f_2\sigma_{zx} + c_3\sigma_{xy}\\
s_{yz} &= c_1\tau_x + f_1\tau_y + e_1\tau_z + b_1\sigma_{yz} + c_3\sigma_{zx} + c_2\sigma_{xy}\\
s_{zx} &= e_2\tau_x + c_2\tau_y + f_2\tau_z + c_3\sigma_{yz} + b_2\sigma_{zx} + c_1\sigma_{xy}\\
s_{xy} &= f_3\tau_x + e_3\tau_y + c_3\tau_z + c_2\sigma_{yz} + c_1\sigma_{zx} + b_3\sigma_{xy}
\end{aligned}\right\} \quad . \quad (12)$$

The first three formulæ give the tensile stresses, the last three the shear forces. We will note once again that there are only fifteen different numerical coefficients in these formulæ.

It is also useful to add that the six stresses being the differential coefficients with regard to $\tau_x, \tau_y, \tau_z, \sigma_{yz}, \sigma_{zx}, \sigma_{xy}$, of the quantity multiplied by $\frac{\Delta V}{2}$ in the expression for the internal work of the parallelopiped under consideration, and that, moreover, this quantity being a homogeneous function of the second degree in $\tau_x, \tau_y, \tau_z, \sigma_{yz}, \sigma_{zx}, \sigma_{xy}$, we could write the expression for the internal work of the element in the form

$$\frac{\Delta V}{2}\left(t_x\tau_x + t_y\tau_y + t_z\tau_z + s_{yz}\sigma_{yz} + s_{zx}\sigma_{zx} + s_{xy}\sigma_{xy}\right).$$

This can be easily verified.

**6. Variation of the Stresses from Point to Point in a Solid.**—We have seen that the stresses acting on the faces A, B, C, A′, B′, C′ of the above parallelopiped are as follows :—

upon A, $t_x, s_{xy}, s_{zx}$ ;

and upon A′, $t_x + \frac{dt_x}{dx}\Delta x,\ s_{xy} + \frac{ds_{xy}}{dx}\Delta x,\ s_{zx} + \frac{ds_{zx}}{dx} . \Delta x$ ;

upon B, $t_y, s_{xy}, s_{yz}$ ;

and upon B′, $t_y + \frac{dt_y}{dy}\Delta y,\ s_{xy} + \frac{ds_{xy}}{dy}\Delta y,\ s_{yz} + \frac{ds_{yz}}{dy}\Delta y$ ;

upon C, $t_z, s_{zx}, s_{yz}$ ;

and upon C′, $t_z + \frac{dt_z}{dz} . \Delta z,\ s_{zx} + \frac{ds_{zx}}{dz}\Delta z,\ s_{yz} + \frac{ds_{yz}}{dz} . \Delta z$.

Between these stresses applied at the surface of the parallelopiped and external forces such as gravity which act on the whole mass, there must be equilibrium.

Let us call $X_0$, $Y_0$, $Z_0$ the components, parallel to the axes, per unit mass of the external force applied to the element of volume $\Delta V = \Delta x\Delta y\Delta z$, and $\rho$ the density of the body at the point $(x, y, z)$ ; the effective values of the external force applied to the parallelopiped will then be

$$X_0\rho\Delta x\Delta y\Delta z,\ Y_0\rho\Delta x\Delta y\Delta z,\ Z_0\rho\Delta x\Delta y\Delta z.$$

Now, for the equilibrium of the parallelopiped, the sum of the components parallel to each axis of all the forces applied to it must be zero. If we consider, for example, those parallel to the axis of $x$, we have the following components :—

on face A, $- t_x\Delta y\Delta z$, and on face A′, $\left(t_x + \frac{dt_x}{dx}\Delta x\right)\Delta y\Delta z$, on face B, $-s_{xy}\Delta z\Delta x$, and on face B′, $\left(s_{xy}+\frac{ds_{xy}}{dy}\Delta y\right)\Delta z\Delta x$,

on face C, $-s_{zx}\Delta x\Delta y$, and on face C′, $\left(s_{zx}+\frac{ds_{zx}}{dz}\Delta z\right)\Delta x\Delta y$,

and on the whole mass of the parallelopiped $X_0\rho\Delta x\Delta y\Delta z$.

The sum of all these forces equated to zero and divided by the volume $\Delta x\Delta y\Delta z$ gives the first of the three following equations, and the two others are obtained in the same way:—

$$\left.\begin{aligned}\frac{dt_x}{dx}+\frac{ds_{xy}}{dy}+\frac{ds_{zx}}{dz}+\rho X_0=0\\ \frac{ds_{xy}}{dx}+\frac{dt_y}{dy}+\frac{ds_{yz}}{dz}+\rho Y_0=0\\ \frac{ds_{zx}}{dx}+\frac{ds_{yz}}{dy}+\frac{dt_z}{dz}+\rho Z_0=0\end{aligned}\right\} \quad . \quad . \quad (13)$$

For the equilibrium of the parallelopiped, it is also necessary that the sum of the moments of the forces about three rectangular axes shall be zero. Let us take these three axes through the centre of the parallelopiped, and parallel to the axes of the co-ordinates. Then the forces $X_0$, $Y_0$, $Z_0$ will have no moments, because, being uniformly distributed over the whole mass of the parallelopiped, their resultant passes through its centre. Similarly, the stresses applied to each face, are distributed over it uniformly, and can therefore be considered to be applied at its centre. Therefore the normal stresses on the faces do not give any moment about the three axes.

Let us consider now the moments about the axis of $x$ of forces parallel to this axis: the faces A, A′ being joined by it through their centres, the stresses applied there will give no moments. On the faces B, B′ there are only the two forces

$$-s_{yz}\Delta z\Delta x,\text{ and }+\left(s_{yz}+\frac{ds_{yz}}{dy}\Delta y\right)\Delta z\Delta x,$$

which tend to cause the parallelopiped to turn around the axis of $x$, and which, neglecting the very small quantity

$$\frac{ds_{yz}}{dy}\Delta x\Delta y\Delta z,$$

form a couple of which the lever-arm is $\Delta y$, and of which therefore the moment is

$$s_{yz}\Delta x\Delta y\Delta z.$$

Finally on the faces C, C′ there are only the forces

$$-s_{yz}\Delta x\Delta y,\text{ and }+\left(s_{yz}+\frac{ds_{yz}}{dy}\Delta x\Delta y\right)\Delta x\Delta y$$

which form a couple tending to turn the parallelopiped in the contrary sense from the previous couple; and of which the moment is

$$- s_{yz}\Delta x \Delta y \Delta z.$$

These two couples neutralize each other, and there is equilibrium of rotation about the axis of $x$ of the moments parallel to the axis of $x$.

We should get similar results for the other two axes.

Thus the three equations (12) are sufficient for the elastic equilibrium of a body, or of any structure; the six stresses $t_x$, $s_{xy}$, etc., are functions of the three displacements $u$, $v$, $w$, as we see by formulæ (4), (7), and (12); so that integration of the three equations we have obtained will give the three displacements as a function of the co-ordinates; from which the formulæ already quoted will give the values of the stresses at any given point.

When a body is acted upon solely by the forces distributed over its surface, and no force such as gravity acts upon the mass, we have $X_0 = 0$, $Y_0 = 0$, $Z_0 = 0$, and the general equations of elasticity become

$$\left.\begin{aligned} \frac{dt_x}{dx} + \frac{ds_{xy}}{dy} + \frac{ds_{zx}}{dz} = 0 \\ \frac{ds_{xy}}{dx} + \frac{dt_y}{dy} + \frac{ds_{yz}}{dz} = 0 \\ \frac{ds_{zx}}{dx} + \frac{ds_{yz}}{dy} + \frac{dt_z}{dz} = 0 \end{aligned}\right\} \quad . \quad . \quad (14)$$

We might remark that the case of a body which is not acted upon by any external force never occurs, because there is always gravity acting upon it; yet, as by the principle of superposition, we can study separately the effects of forces applied to the surface, and of those such as gravity applied to the interior, we see that formulæ (13) will be useful to determine the stresses due solely to the exterior forces applied to the surface.

**7. Components of the Stresses Exerted upon a Small Face Inclined to the Axes.** Let us imagine in the solid body a small tetrahedron such that the three edges meeting at the apex $(x, y, z)$ are parallel to the axes, and such that the normal to the face opposite the apex makes angles $\alpha$, $\beta$, $\gamma$ with the axes.

If we call $a$ the very small area of this face, the areas of the three other faces which are perpendicular to the axes will be

$$a \cos \alpha, \; a \cos \beta, \; a \cos \gamma,$$

and the stresses applied to these three faces,

$$t_x a \cos a,\ s_{xy} a \cos \beta,\ s_{zx} a \cos \gamma,$$
$$s_{xy} a \cos a,\ t_y a \cos \beta,\ s_{yz}\, a \cos \gamma,$$
$$s_{zx} a \cos a,\ s_{yz} a \cos \beta,\ t_z a \cos \gamma;$$

therefore, if we call X, Y, Z the components parallel to the axes of the stress exerted across the face $a$, and referred to unit surface, since the sum of the components parallel to each axis must be zero, we get the three following equations :—

$$\left.\begin{aligned} X &= t_x \cos a + s_{xy} \cos \beta + s_{zx} \cos \gamma \\ Y &= s_{xy} \cos a + t_y \cos \beta + s_{yz} \cos \gamma \\ Z &= s_{zx} \cos a + s_{yz} \cos \beta + t_z \cos \gamma \end{aligned}\right\} \quad . \quad . \quad (15)$$

These formulæ serve to express the stresses exerted on the faces of a parallelopiped, the edges of which are parallel to the three new rectangular axes.

We will examine only the case, which we shall need later, in which we keep the axis of $x$ and take for axes of $y'$, $z'$, two straight lines making angles $\beta$ with the axes of $y$ and $z$ respectively (Fig. 7, Pl. I). If we consider first that face of the new parallelopiped perpendicular to the axis of $x$, we must put in formulæ (15)

$$\cos a = 1,\ \cos \beta = 0,\ \cos \gamma = 0$$

which gives

$$X = t_x,\ Y = s_{xy},\ Z = s_{zx}.$$

Resolving these forces according to the new axes, we get

$$t_{x'} = t_x$$
$$s_{x'y'} = s_{zx} \sin \beta + s_{xy} \cos \beta,$$
$$s_{z'x'} = s_{zx} \cos \beta - s_{xy} \sin \beta.$$

If we next consider the face of the parallelopiped perpendicular to the axis of $y$, we must put, in equations (15),

$$\cos a = 0,\ \cos \beta = \cos \beta,\ \cos \gamma = \sin \beta,$$

which gives

$$X = s_{xy} \cos \beta + s_{zx} \sin \beta$$
$$Y = t_y \cos \beta + s_{yz} \sin \beta$$
$$Z = s_{yz} \cos \beta + t_z \sin \beta,$$

from which it follows that the components, parallel to the new axes, of the stress exerted across the face under consideration, will be

$$\begin{aligned} s_{x'y'} &= X = s_{xy} \cos \beta + s_{zx} \sin \beta, \\ t_{y'} &= Y \cos \beta + Z \sin \beta = t_y \cos^2 \beta + 2 s_{yz} \sin \beta \cos \beta + t_z \sin^2 \beta, \\ s_{y'z'} &= -Y \sin \beta - Z \cos \beta = (-t_y + t_z) \sin \beta \cos \beta \\ &\qquad + s_{yz} (\cos^2 \beta - \sin^2 \beta). \end{aligned}$$

If we consider finally the face of the parallelopiped perpendicular to the axis of $z'$, we get the expressions for $s_{z'x'}$, $s_{y'z'}$, $t_{z'}$, and, as we have already obtained the two former, we need only give the third, which will be

$$t_{z'} = t_y \sin^2 \beta - 2s_{yz} \sin \beta \cos \beta + t_z \cos^2 \beta.$$

Putting together the previous results, we shall have, in order to express the stresses with regard to the new axes, the following formulæ :—

$$\left.\begin{aligned} t_{x'} &= t_x, \\ t_{y'} &= t_y \cos^2 \beta + 2s_{yz} \sin \beta \cos \beta + t_z \sin^2 \beta, \\ t_{z'} &= t_y \sin^2 \beta - 2s_{yz} \sin \beta \cos \beta + t_z \cos^2 \beta, \\ s_{y'z'} &= ( - t_y + t_z) \sin \beta \cos \beta + s_{yz} (\cos^2 \beta - \sin^2 \beta), \\ s_{z'x'} &= s_{zx} \cos \beta - s_{xy} \sin \beta, \\ s_{x'y'} &= s_{zx} \sin \beta + s_{xy} \cos \beta. \end{aligned}\right\} \quad (16)$$

**8. Simplification of formulæ (12) expressing the stresses, when the body possesses particular elastic properties.**

**Case I. Homogeneous Body having at each Point a Plane of Equal Elastic Properties Parallel to the Plane of *yz*.**

We say that a body has a plane of equal elastic properties parallel to a given plane at each point when each strain parallel to this plane produces an equal stress in the two faces symmetrical with regard to it. Let us now see to what form expressions (12) for the stresses $t_x$, $t_y$, etc., are reduced when the plane of equal elastic properties is parallel to the plane of $yz$, that is to say, perpendicular to the axis of $x$.

Let us consider at a point two faces symmetrical with respect to this plane, and let us denote by $a$, $\beta$, $\gamma$ the angles which the normal to one of them makes with the axes, and therefore $180 - a$, $\beta$, $\gamma$ the angles which the normal to the other one makes with the same axes.

The components X, Y, Z of the force upon the first face will be given by formulæ (15), and those X′, Y′, Z′, upon the second face by the same formulæ (15), substituting $180 - a$ for $a$, that is to say, changing $\cos a$ to $- \cos a$. Now, if the strain of the body is symmetrical with respect to the plane $yz$, it follows from our hypothesis that the forces X′, Y′, Z′ should be equal to and symmetrical with respect to the forces X, Y, Z. We shall therefore get

$$X' = - X, \; Y' = Y, \; Z' = Z.$$

Expressing these conditions through equations (15) and those derived from it by replacing $\cos a$ by $- \cos a$ we get

$$s_{zx} = 0, \; s_{xy} = 0.$$

Now, in order that a strain shall be symmetrical with respect to the plane $yz$, it must be made up solely of the three strains

$\tau_x$, $\tau_y$, $\tau_z$ and of the shear strain $\sigma_{yz}$, which reduces the general expressions (12) for $s_{xy}$, $s_{zx}$ to the following form :—

$$s_{zx} = e_2\tau_x + c_2\tau_y + f_2\tau_z + c_3\sigma_{yz}$$
$$s_{xy} = f_3\tau_x + e_3\tau_y + c_3\tau_z + c_2\sigma_{yz},$$

and since $s_{zx}$, $s_{xy}$ must be zero, whatever be the strains $\tau_x$, $\tau_y$, $\tau_z$, $\sigma_{yz}$, it follows that

$$c_2 = 0,\ e_2 = 0,\ f_2 = 0$$
$$c_3 = 0,\ e_3 = 0,\ f_3 = 0.$$

Thus the expressions for the six stresses become :—

$$\left.\begin{aligned} t_x &= a_1\tau_x + b_3\tau_y + b_2\tau_z + c_1\sigma_{yz} \\ t_y &= b_3\tau_x + a_2\tau_y + b_1\tau_z + f_1\sigma_{yz} \\ t_z &= b_2\tau_x + b_1\tau_y + a_3\tau_z + e_1\sigma_{yz} \\ s_{yz} &= c_1\tau_x + f_1\tau_y + e_1\tau_z + b_1\sigma_{yz} \\ s_{zx} &= b_2\sigma_{zx} + c_1\sigma_{xy} \\ s_{xy} &= c_1\sigma_{zx} + b_3\sigma_{xy} \end{aligned}\right\} \quad . \quad . \quad (17)$$

**9. Case II. Homogeneous Body having at each Point Three Planes of Equal Elastic Properties.**—If the plane of equal elastic properties, instead of being perpendicular to the axis of $x$ is perpendicular to the axis of $y$, it is easily seen that the coefficients which must be zero, will be

$$c_1,\ e_1,\ f_1$$
$$c_3,\ e_3,\ f_3.$$

Thus if the body possesses at each point two planes of equal elastic properties perpendicular to the axes of $x$ and $y$, the nine coefficients

$$c_1,\ e_1,\ f_1$$
$$c_2,\ e_2,\ f_2$$
$$c_3,\ e_3,\ f_3$$

must all be zero. This shows that the plane perpendicular to the axis of $z$ is also a plane of equal elastic properties, and that, therefore, the body possesses at each point three planes of equal elastic properties, perpendicular to the axis of the co-ordinates.

The expressions for the stresses become in this case, changing the coefficients $b_1$, $b_2$ and $b_3$ to $G_1$, $G_2$, $G_3$,

$$\left.\begin{aligned} t_x &= a_1\tau_x + G_3\tau_y + G_2\tau_z, & s_{yz} &= G_1\sigma_{yz} \\ t_y &= G_3\tau_x + a_2\tau_y + G_1\tau_z, & s_{zx} &= G_2\sigma_{zx} \\ t_z &= G_2\tau_x + G_1\tau_y + a_3\tau_z, & s_{xy} &= G_3\sigma_{xy} \end{aligned}\right\} . \quad . \quad (18)$$

The coefficients $G_1$, $G_2$, $G_3$ which, multiplied by the shear strains $\sigma_{yz}$, $\sigma_{zx}$, $\sigma_{xy}$, give the shear stresses, are called *the Glide, shear or rigidity moduli.*

**10. Case III. Homogeneous Body having at each Point an Axis of Elasticity Parallel to the Axis of *x*.**—We say that there is an *axis of symmetry* or of *elasticity* at each point of a homogeneous solid when the body exhibits exactly the same elastic properties in all the planes passing through this axis.

It follows that if we consider a system of rectangular axes, OX′, OY′, OZ′, the axis of elasticity being the axis of $x$, and if we express the six stresses $t_x$, $t_y$, etc., as functions of the three direct strains $\tau_{x'}$, $\tau_{y'}$, $\tau_{z'}$, and of the three shear strains $\sigma_{y'z'}$, $\sigma_{z'x'}$, $\sigma_{x'y'}$, referred to the same axes, the formulæ thus obtained will be independent of the direction of the axes of $y'$ and $z'$.

Now, in order to examine under what conditions this is possible, take, to represent the two movable axes of $y'$ and $z'$, two fixed rectangular axes OY, OZ in the same plane; let these belong to a system of rectangular axes OX, OY, OZ having the axis OX (not represented in the figure, Fig. 7, Pl. I) coincident with the axis OX′.

We see firstly, that every plane passing through this axis will be a plane of equal elastic properties, and that therefore we can start from equations (18).

Further, we see that a direct strain $\tau$ will give two equal stresses $t_y$, $t_z$, whence $G_2 = G_3$, also that the two direct strains $\tau_y$, $\tau_z$ being equal must give equal stresses $t_y$, $t_z$, whence we must also get $a_2 = a_3$.

Thus equations (18) become

$$\left.\begin{aligned} t_x &= a_1\tau_x + G_2(\tau_y + \tau_z), & s_{yz} &= G_1\sigma_{yz} \\ t_y &= G_2\tau_x + a_2\tau_y + G_1\tau_z, & s_{zx} &= G_2\sigma_{zx} \\ t_z &= G_2\tau_x + G_1\tau_y + a_2\tau_z, & s_{xy} &= G_2\sigma_{xy} \end{aligned}\right\} \quad . \quad . \quad (19)$$

Now, to obtain expressions for the stresses $t_{x'}$, $t_{y'}$, etc., referred to the axes of $x'$, $y'$, $z'$, the first of which coincides with the axis of $x$, and the other two of which make respectively the angle $\beta$ with the axes of $y$ and $z$, we will first substitute in the preceding formulæ for $\tau_x$, $\tau_y$ their expressions as a function of $\tau_{x'}$, $\tau_{y'}$, etc., given by formulæ (10); secondly, we will substitute the expressions for $t_x$, $t_y$, $t_z$, etc., thus obtained in formulæ (16) and we shall thus get the following formulæ :—

$$\begin{aligned} t_{x'} = {} & a_1\tau_{x'} + G_2(\tau_{y'} + \tau_{z'}) \\ t_{y'} = {} & G_2\tau_{x'} + [a_2(\cos^4\beta + \sin^4\beta) + 6G_1\cos^2\beta\sin^2\beta]\tau_{y'} \\ & + [2a_2\cos^2\beta\sin^2\beta + G_1(\cos^4\beta - 4\cos^2\beta\sin^2\beta]\tau_{y'} \\ & \qquad\qquad + \sin^4\beta)]\tau_{z'} \\ & + (3G_1 - a_2)(\cos^2\beta - \sin^2\beta)\cos\beta\sin\beta\sigma_{y'z'}, \\ t_{z'} = {} & G_2\tau_{x'} + [2a_2\cos^2\beta\sin^2\beta + G_1(\cos^4\beta - 4\cos^2\beta\sin^2\beta \\ & \qquad\qquad + \sin^4\beta)]\tau_{y'} \\ & + [a_2(\cos^4\beta + \sin^4\beta) + 6G_1\cos^2\beta\sin^2\beta]\tau_z \\ & - (3G_1 - a_2)(\cos^2\beta - \sin^2\beta)\cos\beta\sin\beta\,.\,\sigma_{y'z'}, \end{aligned}$$

$$s_{y'z'} = (3G_1 - a_2)(\cos^2\beta - \sin^2\beta)\cos\beta\sin\beta\,(\tau_{y'} - \tau_{z'})$$
$$+ [G_1(\cos^2\beta - \sin^2\beta)^2 + 2(a_2 - G_1)\cos^2\beta\sin^2\beta]\sigma_{y'z'},$$
$$s_{z'x'} = G_2\sigma_{z'x'}$$
$$s_{x'y'} = G_2\sigma_{x'y'}.$$

Now, these formulæ should be independent of $\beta$ in order that the axis of $x$ may be an axis of elasticity; and we can see that this condition will be completely satisfied if

$$a_2 = 3G_1.$$

We must therefore introduce this relation into formulæ (19) when the axis of $x$ is an axis of elasticity; thus, in this case, the formulæ which express the stresses $t_x$, $t_y$, etc., as a function of the normal and shear strains are as follows :—

$$\left.\begin{aligned} t_x &= a_1\tau_x + G_2(\tau_y + \tau_z), & s_{yz} &= G_1\sigma_{yz} \\ t_y &= G_2\tau_x + G_1(3\tau_y + \tau_z), & s_{zx} &= G_2\sigma_{zx} \\ t_z &= G_2\tau_x + G_1(\tau_y + 3\tau_z), & s_{xy} &= G_2\sigma_{xy} \end{aligned}\right\} \quad . \quad . \quad (20)$$

In these formulæ there are only three coefficients $a_1$, $G_1$, $G_2$.

**11. Case IV. A Body of Constant Elasticity or Isotropic Body.**

If the axis of $y$ is also an axis of elasticity, we shall have

$$G_1 = G_2,\ a_1 = 3G_1$$

and the preceding formulæ will become, when we write G for $G_1$

$$\left.\begin{aligned} t_x &= G(3\tau_x + \tau_y + \tau_z), & s_{yz} &= G\sigma_{yz} \\ t_y &= G(\tau_x + 3\tau_y + \tau_z), & s_{zx} &= G\sigma_{zx} \\ t_z &= G(\tau_x + \tau_y + 3\tau_z), & s_{xy} &= G\sigma_{xy} \end{aligned}\right\} \quad . \quad . \quad (21)$$

in which there is only one coefficient G.

We can easily see that when there are two axes of elasticity parallel to the axes of $x$ and $y$, there will be a third, parallel to the axis of $z$.

We can also demonstrate that in this case every straight line is an axis of elasticity; for, all being equal in respect to the resistances around the axis of $x$, each straight line which is perpendicular to it will possess the same properties as the axes of $y$ and $z$, so that, when these are the axes of elasticity, all the lines passing through the origin of the co-ordinates in their plane will also be axes of elasticity. For the same reason, all the straight lines in the two other planes of the co-ordinates will be axes of elasticity; thus, turning the axes of $y$ and $z$ around the axis of $x$, we see that every straight line will be an axis of elasticity.

If we use $\nu$ to express the cubical dilatation or unit volume strain, we have seen that we may obtain formula (5),

$$\nu = \tau_x + \tau_y + \tau_z\,;$$

thus formulæ (21) may be put in the form

$$\left.\begin{aligned} t_x &= G(2\tau_x + \nu), \; s_{yz} = G\sigma_{yz} \\ t_y &= G(2\tau_y + \nu), \; s_{zx} = G\sigma_{zx} \\ t_z &= G(2\tau_z + \nu), \; s_{xy} = G\sigma_{xy} \end{aligned}\right\} \quad . \quad . \quad . \quad (22)$$

Bodies to which these formulæ apply and which have the same elasticity in every direction are called by Lamé, *bodies of constant elasticity* and by Cauchy and Barré de Saint Venant, *isotropic bodies*.

**12. Equations of Condition Relative to the Surface.**—The small tetrahedron which we considered in paragraph 7, can be chosen so that its oblique face, relative to the axes, shall be at the surface of the solid. Then X, Y, Z will be the components along the axes, of the force applied to the surface, per unit area. If we use $f$ to designate this force, and $a'$, $\beta'$, $\gamma'$, to designate the angles which its direction makes with the axes, the three equations (15) become

$$\left.\begin{aligned} f\cos a' &= t_x \cos a + s_{xy} \cos\beta + s_{zx} \cos\gamma \\ f\cos\beta' &= s_{xy}\cos a + t_y \cos\beta + s_{yz} \cos\gamma \\ f\cos\gamma' &= s_{zx}\cos a + s_{yz} \cos\beta + t_z \cos\gamma \end{aligned}\right\} \quad . \quad (23)$$

in which $a$, $\beta$, $\gamma$ are the angles that the normal to the surface of the body at the point under consideration makes with the axes.

Thus the functions of $x, y, z$ representing the stresses $t_x$, $t_y$, etc., must be such that they satisfy equations (13) whatever the values of the co-ordinates may be; and moreover they must satisfy the three preceding equations for all the points of the surface.

**13. Differential Relations between the Normal and the Shear Strains.**—Since the three normal strains

$$\tau_x = \frac{du}{dx}, \quad \tau_y = \frac{dv}{dy}, \quad \tau_z = \frac{dw}{dz} \quad . \quad . \quad . \quad (24)$$

and the three shear strains

$$\sigma_{yz} = \frac{dv}{dz} + \frac{dw}{dy}, \quad \sigma_{zx} = \frac{dw}{dx} + \frac{du}{dz}, \quad \sigma_{xy} = \frac{du}{dy} + \frac{dv}{dx} \quad . \quad (25)$$

depend upon the three displacements $u$, $v$, $w$, it is evident that we cannot give arbitrarily six functions of $x, y, z$ to represent the three normal strains and the three shear strains, but that these functions must satisfy all the differential relations that we can deduce from equations (24) and (25) by the elimination of $u$, $v$, $w$.

To find these relations, let us first differentiate successively in respect to $y$ and $z$ the first of equations (25). We get

$$\frac{d^2\sigma_{yz}}{dydz} = \frac{d^3v}{dydz^2} + \frac{d^3w}{dy^2dz}.$$

Substituting for $\frac{dv}{dy}, \frac{dw}{dz}$ their values $\tau_y, \tau_z$, we get the first of the following equations, and the two others are deduced in a similar way.

$$\left.\begin{aligned} \frac{d^2\sigma_{yz}}{dydz} &= \frac{d^2\tau_y}{dz^2} + \frac{d^2\tau_z}{dy^2} \\ \frac{d^2\sigma_{zx}}{dzdx} &= \frac{d^2\tau_z}{dx^2} + \frac{d^2\tau_x}{dz^2} \\ \frac{d^2\sigma_{xy}}{dxdy} &= \frac{d^2\tau_x}{dy^2} + \frac{d^2\tau_y}{dx^2} \end{aligned}\right\} \quad . \quad . \quad . \quad (26)$$

If we now differentiate the three equations (25) respectively with regard to $x$, $y$, and $z$, we shall get the equations :—

$$\begin{aligned} \frac{d\sigma_{yz}}{dx} &= \frac{d^2v}{dzdx} + \frac{d^2w}{dxdy} \\ \frac{d\sigma_{zx}}{dy} &= \frac{d^2w}{dxdy} + \frac{d^2u}{dydz} \\ \frac{d\sigma_{xy}}{dz} &= \frac{d^2u}{dydz} + \frac{d^2v}{dzdx} \end{aligned}$$

in which the second members only contain the differential coefficients

$$\frac{d^2u}{dydz}, \frac{d^2v}{dzdx}, \frac{d^2w}{dxdy}.$$

Adding the two latter equations and subtracting the first, we get

$$-\frac{d\sigma_{yz}}{dx} + \frac{d\sigma_{zx}}{dy} + \frac{d\sigma_{xy}}{dz} = \frac{2d^2u}{dydz}.$$

This result, differentiated with respect to $x$ and noting that $\frac{du}{dx} = \tau_x$, leads to the first of the following equations. The other two are obtained in a similar way:—

$$\left.\begin{aligned} 2\frac{d^2\tau_x}{dydz} &= \frac{d}{dx}\left(-\frac{d\sigma_{yz}}{dx} + \frac{d\sigma_{zx}}{dy} + \frac{d\sigma_{xy}}{dz}\right) \\ 2\frac{d\tau_y}{dzdx} &= \frac{d}{dy}\left(\frac{d\sigma_{yz}}{dx} - \frac{d\sigma_{zx}}{dy} + \frac{d\sigma_{xy}}{dz}\right) \\ 2\frac{d\tau_z}{dxdy} &= \frac{d}{dz}\left(\frac{d\sigma_{yz}}{dx} + \frac{d\sigma_{zx}}{dy} - \frac{d\sigma_{xy}}{dz}\right) \end{aligned}\right\} \quad . \quad . \quad (27)$$

Thus in order that the six functions of $x$, $y$, and $z$ shall represent the three normal strains and the three shears, they must satisfy both equations (26) and equations (27).

**14. Determination of the Elastic Equilibrium of a Solid Body Acted upon by any Forces.**—The general and direct solution of this problem depends upon the substitution in equations (13) for the stresses $t_x$, $t_y$ their values as functions of the three normal strains and the three shear strains given by formulæ (12) or by formulæ (17), (18), (20), (21), according as to whether the homogeneous body has absolutely variable elastic properties, or whether it possesses at each point a plane of equal elastic properties, or three of these planes, or an axis of elasticity, or finally whether it is isotropic.

In the differential equations of the first order thus obtained we will substitute for $\tau_x$, $\tau_y$, $\tau_z$, $\sigma_{yz}$, $\sigma_{zx}$, $\sigma_{xy}$ their values as functions of the three displacements, $u$, $v$, $w$, and we shall obtain three differential equations with regard to these three unknowns. We should then integrate them, determining the arbitrary functions of the co-ordinates, so that the three equations of condition (23) relative to the surface shall be satisfied for all the points or the surface of the body. This is the direct method which would have to be followed. But the integration that it necessitates are so far beyond the known powers of analysis that we have not been able, up to the present, to solve even the most simple problems.

Thus to investigate the case which occurs most commonly in practice, it has been necessary to follow the reverse process. As experience led us to assume a certain distribution of stresses in the interior of solids, we have attempted to see whether this distribution would satisfy the equations of elastic equilibrium both in the interior and at the surface.

Now, it happened that, in the simple cases of uniform compression over the whole surface of a body, or of the longitudinal extension of a prism, all these equations were satisfied; while in the more complicated cases, which are always the most important, i.e. in cases of torsion, bending, or shear of a prism or a cylinder, the assumed stresses did not satisfy the equations of equilibrium. For these last cases M. Barré de Saint Venant had a very ingenious idea; he regarded as being known only a part of the stresses, namely, those for which there is more reason to be thought exact, and he left the others in indeterminate form: on expressing the fact that all the conditions of equilibrium must be satisfied, it resulted that the assumed stresses satisfied them effectively; and he obtained between the stresses left indeterminate an equation with partial differentials sufficient to determine them, and which in a large number of cases could be integrated.

We will set forth shortly the solutions obtained, confining ourselves to the principal cases. But first we must make an important observation.

It might be stated that the solutions we shall obtain by the indirect method will only be particular solutions, not giving the true conditions of equilibrium of solids; but it is evident that to each system of exterior forces applied to an elastic system, there corresponds a particular condition of elastic equilibrium, and that this condition is unique. Hence it follows that when we have obtained, for a given body, the stresses making equilibrium with an external system of forces, the solution obtained is unique, and therefore, that it must be the same whatever method may be followed to obtain it.

In all the problems which follow, we will suppose that the bodies under consideration are only acted upon by forces applied to the surface, that is to say, that no force, such as gravity, acting upon the whole mass, is present; as a result, the elastic equilibrium of the interior of the solid can always be expressed by equations (14).

**15. Body of any shape, acted upon over its whole surface by a normal and uniform pressure.**

Let us imagine a homogeneous body of any form and elastic properties, the surface of which is acted upon by a normal pressure of $p$ kilograms per square metre. In this case equations (23), expressing the equilibrium at the surface, must be satisfied by putting

$$a' = 180° + a,\ \beta' = 180° + \beta,\ \gamma' = 180° + \gamma.$$

The equations then become:—

$$(t_x + f)\cos a + s_{xy}\cos\beta + s_{zx}\cos\gamma = 0.$$
$$s_{xy}\cos a + (t_y + f)\cos\beta + s_{yz}\cos\gamma = 0.$$
$$s_{zx}\cos a + s_{yz}\cos\beta + (t_z + f)\cos\gamma = 0.$$

We can see, that they would be satisfied if we had

$$t_x = t_y = t_z = -f$$
$$s_{yz} = s_{zx} = s_{xy} = 0 \quad . \quad . \quad . \quad . \quad (28)$$

Let us then see if such a distribution of stresses is possible in the interior.

We notice first that the three indefinite equations (14) are satisfied by hypothesis (28), for all the differentials of the supposed stresses become zero.

If, further, we substitute in the general formulæ (12) for $t_x$, $t_y$, $t_z$ the value $-f$, and for $s_{yz}$, $s_{zx}$, $s_{xy}$ the value zero, we get for the three normal strains and the three shear strains, values which are constant and proportional to $f$, whence it follows that all the equations (26) and (27) will be satisfied.

Thus the stresses which we have assumed are similar to

those which will occur in the body as a consequence of the uniform external pressure.

If at every point the body possesses three planes of equal elastic properties perpendicular to the axes of the co-ordinates, we see that formulæ (18) will give

$$\sigma_{yz} = 0, \ \sigma_{zx} = 0, \ \sigma_{xy} = 0,$$

that is to say, the shear strains will be everywhere equal to zero.

If the body is isotropic the three first equations (21) become

$$-\frac{f}{G} = 3\tau_x + \tau_y + \tau_z,$$

$$-\frac{f}{G} = \tau_x + 3\tau_y + \tau_z,$$

$$-\frac{f}{G} = \tau_x + \tau_y + 3\tau_z,$$

whence we get

$$\tau_x = \tau_y = \tau_z = -\frac{1}{5}\frac{f}{G}$$

and the volume compression will be

$$-\nu = -(\tau_x + \tau_y + \tau_z) = \frac{3}{5}\frac{f}{G}.$$

We shall then see that what we usually call *coefficient of elasticity*, and designate by the letter E, is given by the relation

$$E = \frac{5}{2}G, \quad . \quad . \quad . \quad . \quad (29)$$

and we shall have

$$\tau_x = \tau_y = \tau_z = -\frac{f}{2E} \quad . \quad . \quad . \quad (30)$$

$$-\nu = \frac{3f}{2E} \cdot \quad . \quad . \quad . \quad . \quad (31)$$

**16. Prism or cylinder stretched by forces uniformly distributed over its bases and directed along its axis (Fig. 8 Pl. I).**

Let us imagine a prism or cylinder which is not acted upon by any forces on its lateral surface, and of which the bases are acted upon by normal forces, uniformly distributed at the rate of $t$ kilograms per square metre. If we take for the axis of $x$ a line parallel to the axis of the prism or cylinder, it is evident that for all points on the lateral surface we shall have

$$f = 0, \ \cos a = 0, \ \cos\gamma = \sin\beta;$$

whence it follows that the equations of condition at the surface become

$$s_{xy} \cos \beta + s_{zx} \sin \beta = 0$$
$$t_y \cos \beta + s_{yz} \sin \beta = 0$$
$$s_{yz} \cos \beta + t_z \sin \beta = 0$$

and as these equations must be satisfied whatever element of the surface may be considered, that is to say, whatever the values of $\beta$ may be, it is necessary to ascertain whether we can have, for all the points on the lateral surface,

$$t_y = 0,\ t_z = 0,\ s_{yz} = 0,\ s_{zx} = 0,\ s_{xy} = 0 \quad . \quad . \quad (32)$$

For all points on the two bases we have

$$f = p\,; \quad \cos \beta = \cos \beta' = 0\,; \quad \cos \gamma = \cos \gamma' = 0\,;$$

and, for one of the bases,

$$\cos a = \cos a' = 1$$

while for the other we have

$$\cos a = \cos a' = -1.$$

It follows that the three equations of condition at the surface become, for the bases of the prism or cylinder,

$$t_x = t,\ s_{xy} = 0,\ s_{zx} = 0.$$

The latter two of these three relations occur in equations (32), so that equations (23) would be satisfied for all points of the surface, if we had, for all points of the body,

$$t_x = t,\ t_y = t_z = s_{yz} = s_{zx} = s_{xy} = 0 \quad . \quad . \quad (33)$$

Now, as all these stresses are constant, all their differentials are zero, and the indefinite equations (14) are evidently satisfied. Further, by expressing by means of equations (12) the conditions (33) and resolving with regard to $\tau_x$, $\tau_y$, $\tau_z$, $\sigma_{yz}$ . . . the six equations that result, we obtain the following :—

$$\left.\begin{array}{l} \tau_x = \dfrac{t}{\mathrm{E}},\ \tau_z = -\,\Pi\dfrac{t}{\mathrm{E}},\ \tau_y = -\,\Pi'\dfrac{t}{\mathrm{E}} \\ \sigma_{yz} = \dfrac{t}{\mathrm{H}_1},\ \sigma_{zx} = \dfrac{t}{\mathrm{H}_2},\ \sigma_{xy} = \dfrac{t}{\mathrm{H}_3} \end{array}\right\} \quad . \quad . \quad (34)$$

where E, $\Pi$, $\Pi'$, $\mathrm{H}_1$, $\mathrm{H}_2$, $\mathrm{H}_3$ are constant coefficients. We will call the quantity E the *normal coefficient of elasticity*, and $\Pi$ and $\Pi'$ the *coefficients of contraction, or Poisson's ratio ;* these express the relations between the transverse contraction of the prism along the axes of $y$ and $z$ and its longitudinal extension.

When the extensions and the shear strains given by formulæ (34) are all constant, they will evidently satisfy the equations of condition (26) and (27), hence it follows that the solution given by hypothesis (33) is the true solution of the problem.

It shows that a prism or cylinder, acted upon solely at its

bases by forces uniformly distributed and acting along its axis, is stretched in this direction and contracts laterally, and that in any given section parallel to the bases there is exerted only a force directed along the axis of the prism and equal to the force applied to its bases.

If the body possesses at each point three planes of equal elastic properties, we must express conditions (33) by means of formulæ (18); in this case the shear strains $\sigma_{yz}$, $\sigma_{zx}$, $\sigma_{xy}$ will be zero for all the points, and we shall have, in order to determine the extension strains $\tau_x$, $\tau_y$, $\tau_z$, the three equations

$$t = a_1\tau_x + G_3\tau_y + G_2\tau_z$$
$$0 = G_3\tau_x + a_2\tau_y + G_1\tau_z$$
$$0 = G_2\tau_x + G_1\tau_y + a_3\tau_z$$

whence, writing for simplicity

$$a_1a_2a_3 - a_1G_1^2 - a_2G_2^2 - a_3G_3^2 + 2G_1G_2G_3 = K \quad . \quad (35)$$

we get

$$\left.\begin{aligned} \tau_x &= \frac{a_2a_3 - G_1^2}{K}t, \quad \tau_y = -\frac{a_3G_3 - G_1G_2}{K}t \\ \tau_z &= -\frac{a_2G_2 - G_1G_3}{K}t \end{aligned}\right\} \quad . \quad . \quad (36)$$

Comparing these expressions for the extension with those given by formulæ (34) we get

$$\left.\begin{aligned} E &= \frac{K}{a_2a_3 - G_1^2}, \quad \Pi = \frac{a_3G_3 - G_1G_2}{a_2a_3 - G_1^2} \\ \Pi' &= \frac{a_2G_2 - G_1G_3}{a_2a_3 - G_1^2} \end{aligned}\right\} \quad . \quad . \quad (37)$$

If the axis of $x$ is an axis of elasticity, we have seen that we get

$$a_2 = a_3 = 3G_1, \; G_2 = G_3,$$

thus in this case we get

$$\left.\begin{aligned} K &= 4G_1(2a_1G_1 - G_2^2) \\ E &= \frac{2a_1G_1 - G_2^2}{2G_1}, \quad \Pi = \Pi' = \tfrac{1}{4}\frac{G_2}{G_1} \end{aligned}\right\} \quad . \quad . \quad (38)$$

And if, further, the body is isotropic, we have

$$a_1 = 3G_1, \; G_2 = G_1 = G,$$

and consequently

$$E = \frac{5}{2}G, \; \Pi = \tfrac{1}{4} \quad . \quad . \quad . \quad . \quad (39)$$

We see that for isotropic bodies the normal modulus of elasticity is equal to two and a half times the Glide modulus of elasticity, and that the lateral contraction in a prism acted upon solely by forces normal to its bases, is equal to a fourth of the longitudinal extension.

**17. Torsion of Homogeneous Cylinders or Prisms.**—Let us imagine a prism or cylinder whose lateral surface is not subjected to any force and whose ends are subjected only to tangential forces tending to twist it.

We propose to determine the law according to which these forces must be distributed over the ends in order that the stresses in each transverse section are exactly equal to the external forces applied on the ends and are distributed in the same way.

We will take the axes of $y$ and $z$ in one of the ends and the axis of $x$ parallel to the axis of the prism; and we will call positive the abscissæ $x$ on the prism side.

On an element of area parallel to the ends, i.e. perpendicular to the axis of $x$, the only stresses must be $s_{zx}$, $s_{xy}$ and these must have the same value in corresponding points of successive sections, i.e. keeping $y$ and $z$ constant and varying $x$. It follows that the stresses $s_{zx}$, $s_{xy}$ will be functions only of $y$ and $z$ and therefore

$$\frac{ds_{zx}}{dx} = 0, \frac{ds_{xy}}{dx} = 0.$$

We then have for every element of the end corresponding to $x = 0$

$$\cos a = -1, \cos \beta = 0, \cos \gamma = 0,$$

and for every element of the other end

$$\cos a = 1, \cos \beta = 0, \cos \gamma = 0.$$

Further, as the external forces applied at the ends are perpendicular to the axis of $x$, we shall have for both ends $\cos a' = 0$; so that equations (23) will give

$$t_x = 0, f \cos \beta' = s_{xy}, f \cos \gamma' = s_{zx}.$$

For all elements of the lateral surface of the prism or cylinder we have

$$f = 0, \cos a = 0, \cos \gamma = \sin \beta,$$

and equations (23) become

$$\begin{aligned} s_{xy} \cos \beta + s_{zx} \sin \beta &= 0 \\ t_y \cos \beta + s_{yz} \sin \beta &= 0 \\ s_{yz} \cos \beta + t_z \sin \beta &= 0. \end{aligned}$$

The two stresses $s_{zx}$, $s_{xy}$ cannot by hypothesis be zero either at the surface or at the interior; but the other four stresses will satisfy the two equations for the lateral surface, and the equation $t_x = 0$ at the ends, if they were zero throughout.

We will see therefore whether we can have at all points

$$t_x = t_y = t_z = s_{yz} = 0; \frac{ds_{zx}}{dx} = \frac{ds_{xy}}{dx} = 0 \quad . \quad . \quad (40)$$

Now in substituting these values in the equations (14) we see that the last two are satisfied identically and that the first becomes

$$\frac{ds_{xy}}{dy} + \frac{ds_{zx}}{dz} = 0; \quad . \quad . \quad . \quad . \quad (41)$$

then the stresses $s_{zx}$, $s_{xy}$ must be distributed in the interior of the prism in such a way that they will satisfy equation (41) for all points of the solid.

But it must follow that for all points on the lateral surface, i.e. for all points on the contour of any section, these stresses satisfy the equation

$$s_{xy} \cos \beta + s_{zx} \sin \beta = 0;$$

as $\beta$ is the angle which the normal to this contour makes with the axis of $y$, we shall have

$$\tan \beta = -\frac{dy}{dz}$$

so that the previous equation becomes

$$s_{xy}dz - s_{zx}dy = 0 \quad . \quad . \quad . \quad . \quad (42)$$

Then the stresses $s_{zx}$, $s_{xy}$ must be functions of $y$ and $z$ only; moreover, they must satisfy equation (41) for every point of any section, and equation (42) for every point on the contour of that section.

If we now differentiate the six equations (12) with regard to $x$, it follows from equations (40) that all the first terms will become zero, so that we shall have

$$\frac{d\tau_x}{dx} = \frac{d\tau_y}{dx} = \frac{d\tau_z}{dx} = 0, \frac{d\sigma_{yz}}{dx} = \frac{d\sigma_{zx}}{dx} = \frac{d\sigma_{xy}}{dx} = 0.$$

From these results, equations (26) and (27) give

$$\left.\begin{aligned} \frac{d^2\sigma_{yz}}{dydz} &= \frac{d^2\tau_y}{dz^2} + \frac{d^2\tau_z}{dy^2} \\ \frac{d^2\tau_x}{dy^2} &= \frac{d^2\tau_x}{dz^2} = 0 \\ \frac{d}{dy}\left(\frac{d\sigma_{zx}}{dy} - \frac{d\sigma_{xy}}{dz}\right) &= 0 \\ \frac{d}{dz}\left(\frac{d\sigma_{zx}}{dy} - \frac{d\sigma_{xy}}{dz}\right) &= 0 \end{aligned}\right\} \quad . \quad . \quad . \quad (43)$$

These last two equations show that the quantity

$$\left(\frac{d\sigma_{zx}}{dy} - \frac{d\sigma_{xy}}{dz}\right)$$

cannot be a function of $y$ or $z$; and as on account of the equations

$$\frac{d\sigma_{zx}}{dx} = \frac{d\sigma_{xy}}{dx} = 0,$$

it cannot be a function of $x$, it follows that it must be constant.

We therefore have

$$\frac{d\sigma_{zx}}{dy} - \frac{d\sigma_{xy}}{dz} = 2\theta \quad . \quad . \quad . \quad (44)$$

$\theta$ being a constant quantity.

Substituting in this equation

$$\sigma_{zx} = \frac{dw}{dx} + \frac{du}{dz}, \; \sigma_{xy} = \frac{du}{dy} + \frac{dv}{dx},$$

we obtain

$$\frac{d^2w}{dxdy} - \frac{d^2v}{dzdx} = 2\theta,$$

and as we have

$$\frac{d\sigma_{yz}}{dx} = 0, \text{i.e. } \frac{d^2v}{dzdx} + \frac{d^2w}{dxdy} = 0,$$

we deduce that

$$\frac{d^2w}{dxdy} = \theta, \; \frac{d^2v}{dzdx} = -\theta.$$

If we integrate these two equations on the assumption that the point coinciding with the origin before strain remains fixed, and that the point infinitely near on the axis of $x$ remains so after strain; i.e. if we integrate on the assumption that for $x = y = z = 0$, we must have

$$v = w = \frac{dv}{dx} = \frac{dw}{dx} = 0,$$

we obtain the results

$$w = \theta xy\,; \; v = -\theta xz,$$

and substituting these values in the expressions for $\sigma_{zx}$, $\sigma_{xy}$, we have

$$\sigma_{zx} = \frac{du}{dz} + \theta y, \; \sigma_{xy} = \frac{du}{dy} - \theta z \quad . \quad . \quad (45)$$

Summing up, we see that the three direct and three shear strains must be functions of $y$ and $z$ only, satisfying the differential equations (43) and (44), and giving for the stresses $s_{zx}$, $s_{xy}$ values satisfying equation (41) whatever may be the values of $y$ and $z$ and equation (42) for all points on the contours of the sections.

We will now examine the case in which the problem can be solved in finite terms; but first we will note that in regarding $s_{zx}$, $s_{xy}$ as functions of $y$ and $z$, equation (41) expresses the condition under which equation (42) can be integrated, so that if after having found two functions for $s_{zx}$, $s_{xy}$ which satisfy equation (41) we substitute them in equation (42), the resulting equation will be integrable.

**18. Torsion Continued. Homogeneous Prism of Elliptic Section with Varying Elastic Properties.**—When the body has varying elastic properties, the stresses are expressed as functions of the strains by means of equations (12); now by putting in the first four $t_x = t_y = t = s_{yz} = 0$, we shall obtain expressions for $\tau_x, \tau_y, \tau_z, \sigma_{yz}$ in terms of $\sigma_{zx}, \sigma_{xy}$, and these expressions will be linear; from this it follows that if the shear strains $\sigma_{zx}, \sigma_{xy}$ are linear functions of $y$ and $z$, the direct strains $\tau_x, \tau_y, \tau_z$ and the shear strain $\sigma_{yz}$ will also be linear functions, and therefore the last two formulæ (12) will give linear functions of $y$ and $z$ for $s_{zx}, s_{xy}$, i.e. expressions of the form

$$\left.\begin{aligned} s_{zx} &= a + by + cz \\ s_{xy} &= a_1 + b_1 y + c_1 z \end{aligned}\right\} \quad . \quad . \quad . \quad . \quad (46)$$

where $a, b, c, a_1, b_1, c_1$ are constant coefficients.

We will now observe that the direct and shear strains being linear functions of $y$ and $z$, the equations (43) and (44) are fully satisfied. Moreover, the given expressions for $s_{zx}, s_{xy}$ will satisfy the indefinite equation (41) provided that we have

$$b_1 + c = 0 \quad . \quad . \quad . \quad . \quad (47)$$

Taking this relation into account and substituting in equation (42) for $s_{zx}, s_{xy}$ their expressions as functions of $y$ and $z$, we obtain the differential equation

$$a_1 dz - a dy - c(ydz + zdy) + c_1 zdz - bydy = 0$$

which must be satisfied for the whole contour of the sections. This equation must therefore be the integral of the last equation, i.e.

$$a_1 z - ay - cyz + \tfrac{1}{2}c_1 z^2 - \tfrac{1}{2}by^2 = d \quad . \quad . \quad (48)$$

$d$ being an arbitrary constant.

We see that the contour of the section must be a curve of the second degree; it could be a parabola or a hyperbola, but then the prism would have an infinite section; therefore we have only to consider the remaining case of the elliptic section.

Thus, when we have a homogeneous prism of elliptic section, but of varying elastic properties, and we wish to twist it about a line parallel to its axis, so that the stresses in any section are equal to the tangential forces applied on its ends, we will write the equation of the section of the prism referred to two rectangular axes passing through the point in which the axis of rotation intersects the section. We will then compare that equation with equation (48) so that the latter represents the same ellipse; this will be done by putting the coefficients of equation (48) proportional to those of the equation of the given ellipse. We shall thus obtain five relations between the six coefficients $a, a_1, b, c,$

$c_1$, $d$, and by combining them with relation (47), we shall have six relations between the six coefficients of the equations (46) and the arbitrary constant $d$.

To find another relation, which added to the other six will complete the number necessary for the determination of the unknowns, we will note that if we call $T_M$ the sum of the moments about the axis of torsion of all the forces applied on one end, i.e. the twisting moment of the prism, we shall have

$$T_M = \Sigma(s_{zx}y - s_{xy}z)dA \quad . \quad . \quad . \quad (49)$$

$dA$ being an element of area of the end section and the summation being extended over the whole base. Substituting in this formula for $s_{zx}$, $s_{xy}$ their expressions (46), we obtain

$$T_M = a\Sigma y dA - a_1\Sigma z dA + b\Sigma y^2 dA - c_1\Sigma z^2 dA + 2c\Sigma yz dA \quad (50)$$

which is the final relation between the unknown quantities.

In conclusion, when a prism has variable elastic properties in all directions, the problem of torsion can only be solved in finite terms when the section is an ellipse.

**19. Torsion Continued. Homogeneous prism having at each point a plane of symmetry parallel to the ends or three planes of symmetry one of which is parallel to the ends.**

When the prism has at each point a plane of symmetry perpendicular to its axis and therefore to the axis of $x$, the stresses will be expressed as functions of the direct and shear strains by means of formulæ 17. Now if in the first four we put

$$t_x = t_y = t_z = s_{yz} = 0$$

in accordance with hypothesis (40), we shall get

$$\tau_x = \tau_y = \tau_z = \sigma_{yz} = 0,$$

from which it follows that equations (43) are then identically satisfied. We could then choose for $\sigma_{zx}$, $\sigma_{xy}$ any functions of $y$ and $z$ provided that they satisfy equation (44) and that the stresses $s_{zx}$, $s_{xy}$ deduced by the last two equations (17) satisfy the indefinite equation (41) for all values of $y$ and $z$ and equation (42) for values of $y$ and $z$ corresponding to the contours of the sections.

Now suppose that we choose for $\sigma_{zx}$, $\sigma_{xy}$ any two functions of $y$ and $z$ containing several unknown coefficients; in substituting these functions in equation (44) we shall obtain certain relations between these coefficients which will diminish their number. In substituting them later in the last two formulæ (17) and then introducing the expressions thus obtained for $y$ and $z$ in equation (41), which must be satisfied for all values of $y$ and $z$, we shall obtain new relations between these unknown coefficients, the number of which will be still further diminished; finally in substituting the expressions for $\sigma_{zx}$, $\sigma_{xy}$ in equation (42) and having regard to the relations obtained between the unknown coefficients, we shall obtain a differential equation, directly integrable, which must satisfy the contour of the sections and which will therefore be the differential equation of this contour. We shall integrate this equation and obtain in finite terms the equation of the contour that the section of the prism must have in order that the assumed expressions for $\sigma_{zx}$, $\sigma_{xy}$ may be possible.

This equation for the contour will in general still contain several unknown coefficients, and we might arrange them so that they represent, at least approximately, a given curve.

We see then that for a prism having at every point a plane of symmetry perpendicular to its axis, the problem of torsion can be solved, at least approximately, whatever be the shape of the cross-section.

If the prism possesses at each point three planes of symmetry of which one is perpendicular to the axis of the prism, the solution of the problem will be further simplified, because we shall have in this case, by formulæ (18),

$$s_{zx} = G_2\sigma_{zx},\ s_{xy} = G_3\sigma_{xy},$$

or, from formulæ (45),

$$\left.\begin{aligned} s_{zx} &= G_2\left(\frac{du}{dz} + \theta y\right) \\ s_{xy} &= G_3\left(\frac{du}{dy} - \theta z\right) \end{aligned}\right\} \quad . \quad . \quad . \quad (51)$$

and it is clear that these formulæ satisfy equation (44).

Substituting these expressions for $s_{zx}$, $s_{xy}$ in equations (41) and (42) we get

$$G_3\frac{d^2u}{dy^2} + G_2\frac{d^2u}{dz^2} = 0, \quad . \quad . \quad . \quad (52)$$

$$G_3\left(\frac{du}{dy} - \theta z\right)dz - G_2\left(\frac{du}{dz} + \theta y\right)dy = 0, \quad . \quad . \quad (53)$$

the first of which must hold for all values of $y$ and $z$ and the second only for the contour.

We will now solve this problem in finite terms for two cases which are the most important arising in practice.

**20. Torsion Continued. Prism of Rectangular Section.**—We will assume that the faces of the prism are parallel in pairs to the three planes of symmetry, and will call $b$ the side of the section parallel to the axis of $y$ and $c$ the side parallel to the axis of $z$.

In this case, for the sides parallel to the axis of $y$ we shall have $\frac{dz}{dy} = 0$ for all values of $y$, and for the other sides $\frac{dy}{dz} = 0$ for all values of $z$.

$\therefore$ Equation (53) splits up into the following conditions:—

$\frac{du}{dy} - \theta z = 0$ for $y = \pm\frac{b}{2}$ for all values of $z$ between $\frac{c}{2}$ and $-\frac{c}{2}$,

$\frac{du}{dz} + \theta y = 0$ ,, $z = \pm\frac{c}{2}$ ,, ,, $y$ ,, $\frac{b}{2}$ and $-\frac{b}{2}$,

and we must integrate equation (52) so that both these conditions are satisfied.

Now by putting

$$u = \theta yz + \Sigma A_n(e^{amz} - e^{-amz}) \sin my,$$

equation (52) will be satisfied, as we can prove easily, whatever be the number of terms included in the summation, provided that we have for all these terms

$$a = \sqrt{\frac{G_3}{G_2}};$$

the quantities $A_n$, $m$ may change from one term to another and are determined so that the expression for $u$ satisfies the two conditions into which equation (53) has been divided.

The first of these conditions, i.e. $\frac{du}{dy} - \theta z = 0$ for $y = \pm\frac{b}{2}$, becomes, on substituting the above value of $u$,

$$\Sigma A_n(e^{amz} - e^{-amz})m \cos\frac{mb}{2} = 0;$$

and since this must be satisfied for all values of $z$ between $\frac{c}{2}$ and $-\frac{c}{2}$, we must have for each term

$$\cos\frac{mb}{2} = 0;$$

and, therefore, if $n$ is any whole number

$$\frac{mb}{2} = \frac{2n-1}{2}.\pi, \quad \text{or } m = \frac{(2n-1)\pi}{b}.$$

Thus the expression for $u$ becomes

$$u = \theta yz + \Sigma A_n\left(e^{\frac{a(2n-1)\pi z}{b}} - e^{-\frac{a(2n-1)\pi z}{b}}\right) \sin\frac{(2n-1)\pi y}{b}.$$

Now expressing by means of this formula the second condition that $\frac{du}{dz} + \theta y = 0$ for $z = \pm\frac{c}{2}$, we obtain the equation

$$2\theta y + \Sigma A_n \frac{a(2n-1)\pi}{b}\left(e^{\frac{a(2n-1)\pi c}{2b}} + e^{-\frac{a(2n-1)\pi c}{2b}}\right) \sin\frac{(2n-1)\pi y}{b} = 0 \quad (54)$$

which must be satisfied for all values of $y$ between

$$\frac{b}{2} \text{ and } -\frac{b}{2}.$$

Now in Lagrange's formula

$$\phi(y) = \frac{2}{l}\sum_{n=1}^{n=\infty} \sin\frac{(2n-1)\pi y}{2l}\left(\int_0^l \phi(y')\sin\frac{(2n-1)\pi y'}{2l}\,dy'\right)$$

if we put $\phi(y) = 2\theta y$ and $\phi(y') = 2\theta y'$, it becomes

$$2\theta y - \frac{4\theta}{l}\sum_{n=1}^{n=\infty}\left(\sin\frac{(2n-1)\pi y}{2l}\int_0^l y'\sin\frac{(2n-1)\pi y'}{2l}\,dy'\right) = 0$$

and noting that we have

$$\int_0^l y'\sin\frac{(2n-1)\pi y'}{2l}\,dy' = \pm\frac{4l^2}{(2n-1)^2\pi^2},$$

this formula reduces to the following:—

$$2\theta y - \frac{16l\theta}{\pi^2}\sum_{n=1}^{n=\infty}\left(\pm\frac{\sin\frac{(2n-1)\pi y}{2l}}{(2n-1)^2}\right) = 0$$

in which we should take the positive sign inside the $\Sigma$ for odd values of $n$ and the negative sign for even. Comparing the last formula with (54), we see that they become identical on taking

$$2l = b,$$

$$A_n\frac{a(2n-1)\pi}{b}\left(e^{\frac{a(2n-1)\pi c}{2b}} + e^{-\frac{a(2n-1)\pi c}{2b}}\right) = \mp\frac{16l\theta}{(2n-1)^2\pi^2};$$

thus equation (54) will be satisfied identically for any value of $y$ between $\frac{b}{2}$ and $-\frac{b}{2}$ provided that we take, in general, for any value of $n$

$$A_n = \mp\frac{8b^2\theta}{a(2n-1)^3\pi^3\left[e^{\frac{a(2n-1)\pi c}{2b}} + e^{-\frac{a(2n-1)\pi c}{2b}}\right]},$$

where the upper sign should be taken for odd values of $n$ and the lower for even values.

If we substitute this value of $A_n$ in the general expression for $u$, and if we call respectively hyperbolic sine, cosine, tangent, and cotangent of $x$ the expressions

$$\frac{e^x - e^{-x}}{2},\quad \frac{e^x + e^{-x}}{2},\quad \frac{e^x - e^{-x}}{e^x + e^{-x}},\quad \frac{e^x + e^{-x}}{e^x - e^{-x}},$$

written sinh $x$, cosh $x$, tanh $x$, coth $x$, we obtain finally the formula

$$u = \theta \left[ yz + \frac{8b^2}{a\pi^3} \sum_{n=1}^{n=\infty} \frac{(-1)^n \sinh \dfrac{a(2n-1)\pi z}{b} \sin \dfrac{(2n-1)\pi y}{b}}{(2n-1)^3 \cosh \dfrac{a(2n-1)\pi c}{2b}} \right]. \quad (55)$$

We may note that if we reverse the co-ordinates $y$ and $z$, the moduli $G_2$, $G_3$, and the sides $b$ and $c$ of the section of the prism, and if, moreover, we change $\theta$ to $-\theta$, equation (52) and the contour conditions that we have deduced from equation (53) will not alter at all; it follows from this that making the same changes in equation (55) we shall obtain an expression for $u$ which will still be the integral of equation (52) having regard to the conditions deduced from equation (53). As, moreover, changing $G_2$, $G_3$ means changing $a$ to $\dfrac{1}{a}$, we shall also have

$$u = -\theta \left[ yz + \frac{8c^2 a}{\pi^3} \sum_{n=1}^{n=\infty} \frac{(-1)^n \sinh \dfrac{(2n-1)\pi y}{ac} \cdot \sin \dfrac{(2n-1)\pi z}{c}}{(2n-1)^3 \cosh \dfrac{(2n-1)\pi b}{2ac}} \right] \quad (56)$$

The two values of $u$ thus obtained, although apparently different, both represent the surface into which transverse sections of the prism will change which were plane before strain.

We can now obtain, by means of equations (51), the expressions for the stresses $s_{zx}$, $s_{xy}$, and as we have found two expressions for $u$ we shall also obtain two values for each stress, i.e.

$$\left.\begin{aligned} s_{zx} &= G_2\theta \left[ 2y + \frac{8b}{\pi^2} \sum_{n=1}^{n=\infty} \frac{(-1)^n \cosh \dfrac{a(2n-1)\pi z}{b} \cdot \sin \dfrac{(2n-1)\pi y}{b}}{(2n-1)^2 \cosh \dfrac{a(2n-1)\pi c}{2b}} \right] \\ &= -G_2\theta \frac{8ca}{\pi^2} \sum_{n=1}^{n=\infty} \frac{(-1)^n \sinh \dfrac{(2n-1)\pi y}{ac} \cdot \cos \dfrac{(2n-1)\pi z}{c}}{(2n-1)^2 \cosh \dfrac{(2n-1)\pi b}{2ac}}; \end{aligned}\right\} \quad (57)$$

$$\left.\begin{aligned} s_{xy} &= G_3\theta \frac{8b}{a\pi^2} \sum_{n=1}^{n=\infty} \frac{(-1)^n \sinh \dfrac{a(2n-1)\pi z}{b} \cdot \cos \dfrac{(2n-1)\pi y}{b}}{(2n-1)^2 \cosh \dfrac{a(2n-1)\pi c}{2b}} \\ &= -G_3\theta \left[ 2z + \frac{8c}{\pi^2} \sum_{n=1}^{n=\infty} \frac{(-1)^n \cosh \dfrac{(2n-1)\pi y}{ac} \cdot \sin \dfrac{(2n-1)\pi z}{ac}}{(2n-1)^2 \cosh \dfrac{(2n-1)\pi b}{2ac}} \right] \end{aligned}\right\} \quad (58)$$

In these formulæ we have the quantity $\theta$ which expresses the *angle of torsion* of the prism, i.e. the angle through which one section will have turned after strain with reference to another section at unit distance from it.

This interpretation of $\theta$ is deduced from the two equations

$$w = \theta xy, \; v = -\theta xz.$$

Now, to determine the angle of torsion $\theta$, we will express it as a function of the Twisting Moment, $T_M$, i.e. the sum of moments about the axis of torsion of all the forces applied at one end. If we consider an element of area $dydz$ in one end having co-ordinates $y$, $z$, the effective forces applied on this element will be $s_{zx}dydz$ and $s_{xy}dydz$, and their moments about the axis of torsion will be $ys_{zx}dydz$, $-zs_{xy}dydz$; we thus have

$$T_M = \int_{-\frac{b}{2}}^{+\frac{b}{2}} dy \int_{-\frac{c}{2}}^{+\frac{c}{2}} (s_{zx} \cdot y - s_{xy} \cdot z) dz.$$

Substituting in this formula the first or second expressions for $s_{zx}$, $s_{xy}$ given in (57), (48), we obtain

$$\left.\begin{aligned} T_M &= G_2\theta b^3 c \left[\frac{1}{3} - \frac{64b}{\pi^5 ac} \sum_{n=1}^{n=\infty} \frac{\tanh \dfrac{a(2n-1)\pi c}{2b}}{(2n-1)^5}\right] \\ &= G_3\theta bc^3 \left[\frac{1}{3} - \frac{64ac}{\pi^5 b} \sum_{n=1}^{n=\infty} \frac{\tanh \dfrac{(2n-1)\pi b}{2ac}}{(2n-1)^5}\right] \end{aligned}\right\} . \qquad (59)$$

By means of formulæ (59) we can calculate the angle of torsion $\theta$ when the twisting moment $T_M$ is given.

Formulæ (59) can be put into a more convenient form for calculation as follows. We have

$$-\sum_{n=1}^{n=\infty} \frac{\tanh \dfrac{a(2n-1)\pi c}{2b}}{(2n-1)^5} = \sum_{n=1}^{n=\infty} \frac{1 - \tanh \dfrac{a(2n-1)\pi c}{2b}}{(2n-1)^5} - \sum_{n=1}^{n=\infty} \frac{1}{(2n-1)^5}$$

and we can easily show that

$$\frac{64}{\pi^5} = \cdot 20914, \quad \sum_{n=1}^{n=\infty} \frac{1}{(2n-1)^5} = 1 \cdot 00452,$$

$$\text{and } \therefore \frac{64}{\pi^5} \sum_{n=1}^{n=\infty} \frac{1}{(2n-1)^5} = \cdot 21008,$$

so that formulæ (59) become

$$\left.\begin{aligned} T_M &= G_2\theta b^3 c\left[\frac{1}{3} - \cdot 21008\frac{b}{ac} + \cdot 20914\frac{b}{ac}\sum_{n=1}^{n=\infty}\frac{1 - \tanh\dfrac{a(2n-1)\pi c}{2b}}{(2n-1)^5}\right] \\ &= G_3\theta bc^3\left[\frac{1}{3} - \cdot 21008\frac{ac}{b} + \cdot 20914\frac{ac}{b}\sum_{n=1}^{n=\infty}\frac{1 - \tanh\dfrac{(2n-1)\pi b}{2ac}}{(2n-1)^5}\right] \end{aligned}\right\} \quad (60)$$

The first expression is more or less convergent than the second according as $ac$ is greater or less than $b$, so that to facilitate calculation we should employ one or the other.

As we have

$$\tanh x = \frac{e^x - e^{-x}}{e^x + e^{-x}} = \frac{e^{2x} - 1}{e^{2x} + 1}, \text{ and } 1 - \tanh x = \frac{2}{e^{2x} + 1},$$

we see that when $x$ is large, the quantity $e^{2x}$ is also large, and that $(1 - \tanh x)$ is small; so that to use formulæ (60) it will in general suffice to calculate a very small number of terms.

If the material has equal glide moduli in the directions of $y$ and $z$, we shall have $G_2 = G_3$ and $a = 1$, which will simplify the formulæ somewhat. This is the condition which we must generally assume in practice because we do not usually know the moduli $G_2$, $G_3$ but only a mean value, so that we must assume each to be equal to this known value.

If, in addition, the section is a square, $b = c$ and therefore

$$T_M = G_2\theta b^4\left[\frac{1}{3} - \cdot 21008 + \cdot 20914\sum_{n=1}^{n=\infty}\frac{1 - \tanh\dfrac{(2n-1)\pi}{2}}{(2n-1)^5}\right] \quad . \quad (61)$$

The quantity inside the brackets is purely numerical, and can be calculated at once; we may observe that to obtain results correct to five places of decimal it is sufficient to take account only of the first term of the summation, so that we get

$$T_M = \cdot 84368 G_2\theta\frac{b^4}{6}.$$

Now $\dfrac{b^4}{6}$ is the polar moment of inertia $I_P$ of the section about its centre, so that we have

$$T_M = \cdot 84368 G_2\theta I_P \quad . \quad . \quad . \quad . \quad (62)$$

in place of $T_M = G_2\theta I_P$, which is given by the old theory that plane sections remain plane after strain.

**21. Torsion Continued. Prism of any Cross-section.**—We will follow here the converse method of that which we have used for the rectangular section, i.e. we will assume as given the function $u$ representing the curved surface into which plane sections perpendicular to the axis before strain are changed, and we will find the corresponding form of the contour of the sections.

From the infinite number of forms which we might give to the function $u$ let us choose the following:—

$$u = Cf(\beta y + \gamma z),$$

C, $\beta$, and $\gamma$ being constant coefficients.

We then have

$$\frac{du}{dy} = C\beta f'(\beta y + \gamma z), \quad \frac{d^2u}{dy^2} = C\beta^2 f''(\beta y + \gamma z),$$

$$\frac{du}{dz} = C\gamma f'(\beta y + \gamma z), \quad \frac{d^2u}{dz^2} = C\gamma^2 f''(\beta y + \gamma z),$$

and on substituting in equation (52) we see that it will be satisfied if

$$G_3\beta^2 + G_2\gamma^2 = 0,$$

from which, by making $\beta = 1$, which does not diminish the general scope of the function, we deduce

$$\gamma = \pm\sqrt{\frac{G_3}{G_2}}\sqrt{-1}.$$

Thus the function $u$ can be put into the form

$$u = Cf\left(y \pm \sqrt{-1}\,.\,y\sqrt{\frac{G_3}{G_2}}\right).$$

Developing the right-hand side of this equation and separating out the terms containing $\sqrt{-1}$ and those independent of it, we obtain

$$u = C[\phi(y, z) \pm \sqrt{-1}\psi(y, z)],$$

$\phi$ and $\psi$ being two functions which we can easily determine.

Now we know that if several functions of $y$ and $z$ satisfy equation (52), on adding them, after multiplying each by a constant, we obtain a sum which will also satisfy it.

Then since the two functions

$$\phi(yz) + \sqrt{-1}\psi(y, z),$$

$$\phi(yz) - \sqrt{-1}\psi(y, z),$$

both satisfy equation (52), by adding them after multiplying both by $\frac{1}{2}B_1$ or the first by $\frac{1}{2}C_1\sqrt{-1}$ and the second by $\frac{1}{2}C_1\sqrt{-1}$ we obtain the two functions

$$B_1\phi(y, z), \; C_1\psi(y, z),$$

which satisfy equation (52) and whose sum does so also. We can thus take for $u$ a very general expression of the form

$$u = \Sigma[B_1\phi(y, z) + C_1\psi(y, z)] \quad . \quad . \quad . \quad (63)$$

As we can introduce into this expression any number of terms, giving to each an arbitrary coefficient, and as these coefficients remain in the equation for the contour, it is clear that we could choose them so that they represent, at least approximately, any form of curve.

Take for example

$$f\left(y \pm \sqrt{-1}\,.\,z\sqrt{\frac{G_3}{G_2}}\right) = (y \pm az\sqrt{-1})^n.$$

Then on expanding we shall find

$$\left.\begin{aligned}\phi_n(y, z) &= y^n - \frac{n(n-1)}{1\,.\,2}a^2z^2y^{n-2} + \frac{n(n-1)(n-2)(n-3)}{1\,.\,2.\,3\,.\,4}\,.\,a^4z^4y^{n-4} - \ldots \\ \psi_n(y, z) &= na\Big[zy^{n-1} - \frac{(n-1)(n-2)}{2\,.\,3}\,.\,a^2z^3y^{n-3} \\ &\qquad + \frac{(n-1)(n-2)(n-3)(n-4)}{2.3.4.5}a^4z^5y^{n-5} - \ldots\Big]\end{aligned}\right\} \quad (64).$$

The subscript $n$ is given to the symbols $\phi$, $\psi$ to indicate the degree represented by these two functions.

If now we take

$$u = \Sigma[B_n\phi_n(y, z) + C_n\psi_n(y, z)], \quad . \quad . \quad (65)$$

$B_n$ and $C_n$ being arbitrary constants, and if we substitute in equation (53) written in the form

$$a^2\frac{du}{dy}\,.\,dz - \frac{du}{dz}dy - \theta(a^2zdz + ydy) = 0,$$

we obtain

$$\Sigma B_n\left(a^2\frac{d\phi_n}{dy}\,.\,dz - \frac{d\phi_n}{dz}\,.\,dy\right) + \Sigma C_n\left(a^2\frac{d\psi_n}{dy}\,.\,dz - \frac{d\psi_n}{dz}\,.\,dy\right)$$
$$- \theta(a^2zdz + ydy) = 0.$$

But we can easily see that we have

$$a^2\frac{d\phi_n}{dy}\,.\,dz - \frac{d\phi_n dy}{dz} = ad\psi_n$$
$$a^2\frac{d\psi_n}{dy}\,.\,dz - \frac{d\psi_n dy}{dz} = -\,ad\psi_n,$$

so that the integral of the preceding differential equation will be

$$a\Sigma[B_n\psi_n(y, z) - C_n\phi_n(y, z)] - \frac{\theta}{2}(a^2z^2 + y^2) = D \quad . \quad (66)$$

D being an arbitrary constant.

Such then is the general equation for the contour of sections which under torsion will deform according to the curved surface represented by equation (65).

If we successively put $n = 1$, $n = 2$, $n = 3$, etc., in formula (65), we obtain

$$\left.\begin{aligned} u = {} & B_1 y + B_2(y^2 - a^2z^2) + B_3(y^3 - 3a^2z^2y) \\ & + B_4(y^4 - 6a^2z^2y^2 + a^4z^4) + \ldots + C_1 az + C_2 \,.\, 2azy \\ & + C_3(3azy^2 - a^3z^3) + C_4(4azy^3 - 4a^3z^3y) + \ldots \end{aligned}\right\}, \quad (67)$$

and equation (66), which represents the contour of the section, becomes, on dividing by $a$,

$$\left(\begin{aligned} & B_1 az + B_2 \,.\, 2azy + B_3(3azy^2 - a^3z^3) + B_4(4azy^3 - 4a^3z^3y) \\ & + \ldots - C_1 y - C_2(y^2 - a^2z^2) - C_3(y^3 - 3a^2z^2y) \\ & - C_4(y^4 - 6a^2y^2z^2 + a^4z^4) - \ldots - \frac{\theta}{2a}(y^2 + a^2z^2) \end{aligned}\right) = \frac{D}{a} \quad (68)$$

Taking in the first part a sufficiently large number of terms and determining their coefficients, $B_1, B_2 \ldots C_1, C_2$, etc., it is clear that this equation could represent, at least approximately, any contour.

**22. Torsion Continued. Prism with an Elliptic Section.**— When the section of the prism is an ellipse, the solution of the torsion problem is comprised within the case dealt with in the previous paragraph. Suppose, for example, that the torsion takes place about the axis of the prism, and that planes passing through this axis and through the axes of the ellipse are planes of equal elastic properties. If we call $b$, $c$ the semi-axes of the ellipse, its equation will be

$$\frac{y^2}{b^2} + \frac{z^2}{c^2} = 1 \quad . \quad . \quad . \quad . \quad (69)$$

Now equation (68) reduced to terms of the second degree only becomes

$$B_2 \,.\, 2azy - C_2(y^2 - a^2z^2) - \frac{\theta}{2a}(y^2 + a^2z^2) = \frac{D}{a},$$

and in order that this may be identical with the equation to the ellipse, the coefficients $B_2$, $C_2$, D must be such that

$$B_2 = 0; \quad -\frac{aC_2 + \dfrac{\theta}{2}}{D} = \frac{1}{b^2}; \quad \frac{a^2\left(aC_2 - \dfrac{\theta}{2}\right)}{D} = \frac{1}{c^2},$$

from which we deduce on eliminating D,

$$\frac{aC_2 + \dfrac{\theta}{2}}{aC_2 - \dfrac{\theta}{2}} = -\frac{a^2c^2}{b^2},$$

and therefore

$$C_2 = -\frac{\theta}{2a} \cdot \frac{b^2 - a^2c^2}{b^2 + a^2c^2}.$$

The expression for $u$ for the elliptic contour considered is obtained from formula (68) by putting all the arbitrary coefficients equal to zero except $C_2$, which must have the value found above, so that we shall have

$$u = -\theta \cdot \frac{b^2 - a^2c^2}{b^2 + a^2c^2} \cdot yz \quad . \quad . \quad . \quad (70)$$

On introducing this value of $u$ into formulæ (51), and noting that $a^2 = \frac{G_3}{G_2}$, we obtain

$$\left.\begin{aligned} s_{zx} &= G_2\theta\left(-\frac{b^2 - a^2c^2}{b^2 + a^2c^2} + 1\right)y = \frac{2G_2G_3c^2}{G_2b^2 + G_3c^2} \cdot \theta y \\ s_{xy} &= G_3\theta\left(-\frac{b^2 - a^2c^2}{b^2 + a^2c^2} - 1\right)z = -\frac{2G_2G_3b^2}{G_2b^2 + G_3c^2} \cdot \theta z \end{aligned}\right\} \quad (71)$$

We will now find the twisting moment $T_M$, which in accordance with § 18 is given by the formula

$$T_M = \iint(s_{zx}y - s_{xy}z)dydz,$$

extending the integration over the whole section. If we substitute in this formula the values given by equations (71), and note that

$$\iint y^2dydz = I_z; \ \iint z^2dydz = I_y,$$

$I_z$, $I_y$ being the moments of inertia of the ellipse about its axes $z$, $y$, we obtain the result

$$T_M = \frac{2G_2G_3(c^2I_z + b^2I_y)}{G_2b^2 + G_3c^2} \cdot \theta \quad . \quad . \quad (72)$$

From this we can deduce the angle of torsion $\theta$ when the twisting moment $T_M$ is given.

We may note that formula (70) gives $u = 0$ for all values of $y$ and $z$ when we have between $b$ and $c$ the relation

$$b^2 - a^2c^2 = 0, \quad \text{or} \ \frac{b^2}{G_3} = \frac{c^2}{G_2} \quad . \quad . \quad . \quad (73)$$

i.e. when the squares of the semi-axes are proportional to the Glide moduli in their direction; in this case the plane sections remain plane after strain.

When $G_2 = G_3$, which arises when the axis of $x$ is an axis of elasticity or the body is isotropic, formulæ (70), (71), (72) become

$$\left.\begin{aligned} u &= -\theta \cdot \frac{b^2 - c^2}{b^2 + c^2} \cdot yz \\ s_{zx} &= \frac{2G_2 c^2 \theta}{b^2 + c^2} \cdot y \\ s_{xy} &= -\frac{2G_2 b^2 \theta}{b^2 + c^2} \cdot z \\ T_M &= \frac{2G_2(c^2 I_z + b^2 I_y)}{b^2 + c^2} \cdot \theta \end{aligned}\right\} \quad . \quad . \quad (74)$$

And we see from the first of these formulæ that in the circular section plane sections remain plane after strain.

**23. Bending of Homogeneous Prisms or Cylinders.**—Imagine a prism or cylinder ABCD (Fig. 8, Pl. I) referred to three rectangular axes OX, OY, OZ which coincide before strain with the following lines :—

OX with the axis of the body.

OY, OZ with the principal axes of the base AB.

We will assume that the three axes move with the body so that the following conditions are satisfied :—

1. The centroid of the base AB coincides always with the origin so that calling $u$, $v$, $w$ the displacements of any point parallel to the axes, we have

$$u = v = w = 0$$
$$\text{for} \quad x = y = z = 0.$$

2. The element of the base AB corresponding to $y = z = 0$, remains tangential to the plane of $y$, $z$, which means that for $x = y = z = 0$, we have

$$\frac{du}{dy} = \frac{du}{dz} = 0\ ;$$

3. The principal axis of the base AB which coincides with the axis of $y$ before bending is tangential to this axis after bending, i.e. for

$$x = y = z = 0$$

we have
$$\frac{dw}{dy} = 0.$$

With these conditions the axes of the co-ordinates are regarded as moving with the body, so that $u$, $v$, $w$ represent only relative displacements, i.e. those due to strain.

Now suppose that the lateral surface of the body is not subjected to any load and that the ends are subjected to forces in the following manner :—

(i) The base CD by a force whose tangential components are F cos $\beta'$, F cos $\gamma'$ distributed in a manner which we will determine later, and whose normal component F cos $\alpha'$ is so distributed that at each point it is proportional to the co-ordinate $z$, so that we have

$$\mathrm{F}\cos\alpha' = mz\,;\ \mathrm{F}\cos\beta' = \phi(y, z)\,;\ \mathrm{F}\cos\gamma' = \psi(y, z).$$

(ii) The base AB by a force whose tangential components correspond exactly with those on the base CD but act in the opposite direction, and whose normal component is directed in an opposite sense to that applied on the other end and is proportional to the co-ordinate $z$, so that we have

$$\mathrm{F}\cos\alpha' = -(m + \mathrm{C}a)z\,;\ \mathrm{F}\cos\beta' = -\phi(y, z)\,;\ \mathrm{F}\cos\gamma' = -\psi(y, z)$$

C being a new arbitrary constant and $a$ the length of the prism.

Now let us see if the internal stresses can satisfy the following conditions :—

1. That the shear stresses $s_{zx}$, $s_{xy}$ are the same at the ends and at any section, so that we have everywhere

$$s_{xy} = \phi(y, z),\ s_{zx} = \psi(y, z).$$

2. That the normal stresses $t_x$ vary over different points of the same section as the transverse co-ordinate $z$ and follow a linear relation with regard to $x$ for different sections.

3. That $t_y = t_z = s_{yz} = 0$ throughout.

If these conditions are satisfied, we shall have for any point in the body

$$\left.\begin{aligned} t_x &= [m + \mathrm{C}(a - x)]z \\ s_{zx} &= \psi(y, z) \\ s_{xy} &= \phi(y, z) \\ t_y &= t_z = s_{yz} = 0 \end{aligned}\right\} \quad . \quad . \quad . \quad (75)$$

and therefore

$$\frac{ds_{zx}}{dx} = \frac{ds_{xy}}{dx} = 0 \quad . \quad . \quad . \quad . \quad (76)$$

From equations (75) and (76) we see that the two last relations (14) are identically satisfied, whilst the first takes the following form :—

$$\frac{ds_{xy}}{dy} + \frac{ds_{zx}}{dz} = \mathrm{C}z \quad . \quad . \quad . \quad . \quad (77)$$

Now let us see what equations (23) with regard to the surface become in this case.

For the lateral surface we have throughout

$$\mathrm{F} = 0\,;\ \cos\alpha = 0\,;\ \cos\gamma = \sin\beta,$$

and therefore the last two equations (23) are identically satisfied, while the first reduces to

$$s_{xy}\cos\beta + s_{zx}\sin\beta = 0.$$

Now, since $\beta$ is the angle which the normal to the contour of the sections makes with the axis of $y$, we have

$$\tan\beta = -\frac{dy}{dz},$$

so that the above equation becomes

$$s_{xy}dz - s_{zx}dy = 0 \quad . \quad . \quad . \quad . \quad (78)$$

For the base CD we have

$$\cos a = 1;\ \cos\beta = \cos\gamma = 0;$$

$$\mathrm{F}\cos a' = mz;\ \mathrm{F}\cos\beta' = \phi(y, z);\ \mathrm{F}\cos\gamma' = \psi(y, z),$$

and, noting that for $x = a$ we have

$$t_x = mz,$$

and having regard to the values of $s_{xy}$, $s_{zx}$, we see that equations (23) are identically satisfied.

For the base AB we have

$$\cos a = -1;\ \cos\beta = \cos\gamma = 0;$$

$$\mathrm{F}\cos a' = -(m + \mathrm{C}a)z;\ \mathrm{F}\cos\beta' = -\phi(y, z);\ \mathrm{F}\cos\gamma' = -\psi(y, z),$$

and since by putting $x = 0$ we have

$$t_x = (m + \mathrm{C}a)z,$$

it follows that equations (23) are still identically satisfied.

Now, if we assume that the body has three planes of equal elastic properties perpendicular to the axes of the co-ordinates, it is clear that the first four equations (18), and the differential coefficients with regard to $x$ of the two last, will give us in view of formulæ (75)

$$\left.\begin{array}{l} \tau_x = \dfrac{[m + \mathrm{C}(a - x)]z}{\mathrm{E}},\ \tau_y = -\Pi\tau_x,\ \tau_z = -\Pi'\tau_x \\ \sigma_{yz} = 0,\ \dfrac{d\sigma_{zx}}{dx} = \dfrac{d\sigma_{xy}}{dx} = 0 \end{array}\right\} \quad . \quad (79)$$

the coefficients E, $\Pi$, $\Pi'$ having the same meaning and value as in formulæ (37).

These expressions for the tensile and shear strains satisfy equations (26) and the first of equations (27) identically, and reduce the two last to the following:—

$$\frac{d}{dy}\left(-\frac{d\sigma_{zx}}{dy} + \frac{d\sigma_{xy}}{dz}\right) = \frac{2\Pi\mathrm{C}}{\mathrm{E}}$$

$$\frac{d}{dz}\left(\frac{d\sigma_{zx}}{dy} - \frac{d\sigma_{xy}}{dz}\right) = 0.$$

As, moreover, we deduce also from the two last equations (79)

$$\frac{d}{dx}\left(\frac{d\sigma_{zx}}{dy} - \frac{d\sigma_{xy}}{dz}\right) = 0,$$

it follows that the function

$$\frac{d\sigma_{zx}}{dy} - \frac{d\sigma_{xy}}{dz}$$

cannot be a function of $z$ or $x$, but only of $y$, so that we shall have

$$-\frac{d\sigma_{zx}}{dy} + \frac{d\sigma_{xy}}{dz} = \frac{2\Pi C}{E} y \quad . \quad . \quad . \quad (80)$$

We will not add any arbitrary constant, because it follows from the problem of torsion that the added constant would correspond to a uniform torsion of the body, and we have assumed that this does not occur in the present case.

We see then that assumptions (75) and (76) as to the internal stresses will hold provided that the distribution of the forces on the ends is such that the forces satisfy equations (77) and (80) for every point in the ends and equation (78) for any point on the contour.

We will now express the stresses $s_{xy}$, $s_{zx}$ by means of a single unknown function so that they satisfy the indefinite equation (80). Thus it will suffice to determine this function from equations (77) and (78).

Now, if in equation (80) we substitute for $\sigma_{zx}$, $\sigma_{xy}$ their expressions (25) we get

$$\frac{d^2v}{dzdx} - \frac{d^2w}{dxdy} = \frac{2\Pi C}{E} \, . \, y,$$

but since $\sigma_{yz} = 0$, we shall also have

$$\frac{d\sigma_{yz}}{dx} = 0, \text{ or } \frac{d^2v}{dzdx} + \frac{d^2w}{dxdy} = 0.$$

So that combining this with the previous equation we get

$$\frac{d^2v}{dzdx} = \frac{\Pi C}{E} \, . \, y$$

$$\frac{d^2w}{dxdy} = -\frac{\Pi C}{E} \, . \, y.$$

Moreover the first three equations (79) give

$$\frac{du}{dx} = \frac{[m + C(a - x)] \, . \, z}{E},$$

$$\frac{dv}{dy} = -\frac{\Pi[m + C(a - x)]z}{E},$$

$$\frac{dw}{dz} = -\frac{\Pi'[m + C(a - x)]z}{E},$$

so that we get by differentiation

$$\frac{d^2u}{dz\,.\,dx} = \frac{m + \mathrm{C}(a - x)}{\mathrm{E}}; \quad \frac{d^2u}{dx\,.\,dy} = 0;$$

$$\frac{d^2v}{dx\,.\,dy} = \frac{\Pi \mathrm{C}z}{\mathrm{E}}; \quad \frac{d^2w}{dzdx} = \frac{\Pi' \mathrm{C}z}{\mathrm{E}},$$

and the last two equations (79) give

$$\frac{d^2w}{dx^2} + \frac{d^2u}{dzdx} = 0,$$

$$\frac{d^2u}{dxdy} + \frac{d^2v}{dx^2} = 0;$$

from which we deduce by the aid of the previous results,

$$\frac{d^2v}{dx^2} = 0; \quad \frac{d^2w}{dx^2} = -\frac{m + \mathrm{C}(a - x)}{\mathrm{E}}.$$

Collecting together these results, we get :—

$$\frac{d^2v}{dx^2} = 0; \quad \frac{d^2v}{dxdy} = \frac{\Pi \mathrm{C}z}{\mathrm{E}}; \quad \frac{d^2v}{dzdx} = \frac{\Pi \mathrm{C}y}{\mathrm{E}};$$

$$\frac{d^2w}{dx^2} = -\frac{m + \mathrm{C}(a - x)}{\mathrm{E}}; \quad \frac{d^2w}{dxdy} = -\frac{\Pi \mathrm{C}y}{\mathrm{E}}; \quad \frac{d^2w}{dzdx} = \frac{\Pi' \mathrm{C}z}{\mathrm{E}};$$

i.e. we now know the three differential coefficients with regard to $x$, $y$, $z$ as well as of $\frac{dv}{dx}$ and $\frac{dw}{dx}$.

We therefore deduce on integrating

$$\frac{dv}{dx} = \mathrm{C}' + \frac{\Pi \mathrm{C}yz}{\mathrm{E}},$$

$$\frac{dw}{dx} = \mathrm{C}'' - \frac{mx - \frac{1}{2}\mathrm{C}(a - x)^2}{\mathrm{E}} - \frac{\Pi \mathrm{C}y^2}{2\mathrm{E}} + \frac{\Pi' \mathrm{C}z^2}{2\mathrm{E}},$$

$\mathrm{C}'$ and $\mathrm{C}''$ being arbitrary constants.

Substituting these values in the expressions for $\sigma_{xy}$, $\sigma_{zx}$, we get

$$\sigma_{xy} = \frac{du}{dy} + \mathrm{C}' + \frac{\Pi \mathrm{C}yz}{\mathrm{E}}$$

$$\sigma_{zx} = \frac{du}{dz} + \mathrm{C}'' - \frac{mx - \frac{1}{2}\mathrm{C}(a - x)^2}{\mathrm{E}} - \frac{\Pi \mathrm{C}y^2}{2\mathrm{E}} + \frac{\Pi' \mathrm{C}z^2}{2\mathrm{E}};$$

but we saw at the beginning of this paragraph that for $x = y = z = 0$ we must have $\frac{du}{dy} = \frac{du}{dz} = 0$, so that putting $x = y = z = 0$, and writing the new values of $\sigma_{xy}$, $\sigma_{zx}$ as $\sigma_{xy}^0$, $\sigma_{zx}^0$, we obtain

$$\sigma_{xy}^0 = \mathrm{C}'; \quad \sigma_{zx}^0 = \mathrm{C}'' + \frac{\mathrm{C}a^2}{2\mathrm{E}},$$

so that eliminating $\mathrm{C}'$, $\mathrm{C}''$ we may write

$$\sigma_{xy} = \frac{du}{dy} + \sigma_{xy}{}^0 + \frac{\Pi C yz}{E}$$

$$\sigma_{zx} = \frac{du}{dz} + \sigma_{zx}{}^0 - \frac{mx + \frac{1}{2}C(2ax - x^2)}{E} - \frac{\Pi C y^2}{2E} + \frac{\Pi' C z^2}{2E}.$$

Thus the expressions for the stresses become

$$\left.\begin{aligned} s_{xy} &= G_3\left(\frac{du}{dy} + \sigma_{xy}{}^0 + \frac{\Pi C yz}{E}\right) \\ s_{zx} &= G_2\left(\frac{du}{dz} + \sigma_{zx}{}^0 - \frac{mx + \frac{1}{2}C(2ax - x^2)}{E} - \frac{\Pi C y^2}{2E} + \frac{\Pi' C z^2}{2E}\right) \end{aligned}\right\} \quad (81)$$

These depend only on the function $u$ and, substituting in equation (77), become

$$G_3\frac{d^2u}{dy^2} + G_3\frac{d^2u}{dz^2} = \left(1 - \frac{\Pi G_3}{E} - \frac{\Pi' G_2}{E}\right)Cz \quad . \quad . \quad (82)$$

**24. Bending Continued. Determination of the Constants C and *m*.**—Now that we have shown the possibility of solving the problem of bending, we will determine the meaning of the constants C and $m$.

If in the transverse section corresponding to the abscissa $x$, we consider an element of area $dA = dydz$, the normal force applied to this element will be

$$t_x dA = [m + C(a - x)]zdydz;$$

and the moment of this force about the principal axis parallel to $y$ will be

$$t_x . zdA = [m + C(a - x)]z^2dydz.$$

The sum of the moments about this principal axis of the forces over the section will be

$$\begin{aligned} \Sigma t_x zdA &= [m + C(a - x)]\iint z^2dydz \\ &= [m + C(a - x)]I, \end{aligned}$$

I being the moment of inertia about the principal axis which we will henceforth call the *neutral axis*.

If we suppose the body cut through at the section GH, it is clear that the equilibrium of the portion CDHG will not be altered provided that we apply to the section GH external forces equal to the stresses which act on it. Then the portion CDHG must be in equilibrium under the action of the external forces applied on the end CD and the stresses over the section GH. These will be as follows :—

On the end CD, normal forces $mz$ and tangential forces $\phi(y, z)$, $\psi(y, z)$.

On the section GH, normal stresses $- [m + C(a - x)]z$ and shear stresses $- \phi(y, z)$, $- \psi(y, z)$.

We see in the first place that the shear forces being equal and opposite on the two sections, the conditions of equilibrium, both for translation and rotation about the axis of $x$, are identically satisfied; moreover, the condition that the normal force over CD and GH must be zero leads to the equation

$$m\iint z\,dy\,dz - [m + C(a - x)]\iint z\,dy\,dz = - C(a - x)\iint z\,dy\,dz = 0,$$

the double integration being extended over the whole section. This condition is identically satisfied because, since the distances $z$ are those from a line passing through the centroid, the double integral is zero.

We have still to express the condition that the sum of the moments of all the forces applied to the portion CDHG, about two lines parallel to the axes of $y$ and $z$, is zero. We will take for these lines the principal axes of the section CD, so that the shear stresses on this section will have zero moment.

This being the case we note that the normal forces $mz$ applied to the end CD have zero resultant, as we have seen, and so are statically equivalent to two couples having moments about the axes of $y$ and $z$ equal respectively to

$$m\iint z^2\,dy\,dz\,;\quad m\iint yz\,dy\,dz,$$

the distances $y$ and $z$ being measured from the principal axes of the end CD; but we know that by a property of principal axes, the product moment

$$= \iint yz\,dy\,dz = 0,$$

so that the couple about the axis of $z$ is zero, and we have to consider only the couple about the axis of $y$ whose moment is equal to

$$m\iint z^2\,dy\,dz = mI,$$

I being the moment of inertia about the neutral axis (parallel to the axis of $y$).

In the same way, the normal stresses on the section GH are equivalent to a couple whose moment will be

$$- [m + C(a - x)]I.$$

Thus for equilibrium of moments about the principal axes of the section GH we shall have the two equations

$$- C(a - x)I + (a - x)\iint\psi(y, z)\,dy\,dz = 0,$$
$$(a - x)\iint\phi(y, z)\,dy\,dz = 0.$$

$\therefore$ dividing by $(a - x)$

$$\iint\psi(y, z)\,dy\,dz = CI$$
$$\iint\phi(y, z)\,dy\,dz = 0.$$

The second equation signifies that the sum of the shear

stresses parallel to the axis of $y$ must be zero and the first that the sum of the shear stresses parallel to the axis of $z$ must be equal to CI. Calling this sum the Shearing Force S, we have

$$S = CI$$

or

$$C = \frac{S}{I}.$$

To determine $m$ we note that if we call $M_1$ the sum of the moments of the normal forces on the base CD about the principal axis parallel to the axis of $y$, we shall have

$$M_1 = mI$$

or

$$m = \frac{M_1}{I}.$$

With these results, formulæ (75), (79), (81), and (82) become

$$\left.\begin{aligned} t_x &= \frac{M_1 + S(a-x)}{I}\,.\,z \\ \tau_x &= \frac{M_1 + S(a-x)}{EI}\,.\,z \end{aligned}\right\} \quad . \quad . \quad . \quad (83)$$

$$\left.\begin{aligned} s_{xy} &= G_3\left(\frac{du}{dy} + \sigma_{xy}{}^0 + \frac{\Pi S y z}{EI}\right) \\ s_{zx} &= G_2\left(\frac{du}{dz} + \sigma_{zx}{}^0 - \frac{M_1 x + \frac{1}{2}S(2ax - x^2)}{EI} - \frac{\Pi S y^2}{2EI} + \frac{\Pi' S y^2}{2EI}\right) \end{aligned}\right\} \quad (84)$$

$$G_3\frac{d^2u}{dy^2} + G_2\frac{d^2u}{dz^2} = \left(1 - \frac{\Pi G_3}{E} - \frac{\Pi' G_2}{E}\right)\frac{Sz}{I} \quad . \quad (85)$$

**25. Bending Continued. Application to Rectangular Prism.** —If $b$ and $c$ are the sides of the section parallel respectively to the axes of $y$ and $z$, the origin will be at the centre of one end of the prism.

If we now put

$$u = u' - \sigma_{xy}{}^0 y - \left[\sigma_{zx}{}^0 - \frac{M_1 x + \frac{1}{2}S(2ax - x^2)}{EI} - \frac{\Pi S b^2}{EI\,.\,12} + \frac{1}{G_2}\cdot\frac{Sc^2}{8I}\right]z - \frac{\Pi S y^2 z}{2EI} + \left(1 - \frac{\Pi' G_2}{E}\right)\frac{Sz^3}{6G_2 I},$$

$u'$ being an unknown function of $y$ and $z$, formulæ (84) and equation (85) become

$$\left.\begin{aligned} s_{xy} &= G_3\frac{du'}{dy}, \\ s_{zx} &= G_2\left[\frac{du'}{dz} - \frac{\Pi S}{EI}\left(y^2 - \frac{b^2}{12}\right) + \frac{S}{2G_2 I}\left(z^2 - \frac{c^2}{4}\right)\right] \end{aligned}\right\} . \quad (86)$$

$$G_3\frac{d^2u'}{dy^2} + G_2\frac{d^2u'}{dz^2} = 0 \quad . \quad . \quad . \quad . \quad . \quad (87)$$

The equation $\qquad s_{xy}dz - s_{zx}dy = 0,$

which applies to the contour of the sections, divides into two others, because on the sides $c$ we have throughout $dy = 0$, and therefore $s_{xy} = 0$, and on the sides $b$, $dz = 0$, so that $s_{zx} = 0$. We therefore have the following conditions :—

$$\text{For } y = \pm\frac{b}{2},\ s_{xy} = 0 \text{ or } \frac{du'}{dy} = 0 \text{ for all values of } z \quad . \quad (88)$$

$$\text{For } z = \pm\frac{c}{2},\ s_{zx} = 0 \text{ or } \frac{du'}{dz} - \frac{\Pi S}{EI}\left(y^2 - \frac{b^2}{12}\right) = 0 \text{ for all values of } y \quad (89)$$

Now, suppose that we have

$$u' = \sum A_n\left(e^{amz} - e^{-amz}\right)\cos my,$$

in which the summation may cover any number of terms and the quantities $A_n$, $m$ may be different for different terms.

We will note in the first place that this expression satisfies equation (87) provided that we take

$$G_3 - a^2G_2 = 0,$$

$$\text{i.e. } a = \pm\sqrt{\frac{G_3}{G_2}}.$$

We see, moreover, that, taking the differential coefficient of $u'$ with regard to $y$ and equating to zero after having put $y = \pm\frac{b}{2}$, according to condition (88) we obtain

$$\pm\sum mA_n\left(e^{amx} - e^{-amx}\right)\sin\frac{mb}{2} = 0$$

from which we deduce

$$\frac{mb}{2} = n\pi, \text{ i.e. } m = \frac{2n\pi}{b},$$

$n$ being any integer. Then taking this value for $m$ we have

$$u' = \sum A_n\left(e^{a\frac{2n\pi z}{b}} - e^{-a\frac{2n\pi z}{b}}\right)\cos\frac{2n\pi y}{b}.$$

Taking the differential coefficient with regard to $z$ and putting $z = +\frac{c}{2}$ and then substituting in equation (89) we get

$$\frac{\Pi S}{EI}\left(\frac{b^2}{12} - y^2\right) + \frac{2\pi a}{b}\sum nA_n\left(e^{a\frac{n\pi c}{b}} + e^{-a\frac{n\pi c}{b}}\right)\cos\frac{2n\pi y}{b} = 0,$$

and we have now to determine the number of the terms in the summation and the coefficient $A_n$ in order that this equation may be satisfied.

Now, by Lagrange's formula

$$\phi(y) - \frac{1}{l}\int_0^l \phi(y')dy' = \frac{2}{l}\sum_{n=1}^{n=\infty} \cos\frac{n\pi y}{l}\int_0^l \phi(y')\cos\frac{n\pi y'}{l}dy',$$

the symbol $\phi(y)$ representing any function of $y$. If we take

$$\phi(y) = y^2,\ \phi(y') = y'^2$$

we shall have at once

$$\int_0^l y'^2 dy' = \frac{l^3}{3},$$

$$\int_0^l y'^2 \cos\frac{n\pi y'}{l}dy' = -\frac{(-1)^{n-1}2l^3}{n^2\pi^2}$$

and Lagrange's formula becomes

$$\frac{l^3}{3} - y^2 - \frac{4l^2}{\pi^2}\sum_{n=1}^{n=\infty}\frac{(-1)^{n-1}\cos\frac{n\pi y}{l}}{n^2} = 0.$$

Comparing this equation with the one which we have deduced from equation (89), we see that it will be identically satisfied by taking $l = \frac{b}{2}$ so that

$$\frac{2\pi a\mathrm{EI}}{\Pi \mathrm{S}b} n\mathrm{A}_n\left(e^{\frac{an\pi c}{b}} + e^{-\frac{an\pi c}{b}}\right) = -\frac{(-1)^{n-1}b^2}{\pi^2 n^2}$$

$$\text{i.e. } \mathrm{A}_n = -\frac{\Pi \mathrm{S}b^3}{2\pi^3 a\mathrm{EI}} \times \frac{(-1)^{n-1}}{n^3\left(e^{\frac{an\pi c}{b}} + e^{-\frac{an\pi c}{b}}\right)}.$$

Substituting this value of $\mathrm{A}_n$ in the general expression for $u'$, we have

$$u' = -\frac{\Pi \mathrm{S}b^3}{2\pi^3 a\mathrm{EI}}\sum_{n=1}^{n=\infty}\frac{(-1)^{n-1}\sinh\frac{a2n\pi z}{b}\,.\cos\frac{2n\pi y}{b}}{n^3\cosh\,.\,\frac{an\pi c}{b}}\quad. \tag{90}$$

This is the general integral of equation (87) having regard to equations (88) and (89).

Taking the partial differential coefficients of this expression with regard to $y$ and $z$ and substituting in equations (86) for the shear stresses, we have

$$\left.\begin{aligned} s_{xy} &= \frac{\mathrm{G}_3\,.\,\Pi \mathrm{S}b^2}{\pi^2 a\mathrm{EI}}\sum_{n=1}^{n=\infty}\frac{(-1)^{n-1}\sinh\frac{a\,.\,2n\pi z}{b}\sin\frac{2n\pi y}{b}}{n^2\cosh\frac{an\pi c}{b}} \\ s_{zx} &= -\frac{\mathrm{S}}{2\mathrm{I}}\left(\frac{c^2}{4} - z^2\right) + \frac{\Pi \mathrm{G}_2\mathrm{S}}{\mathrm{EI}}\left[\frac{b^2}{12} - y^2 - \frac{b^2}{\pi^2}\sum_{n=1}^{n=\infty}\frac{(-1)^{n-1}\cosh\frac{a\,.\,2n\pi z}{b}\,.\cos\frac{2n\pi y}{b}}{n^2\cosh\frac{a\,.\,n\pi c}{b}}\right] \end{aligned}\right\} \text{(9}$$

If we require the shear stresses at the centre of the section we must put $y = z = 0$, this giving

$$\left.\begin{aligned} s_{xy} &= 0 \\ s_{zx} &= -\frac{Sc^2}{8I}\left[1 - \frac{2\Pi G_2 b^2}{3Ec^2}\left(1 - \frac{12}{\pi^2}\sum_{n=1}^{n=\infty}\frac{(-1)^{n-1}}{n^2\cosh.\dfrac{an\pi c}{b}}\right)\right] \end{aligned}\right\} \qquad (92)$$

**26. Bending Continued. Application to a Prism of Elliptic Cross-section.**—Now let us make

$$u = -\sigma_{xy}{}^0 y - \left[\sigma_{zx}{}^0 - \frac{M_1 x + \frac{1}{2}S(2ax - x^2)}{EI}\right]z + (1 - \Pi G_3 - \Pi' G_2)\frac{Sz^3}{6G_2 I}$$

$$+ Kz + B\left(y^2 z - \frac{G_3}{3G_2}z^3\right),$$

K and B being unknown coefficients.

We will note in the first place that since for

$$x = y = z = 0 \text{ we must have } \frac{du}{dy} = \frac{du}{dz} = 0,$$

we must also have $\sigma_{xy}{}^0 = 0$; $\sigma_{zx}{}^0 = K$, so that the expression for $u$ becomes

$$\frac{M_1 x + \frac{1}{2}S(2ax - x^2)}{EI}z + \left(1 - \frac{\Pi G_3}{E} - \frac{\Pi' G_2}{E}\right)\frac{Sz^3}{6G_2 I} + B\left(y^2 z - \frac{G_3}{3G_2}z^3\right); \quad (93)$$

this satisfies equation (85) and reduces equations (84) to the following form:—

$$s_{xy} = 2G_3\left(B + \frac{\Pi S}{2EI}\right)yz,$$

$$s_{zx} = G_2\left[\sigma_{zx}{}^0 + \left(B - \frac{\Pi S}{2EI}\right)y^2\right] + \left[\frac{S}{2I} - G_3\left(B + \frac{\Pi S}{2EI}\right)\right]z^2.$$

Putting for simplification

$$B + \frac{\Pi S}{2EI} = P \quad . \quad . \quad . \quad . \quad (94)$$

these two formulæ become

$$s_{xy} = 2G_3 P yz$$

$$s_{zx} = G_2\left[\sigma_{zx}{}^0 + \left(P - \frac{\Pi S}{EI}\right)y^2\right] + \left[\frac{S}{2I} - G_3 P\right]z^2;$$

so that the contour equation

$$s_{xy}dz - s_{zx}dy = 0$$

gives on substitution

$$2G_3 P yz dz - \left[G_2\sigma_{zx}{}^0 + G_2\left(P - \frac{\Pi S}{EI}\right)y^2 + \left(\frac{S}{2I} - G_3 P\right)z^2\right]dy = 0.$$

We can make this equation homogeneous by putting

$$G_2\sigma_{zx}{}^0 + \left(\frac{S}{2I} - G_3 P\right)z^2 = \left(\frac{S}{2I} - G_3 P\right)z'^2;$$

because we deduce therefrom that $zdz = z'dz'$ and the equation becomes

$$2G_3Pyz'dz' - \left[\left(\frac{S}{2I} - G_3P\right)z'^2 + G_2\left(P - \frac{\Pi S}{EI}\right)y^2\right]dy = 0.$$

To integrate this equation, we must separate the variables; this can be done by introducing into the calculation a new variable $t$ related to $z'$, $y$ by the equation

$$z' = ty;$$

then substituting $ty$ for $z'$ in the above equation and noting that $dz' = tdy + ydt$, we obtain, after dividing by $y^2$,

$$2G_3Pt(tdy + ydt) - \left[\left(\frac{S}{2I} - G_3P\right)t^2 + G_2\left(P - \frac{\Pi S}{EI}\right)\right]dy = 0,$$

or $$2G_3Pytdt - \left[\left(\frac{S}{2I} - 3G_3P\right)t^2 + G_2\left(P - \frac{\Pi S}{EI}\right)\right]dy = 0,$$

and therefore

$$\frac{dy}{y} = \frac{2G_3Ptdt}{\left(\frac{S}{2I} - 3G_3P\right)t^2 + G_2\left(P - \frac{\Pi S}{EI}\right)}.$$

This equation can easily be integrated and gives, if C is an arbitrary constant,

$$\log Cy = \frac{G_3P}{\left(\frac{S}{2I} - 3G_3P\right)} \log\left[\left(\frac{S}{2I} - 3G_3P\right)t^2 + G_2\left(P - \frac{\Pi S}{EI}\right)\right];$$

putting $C^{\frac{S}{2G_3PI} - 3} = C'$ and expressing as powers, this formula becomes

$$C'y^{\frac{S}{2G_3PI} - 3} = \left(\frac{S}{2I} - 3G_3P\right)t^2 + G_2\left(P - \frac{\Pi S}{EI}\right);$$

multiplying each side by $y^2$ and substituting $z'^2$ for $t^2y^2$, we have

$$C'y^{\frac{S}{2G_3PI} - 1} = \left(\frac{S}{2I} - 3G_3P\right)z'^2 + G_2\left(P - \frac{\Pi S}{EI}\right)y^2.$$

Finally from the relation which we have established between $z^2$ and $z'^2$, we deduce

$$z'^2 = z^2 + \frac{G_2\sigma_{zx}^0}{\frac{S}{2I} - G_3P};$$

so that the preceding equation becomes

$$C'y^{\frac{S}{2G_3PI} - 1} = \left(\frac{S}{2I} - 3G_3P\right)z^2 + G_2\left(P - \frac{\Pi S}{EI}\right)y^2 + G_2\sigma_{zx}^0 \cdot \frac{\frac{S}{2I} - 3G_3P}{\frac{S}{2I} - G_3P} \quad (95)$$

This then is the general equation of the section of cylinders for which the curved form of cross-section after flexure is represented by equation (93).

Differentiating equation (95) we obtain

$$\left[\left(\frac{S}{2G_3PI} - 1\right)C'y^{\frac{S}{2G_3PI} - 2} - 2G_2\left(P - \frac{\Pi S}{EI}\right)y\right]dy = 2\left(\frac{S}{2I} - 3G_3P\right)zdz,$$

from which it follows that if $\frac{S}{2G_3PI} - 2$ is positive, i.e. if

$$\frac{S}{2G_3PI} - 1 > 1 \quad . \quad . \quad . \quad . \quad . \quad (96)$$

we shall also have

$$\frac{dz}{dy} = 0 \text{ for } y = 0,$$

and

$$\frac{dy}{dz} = 0 \text{ for } z = 0;$$

this shows that the value of $z$ corresponding to $y = 0$ is a maximum or a minimum as is also the value of $y$ corresponding to $z = 0$.

Condition (96) being regarded as satisfied, it is clear that the quantity $\frac{S}{2G_3PI} - 1$ can always be put in the form $\frac{2a}{2\beta}$, $a$ and $\beta$ being two integers, of which the second is less than the first; the term $C'y^{\frac{S}{2G_3PI} - 1}$ will therefore not change sign in changing the sign of $y$. The curve represented by equation (96) will therefore be symmetrical about each of the axes of $y$ and $z$.

Now, let us call $\pm b$ the values of $y$ corresponding to $z = 0$ and $\pm c$ the values of $z$ corresponding to $y = 0$. Let us divide equation (95) by its last term and put

$$\frac{\frac{S}{2I} - G_3P}{G_2\sigma_{zx}^0} = -\frac{1}{c^2},$$

$$\frac{\frac{S}{2I} - G_3P}{\sigma_{zx}^0} \cdot \frac{P - \frac{\Pi S}{EI}}{\frac{S}{2I} - 3G_3P} = -\frac{K}{b^2} \quad . \quad . \quad . \quad (97)$$

and

$$\frac{\frac{S}{2I} - G_3P}{G_2\sigma_{zx}^0} \cdot \frac{C'}{\frac{S}{2I} - 3G_3P} = C'',$$

K being a constant coefficient. Equation (95) then becomes

$$C'' y^{\frac{S}{2G_3PI} - 1} + K\left(\frac{y}{b}\right)^2 + \left(\frac{z}{c}\right)^2 = 1,$$

and since for $z = 0$ we must have $y = b$, it follows that

$$C'' = \frac{1 - K}{b^{\frac{S}{2G_3PI} - 1}};$$

so that eliminating C″ and putting for simplification

$$\frac{S}{2G_3PI} - 1 = n, \quad . \quad . \quad . \quad . \quad (98)$$

the equation for the contour of the sections of the prism or cylinder becomes

$$(1 - K)\left(\frac{y}{b}\right)^n + K\left(\frac{y}{b}\right)^2 + \left(\frac{z}{c}\right)^2 = 1 \quad . \quad . \quad (99)$$

Now, from equation (98) and (97) we deduce

$$\left.\begin{aligned} P &= \frac{S}{2G_3I(n+1)} \\ \sigma_{zx}{}^0 &= -\frac{Snc^2}{2G_2I(n+1)} \\ K &= \frac{1 - (n+1)\dfrac{2\Pi G_3}{E}}{n - 2} \cdot \frac{G_2 b^2}{G_3 c^2} \end{aligned}\right\} \quad . \quad . \quad (100)$$

and the expressions for the shear stresses become

$$\left.\begin{aligned} s_{xy} &= \frac{S}{(n+1)I} \cdot yz, \\ s_{zx} &= -\frac{S}{2(n-1)I}\left[n(c^2 - z^2) - (n-2)K\frac{c^2}{b^2} \cdot y^2\right] \end{aligned}\right\} \quad (101)$$

It is clear that equation (99) may represent an infinite number of curves with semi-axes $b$ and $c$, for we can give $n$ and therefore K an infinite number of values. Putting $K = 1$, the equation becomes

$$\frac{y^2}{b^2} + \frac{z^2}{c^2} = 1$$

and represents an ellipse. The last of the equations (100) then gives

$$1 = \frac{1 - (n+1)\dfrac{2\Pi G_3}{E}}{(n - 2)} \cdot \frac{G_2 b^2}{G_3 c^2}$$

from which we derive

$$n = \frac{2 + \left(1 - \dfrac{2\Pi G_3}{E}\right)\dfrac{G_2 b^2}{G_3 c^2}}{1 + \dfrac{2\Pi G_3}{E} \cdot \dfrac{G_2 b^2}{G_3 c^2}} \quad . \quad . \quad (102)$$

Thus for prisms or cylinders of elliptic cross-section as well as for those of rectangular cross-section, the problem of bending can be completely solved when the body has at each point three planes of equal elastic properties, perpendicular respectively to the axis of the body and to the principal axes of its ends.

**27. Bending Continued. Form of the Prism or Cylinder after Bending.**—The determination of the form of the prism or cylinder after bending requires that we shall find the expression for the displacements $u$, $v$, $w$ in terms of $x$, $y$, $z$.

As regards $u$, we have seen that we can only find it for very special sections; on the other hand, we shall show later that the expressions for $v$ and $w$ can be found in terms of the coordinates for all forms of cross-section.

We will discuss in the first place the expression for $u$ obtained in the cases already studied in order to obtain therefrom some important properties; take, for example, formula (93) for cylinders whose sections are represented by equation (99).

It is clear that if we consider $x$ constant, formula (93) gives the displacements parallel to the axis of $x$ of the different points of a section of the cylinder, i.e. it represents the curved surface that a section initially plane will become after bending. Now formula (93) can be put in the form

$$u = u' + u'',$$

in which

$$u' = \frac{M_1 x + \frac{1}{2}S(2ax - x^2)}{EI} \cdot z \qquad . \quad . \quad (103)$$

$$u'' = \left(1 - \frac{\Pi G_3}{E} - \frac{\Pi' G_2}{E}\right)\frac{Sz^3}{6G_2 I} + B\left(y^2 z - \frac{G_3}{3G_2} z^3\right) \quad . \quad (104)$$

Regarding $x$ as constant, i.e. considering points in the same cross-section, the first of these equations will represent a plane parallel to the axis of $y$, and the second a surface of the third order. This plane and surface are tangential to each other at the centroid of the section because for $y = z = 0$ we have

$$\frac{du''}{dy} = 0, \frac{du''}{dz} = 0.$$

From this it follows that equation (104) represents the curved surface into which the plane surface corresponding to the abscissa $x$ becomes changed after bending, this surface being referred to the plane tangential to its centre point. As the expression for $u'$ is independent of $x$, we see that all plane sections will adopt the same curved form after bending.

We should arrive at similar conclusions for prisms with rectangular ends.

Now let us pass to the determination of the displacements $v$ and $w$.

If we integrate with reference to $x$ the equation

$$\frac{d^2v}{dzdx} - \frac{d^2w}{dxdy} = \frac{2\Pi C}{E}y,$$

we obtain, on substituting for C its value $\frac{S}{I}$ and calling $\phi(y, z)$ an arbitrary function of $y$ and $z$,

$$\frac{dv}{dz} - \frac{dw}{dy} = \frac{2\Pi S}{EI}xy + 2\phi(y, z);$$

but, as the shear strain $\sigma_{yz}$ must be zero, we also have

$$\frac{dv}{dz} + \frac{dw}{dy} = 0,$$

so that, combining this equation with the previous one, we have

$$\frac{dv}{dz} = \frac{\Pi S}{EI}xy + \phi(y, z),$$

$$\frac{dw}{dy} = -\frac{\Pi S}{EI}xy - \phi(y, z).$$

Differentiating the first of these equations with regard to $y$, and the second with regard to $z$, we have

$$\frac{d^2v}{dydz} = \frac{\Pi S}{EI}x + \frac{d\phi(y, z)}{dy},$$

$$\frac{d^2w}{dydz} = -\frac{d\phi(y, z)}{dz}.$$

Now the two formulæ

$$\frac{dv}{dy} = -\frac{\Pi[m + C(a - x)]z}{E}$$

$$\frac{dw}{dz} = -\frac{\Pi'[m + C(a - x)]z}{E}$$

give, on differentiating the first with regard to $z$ and the second with regard to $y$, and substituting for $m$ and C their values,

$$\frac{d^2v}{dydz} = -\frac{\Pi[M_1 + S(a - x)]}{EI}$$

$$\frac{d^2w}{dydz} = 0.$$

Comparing these expressions with those obtained above, we see that we have

$$\frac{d\phi(y, z)}{dy} = -\frac{\Pi(M_1 + Sa)}{EI}$$

$$\frac{d\phi(y, z)}{dz} = 0,$$

from which we deduce by integrating

$$\phi(y, z) = -\frac{\Pi(\mathrm{M}_1 + \mathrm{S}a)}{\mathrm{EI}}y + \mathrm{D},$$

D being an arbitrary constant.

We can now substitute this value of the function $\phi(y, z)$ in the expressions for $\frac{dv}{dz}, \frac{dw}{dy}$, thus obtaining

$$\frac{dv}{dz} = \frac{\Pi \mathrm{S}}{\mathrm{EI}}xy - \frac{\Pi(\mathrm{M}_1 + \mathrm{S}a)}{\mathrm{EI}}y + \mathrm{D},$$

$$\frac{dw}{dy} = -\frac{\Pi \mathrm{S}}{\mathrm{EI}}xy + \frac{\Pi(\mathrm{M}_1 + \mathrm{S}a)}{\mathrm{EI}}y - \mathrm{D};$$

and as for $x = y = z = 0$ we must have $\frac{dw}{dz} = 0$, it follows that the constant D must be zero

Since we have given in § 23 expressions for $\frac{dv}{dx}, \frac{dv}{dy}$ $\frac{dw}{dx}$ and $\frac{dw}{dz}$, it follows that we now know the three differential coefficients of $v$ and $w$ with regard to $x, y, z$. These differential coefficients are as follows:—

$$\frac{dv}{dx} = \sigma_{xy}{}^0 + \frac{\Pi \mathrm{S}yz}{\mathrm{EI}}$$

$$\frac{dw}{dx} = \sigma_{zx}{}^0 - \frac{\mathrm{M}_1 x + \frac{1}{2}\mathrm{S}(2ax - x^2)}{\mathrm{EI}} - \frac{\Pi \mathrm{S}y^2}{2\mathrm{EI}} + \frac{\Pi' \mathrm{S}z^2}{2\mathrm{EI}}$$

$$\frac{dv}{dy} = -\frac{\Pi[\mathrm{M}_1 + \mathrm{S}(a - x)]z}{\mathrm{EI}}$$

$$\frac{dw}{dy} = \frac{\Pi[\mathrm{M}_1 + \mathrm{S}(a - x)]y}{\mathrm{EI}}$$

$$\frac{dv}{dz} = -\frac{\Pi[\mathrm{M}_1 + \mathrm{S}(a - x)]y}{\mathrm{EI}}$$

$$\frac{dw}{dz} = -\frac{\Pi'[\mathrm{M}_1 + \mathrm{S}(a - x)]z}{\mathrm{EI}}.$$

We can now complete the integration, and since for $x = y = z = 0$ we must have $v = w = 0$, we shall not have to add an arbitrary constant, so that we shall get

$$\left.\begin{aligned} v &= \sigma_{xy}{}^0 x - \frac{\Pi[\mathrm{M}_1 + \mathrm{S}(a - x)]yz}{\mathrm{EI}} \\ w &= \sigma_{zx}{}^0 . x - \frac{\mathrm{M}_1 x^2 + \mathrm{S}\left(ax - \frac{x^3}{3}\right)}{2\mathrm{EI}} + \frac{[\mathrm{M}_1 + \mathrm{S}(a - x)](\Pi y^2 - \Pi' z^2)}{2\mathrm{EI}} \end{aligned}\right\} \quad (105)$$

Such are the general expressions in finite form for the transverse displacements $v$ and $w$; it is very remarkable that we have been able to obtain these without making any restriction upon the form of the section of the prism or cylinder, but it is necessary to point out that this general condition is only apparent, because the shear strains $\sigma^0_{xy}$, $\sigma^0_{zx}$, occurring in the expressions for $v$ and $w$ depend upon the forms of the sections.

If we wish to determine the curve taken up by the axis of the body after bending, we put $y = z = 0$ for all values of $x$ in equations (105), and we obtain

$$\left.\begin{aligned} v &= \sigma^0_{xy} \cdot x \\ w &= \sigma^0_{zx} \cdot x - \frac{M_1 x^2 + S\left(ax - \dfrac{x^3}{3}\right)}{2EI} \end{aligned}\right\}, \quad . \quad . \quad (106)$$

which give the curve into which the axis bends; the first of these equations shows that this curve is plane.

When the section of the prism or cylinder is symmetrical about the principal axis parallel to the axis of $z$, it is clear that the axis, in bending, must remain in a vertical plane, so that we shall have $v = 0$, and therefore $\sigma^0_{xy} = 0$, for all values of $x$.

**28. Prisms Subjected to Different Kinds of Forces.**—We have studied the elastic equilibrium of a straight prism subjected to the following external forces:—

1. A force uniformly distributed both over its surface and over its ends;
2. A force uniformly distributed only over the ends and perpendicular to them;
3. Tangential forces distributed over the ends according to a certain law depending upon the form of the section and producing a uniform torsion;
4. Tangential forces distributed over the ends according to a certain law depending upon the form of the sections, and normal forces distributed linearly over the two ends; all these forces resulting in a bending action in the plane determined by the axis of the prism and by a principal axis of the ends.

Now, we have seen that for each of these cases the stresses induced are linear functions of the external forces; from this it follows that if all the external forces considered above were applied simultaneously, the stresses $t_x$, $t_y$, $t_z$, $s_{yz}$, $s_{zx}$, $s_{xy}$ would be the sums of the values which they would have if each of these external forces were separately applied.

Consider, for example, the case which we shall have to deal with continually in the subsequent treatment in this book, viz. the case in which the prism, referred to three rectangular axes as in the problem of bending, has at each point three planes of equal elastic properties perpendicular to the axes of the coordinates and subjected to the following forces :—

1. On the end corresponding to $x = a$ ($a$ being the length of the prism), to a normal force P uniformly distributed, and on the end corresponding to $x = 0$, to a normal force $-$ P also uniformly distributed ;
2. On the end corresponding to $x = a$, to normal forces distributed according to the linear function
$$\frac{M_{1X}}{I_Y} \cdot z,$$
and to tangential forces distributed as in the problem of bending and giving zero resultant parallel to the axis of $y$ and a resultant equal to $- S_Z$ parallel to the axis of $z$ ; further, on the end corresponding to $x = 0$, to normal forces distributed according to the linear function
$$-\frac{M_{1Y} + S_Z a}{I_Y} \cdot z,$$
and to tangential forces equal to those on the other end but in the opposite direction ;
3. On the base corresponding to $x = a$, to normal forces distributed according to the linear function
$$-\frac{M_{1Z} y}{I_Z},$$
and to tangential forces distributed as in the problem of bending when the neutral axis is parallel to the axis of $z$, these forces giving a zero resultant parallel to the axis of $z$, and a sum $- S_Y$ parallel to the axis of $y$ ; further, on the end corresponding to $x = 0$, to normal forces distributed according to the linear function
$$\frac{M_{1Z} - S_Y a}{I_Z} y,$$
and to tangential forces equal to those applied at the other end but opposite in direction.
4. Finally, on the end corresponding to $x = a$, to tangential forces distributed as in the problem of torsion and giving a twisting moment $T_x$ about the axis of the prism, and, on the end corresponding to $x = 0$, to tangential forces exactly equal to those on the other end but in the opposite direction.

Now we have seen that if these four systems of forces are separately applied to the prism, we have at any point
$$t_y = t_z = s_{yz} = 0.$$

Further, it follows from the problem of simple tension that the forces of the first system acting alone will give at any point

$$t_x = \frac{P}{A}, \; s_{xy} = s_{zx} = 0.$$

Also it follows from the problem of bending that the forces of the second system, i.e. those producing bending in the plane $xz$, acting alone, will give, for the normal force on a plane element of the section corresponding to $x$,

$$t_x = \frac{M_{1Y} + S_Z(a - x)}{I_Y} \, . \, z,$$

or

$$t_x = \frac{B_Y \, . \, z}{I_Y},$$

$B_Y$ being the *bending moment* about the principal axis parallel to the axis of $y$. It further follows that the shear stresses $s_{xy}$, $s_{zx}$ will be functions of $y$ and $z$ which can be represented by

$$s_{xy} = \phi_1(y, z)S_Z; \; s_{zx} = \phi_2(y, z)S_Z,$$

where the functions $\phi_1(y, z)$, $\phi_2(y, z)$ depend only on the form of the section of the prism or cylinder and not on the magnitude of the forces applied on the ends.

To simplify the notation we will write these two equations in the form

$$s_{xy} = K_1 \frac{S_Z}{A}; \; s_{zx} = K_2 \frac{S_Z}{A},$$

A being the area of the section and $K_1$, $K_2$ functions of $y$ and $z$, which are the previous functions multiplied by A.

The forces of the third system, acting alone, would produce bending in the plane of $xy$, and will give

$$t_x = -\frac{B_Z}{I_Z} \, . \, y$$

and

$$s_{xy} = K_3 \frac{S_Y}{A}; \; s_{zx} = K_4 \frac{S_Y}{A}.$$

Finally, the forces of the fourth system acting alone would produce torsion about the axis of the prism, and the stresses will be

$$t_x = 0,$$
$$s_{xy} = \psi_1(y, z)T_M,$$
$$s_{zx} = \psi_2(y, z)T_M,$$

where $\psi_1(y, z)$ and $\psi_2(y, z)$ are functions of $y$ and $z$ depending upon the shape of the section but not upon the magnitude of the forces applied on the ends; for simplification we will write

$$s_{xy} = C_1 \frac{T_M}{I_P}; \; s_{zx} = C_2 \frac{T_M}{I_P},$$

$I_P$ being the polar moment of inertia of the section.

Thus, from the principle of superposition, when all four systems of forces are acting simultaneously upon the prism or cylinder, we shall have

$$t_y = t_z = s_{yz} = 0, \quad . \quad . \quad . \quad . \quad . \quad (107)$$

and

$$\left.\begin{aligned} t_x &= \frac{P}{A} + \frac{B_Y z}{I_Y} - \frac{B_Z y}{I_Z}, \\ s_{xy} &= K_1 \frac{S_Z}{A} + K_3 \frac{S_Y}{A} + C_1 \frac{T_M}{I_P}, \\ s_{zx} &= K_2 \frac{S_Z}{A} + K_4 \frac{S_Y}{A} + C_2 \frac{T_M}{I_P}. \end{aligned}\right\} \quad . \quad . \quad . \quad (108)$$

These are the general formulæ covering all the special cases which we shall want to use later in the book. The functions $K_1$—$K_4$, $C_1$, $C_2$ cannot be determined exactly for several sections which have to be considered in practice; but for rectangular, circular, and elliptic sections we have seen that they can be completely determined.

**29. Equations of Strength or Stability.**—In order that all the results and formulæ given up to the present may be exact, the forces which we have assumed to be applied to the bodies considered must be within certain limits, depending upon the dimensions of the body and the material of which it is composed. These elastic limits are the values of the forces producing strains which do not disappear completely even after the external forces are removed. The values of these limits can only be found by direct experiment for each material, each form of body, and each method of application of the external forces.

But as we have not had up to the present, and shall not have for a long time, sufficient experimental results for the large variety of cases arising in practice, we have attempted by general though not rigorously exact methods, to determine these limits, at least for isotropic bodies.

Precisely because the methods are not rigorously exact, two methods have been employed by different writers. We will give these and explain them briefly.

1. Some writers, particularly Navier and Barré de Saint-Venant, have adopted the principle that in all isotropic bodies of the same material the limits of elastic resistance are reached, whatever be the method of application of the external forces, whenever the maximum direct strain reaches a given value.

This principle does not appear to be exact; in fact, it may happen that the limits of elastic resistance at a point do not depend only on the value of the maximum direct strain at this point, but also upon the manner in which the strain varies in different directions. For example, if a prism is subjected to a uniform tension over its whole surface, the strain at any point is the same in all directions; but, if the tension is uniformly distributed on its ends only, the maximum strain occurs parallel to the sides of the prism, while in directions making increasing angles with the sides the strain progressively diminishes and changes to a contraction, as we have already seen. Now, it is not very clear that in these two very different cases, the limit of elastic resistance should be reached when the maximum direct strain reaches the same value.

2. Other writers, particularly M. Clebsch, have assumed that in isotropic bodies of the same material the limits of elastic resistance are reached when the maximum stress at a point reaches a given value, whatever be the distribution of the external forces.

This principle, like the preceding one, does not appear to be rigorously exact; it may happen that the limits of elastic resistance depend not only on the value of the maximum stress, but also upon the manner in which the stress varies around the point under consideration when we vary the direction of the element of area. In a prism subjected to uniform tension over its whole surface, for example, the maximum stress at any point is the same in all directions, whilst if the prism is subjected only to an axial tension, the stress has a maximum value in a direction parallel to the ends, but diminishes progressively to zero as the direction changes until it is perpendicular to the ends.

Now we have seen that the maximum value of the stress at a point will be the same in these two cases provided that the stress on the ends is the same in each. But as in the second case the external forces are applied on the ends only, whereas in the first case they are also applied on the lateral surface, it is not quite clear that the limits of elastic resistance will be reached for the same value of the maximum stress in the two cases.

Thus neither of the two principles proposed appears to be exact; but as we often have to use them in the absence of anything better, we will give the formulæ for their application, generalising the first, as M. Barré de Saint-Venant has done, to render it applicable to bodies having three planes of equal elastic properties.

We will note in this connection that if, for a body with absolutely variable elastic properties, we call $\tau_r$ the tensile strain in any direction and $\tau_{lr}$ the strain in the same direction corresponding to the elastic limit, $\tau_r$ must always be less than $\tau_{lr}$; moreover, the direction in which we should most expect the elastic limit to be reached is that in which the ratio $\frac{\tau_r}{\tau_{lr}}$ is greatest, while remaining less than unity.

As for the second principle, we cannot generalise in the same way, for if we call $\mathfrak{f}$ the tensile stress in any direction and $t_l$ the elastic limit normal stress at the same point, we cannot take $\frac{\mathfrak{f}}{t_l}$ as a measure of the working factor, since the stress $\mathfrak{f}$ will not in general be perpendicular to the area under consideration.

**30. Equations of Strength Deduced from the Maximum Strains.**—We found in § 3 the formula

$$\tau_r = \tau_x \cos^2 a + \tau_y \cos^2 \beta + \tau_z \cos^2 \gamma + \sigma_{yz} \cos \beta \cos \gamma + \sigma_{zx} \cos \gamma \cos a + \sigma_{xy} \cos a \cos \beta.$$

Now suppose that the body has at every point three planes of equal elastic properties parallel to the co-ordinate planes, and that the strains at the elastic limit in the directions perpendicular to these planes are

$$\tau_{lx}, \tau_{ly}, \tau_{lz};$$

the strain $\tau_{lr}$ corresponding to the elastic limit in a direction making angles $a$, $\beta$, $\gamma$ with the axes must be a function of these three angles having the values $\tau_{lx}$, $\tau_{ly}$, $\tau_{lz}$ when the line $r$ is taken successively parallel to the axes.

Now we will suppose for simplification that we have

$$\tau_{lr} = \tau_{lx} \cos^2 a + \tau_{ly} \cos^2 \beta + \tau_{lz} \cos^2 \gamma, \qquad (109)$$

and we will find the values of $a$, $\beta$, $\gamma$ corresponding to the maximum value of the ratio $\frac{\tau_r}{\tau_{lr}}$.

As we have the equation

$$\cos^2 a + \cos^2 \beta + \cos^2 \gamma = 1, \quad . \quad . \quad (110)$$

we can regard one of the angles—e.g. $a$—as a function of the two others, and therefore the equations for determining the maximum and minimum values of $\frac{\tau_r}{\tau_{lr}}$ will be

$$\frac{d\frac{\tau_r}{\tau_{lr}}}{d\cos a} \cdot \frac{d\cos a}{d\cos \beta} + \frac{d\frac{\tau_r}{\tau_{lr}}}{d\cos \beta} = 0;$$

$$\frac{d\frac{\tau_r}{\tau_{lr}}}{d\cos a} \cdot \frac{d\cos a}{d\cos \gamma} + \frac{d\frac{\tau_r}{\tau_{lr}}}{d\cos \gamma} = 0.$$

Now equation (110) gives

$$\cos a \frac{d \cos a}{d \cos \beta} + \cos \beta = 0,$$

$$\cos a \frac{d \cos a}{d \cos \gamma} + \cos \gamma = 0,$$

so that, eliminating from the two previous equations the ratios $\frac{d \cos a}{d \cos \beta}$, $\frac{d \cos a}{d \cos \gamma}$, we obtain

$$\frac{1}{\cos a} \cdot \frac{d \frac{\tau_r}{\tau_{lr}}}{d \cos a} = \frac{1}{\cos \beta} \cdot \frac{d \frac{\tau_r}{\tau_{lr}}}{d \cos \beta} = \frac{1}{\cos \gamma} \cdot \frac{d \frac{\tau_r}{\tau_{lr}}}{d \cos \gamma} = k$$

i.e.

$$\left.\begin{aligned} \frac{d \frac{\tau_r}{\tau_{lr}}}{d \cos a} &= k \cos a \\ \frac{d \frac{\tau_r}{\tau_{lr}}}{d \cos \beta} &= k \cos \beta \\ \frac{d \frac{\tau_r}{\tau_{lr}}}{d \cos \gamma} &= k \cos \gamma \end{aligned}\right\} \quad . \quad . \quad . \quad (111)$$

Adding one to the other of these three equations after having multiplied by $\cos a$, $\cos \beta$, $\cos \gamma$ respectively, and having regard to equation (110), we shall find

$$\cos a \cdot \frac{d \frac{\tau_r}{\tau_{lr}}}{d \cos a} + \cos \beta \cdot \frac{d \frac{\tau_r}{\tau_{lr}}}{d \cos \beta} + \cos \gamma \cdot \frac{d \frac{\tau_r}{\tau_{lr}}}{d \cos \gamma} = k.$$

As $\frac{\tau_r}{\tau_{lr}}$ is a function homogeneous with reference to $\cos a$, $\cos \beta$, $\cos \gamma$, we know that the left-hand side of this equation is equal to the function $\frac{\tau_r}{\tau_{lr}}$ itself multiplied by its degree, so that since this degree is zero we shall have $k = 0$, and equations (111) will become, after differentiating and then multiplying by $\tau_r$,

$$\frac{d\tau_r}{d \cos a} - \frac{\tau_r}{\tau_{lr}} \cdot \frac{d\tau_{lr}}{d \cos a} = 0$$

$$\frac{d\tau_r}{d \cos \beta} - \frac{\tau_r}{\tau_{lr}} \cdot \frac{d\tau_{lr}}{d \cos \beta} = 0$$

$$\frac{d\tau_r}{d \cos \gamma} - \frac{\tau_r}{\tau_{lr}} \cdot \frac{d\tau_{lr}}{d \cos \gamma} = 0,$$

i.e. substituting for $\tau_r$ and $\tau_{lr}$ their values given by expressions (8) and (108),

$$\left.\begin{aligned} 2\left(\tau_x - \tau_{lx} \cdot \frac{\tau}{\tau_{lr}}\right)\cos a + \sigma_{xy}\cos\beta + \sigma_{zx}\cos\gamma = 0 \\ \sigma_{xy}\cos a + 2\left(\tau_y - \tau_{ly} \cdot \frac{\tau_r}{\tau_{lr}}\right)\cos\beta + \sigma_{yz}\cos\gamma = 0 \\ \sigma_{zx}\cos a + \sigma_{yz}\cos\beta + 2\left(\tau_z - \tau_{lz} \cdot \frac{\tau_r}{\tau_{lr}}\right)\cos\gamma = 0 \end{aligned}\right\} . \qquad (112)$$

By eliminating from these three equations the ratios

$$\frac{\cos\beta}{\cos a}, \frac{\cos\gamma}{\cos a},$$

we obtain

$$\left.\begin{aligned} 8\left(\tau_x - \tau_{lx}\frac{\tau_r}{\tau_{lr}}\right)\left(\tau_y - \tau_{ly}\frac{\tau_r}{\tau_{lr}}\right)\left(\tau_z - \tau_{lz}\frac{\tau_r}{\tau_{lr}}\right) - 2\sigma^2_{yz}\left(\tau_x - \tau_{lx}\frac{\tau_r}{\tau_{lr}}\right) \\ - 2\sigma^2_{zx}\left(\tau_y - \tau_{ly}\frac{\tau_r}{\tau_{lr}}\right) - 2\sigma^2_{xy}\left(\tau_z - \tau_{lz}\frac{\tau_r}{\tau_{lr}}\right) + 2\sigma_{yz}\sigma_{zx}\sigma_{xy} = 0 \end{aligned}\right\} . \qquad (113)$$

This is an equation of the third degree in $\frac{\tau_r}{\tau_{lr}}$, and the maximum root will give the maximum value of this ratio. When this maximum root has been found we shall obtain the corresponding values of the angles $a$, $\beta$, $\gamma$ by means of equations (112) combined with equation (110).

We see that equation (113), being of an odd degree, must have a real root, and we can even prove that all the roots are real.

If we write it in the form

$$\left.\begin{aligned} \left(\frac{\tau_x}{\tau_{lx}} - \frac{\tau_r}{\tau_{lr}}\right)\left[4\left(\frac{\tau_y}{\tau_{ly}} - \frac{\tau_r}{\tau_{lr}}\right)\left(\frac{\tau_z}{\tau_{lz}} - \frac{\tau_r}{\tau_{lr}}\right) - \frac{\sigma^2_{yz}}{\tau_{ly}\tau_{lz}}\right] \\ - \frac{\sigma^2_{zx}}{\tau_{lz}\tau_{lx}}\left(\frac{\tau_y}{\tau_{ly}} - \frac{\tau_r}{\tau_{lr}}\right) - \frac{\sigma_{xy}}{\tau_{lx}\tau_{ly}}\left(\frac{\tau_z}{\tau_{lz}} - \frac{\tau_r}{\tau_{lr}}\right) - \frac{\sigma_{yz}\sigma_{zx}\sigma_{xy}}{\tau_{lx}\tau_{ly}\tau_{lz}} = 0, \end{aligned}\right\} . \qquad (114)$$

and call $a$ and $b$ the values of $\frac{\tau_r}{\tau_{lr}}$ which make the quantity in square brackets zero, we can easily see that $a$ and $b$ are real quantities, and that one of them, e.g. $a$, is the maximum and the other, $b$, is the minimum value of each of the ratios $\frac{\tau_y}{\tau_{ly}}, \frac{\tau_z}{\tau_{lz}}$.

Putting $a = \frac{\tau_r}{\tau_{lr}}$ in equation (114) we have

$$- \frac{\sigma^2_{zx}}{\tau_{lz}\tau_{lx}}\left(\frac{\tau_y}{\tau_{ly}} - a\right) - \frac{\sigma^2_{xy}}{\tau_{lx}\tau_{ly}}\left(\frac{\tau_z}{\tau_{lz}} - a\right) + \frac{\sigma_{yz}\sigma_{zx}\sigma_{xy}}{\tau_{lx}\tau_{ly}\tau_{lz}} = 0$$

and the left-hand side is a perfect square, since we have assumed

$$4\left(\frac{\tau_y}{\tau_{ly}} - a\right)\left(\frac{\tau_z}{\tau_{lz}} - a\right) - \frac{\sigma^2_{yz}}{\tau_{ly}\tau_{lz}} = 0.$$

It represents therefore a quantity which is always positive, the first two terms being necessarily positive.

We shall see in the same way that if we substitute $b$ for $\frac{\tau_r}{\tau_{lr}}$, the left-hand side of equation (114) reduces to a perfect square prefaced with a negative sign, and represents therefore a quantity which is always negative. By giving to $\frac{\tau_r}{\tau_{lr}}$ values $+\infty$, $a$, $b$, $-\infty$ we obtain values having signs $- + - +$, thus showing that the three roots of equation (113) are real.

Now let us apply this equation to the case considered in previous paragraphs, assuming the prism to have at each point three planes of equal elastic properties perpendicular to the axes of the co-ordinates.

We have seen that even when this prism is subjected simultaneously to the different systems of forces considered in § 28, we have

$$\sigma_{yz} = 0,\ \tau_y = -\Pi\tau_x,\ \tau_z = -\Pi'\tau_x,$$

so that equation (113) becomes

$$4\left(\tau_x - \tau_{lx}\frac{\tau_r}{\tau_{lr}}\right)\left(\Pi\tau_x + \tau_{ly}\frac{\tau_r}{\tau_{lr}}\right)\left(\Pi'\tau_x + \tau_{lz}\frac{\tau_r}{\tau_{lr}}\right) + \sigma^2_{zx}\left(\Pi\tau_x + \tau_{ly}\frac{\tau_r}{\tau_{lr}}\right)$$
$$+ \sigma^2_{xy}\left(\Pi'\tau_x + \tau_{lz}\frac{\tau_r}{\tau_{lr}}\right) = 0.$$

If, moreover, the elastic properties are the same in directions parallel to the axes of $y$ and $z$, we have

$$\Pi = \Pi',\ \tau_{ly} = \tau_{lz},$$

and the above equation becomes divisible by the factor

$\Pi\tau_x + \tau_{ly}\frac{\tau_r}{\tau_{lr}}$, and reduces to

$$4\left(\tau_x - \tau_{lx}\frac{\tau_r}{\tau_{lr}}\right)\left(\Pi\tau_x + \tau_{ly}\frac{\tau_r}{\tau_{lr}}\right) + \sigma^2_{zx} + \sigma^2_{xy} = 0,$$

of which the roots are

$$\frac{\tau_r}{\tau_{lr}} = \frac{1}{2}\left(\frac{\tau_x}{\tau_{lx}} - \frac{\Pi\tau_x}{\tau_{ly}}\right) \pm \sqrt{\frac{1}{4}\left(\frac{\tau_x}{\tau_{lx}} + \frac{\Pi\tau_x}{\tau_{ly}}\right)^2 + \frac{\sigma^2_{zx} + \sigma^2_{xy}}{4\tau_{lx}\tau_{ly}}}. \quad (115)$$

Now in order that the elastic limit may not be exceeded, it is necessary, according to the principle of M. Barré de Saint-Venant, that the ratio $\frac{\tau_r}{\tau_{lr}}$ shall always be less than 1, so that we have the condition

$$\frac{1}{2}\left(\frac{\tau_x}{\tau_{lx}} - \frac{\Pi\tau_x}{\tau_{ly}}\right) + \sqrt{\frac{1}{4}\left(\frac{\tau_x}{\tau_{lx}} + \frac{\Pi\tau_x}{\tau_{ly}}\right)^2 + \frac{\sigma^2_{zx} + \sigma^2_{xy}}{4\tau_{lx}\tau_{ly}}} < 1. \quad (116)$$

To apply this formula easily, it is convenient to introduce stresses instead of strains.

If $E_x$, $E_y$ are the elastic moduli in the direction $x$ and the two directions $y$, $z$, G the glide modulus in the two directions $y$ and $z$ (i.e. the value of $G_2$, $G_3$ in equations 18), and $t_{lx}$, $t_{ly}$ the limit tensile stresses in the directions of $x$ and $y$ corresponding to the limit strains $\tau_{lx}$, $\tau_{ly}$, we shall have

$$\tau_x = \frac{t_x}{E_x},\ \sigma_{zx} = \frac{s_{zx}}{G},\ \sigma_{xy} = \frac{s_{xy}}{G},$$

$$\tau_{lx} = \frac{t_{lx}}{E_x},\ \tau_{ly} = \frac{t_{ly}}{E_y},$$

and formula (116) will become

$$\frac{t_x}{2t_{lx}}\left(1 - \frac{\Pi t_{lx}E_y}{t_{ly}E_x}\right) + \sqrt{\frac{t_x^2}{4t^2_{lx}}\left(1 + \frac{\Pi t_{lx}E_y}{t_{ly}E_x}\right)^2 + \frac{E_xE_y}{4G^2}\left(\frac{s^2_{zx} + s^2_{xy}}{t_{lx}t_{ly}}\right)} < 1 \quad (117)$$

Writing for simplification

$$\frac{\Pi t_{lx}E_y}{t_{ly}E_x} = \chi,\ \frac{4G^2}{E_xE_y}\cdot t_{lx}\cdot t_{ly} = s_l^2 \quad . \quad . \quad (118)$$

this formula becomes

$$\frac{1-\chi}{2}\cdot\frac{t_x}{t_{lx}} + \sqrt{\left(\frac{1+\chi}{2}\right)^2\frac{t_x^2}{t_{lx}^2} + \frac{s^2_{zx} + s^2_{xy}}{s_l^2}} < 1 \quad . \quad (119)$$

We see that it gives

$$\frac{\sqrt{s^2_{zx} + s^2_{xy}}}{s_l} < 1$$

when the tensile stress $t_x$ is zero; from which it follows that $s_l$ is the value that the shear stress $\sqrt{s^2_{zx} + s^2_{xy}}$ must never exceed, or in other words, $s_l$ is the *limiting shear stress.*

If for a body having at each point three planes of equal elastic properties, as we have supposed above, experiment has given

$$\chi = \frac{1}{3},\ \frac{s_l}{t_{lx}} = \frac{3}{4},$$

formula (119) will become, on multiplying each side by $t_{lx}$,

$$\tfrac{1}{3}t_x + \tfrac{2}{3}\sqrt{t_x^2 + 4(s^2_{zx} + s^2_{xy})} < t_{lx} \quad . \quad . \quad (120)$$

For an isotropic body we have

$$\Pi = \frac{1}{4};\ t_{lx} = t_{ly} = t_l;\ E_x = E_y = E;\ G = \frac{2E}{5},$$

and therefore

$$\chi = \frac{1}{4}, \; s_t = \frac{4}{5}t_t,$$

so that formula (119) becomes, multiplying each side by $t_t$,

$$\frac{3}{8}t_x + \frac{5}{8}\sqrt{t_x^2 + 4(s^2_{zx} + s^2_{xy})} < t_t . \qquad (121)$$

The result $s_t = \frac{4}{5}t_t$ shows that for isotropic bodies the shear strength is $\frac{4}{5}$ of the tensile strength.

**31. Equations of Strength Deduced from Maximum Stresses.**—We have found in § 7 that the components parallel to the axes of a stress acting upon a plane element whose normal makes angles $\alpha, \beta, \gamma$ with the axes are expressed by the formulæ

$$\left.\begin{aligned} X &= t_x \cos\alpha + s_{xy}\cos\beta + s_{zx}\cos\gamma \\ Y &= s_{xy}\cos\alpha + t_y\cos\beta + s_{yz}\cos\gamma \\ Z &= s_{zx}\cos\alpha + s_{yz}\cos\beta + t_z\cos\gamma \end{aligned}\right\} \qquad (122)$$

Now let us assume a plane element on which there is a stress F normal to it; if we call $\alpha, \beta, \gamma$ the angles which its normal makes with the axes, the components of F will be

$$X = F\cos\alpha, \; Y = F\cos\beta, \; Z = F\cos\gamma,$$

and equations (122) will become

$$\left.\begin{aligned} (t_x - F)\cos\alpha + s_{xy}\cos\beta + s_{zx}\cos\gamma &= 0 \\ s_{xy}\cos\alpha + (t_y - F)\cos\beta + s_{yz}\cos\gamma &= 0 \\ s_{zx}\cos\alpha + s_{yz}\cos\beta + (t_z - F)\cos\gamma &= 0 \end{aligned}\right\} \qquad (123)$$

These three equations combined with the following

$$\cos^2\alpha + \cos^2\beta + \cos^2\gamma = 1$$

serve to determine the unknowns, $\cos\alpha$, $\cos\beta$, $\cos\gamma$, F.

Now if we first eliminate the ratios $\frac{\cos\beta}{\cos\alpha}$, $\frac{\cos\gamma}{\cos\alpha}$ between the three equations (121), we obtain

$$(t_x - F)(t_y - F)(t_z - F) - (t_x - F)s^2_{yz} - (t_y - F)s^2_{zx} - (t_z - F)s^2_{xy} + 2s_{yz}\,s_{zx}\,s_{xy} = 0 . \qquad (124)$$

This is an equation of the third degree, whose three roots, which we will call $F_1$, $F_2$, $F_3$, are real, as we can prove by the same reasoning as we adopted with regard to equation (113) of the previous paragraph.

Suppose that one of the roots of equation (124), e.g. $F_1$, is found, and we take the corresponding direction as the axis of $x$; then the stress on the element perpendicular to the axis of $x$ will be normal to this element, and we shall have

$$t_x = F_1, \; s_{xy} = 0, \; s_{zx} = 0;$$

thus equations (123) will become

$$(F_1 - F)\cos\alpha = 0,$$
$$(t_y - F)\cos\beta + s_{yz}\cos\gamma = 0,$$
$$s_{yz}\cos\beta + (t_z - F)\cos\gamma = 0.$$

In order that the first may be satisfied for $F = F_2$ or $F = F_3$, we must have $\cos\alpha = 0$; from this it follows that the normals to the two elements on which the stresses $F_2$, $F_3$ act are perpendicular to the axis of $x$. Thus as $F_1$, $F_2$, $F_3$ may represent the three roots of equation (124) in any order, it must follow that the three elements on which the normal stresses act must be mutually at right angles.

Then if we take axes of co-ordinates in the directions in which the three stresses $F_1$, $F_2$, $F_3$ act, we shall have

$$s_{yz} = 0,\ s_{zx} = 0,\ s_{xy} = 0,$$

and equations (122) will become

$$X = t_x\cos\alpha,\ Y = t_y\cos\beta,\ Z = t_z\cos\gamma.$$

In this case the stress F on the element whose normal makes angles $\alpha$, $\beta$, $\gamma$ with the axes will be the resultant of the three stresses X, Y, Z, so that we shall have

$$F^2 = t_x^2\cos^2\alpha + t_y^2\cos^2\beta + t_z^2\cos^2\gamma \quad . \quad . \quad (125)$$

Now starting from the centre of this element take on its normal a length containing the same number of units as there are in the ratio $\frac{1}{F}$ taken positively, and call $x$, $y$, $z$ the projections of this length on three rectangular axes leading from the point considered in the directions of the three stresses F (called principal stresses). We shall thus have

$$x = \frac{\cos\alpha}{F},\ y = \frac{\cos\beta}{F},\ z = \frac{\cos\gamma}{F},$$

and equation (125) will become

$$t_x^2x^2 + t_y^2y^2 + t_z^2z^2 = 1 \quad . \quad . \quad . \quad (126)$$

This represents an ellipsoid of which the principal axes are in the directions of the principal stresses $F_1$, $F_2$, $F_3$, and whose radii, measured from the centre, are numerically equal to the reciprocals of the stresses.

The maximum and minimum stresses correspond to the minimum and maximum axes of the ellipsoid, and will therefore be two of the three principal stresses $F_1$, $F_2$, $F_3$, i.e. two of the three stresses normal to the plane elements on which they act.

Now take equation (124) and suppose that for the three elements perpendicular to the axes we have, as for the prism subjected separately or simultaneously to tension, torsion, or bending,

$$t_y = 0,\ t_z = 0,\ s_{yz} = 0;$$

that equation will then be divisible by F and will give

$$(t_x - F)F + s^2_{zx} + s^2_{xy} = 0,$$

from which we deduce

$$F = \frac{t_x}{2} \pm \sqrt{\frac{t_x^2}{4} + s^2_{zx} + s^2_{xy}}.$$

The maximum tensile stress will therefore be equal to

$$\frac{t_x}{2} + \sqrt{\frac{t_x^2}{4} + s^2_{zx} + s^2_{xy}}, \quad . \quad . \quad . \quad (127)$$

so that if $t_l$ is the maximum stress per square metre that the body under consideration may carry, we shall have

$$\frac{t_x}{2} + \frac{1}{2}\sqrt{t_x^2 + 4(s^2_{zx} + s^2_{xy})} < t_l \quad . \quad . \quad (128)$$

If we now compare this condition with condition (121) found for the same case at the end of the previous paragraph, we see at once that, taking $t_l$ the same in both cases, condition (128) will always be satisfied when condition (121) is satisfied. For we have clearly

$$\frac{3}{8}t_x + \frac{5}{8}\sqrt{t_x^2 + 4(s^2_{zx} + s^2_{xy})} = \frac{t_x}{2} + \frac{1}{2}\sqrt{t_x^2 + 4(s^2_{zx} + s^2_{xy})}$$
$$+ \frac{1}{8}\left(\sqrt{t_x^2 + 4(s^2_{zx} + s^2_{xy})} - t_x\right),$$

and therefore

$$\frac{t_x}{2} + \frac{1}{2}\sqrt{t_x^2 + 4(s^2_{zx} + s^2_{xy})} < \frac{3}{8}t_x + \frac{5}{8}\sqrt{t_x^2 + 4(s^2_{zx} + s^2_{xy})} < t_l.$$

We see therefore that for prisms subjected to forces in the manner assumed, it will always be safer to adopt the maximum strain condition than the maximum stress condition; the more so because the first can be generalised to make it applicable to bodies having an axis of elasticity, or even three planes of equal elastic properties, while the second is applicable to isotropic bodies only.

## CHAPTER IV.

### APPROXIMATE APPLICATIONS.

**1. Experimental Law.**—In the problems that we have analysed in the previous chapter, we have seen that we can determine the stresses which occur in a bar subjected to pull, bending, or torsion, only when it is acted upon by forces applied at its ends and distributed according to a known law.

Now it never happens in practice that we know the law according to which the forces applied at the ends of a prism are distributed; so that, strictly speaking, the solutions which we have given in the previous chapter would be practically useless, if the knowledge of this law were absolutely essential for the determination, even approximately, of the stresses occurring internally. Fortunately we can proceed by means of the following result which experience has proved and which is of the greatest importance.

*When a bar, which is very long in comparison with its transverse dimensions, is acted upon by forces applied at its ends, the stresses at any section not very near to an end remain practically the same if the distribution of the external forces is changed, provided that the resultants and the resultant moments of the forces applied at each end with reference to three fixed rectangular axes remain constant. The influence of the distribution of the external forces is felt only on the sections very near to the ends, but it diminishes very rapidly and is altogether negligible at a small distance comparable with the transverse dimensions of the bar.*

It follows then that except for two short elements near to the ends, the distribution of the stresses in the interior of the prism follows the laws which we have proved in the problems upon tension, torsion, and bending.

But this result also can be generalised for the most frequent cases that occur in practice.

Suppose that the bar ABDC (Fig. 8, Pl. I), which is very long compared with its transverse dimensions, is acted upon not only by forces applied at its ends and distributed according to a known law, but also by its weight and by forces distributed along the whole of its surface according to a continuous law. Let us consider a section GH, not very near to the base, and let $T_X$, $S_Y$, $S_Z$ be the components, acting along the axis of the bar and along the principal axes of the section GH, of all the forces acting on the bar to the right of this section, and let $T_M$, $B_Y$, $B_Z$ be the torsion and bending moments about the same axes.

Now take to the right and left of GH two short lengths GN, GL, the lengths of which are comparable with the transverse dimensions of the bar, and therefore very small compared with its total length. If the section GH is sufficiently far from the two ends, it is clear that the weight of the two short lengths GN, GL and the forces applied over their surface will be negligible compared with the weight of the portion GC and with the forces applied over its surface and on the end CD. So that if we suppose that we have destroyed the effect of the weight on the length LN and have taken away the external forces applied on its surface, and have then applied to the sections at L and N external forces in such a manner that we still have about the section GH the forces $T_X$, $S_Y$, $S_Z$ and the moments $T_M$, $B_Y$, $B_Z$, the conditions of elastic equilibrium of the length LN will have changed very little and the stresses across the section GH will have remained approximately the same as before.

Now if the length LN is weightless and no force is applied to its surface, it may be regarded as a bar acted upon only by forces applied at its ends; for the forces applied to the lengths AL, NC only act on the length LN through the stresses on the sections at L, N. It follows then from the experimental law enunciated above that the stresses on the section GH, whose distances from the sections at L and N are comparable with the transverse dimensions of the bar, will be distributed approximately according to the laws of simple pull, torsion, and bending found by M. Barré de Saint-Venant, and of which we have previously given proofs.

Therefore we have for any point of the section GH

$$t_y = 0, \; t_z = 0, \; s_{yz} = 0,$$

and at a point of which $y$, $z$ are the transverse co-ordinates, i.e. the distances from the principal axes,

$$\left.\begin{aligned} t_x &= \frac{T_X}{A} + \frac{B_Y}{I_Y} \, . \, z - \frac{B_Z}{I_Z} \, . \, y, \\ s_{xy} &= \frac{F_1 S_Z}{A} + \frac{H_1 S_Y}{A} + \frac{K_1 T_M}{I_P}, \\ s_{zx} &= F_2 \, . \, \frac{S_Z}{A} + H_2 \, . \, \frac{S_Y}{A} + K_2 \, . \, \frac{T_M}{I_P}, \end{aligned}\right\} \quad . \quad . \quad (1)$$

$F_1$, $H_1$, $K_1$, $F_2$, $H_2$, $K_2$ being functions of $y$ and $z$ dependent on the form of the section, but independent of the external forces. The functions will be determined in each case according to the method we adopted in dealing with problems of torsion and bending.

The above are the formulæ which give the approximate distribution of stresses at any section of a bar when it is acted upon by its weight and by forces distributed over its ends and lateral surface according to any arbitrary but continuous law.

These formulæ will still be applicable when the bar is loaded in a discontinuous manner, e.g. by weights applied at different points, provided that in that case we do not consider sections very close to the points of application of the weights.

**2. Long Slender Bodies Curved in any Manner.**—Consider a long slender body whose axis (i.e. the line joining the centroids of its sections) follows any line, straight or curved, the radius of curvature in the latter alternative being very great compared with the transverse dimensions.

We will assume that the locus of the principal axes of the normal sections of the body is composed of two continuous ruled surfaces, perpendicular to each other and such that the angle of two straight generatrices is very small in comparison with their distance measured along the axis of the body.

We will also assume that two sections very close together possess differences in dimensions which are very small compared with the distance apart of the sections.

It is now clear that if we consider a portion of such a body, of which the length is short but comparable with the dimensions of the sections, it can be regarded as approximately a prism or cylinder, and we can therefore apply formulæ (1) given in the previous paragraph to determine the stresses at any point.

These formulæ will give approximately the distribution of the stresses not only for cylindrical or prismatic bodies but also for solid bodies of the kind defined above.

It is to be understood that the same restrictions hold for these bodies as for cylinders or prisms, i.e. that when external forces applied to the ends of a body are not distributed according to the law represented in formulæ (1), or when those applied to the lateral surface are not applied in a continuous manner but are isolated at different points, formulæ (1) will give approximately the true distribution of the stresses for all the sections except those very near the ends or the points of application of the isolated loads.

**3. Relations Between the Resultant Shear Forces $S_Y$, $S_Z$ and the Resultant Bending Moments** $B_Y$, $B_Z$.—Let us consider in a body satisfying the conditions enunciated above two sections infinitely close together, the length from the origin of one, measured along the curved axis, being $s$, and of the other $s + ds$.

Let $B_Y$, $B_Z$ be the bending moments about the principal axes and $S_Y$, $S_Z$ be the shearing forces parallel to these axes of the first section.

Further, let $pds$ be the component normal to the section of the force due to the weight of the portion between the two sections and to the external transverse forces applied over the same length, and let $y$, $z$ be the co-ordinates with reference to the principal axes of the point of application of this component.

Since by hypothesis the principal axes of the two sections may be regarded as parallel to each other, the bending moments about the principal axes of the second section, neglecting small quantities of the second order, will be

$$B'_Y = B_Y - S_Z ds - pds\,.\,z,$$
$$B'_Z = B_Z + S_Y ds + pds\,.\,y.$$

Now if no finite force is applied between the two sections under consideration, and if, therefore, in the neighbourhood of these sections the bending moments are continuous functions of $s$, we shall have in accordance with the principles of the differential calculus

$$B'_Y = B_Y + \frac{dB_Y}{ds}ds$$

$$B'_Z = B_Z + \frac{dB_Z}{ds}ds.$$

By comparing these results with the above and dividing by $ds$ we get

$$\frac{dB_Y}{ds} = -S_Z - pz$$

$$\frac{dB_Z}{ds} = S_Y + py.$$

Now the terms $pz$, $py$ are zero in the following two cases :—

1. When the force applied to the element comprised between the sections considered is normal to the axis of the body, because then the component $pds$ is zero.

2. When the co-ordinates $y$, $z$ are zero, i.e. when the force applied to the element is uniformly distributed over the cross-section.

The above equations then become

$$\frac{dB_Y}{ds} = -S_Z; \quad \frac{dB_Z}{ds} = S_Y,$$

so that the *differential coefficients of the bending moments are equal to the shearing forces.*

Beyond these two special cases, this rule does not rigorously apply : nevertheless for very slender bodies, the co-ordinates $y$, $z$ are necessarily very small, and therefore the products $pz$, $py$ can in general be neglected, so that the above rule will hold at least approximately for these bodies.

**4. Formula for Internal Work.**—We will now deduce a formula for giving the internal work of a body satisfying the conditions assumed in the above two paragraphs.

With this in view we will recall the formula which we found in § 5, chap. iii., to express the internal work of the element of volume $\delta v = \Delta x \Delta y \Delta z$, i.e.

$$\tfrac{1}{2}(t_x\tau_x + t_y\tau_y + t_z\tau_z + s_{yz}\sigma_{yz} + s_{zx}\sigma_{zx} + s_{xy}\sigma_{xy})\Delta x\Delta y\Delta z.$$

If we take as origin a point on the axis of the body, and for axes of $y$ and $z$ the principal axes of the transverse section through this point, and if we consider the element of volume comprised between the section passing through the origin and that having abscissa $\Delta x$ and parallel to the first, we have seen in § 1 that for every point in this element we may assume

$$t_y = 0, \quad t_z = 0, \quad s_{yz} = 0,$$

from which it follows that for each element of volume $\Delta x\Delta y\Delta z$ taken between the two sections considered, the expression for the internal work will be

$$\tfrac{1}{2}(t_x\tau_x + s_{xy}\sigma_{xy} + s_{zx}\sigma_{zx})\Delta x\Delta y\Delta z.$$

Therefore the internal work of the whole volume comprised between the two sections will be given by the expression

$$\tfrac{1}{2}\Delta x\iint(t_x\tau_x + s_{xy}\sigma_{xy} + s_{zx}\sigma_{zx})\Delta y\Delta z,$$

taking the double integral over the whole section.

Adopting the language of the infinitesimal caculus we may change $\Delta$ to $d$ and our formula becomes

$$W_i = \tfrac{1}{2}ds\iint(t_x\tau_x + s_{xy}\sigma_{xy} + s_{zx}\sigma_{zx})dydz \quad . \quad . \quad (2)$$

If we assume the body is homogeneous and has three planes of equal elastic properties perpendicular to the axes of the co-ordinates, we shall have in accordance with formulæ (18) of the previous chapter

$$\begin{aligned} t_x &= a_1\tau_x + G_3\tau_y + G_2\tau_z, & 0 &= G_1\sigma_{yz}, \\ 0 &= G_3\tau_x + a_2\tau_y + G_1\tau_z, & s_{zx} &= G_2\sigma_{zx}, \\ 0 &= G_2\tau_x + G_1\tau_y + a_3\tau_z, & s_{xy} &= G_3\sigma_{xy}, \end{aligned}$$

from which we deduce, calling E the elastic modulus in direct stress of the material,

$$\tau_x = \frac{t_x}{E}, \qquad \sigma_{zx} = \frac{s_{zx}}{G_2}, \qquad \sigma_{xy} = \frac{s_{xy}}{G_3}$$

and formula (2) becomes

$$W_i = \tfrac{1}{2}ds\iint\left(\frac{t_x^2}{E} + \frac{s^2_{zx}}{G_2} + \frac{s^2_{xy}}{G_3}\right)dydz \quad . \quad . \quad (3)$$

We can substitute in this formula for $t_x$, $s_{zx}$, $s_{xy}$ their expressions (1): then if we note that the ordinates $y$, $z$ are with reference to the principal axes and that therefore we have

$$\iint ydydz = 0, \quad \iint zdydz = 0, \quad \iint yzdydz = 0$$
$$\iint dydz = A, \quad \iint z^2dydz = I_Y, \quad \iint y^2dydz = I_Z$$

we obtain at once

$$\tfrac{1}{2}ds\iint t_x^2dydz = \frac{ds}{2E}\left(\frac{T^2}{A} + \frac{B_Z^2}{I} + \frac{B_Y^2}{I_Y}\right) \quad . \quad . \quad (4)$$

Moreover, if we put

$$\iint\left(\frac{F_1^2}{G_3} + \frac{F_2^2}{G_2}\right)dydz = \frac{FA}{G_2}, \qquad \iint\left(\frac{H_1K_1}{G_3} + \frac{H_2K_2}{G_2}\right)dydz = \frac{L}{G_2}$$
$$\iint\left(\frac{H_1^2}{G_3} + \frac{H_2^2}{G_2}\right)dydz = \frac{HA}{G_2}, \qquad \iint\left(\frac{K_1F_1}{G_3} + \frac{K_2F_2}{G_2}\right)dydz = \frac{N}{G_2}$$
$$\iint\left(\frac{K_1^2}{G_3} + \frac{K_2^2}{G_2}\right)dydz = \frac{KI_P}{G_2}, \qquad \iint\left(\frac{F_1H_1}{G_3} + \frac{F_2H_2}{G_2}\right)dydz = \frac{Q}{G_2}$$

it follows that

$$\tfrac{1}{2}ds\iint\left(\frac{s^2_{xy}}{G_3} + \frac{s^2_{zx}}{G_2}\right)dydz$$
$$= \frac{ds}{2G_2}\left(\frac{FS_Z^2}{A} + \frac{HS_Y^2}{A} + \frac{KT_M^2}{I_P} + \frac{LS_YT_M}{AI_P} + \frac{NS_ZT_M}{AI_P} + \frac{QS_YS}{A^2}\right) \quad (5)$$

The coefficients F, H, K, L, N, Q depend only upon the form of the section and on the relation between the glide moduli $G_2$, $G_3$, and remain therefore always the same for a given section and material, whatever may be the forces applied to the body and the curve taken by its axis.

To obtain the total internal work for the very small element considered, we should add formulæ (4) and (5), of which the first gives the internal work due to the direct stresses, and the second that due to the shear stresses

**5. Simplification of Formula (5) due to the Distribution of the External Forces.**—It seldom happens in ordinary construction that the portions of the structure are subjected to torsion; i.e. $T_M$ is nearly always zero. In that case formula (5) becomes

$$\tfrac{1}{2}ds\iint\left(\frac{s^2_{xy}}{G_3} + \frac{s^2_{zx}}{G_2}\right)dydz = \frac{ds}{2G_2}\left(\frac{FS_Z^2}{A} + \frac{HS_Y^2}{A} + \frac{QS_ZS_Y}{A^2}\right) \quad (6)$$

It also happens nearly always that for bodies whose axes are curved in a plane the principal axis $y$ is perpendicular to the plane of the curve, and that the external forces are distributed symmetrically with reference to this plane. In this case (the only one which will arise in the applications that we shall give in this book) we have for every section, not only $T_M = 0$, but also

$$B_Y = 0 \text{ and } S_Z = 0$$

so that formulæ (4) and (5) simplify very considerably, and the total internal work of the small element will be

$$dW_i = \frac{ds}{2E}\left(\frac{T_X^2}{A} + \frac{B_Z^2}{I_Z}\right) + \frac{ds}{2} \cdot \frac{FS_Z^2}{AG_2}. \qquad (7)$$

Such is the formula of which we shall make use in all our applications.

**6. Simplification of Formula (5) on Account of the Form of the Sections.**—Formula (5) also becomes simplified when among the quantities $S_Y$, $S_Z$, $T_M$ there are no zero values, provided that the body is symmetrical about one of its principal axes.

But as we shall never have need in the course of this book to remember these simplifications, because the coefficients L, N, Q do not enter into formula (7), we will only give the principal results without proving them.

Let us assume in the first place that the section is symmetrical about the axis of $z$, and let $F_1, H_1, K_1, F_2, H_2, K_2$ be the values of the coefficients which occur in the expressions for $s_{xy}$, $s_{zx}$ for a point in the section. We can prove that for the point symmetrical with the first about the axis of $z$, the coefficients will be $-F_1, H_1, K_1, F_2, -H_2, -K_2$.

It follows that the quantities

$$\frac{K_1F_1}{G_3} + \frac{K_2F_2}{G_2} \text{ and } \frac{F_1H_1}{G_3} + \frac{F_2H_2}{G_2}$$

will have values which are numerically equal but opposite in sign for the two points considered, and that therefore the coefficients N, Q will be zero.

Then if the section is symmetrical about the principal axis $z$ the right-hand side of formula (5) becomes

$$\frac{ds}{2G_2}\left(\frac{FS_Z^2}{A} + \frac{HS_Y^2}{A} + \frac{KT_M^2}{I_P} + \frac{LS_Y\,T_M}{AI_P}\right) \qquad (8)$$

If, on the other hand, the section were symmetrical with regard to the principal axis $y$, the coefficients L, Q would become zero and formula (5) would become

$$\frac{ds}{2G_2}\left(\frac{FS_Z^2}{A} + \frac{HS_Y^2}{A} + \frac{KT_M^2}{I_P} + \frac{NT_M}{AI_P}\right) \qquad (9)$$

We see that the coefficient Q is zero in both cases, i.e. it is zero for every section having an axis of symmetry, because we know that an axis of symmetry is also a principal axis.

Thus if a section is symmetrical about one axis, and if, moreover, the twisting moment $T_M$ is zero, the right-hand side of formula (5) becomes

$$\frac{ds}{2G_2}\left(\frac{FS_Z^2}{A} + \frac{HS_Y^2}{A}\right). \quad . \quad . \quad . \quad . \quad (10)$$

If the twisting moment is not zero, but if the two principal axes of a section are axes of symmetry, it follows from what we have said above that the three coefficients L, N, Q are all zero and that therefore the right-hand side of formula (5) becomes

$$\frac{ds}{2G_2}\left(\frac{FS_Z^2}{A} + \frac{HS_Y^2}{A} + \frac{KT_M^2}{I_P}\right) \quad . \quad . \quad . \quad . \quad (11)$$

We will note that in formula (5) and in all those that we have deduced therefrom, the coefficients F, H, K, L, N, Q depend not only upon the shape of the section but also upon the relation between $G_2$ and $G_3$.

When the elastic properties are the same in the directions of the two principal axes we have $G_2 = G_3 = G$, and the coefficients F, H, etc. depend only on the form of the section.

This is the only case we shall consider in the applications. If the body is isotropic we have, as we have shown,

$$G = \frac{2}{5}E.$$

**7. Strains in Long Slender Bodies.**—Let us consider a body AA′B′B (Fig. 9, Pl. I) satisfying the conditions set out in § 2, and let us determine the relative displacements of its various points, i.e. the strains of the body due to external forces.

We will refer the body to three rectangular axes such that before strain the axis of $x$ is perpendicular to the section AA′ and the axes of $y$ and $z$ coincide with the principal axes of this section; then assume that after strain the axis of $x$ is normal to the centre of the section AA′ and that the axis of $y$ is tangential to the principal axis with which it coincided before strain.

If we knew how to express the internal work of the body as a function of all the elemental forces applied to it, except those applied to the centre of the section AA′, (because this central element must be regarded as fixed during strain), it is clear from what we have proved that by taking the differential coefficients of the internal work of the whole body, with regard to the components parallel to the axes of the elemental force applied at a point, we should find the values of the displacements of this point in the directions of the three axes.

To find the displacements of a point at which no force is applied, we will assume to be applied there three arbitrary forces $F_x$, $F_y$, $F_z$, parallel to the axes ; we will then express the internal work of the whole body as a function of the forces that are actually applied and of $F_x$, $F_y$, $F_z$, and then take the differential coefficients with regard to these arbitrary forces and make them equal to zero.

But, for the rigorous application of this procedure, we should have to be able to solve the problem of determining exactly the stresses which are developed in the body AA′B′B under the action of any external forces, and this is beyond the scope of known analysis, for we have seen in the preceding chapter that this problem can only be solved at the present time for prisms or cylinders, the lateral faces of which are free and the bases of which are subjected to forces distributed in a particular manner.

Fortunately this rigorous solution is not at all necessary in practice, when we are considering bodies very slender in comparison with their length, because we can then proceed in the following manner and obtain results which are approximate but always sufficiently exact.

Let us assume that at the section CC′ external forces are applied, the sums of whose components normal to the section and parallel to the principal axes are $F_1$, $F_2$, $F_3$ ; and whose moments about this normal and these axes are $M_1$, $M_2$, $M_3$. The formulæ given in the preceding paragraphs show that by expressing the internal work of the whole body as a function of these external forces we obtain a formula of the second degree in $F_1$, $F_2$, $F_3$, $M_1$, $M_2$, $M_3$. Now we can say that *the differential coefficients of this formula with regard to* $F_1$, $F_2$, $F_3$ *give to a sufficiently accurate degree of approximation the displacements of the centre of the section in the three directions, whilst those with regard to* $M_1$, $M_2$, $M_3$ *give the angular rotations of the sections about the three axes.*

In fact, it follows from the theorem of the differential coefficients of internal work that, if the forces applied to the section CC′ are so distributed that the section remains plane and without change of its shape during strain, the law enunciated above is rigorously true.

Moreover, the formula which we use to express the internal work contains the three components $F_1$, $F_2$, $F_3$ and the three moments $M_1$, $M_2$, $M_3$, but does not depend at all upon the manner in which the forces applied to the section CC′ are distributed; from this it follows that the differential coefficients of this formula with regard to $F_1$, $F_2$, $F_3$, $M_1$, $M_2$, $M_3$ give the displacements of the centre and the angular rotations of this section, just as if the external forces which are applied to it were distributed in such a manner that it remained plane after strain.

But we have seen that, whatever be the distribution of the external forces applied at the section CC′, provided that the sums of their components are $F_1$, $F_2$, $F_3$ and of their moments $M_1$, $M_2$, $M_3$, the internal stresses, and, therefore, the relative displacements of its various points, remain practically the same for the whole body except for a very small layer having for base the section CC′. If, then, the thickness of this layer is negligible in comparison with the length of the portions AA′C′C, CC′B′B, the absolute displacements of the points in the section CC′ cannot be sensibly altered, whatever be the relative displacements in this thin layer; i.e. they will be the same whatever be the distribution of the external forces over the section CC′.

What we have proved above for any section clearly holds also for the extreme sections AA′, BB′.

**8. Smallness of the Internal Work due to Shearing Forces, when there is no Torsion.**—We have seen that when there is no torsion and all the sections of the body have an axis of symmetry, the internal work of an infinitely thin slice is expressed by

$$\frac{ds}{2E}\left(\frac{T_X^2}{A}+\frac{B_Z^2}{I_Z}+\frac{B_Y^2}{I_Y}\right)+\frac{ds}{2G_2}\left(\frac{FS_Z^2}{A}+\frac{HS_Y^2}{A}\right) \quad . \quad (12)$$

the slice considered being comprised between sections whose abscissæ are $s$ and $s + ds$ measured along the axis of the body.

Now we will show that generally the three terms

$$\frac{ds}{2E}\cdot\frac{T_X^2}{A},\quad \frac{ds}{2G_2}\cdot\frac{FS_Z^2}{A},\quad \frac{ds}{2G_2}\cdot\frac{HS_Y^2}{A}$$

will be very small compared with those containing the bending moments.

Calling $\frac{I_Z}{A} = g_z^2$ and $\frac{I_Y}{A} = g_y^2$, where $g_z$ and $g_y$ are the radii of gyration about the principal axes, and assuming that the body is isotropic so that $G_2 = \frac{2}{5}E$, then expression (12) becomes

$$\frac{ds}{2EA}\left\{\left(T_X^2+\frac{B_Y}{g_y^2}+\frac{B_Z}{g_z^2}\right)+\left(\frac{5F}{2}\cdot S_Z^2+\frac{5H}{2}\cdot S_Y^2\right)\right\}.$$

Now consider all the external forces applied to the body to the right of the slice under consideration, and resolve each into three components, one normal to the section and the others parallel to the principal axes.

The resultant of these normal components will be $T_X$, and its line of action will cut the section at a point whose co-ordinates we will call $y_x$, $z_x$.

The resultant of all the components parallel to the axis of $y$ will be $S_Y$ and will act at a distance from the section which we will call $x_y$, and the corresponding resultant parallel to the axis of $z$ will be $S_Z$ at distance $x_z$.

Now by assuming that $T_X$ is a tension and that $S_Y$, $S_Z$ are directed towards the negative co-ordinates $y$ and $z$, we have clearly

$$B_Y = T_X z_x + S_Z x_z,$$
$$B_Z = -T_X y_x - S_Y x_y.$$

Substituting these values in the above expression, we have

$$\frac{ds}{2EA}\left[T_X^2\left(1+\frac{y_x^2}{g_y^2}+\frac{z_x^2}{g_z^2}\right)+2T_XS_Y\,.\frac{y_x x_y}{g_y^2}+2T_XS_Z\frac{z_x x_z}{g_z^2}\right.$$
$$\left.+\,S_Z^2\left(\frac{5F}{2}+\frac{x_z^2}{g_z^2}\right)+S_Y^2\left(\frac{5H}{2}+\frac{x_y^2}{g_y^2}\right)\right].$$

If the body is long and slender, it will happen in general that the lever arms $y_x$, $z_x$, $x_z$, $x_y$ will be very great compared with its transverse dimensions, and as the radii of gyration are always less than the corresponding breadth of the section, it follows that the quantities $\frac{y_x}{g_y}, \frac{z_x}{g_z}, \frac{x_z}{g_z}, \frac{x_y}{g_y}$ will be large numbers. We can therefore neglect, in comparison with the squares of these, the quantities 1, $\frac{5F}{2}, \frac{5H}{2}$, since, as we shall see later, the latter are relatively small.

Thus the last expression becomes

$$\frac{ds}{2EA}\left[T_X^2\left(\frac{y_x^2}{g_y^2}+\frac{z_x^2}{g_z^2}\right)+2T_XS_Y\,.\frac{y_x x_y}{g_y^2}+2T_XS_Z\,.\frac{z_x x_z}{g_z^2}+\frac{S_Z^2\,.\,x_z^2}{g_z^2}+\frac{S_Y^2\,.\,x_y^2}{g_y^2}\right]$$

which is equal to

$$\frac{ds}{2E}\left(\frac{B_Z^2}{I_Z}+\frac{B_Y^2}{I_Y}\right) \quad . \quad . \quad . \quad . \quad (13)$$

We see therefore, as we wished to prove, that we can generally neglect in the formula for the internal work in a thin slice of the body the three terms containing $T_X$, $S_Y$, $S_Z$ in comparison with those containing the bending moments.

The reasoning by which we have arrived at this result is not applicable to sections very near to the ends of the body. Nevertheless we should always remember that in practical applications we do not have to consider the internal work of separate slices, but that of the whole body, so that the error which the two short lengths at the ends introduce will be negligible, and we can therefore use the simpler formula (13) instead of the more exact form (12).

It may, and often does, happen that the lever-arms or eccentricities of the resultant $T_X$ about the principal axes are zero or very small. In this case $\frac{y_x^2}{g_y^2}$ and $\frac{z_x^2}{g_z^2}$ are not very great, and we cannot neglect unity in comparison with them; but if the arms $x_y$, $x_z$ are very great in comparison with the radii of gyration, we can easily see that as before we may neglect $\frac{5F}{2}S_Z^2$, $\frac{5H}{2}S_Y^2$ in comparison with the terms $\frac{B_Y^2}{g_z^2}$, $\frac{B_Z^2}{g_y^2}$. Moreover, the term $T_X$ is of the same order of magnitude as the terms $\frac{5F}{2}S_Z^2$, $\frac{5H}{2}S_Y^2$, and should be neglected at the same time as the latter; so that even when $y_x$, $z_x$ are zero or very small formula (12) reduces to the much simpler form (13).

**9. Curved Form Adopted by Sections after Strain due to Normal Stresses only.**—We will now show that if the shear stresses $s_{xy}$, $s_{zx}$ are zero, plane sections remain plane after strain.

If we consider two molecules $m$, $m'$ (Fig. 9, Pl. I) infinitely close, and such that before strain the straight line joining them is perpendicular to the section CC′, i.e. parallel to the element $nn'$ of the axis of the body, it is clear that after strain these two lines will still be parallel; further, if the shear stresses are zero, the shear strains will also be zero, and the line $mm'$ must be normal to CC′ after strain, i.e. perpendicular to the element of the section corresponding to $m$.

Then since after strain the lines $mm'$, $nn'$ are parallel, it is clear that all the elements of the section must be perpendicular to the line $nn'$, and that therefore the section must be plane and perpendicular to the axis of the body.

The same result can be deduced from the formula

$$t_x = \frac{T_X}{A} + \frac{B_Y}{I_Y}z \,.\, - \frac{B_Z}{I_Z}\cdot y,$$

for, putting for $t_x$ its value $E\tau_x = E\frac{du}{dx}$, we have

$$du = \frac{dx}{E}\left(\frac{T_X}{A} + \frac{B_Y}{I_Y}\cdot z - \frac{B_Z}{I_Z}\cdot y\right).$$

Now suppose that the body is referred to three rectangular axes with origin at the centre of CC′, these axes being respectively normal to the section and parallel to the principal axes, then the above formula will give the amount by which a point at distance $dx$ from CC′ and having co-ordinates $y$, $z$ will stretch from the plane of $yz$, and $dx + du$ will give the final distance of the point from this plane. If we apply this formula to all the points in a section infinitely close to CC′, the quantities $\frac{T_X}{A}$, $\frac{B_Z}{I_Z}$, $\frac{B_Y}{I_Y}$ will be constant, and the quantity $dx$ should also be regarded as constant if the transverse dimensions of the body are very small in comparison with its radius of curvature.

In this case $dx + du$ will be a linear function of $y$ and $z$; this shows that, neglecting the very small variations in $y$ and $z$ due to lateral strains, the section will remain plane after strain.

**10. Corollary.**—We have seen that the theorem of the differential coefficients of internal work given in § 4 will hold rigorously if the sections remain plane after strain; we have also seen in § 5 that the internal work due to shear stresses is negligible compared with that due to normal stresses for long slender bodies; finally, in the last paragraph we have shown that, when under normal stresses only, plane sections will remain plane after strain. It follows, therefore, that the theorem given in § 4 becomes rigorously applicable when we neglect the internal work due to shear stresses in comparison with that due to normal stresses.

**11. Definitions and Notation for Structures that Usually Occur in Practice.**—In general the elastic structures that we have to consider in practice satisfy the following conditions:—

1. All members have transverse dimensions which are very small in comparison with their length.
2. The centre lines or axes of all the members are in one plane, which we will call the *plane of the structure.*
3. All the transverse sections of the different members have a principal axis in this plane.
4. All the stresses applied to the structure act within this same plane, or are so distributed on both sides of it that the twisting moment $T_M$, the bending moment $B_Z$ about the principal axis in this plane, and the shearing force $S_Y$ perpendicular to this plane, are zero.

For this kind of structure we have only to consider the normal force $T_X$, the bending moment $B_Y$ about the principal axis normal to the plane of the structure, and the shearing force $S_Z$ parallel to this plane. As throughout the following treatment we shall only consider structures satisfying the above conditions, we will for simplification drop the suffixes used in the previous treatment, and adopt the following notation :—

P = normal pressure or thrust;
S = shearing force parallel to the plane of the structure;
B = bending moment about the principal axis normal to this plane;
I = moment of inertia about the principal axis normal to this plane;
A = area of section.

Also we will call the principal axis normal to the plane of the structure the *neutral axis.*

To express the internal work of an infinitely thin slice of length or thickness $dl$ we must use formula (7), which now becomes

$$dW_i = \frac{dl}{2E}\left(\frac{P^2}{A} + \frac{B^2}{I}\right) + \frac{dl}{2G_2} \cdot \frac{FS^2}{A} \quad . \quad . \quad (14)$$

Following from what we have proved in § 5, the first and last terms of the right-hand side are usually very small in comparison with the second.

In future we will call

$$\text{internal work due to thrust} = \frac{dl}{2E} \cdot \frac{P^2}{A},$$

$$\text{internal work due to bending} = \frac{dl}{2E} \cdot \frac{B^2}{I},$$

$$\text{internal work due to shear} = \frac{dl}{2G_2} \cdot \frac{FS^2}{A},$$

and thus in general the internal work due to thrust and shear will be negligible compared with that due to bending.

**12. Transformation of one of the Differential Equations of Elastic Equilibrium of Solid Bodies.**—Let us imagine a prism or cylinder similar to that which we have studied in the problem of bending in the previous chapter, and which is considered as subjected only to normal forces and to tangential forces applied at its ends.

We have seen that for this prism we must have at any point on the contour the relation

$$s_{xy}dz - s_{zx}dy = 0 \quad . \quad . \quad . \quad (15)$$

We have also seen that at any point in the interior of a body in elastic equilibrium, we must have, neglecting the effect of weight, the equation

$$\frac{dt_x}{dx} + \frac{ds_{xy}}{dy} + \frac{ds_{zx}}{dz} = 0 \, . \quad . \quad . \quad . \quad (16)$$

We shall now see that, having regard to equation (15) which must hold at the contour of the sections, for the prism or cylinder this equation can be put into a form that it is very important to remember.

Consider the section GH (Fig. 8, Pl. I) at distance $x$ from the end AB; then from what we have shown in the problem of bending, we shall have

$$t_x = \frac{B_0 + S(a - x)}{I} \cdot z,$$

$B_0$ being the bending moment on the base CD and S the shearing force on this base and on all sections of the body.

From this formula we get

$$\frac{dt_x}{dx} = -\frac{Sz}{I}$$

so that equation (16) becomes

$$\frac{Sz}{I} = \frac{ds_{xy}}{dy} + \frac{ds_{zx}}{dz}.$$

Now let us suppose that ABA′B′ (Fig. 10, Pl. I) represents the section GH, and let us integrate the last equation, after having multiplied by $dydz$, extending the integration over the whole section comprised between the line C′C″ and the contour C′BC″.

We shall have in the first place

$$\frac{S}{I}\iint z\,dydz = \iint \frac{ds_{xy}}{dy}dydz + \iint \frac{ds_{zx}}{dz}dydz:$$

but if we draw D′D″ parallel to the axis of $y$ and E′E″ very close to it, we have, on integrating first for the strip D′D″E″E′,

$$\iint \frac{ds_{xy}}{dy}dydz = \int dz\int \frac{ds_{xy}}{dy}dy = \int (s_{xy}'' - s_{xy}')dz$$

where $s_{xy}'$ and $s_{xy}''$ are the values of the shear stress $s_{xy}$ at D′ and D″.

In the same way if we draw D′F′, E′H′ perpendicular to C′C″ and integrate for the strip D′F′H′E′, we obtain

$$\iint \frac{ds_{zx}}{dz}dydz = \int dy\int \frac{ds_{zx}}{dz}dz = \int (s_{zx}' - s_{zx})dy,$$

$s_{zx}$ and $s_{zx}'$ being the values of $s_{zx}$ for the points F′, D′. Thus the above equation becomes

$$\frac{S}{I}\iint z\,dydz = \int (s_{xy}'' - s_{xy}')dz + \int (s_{zx}' - s_{zx})dy \qquad (17)$$

Now we may write

$$\int (s_{xy}'' - s_{xy}')dz = \int s_{xy}''dz'' - \int s_{xy}'dz',$$

extending the integrals $\int s_{xy}''dz''$, $\int s_{xy}'dz'$ respectively of the points C″ and C′ up to the highest point of the contour, so that $dz'$, $dz''$ will always be taken positive; if, on the other hand, we traverse the whole arc C′BC″ continuously from C′, it is clear that the differentials $dz''$ will be negative and that it will be necessary therefore to change their sign, giving

$$\int (s_{xy}'' - s_{xy}')dz = - \left(\int s_{xy}''dz'' + \int s'_{xy}dz'\right),$$

or simply

$$\int (s_{xy}'' - s_{xy}')dz = - \int s_{xy}'dz',$$

extending the right-hand integral over the whole arc C′BC″.
We can also write

$$\int (s_{zx}' - s_{zx})dy = \int s_{zx}'dy' - \int s_{zx}dy,$$

extending the first integral on the right-hand side over the whole arc C′BC″ and the second over the line C′C″.

Thus equation (17) can be written

$$\frac{S}{I}\iint zdydz = - \int (s_{xy}'dz' - s_{zx}'dy') - \int s_{zx}dy,$$

and as equation (15) gives for any point on the contour

$$s_{xy}'dz' - s_{zx}'dy' = 0,$$

it follows that

$$\frac{S}{I}\iint zdydz = - \int s_{zx}dy \quad . \quad . \quad . \quad (18)$$

It must again be noted that the integral on the left-hand side must be extended over the whole area between the arc C′B′C″ and the line C′C″, whilst that on the right-hand side must be taken over the line C′C″.

We can further reduce the double integral to a simple integral because the integration with respect to $y$ can be effected, so that if $b$ is the breadth D′D″ of the section at distance $z$ from the neutral axis we have

$$\iint zdydz = \int bzdz.$$

The quantity $b$ should be considered as a function of $z$ and the integration extended between the limits $z_1$, corresponding to the line C′C″, and the maximum value of $z$, which is the neutral axis depth $n$. Thus equation (18) becomes

$$\int s_{zx}dy = - \frac{S}{I}\int_{z_1}^{n} bzdz \quad . \quad . \quad . \quad (19)$$

**13. Alternative Derivation of Formula (19).**—In the preceding paragraph we have derived formula (19) from the general equations of elasticity; but the authors of this treatment have given another proof which we will reproduce here.

Imagine that we have an isolated element of volume comprised between the lateral surface corresponding to the arc C′BC″, the plane parallel to the axis of the prism containing the line C′C″, and the two sections corresponding to abscissæ $x$ and $x + dx$.

One of the necessary conditions for the equilibrium of this element of volume is that the sum of components parallel to the axis of $x$ of all the forces applied to it must be zero.

Now on the face corresponding to the section having the abscissa $x$, we have normal stresses $t_x$, whose sum extended over this face is equal to $\int_{z_1}^{n} t_x b dz$; on the opposite face we have stresses $-\left(t_x + \frac{dt_x}{dx}dx\right)$ or $-\left(t_x - \frac{\mathrm{S}}{\mathrm{I}}z dx\right)$ of which the sum is

$$-\int_{z_1}^{n}\left(t_x - \frac{\mathrm{S}}{\mathrm{I}}z dx\right)b dz = -\int_{z_1}^{n} t_x b dz + dx\frac{\mathrm{S}}{\mathrm{I}}\int_{z_1}^{n} b z dz.$$

By hypothesis, no external force acts upon the curved surface of the element under consideration, and on the rectangular face of which the sides are CC′ and $dx$ we have parallel to the axis of $x$ forces $s_{zx}$ whose sum taken over the whole face is

$$\int s_{zx} dy dx = dx\int s_{zx} dy.$$

Adding these three sums and equating to zero the result divided by $dx$ we obtain the equation

$$\frac{\mathrm{S}}{\mathrm{I}}\int_{z_1}^{n} b z dz + \int s_{zx} dy = 0;$$

this is exactly the same as formula (19).

**14. Application of Formula (19) to any Solid Bodies whose Breadths are Small Compared with their Depths.**—We have proved formula (19) for cylinders acted upon only by forces applied at their ends; but it applies approximately, as do all the others that we have found for prisms, to long slender bodies defined in § 11.

The section of these bodies may have any form, but we will consider the case (arising most frequently in practice) in which the sections are very narrow at the neutral axis compared with their depth.

The line C′C″ parallel to the neutral axis being very short compared with the depth BB′, (Fig. 10, Pl. I), it is clear that the shear stress $s_{zx}$ parallel to the principal axis BB′ will vary very little across the line C′C″ and as an approximation may be regarded as constant. We therefore have

$$\int s_{zx} dy = b_c s_{zx}$$

where $b_c$ is the breadth at the line C′C″, and equation (19) becomes

$$s_{zx} = -\frac{S}{b_c I}\int_{z_1}^{n} bz dz$$

or

$$s_{zx} = \frac{S}{b_c I}\int_{z_1}^{n} bz dz \quad . \quad . \quad . \quad . \quad (20)$$

neglecting the negative sign, which will not affect the results.

As the maximum numerical value of the integral $\int_{z_1}^{n} bzdz$ occurs for $z = 0$, it follows that the maximum shear stress occurs at the neutral axis and is equal to

$$s_{zx}^{0} = \frac{S}{b_0 I}\int_{0}^{n} bz dz.$$

Now $\int_{0}^{n} bzdz$ represents the first moment about the neutral axis of the area above or below this axis, and calling this moment $M_0$ we have

$$s_{zx}^{0} = \frac{SM_0}{b_0 I} \quad . \quad . \quad . \quad . \quad (21)$$

The expression for $s_{zx}$ may now be put into another form, for we may write

$$\int_{z}^{n} bz dz = \int_{0}^{n} bz dz - \int_{0}^{z} bz dz = M_0 - \int_{0}^{z} bz dz,$$

and therefore

$$s_{zx} = \frac{S}{bI}\left(M_0 - \int_{0}^{z} bz dz\right).$$

It will be seen that for sections very narrow compared with their depth the shear stress $s_{xy}$ will always be very small and may be regarded as zero. For every straight line C′C″ this stress in general reaches its maximum at the contour points C′, C″; now for these points equation (15) gives

$$s_{xy} = s_{zx}\frac{dy}{dz},$$

and as in sections of the form under consideration the value of $\frac{dy}{dz}$ is generally very small, it follows that the shear stress $s_{xy}$ will be very small compared with $s_{zx}$.

This reasoning will not be exact for the highest points of the contour, for which the value of $\frac{dy}{dz}$ is not very small, but for these points it follows from formula (20) that the stress $s_{zx}$ will itself be very small, so that the stress $s_{xy}$ will also be small.

**15. Curved Surface Assumed after Bending by Sections very Narrow Compared with their Depth.**—Equation (20) enables us to find at once the equation of the curved surface assumed after bending by sections very narrow compared with their depth. For if we call $\sigma_{zx}$ the shear strain caused by the shear stress $s_{zx}$, and if we suppose the body to have three planes of symmetry respectively perpendicular to the normal to the section, and to its principal axes, we shall have in accordance with formulæ (19) of chapter iii.,

$$s_{zx} = G_2 \sigma_{zx},$$

and as we have

$$\sigma_{zx} = \frac{dw}{dx} + \frac{du}{dz},$$

we derive from equation (20), taking the right-hand side as of negative sign,

$$\frac{du}{dz} = -\frac{dw}{dx} - \frac{S}{G_2 b_c I}\int_{z_1}^{n} bzdz \quad . \quad . \quad . \quad (22)$$

Now $w$ is the displacement of a point in the section parallel to the axis of $z$, and since at a point having an ordinate $z$ we have a tensile stress

$$t_x = \frac{T_x}{A} + \frac{M_1 + S(a-x)}{I} \cdot z,$$

we shall have at this point a lateral contraction equal to

$$\frac{\Pi' t_x}{E} = \frac{\Pi'}{E}\left(\frac{T_x}{A} + \frac{M_1 + S(a-x)}{I} \cdot z\right),$$

and therefore $-w = \frac{\Pi'}{E}\left(\frac{T_x}{A}\int_0^z dz - \frac{M_1 + S(a-x)}{I}\int_0^z zdz\right).$

Effecting these integrations and then differentiating with regard to $x$, we get

$$\frac{dw}{dx} = \frac{\Pi' S}{EI} \cdot \frac{z^2}{2},$$

so that equation (22) becomes

$$\frac{du}{dz} = -\frac{\Pi' S}{EI}\cdot\frac{z^2}{2} - \frac{S}{G_2 bI}\int_{z_1}^{n} bz\,dz \quad . \quad . \quad (23)$$

This is the differential equation of the cylindrical surface assumed by the section after bending.

It will be noted that for $z = 0$ we obtain

$$\left(\frac{du}{dz}\right)_0 = -\frac{S}{G_2 b_0 I}\int_0^n bz\,dz,$$

i.e. by formula (21)

$$\left(\frac{du}{dz}\right)_0 = -\frac{s_{zx}^{\,0}}{G_2}.$$

Now $\frac{s_{zx}^{\,0}}{G_2}$ is the value of the central shear strain $\sigma_{zx}^{\,0}$, or the cosine of the angle (nearly a right angle) which the central element of the section makes with the axis of the body after bending; from this it follows that the plane of $(y, z)$, to which equation (23) is referred, is perpendicular to this axis.

**16. Internal Work for Bodies Defined in § 11. Formula giving the Coefficient F.**

We have seen that for bodies defined in § 11 the internal work of a slice comprised between two infinitely near sections is expressed by formula (14), where F represents a geometric coefficient depending only on the form of the section and on the ratio of the two glide moduli $G_2$, $G_3$. This coefficient is expressed by the formula

$$\frac{FA}{G_2} = \iint\left(\frac{F_1^2}{G_3} + \frac{F_2^2}{G_2}\right)dy\,dz.$$

Now it follows from what we have said in § 5 that if the shear force $S_Y$ and the twisting moment $T_M$ are zero, we have

$$s_{xy} = \frac{F_1 S_Z}{A}, \quad s_{zx} = \frac{F_2 S_Z}{A}.$$

For sections which are very narrow in the direction of the neutral axis, we have approximately

$$s_{xy} = 0, \quad s_{zx} = -\frac{S}{bI}\int_z^n bz\,dz,$$

and therefore

$$F_1 = 0, \quad F_2 = -\frac{A}{bI}\int_z^n bz\,dz.$$

Then substituting in the above formula, we get

$$\frac{FA}{G_2} = \frac{A^2}{G_2 I^2}\iint\left(\int_z^n bz\,dz\right)^2\frac{dy\,dz}{b^2}.$$

The first integration can be effected with regard to $y$, extending it over the whole straight line $C'C'' = b$; then the second integration should be extended over the whole section, i.e. from $-(d-n)$ to $n$, $d$ being the total depth of the section. Then dividing throughout by $\frac{A}{G_2}$ and changing $G_2$ to G, we get

$$F = \frac{A}{I^2}\int_{-(d-n)}^{n}\left(\int_z^n bz\,dz\right)^2\frac{dz}{b} \quad . \quad . \quad . \quad (24)$$

This is the general formula which we must use to determine the coefficient F.

**17. Application of Formulæ (20), (21), (23) and (24) to the Simplest and most Usual Sections.**

I. Rectangular Section.—Consider a rectangular section (Fig. 11, Pl. I) whose neutral axis is parallel to the side $b$ which is assumed to be quite small compared with the side $c$. We shall have

$$b \text{ constant for all values of } z;$$

$$n = d - n = \frac{c}{2}; \quad A = bc, \quad M_0 = \frac{bc^2}{8}; \quad I = \frac{bc^3}{12}.$$

Thus formula (20) becomes

$$s_{zx} = \frac{1}{2}\frac{S}{I}\left(\frac{c^2}{4} - z^2\right) = \frac{3}{2}\frac{S}{A}\left(1 - \frac{4z^2}{c^2}\right) \quad . \quad . \quad (25)$$

Formula (21) becomes

$$s_{zx}{}^0 = \frac{3S}{2A} \quad . \quad . \quad . \quad . \quad (26)$$

Formula (23) becomes

$$\frac{du}{dz} = -\frac{\Pi' S z^2}{EI\,.\,2} - \frac{S}{2G_2 I}\cdot\left(\frac{c^2}{4} - z^2\right),$$

from which, on integrating and noting that for $z - 0$ we must have $u = 0$, we find

$$u = -\frac{\Pi' S}{EI}\cdot\frac{z^3}{6} - \frac{S}{2G_2 I}\left(\frac{c^2 z}{4} - \frac{z^3}{3}\right)$$

i.e.

$$u = -\frac{3S}{2G_2 b}\left[\frac{z}{c} - \frac{4}{3}\left(1 - \frac{\Pi' G_2}{E}\right)\frac{z^3}{c^3}\right] \quad . \quad . \quad (27)$$

Finally formula (24) gives us

$$F = \frac{Ab}{I^2}\int_{-\frac{c}{2}}^{+\frac{c}{2}}\left(\int_z^{\frac{c}{2}} z\,dz\right)^2 dz,$$

from which we find, on effecting the integrations,

$$F = \frac{6}{5}* \quad . \quad . \quad . \quad . \quad . \quad (28)$$

* We can obtain the exact value of F for any value of $\frac{b}{c}$ by using the exact expressions for $s_{xy}$, $s_{zx}$ which we found in the problem of bending. We then have

II. Elliptic Section.—Let $b$ and $c$ (Fig. 12, Pl. I) be the semi-axes of an elliptic section, the first being taken as the neutral axis and being less in length than the second.

The equation of the contour will be

$$\frac{y^2}{b^2}+\frac{z^2}{c^2}=1,$$

and we shall have in equation (24)

$$b=2y=2b\sqrt{1-\frac{z^2}{c^2}},$$

$$n=d-n=c,\quad \mathrm{A}=\pi bc,\quad \mathrm{M}_0=\tfrac{2}{3}bc^2,\quad \mathrm{I}=\frac{\pi bc^3}{4}.$$

Thus we shall have

$$\int_z^c bzdz=2b\int_z^c\left(\sqrt{1-\frac{z^2}{c^2}}\right)zdz=\tfrac{2}{3}bc^2\left(1-\frac{z^2}{c^2}\right)^{\frac{3}{2}}=\frac{2c^2}{3b^2}\cdot y^3 \tag{29}$$

and therefore

$$s_{zx}=\frac{4}{3}\cdot\frac{y^2}{b^2}\frac{\mathrm{S}}{\mathrm{A}},\quad s_{zx}{}^0=\frac{4\mathrm{S}}{3\mathrm{A}} \quad . \quad . \quad . \tag{30}$$

Formula (23) becomes

$$\frac{du}{dz}=-\frac{\Pi'\mathrm{S}}{\mathrm{EI}}\frac{z^2}{2}-\frac{\mathrm{S}}{\mathrm{G}_2\mathrm{I}}\frac{c^2}{3}\left(1-\frac{z^2}{c^2}\right),$$

and integrating so that for $z=0$ we have $u=0$, we get

$$u=-\frac{4\mathrm{S}}{3\mathrm{G}_2\mathrm{A}}\left[z-\left(1-\frac{3\Pi'\mathrm{G}_2}{2\mathrm{E}}\right)\frac{z^3}{3c^2}\right] \quad . \quad . \tag{31}$$

Finally formula (24) gives us

$$\mathrm{F}=\frac{\mathrm{A}}{\mathrm{I}^2}\cdot\frac{2c^4}{9b^4}\int_{-c}^{+c}y^5dz$$

---

$$\mathrm{F}=\frac{6}{5}+\left(\frac{\Pi\mathrm{G}}{\mathrm{E}}\right)^2\frac{b^4}{c^4}\left(\frac{4}{5}-\frac{144b}{\pi^5c}\sum_{n=1}^{n=\infty}\frac{\tanh\frac{n\pi c}{b}}{n^3}+\frac{72b}{\pi^5c}\sum_{n=1}^{n=\infty}\frac{\tanh\frac{n\pi c}{b}}{n^5}\right)$$

assuming that $\mathrm{G}_2=\mathrm{G}_3=\mathrm{G}$.

As the ratio $\left(\frac{\Pi\mathrm{G}}{\mathrm{E}}\right)^2$ is generally small, we see that the value of F is very nearly $=\frac{6}{5}$, even when the side $b$ is not small compared with $c$. For isotropic bodies, for example, we have

$$\Pi=\tfrac{1}{4},\quad \mathrm{G}=\frac{2\mathrm{E}}{5}.$$

$$\therefore\left(\frac{\Pi\mathrm{G}}{\mathrm{E}}\right)^2=\frac{1}{100}$$

so that we may take $\mathrm{F}=\frac{6}{5}$ even if $b=c$.

whence, having regard to the symmetry of the figure about the axis of $y$

$$F = \frac{A}{I^2} \cdot \frac{4c^4}{9b^4}\int_0^c y^5 dz.$$

Now we have

$$dz = -\frac{c}{b^2} \cdot \frac{ydy}{\sqrt{1 - \frac{y^2}{b^2}}},$$

and as $z = 0$ for $y = b$ and $z = c$ for $y = 0$, it follows that

$$\int_0^c y^5 dz = -\frac{c}{b^2}\int_b^0 \frac{y^6 dy}{\sqrt{1 - \frac{y^2}{b^2}}} = \frac{1\,.\,3\,.\,5}{2\,.\,4\,.\,6}\pi c b^5$$

and therefore $F = \frac{10}{9}$.*

III. I AND [ SECTIONS.—Now let us consider the I section shown in Fig. 13, Pl. I, which is symmetrical about the neutral axis YY′. Taking the dimensions given we have at once

$$\left.\begin{aligned} A &= ab - (a - e)b_1 \\ M_1 &= \tfrac{1}{8}[ab^2 - (a - e)b_1^2] \\ I &= \tfrac{1}{12}[ab^3 - (a - e)b_1^3] \end{aligned}\right\} \quad . \quad . \quad . \quad (32)$$

In accordance with formula (20) the shear stress $s_{zx}$ for points in a line $mn$ drawn in the web parallel to the neutral axis will be given by the formula

$$s_{zx} = \frac{S}{eI}(M_1 - \tfrac{1}{2}ez^2) \quad . \quad . \quad . \quad . \quad (33)$$

and the maximum value of this stress will occur at the neutral axis, for which $z = 0$, and will be

$$\overset{0}{s}_{zx} = \frac{SM_1}{eI} \quad . \quad . \quad . \quad . \quad (34)$$

The total shear force acting on the web will be

$$\int_{-\frac{b_1}{2}}^{+\frac{b_1}{2}} s_{zx} e dz = \frac{S}{I}\int_{-\frac{b_1}{2}}^{+\frac{b_1}{2}}(M_1 - \tfrac{1}{2}ez^2)dz = \frac{S}{I}\left(M_1 b_1 - \frac{eb_1^3}{24}\right)$$

$$= \frac{S}{I}\left\{a\left(\frac{b^2 b_1 - b_1^3}{8}\right) + \frac{eb_1^3}{12}\right\}.$$

* For the elliptic as for the rectangular section we can find the value of F whatever be the ratio $\frac{b}{c}$ by means of the expressions for $s_{xy}$, $s_{zx}$ given for cylinders with elliptic section in the problem of bending; the formula which is found, and which we do not think it necessary to reproduce here, gives $A = \frac{10}{9}$ approximately, as we have obtained above, when the ratio $\frac{b}{c}$ is very small.

Moreover, we know that the total shear force over the whole section must be equal to S, so that the shear forces acting over the two flanges will be

$$S - \frac{S}{I}\left\{\frac{a(b^2b_1 - b_1^3)}{8} + \frac{eb_1^3}{12}\right\} = \frac{S}{I}\left\{I - \frac{a(b^2b_1 - b_1^3)}{8} - \frac{eb_1^3}{12}\right\}$$

$$= \frac{S}{I} \cdot \frac{a(b - b_1)^2(2b + b_1)}{24} \qquad (35)$$

putting in the above value for I and reducing.

Now the moment of inertia of the whole section is greater than that which we should obtain by neglecting the web, i.e. greater than

$$I' = \tfrac{1}{12}a(b^3 - b_1^3) = \tfrac{1}{12}a(b - b_1)(b^2 + bb_1 + b_1^2),$$

and as we clearly have $b^2 + bb_1 + b_1^2 > 2bb_1 + b_1^2$, it follows that by putting $I'' = \frac{1}{12}a(b - b_1)(2bb_1 + b_1^2)$ we shall have

$$I > I' > I''.$$

Therefore formula (35) will have a value which is certainly less than

$$\frac{S}{I''}\frac{a(b - b_1)^2(2b + b_1)}{24} = S \cdot \frac{(b - b_1)}{2b_1}.$$

Thus if the flanges are very thin compared with the depth of the section, as is nearly always the case in practice, the ratio $\frac{(b - b_1)}{2b_1}$, will be very small; i.e. the total shear force carried by the flanges will be very small, and may be neglected in comparison with that carried by the web.

Formula (23) becomes between $z = -\frac{b_1}{2}$ and $z = +\frac{b_1}{2}$

$$\frac{du}{dz} = -\frac{\Pi' S z^2}{EI \cdot 2} - \frac{S}{G_2 e I}(M_0 - \tfrac{1}{2}ez^2),$$

and on integrating so that $u = 0$ for $z = 0$, we get

$$u = -\frac{S}{G_2 I}\left[\frac{M_0}{e} \cdot z - \left(1 - \frac{\Pi' G_2}{E}\right)\frac{z^3}{6}\right] \quad . \quad . \quad (36)$$

Now instead of applying formula (24) to the I section, we will endeavour to find directly, for the sake of greater clearness, the expression for the internal work due to the shear stresses.

We have seen that for a very thin slice this expression is for all forms of section

$$dW_i = \frac{ds}{2}\iint\left(\frac{s_{xy}^2}{G_3} + \frac{s_{zx}^2}{G_2}\right)dydz.$$

For the I section we may take not only $s_{zy} = 0$ for all points in the section, but also, from what we have proved above, $s_{xy} = 0$

for all points in the flanges, so that the above expression may be written

$$dW_i = \frac{ds}{2G_2}\iint s_{zx}^2 dydz,$$

the integration being extended over the web of the section only. Performing this integration with regard to $y$ and substituting for $s_{zx}$ its value (33), the formula becomes

$$dW_i = \frac{ds}{2G_2}\cdot\frac{S^2}{eI^2}\int_{-\frac{b_1}{2}}^{+\frac{b_1}{2}}(M_0 - \tfrac{1}{2}ez^2)dz$$

$$= \frac{ds}{2G_2}\cdot\frac{S^2 b_1}{eI^2}(M_0^2 - \tfrac{1}{12}M_0 e b_1^2 + \tfrac{1}{320}e^2 b_1^4),$$

(performing the integration with regard to $z$ also). We can write this in the form

$$dW_i = \frac{QS^2}{2G_2A}ds$$

$$\text{where } Q = \frac{Ab_1}{I^2 e}(M_0^2 - \tfrac{1}{12}M_0 e b_1^2 + \tfrac{1}{320}e^2 b_1^4) \quad . \quad (37)$$

Now $M_0^2 - \frac{1}{12}M_0 e b_1^2 + \frac{1}{320}e^2 b_1^4 = (M_0 - \frac{1}{24}e b_1^2)^2 + \frac{1}{720}e^2 b_1^4$; and as $\frac{1}{720}e^2 b_1^4$ will be comparatively negligible in practice, we have

$$Q = \frac{Ab_1}{I^2 e}\left(M_0 - \frac{e b_1^2}{24}\right)^2 \quad . \quad . \quad . \quad (38)$$

It will be noted that in the present paragraph we have never imposed the condition that the web comes in the centre of the flanges, so that the formulæ apply also to the channel section of Fig. 14.

# CHAPTER V.

## THEORY OF LATTICE GIRDERS.

**1. Definitions.**—We will use the term *lattice girder* to denote a girder composed of two members of **T** section joined to each other by inclined bars as shown in Fig. 2, Pl. II.

The two members of **T** section which are thus connected are called *flanges*. The bars are usually riveted to the flanges and to each other at the points where they cross; thus the lattice girders form rigid structures, i.e. structures subjected only to the deformations due to the elasticity of the materials.

The accurate determination of the stresses which occur in the different members of a lattice girder, and in the different portions of each member, could be effected by means of the theorem of the differential coefficients of internal work, regarding the lattice girders as composite structures. But this method, although leading to equations of the first order, generally results in such a large number of equations and unknowns, that its solution becomes very tedious.

We must, then, be satisfied with an approximate solution, which will in general be more than sufficient in practice.

To explain this solution in a logical manner, we will examine certain theoretical cases.

**2. First Case of a Pin-Jointed Lattice Girder.**—Let us consider the girder represented in Fig. 15, Pl. I, composed of bars pin-jointed at all the nodes, so that DE, EF, CG, GH, etc., are bars pin-jointed to each other and to the diagonal bars.

Suppose that at the point A two forces are applied; one, P, vertical, and the other, $P'' - P'$, acting from A towards D; and at the point B two other forces, of which one, P, is vertical and the other, $P'' + P'$, acts from B towards C.

Now let us imagine that the girder is cut through along the line IL and that forces $T_1$, $T_2$, $T_3$, $T_4$ are applied to the bars cut through, these forces being equal to the forces or stresses to which the bars are subjected. For equilibrium of the portion of the girder to the right of the section IL, we must express the conditions that the sums of the horizontal and vertical forces are zero, and that the sum of the moments about any point is also zero; this gives the following three equations:—

$$\left.\begin{aligned} &T_1 + T_2 + (T_3 + T_4)\cos a + 2P'' = 0 \\ &(T_3 - T_4)\sin a + 2P = 0 \\ &(T_1 - T_2)\frac{h}{2} - 2P\left(x + \frac{h}{2\tan a}\right) - P'h = 0 \end{aligned}\right\} \quad . \quad (1)$$

These three equations are not sufficient to determine the four unknowns $T_1$, $T_2$, $T_3$, $T_4$. But if we suppose that a section NO is taken, and that forces $T_5$, $T_6$ are applied to it equal to the stresses in the bars cut through, it is clear that the four stresses $T_1$, $T_3$, $T_5$, $T_6$ must equilibrate each other; this necessitates that, among other things, the sum of the vertical components of the four stresses shall be zero, i.e. that

$$(T_3 + T_6)\sin a = 0$$

or

$$T_3 = -T_6.$$

By applying similar reasoning to each of the nodes, we see clearly that all the diagonal bars are subjected to equal stresses, but that these stresses are tensile for bars parallel to DG and compressive for those parallel to CE.

It follows that $T_4$ will be a tension, as we have assumed, and $T_3$ a thrust, and that we have

$$T_3 = -T_4.$$

This equation and the previous ones give us all that are required for the determination of the unknowns, and we thus have

$$\left.\begin{aligned} T_1 &= \frac{2P}{h}\left(x + \frac{h}{2\tan a}\right) + P' - P'' \\ T_2 &= -\frac{2P}{h}\left(x + \frac{h}{2\tan a}\right) - P' - P'' \\ T_4 &= \frac{P}{\sin a} \\ T_3 &= -\frac{P}{\sin a} \end{aligned}\right\} \quad . \quad . \quad (2)$$

It will be seen that if we call B the bending moment and S the shearing force about the section passing through the point M, we shall have

$$B = 2P\left(x + \frac{h}{2\tan a}\right) + P'h \quad . \quad . \quad . \quad (3)$$

$$S = 2P$$

and therefore

$$\left.\begin{aligned} T_1 &= \frac{B}{h} - P'' \\ T_2 &= -\frac{B}{h} - P'' \end{aligned}\right\} \quad . \quad . \quad . \quad . \quad . \quad (4)$$

$$T_3 = -T_4 = -\frac{S}{2 \sin a} \quad . \quad . \quad . \quad . \quad (5)$$

Moreover, if A is the cross-sectional area of the bars having the stresses $T_1$, $T_2$, the intensity of stress will be given by

$$t_1 = \frac{T_1}{A} = \frac{B}{Ah} - \frac{P''}{A},$$

$$t_2 = \frac{T_2}{A} = -\frac{B}{Ah} - \frac{P''}{A},$$

and as the moment of inertia of the section composed of the two bars is given by

$$I = \frac{Ah^2}{2},$$

the above formulæ become

$$\left.\begin{aligned} t_1 &= \frac{B\frac{h}{2}}{I} - \frac{2P''}{2A} \\ t_2 &= -\frac{B\frac{h}{2}}{I} - \frac{2P''}{2A} \end{aligned}\right\} \quad . \quad . \quad . \quad . \quad . \quad (6)$$

i.e. they are exactly similar to those which give the distribution of stress in a solid beam.

Further, the sum of the normal components of the two stresses $T_3$, $T_4$ is

$$(T_4 - T_3) \sin a = S.$$

On account of these analogous results we may say, although only approximately, that *in the lattice girders considered above the horizontal bars resist the bending moment and normal thrust and the diagonal bars resist the shearing force.*

**3. Second Case of a Pin-Jointed Lattice Girder.**—Now let us consider the girder represented in Fig. 16, Pl. I, which is composed of a series of bars, DE, EF, FG, etc., CH, HK, etc., pin-jointed to each other and to the diagonal bars. We will suppose that the height AB of the girder is divided into three parts, and that the diagonal bars are arranged as shown in the figure; vertical forces equal to $\frac{2P}{6}$ will be assumed as acting at the points A, B, and vertical forces equal to $\frac{2P}{3}$ at the points L, N, the sum of the vertical forces at the end of the girder being thus 2P.

We will also suppose that at the points A, B horizontal forces $P'' - P'$ and $P'' + P'$ act from A to D and from B to C.

If we resolve each of the four forces applied at the end ALNB of the girder in the directions of the two bars which intersect at its point of application, we see that the diagonal bars from the points A, L, N parallel to CG are subjected to a compressive stress equal to $\frac{2P}{6 \sin a}$, while the bars from the points L, N, B parallel to PI are subjected to a tensile stress also equal to $\frac{2P}{6 \sin a}$.

Moreover, if we consider the two diagonal bars meeting at any node, for example at G, we see, by the same reasoning as that employed in the previous paragraph, that the sum of the vertical components of the stresses in these bars must be zero; from this we deduce that one of the bars will be in compression and the other in tension, the two stresses being numerically equal. Passing thus from one node to another we see that all bars parallel to CG are subjected to a compressive stress equal to $\frac{2P}{6 \sin a}$, and all bars parallel to DI to a tensile stress also equal to $\frac{2P}{6 \sin a}$.

Now if we suppose the girder to be cut through at the section RU, it is clear that the portion of the girder to the right of this section must remain in equilibrium on applying to the bars cut through tensions or compressions equal to those which occur in the structure. But, among the six diagonal bars cut through, the three parallel to CG are subjected to a thrust $\frac{2P}{6 \sin a}$, and the other three to a tension $\frac{2P}{6 \sin a}$; thus the sum of the vertical components of these six forces will be exactly 2P, and the sum of the horizontal components will be zero. Therefore, of the three equations for the equilibrium of the section ARUB, the first, which expresses the condition that the sum of the vertical components of all the forces is zero, is satisfied; the second, expressing the condition that the sum of the horizontal components is zero, will be simply

$$T_1 + T_2 + 2P'' = 0;$$

and the third, expressing the condition that the sum of the moments about any point, e.g. V, is zero, will be

$$(T_1 - T_2)\frac{h}{2} - 2Px - P'h = 0$$

or

$$T_1 - T_2 - \frac{2B}{h} = 0,$$

B being the bending moment $2Px + P'h$.

From these two equations we deduce that

$$T_1 = \frac{B}{h} - P''$$

$$T_2 = -\frac{B}{h} - P''.$$

These are exactly the same as those which we found for the girder considered previously, and lead us to the same conclusion that the horizontal bars resist the bending moment whilst the diagonal bars resist the shear.

4. **Third Case of a Pin-Jointed Lattice Girder.**—In the above two cases of pin-jointed lattice girders, the ordinary static equations were sufficient to determine the stresses in all the members, so that there was no need to consider their elasticity.

We will now consider another case in which the number of bars is more than is strictly necessary to form a firm frame, i.e. in which the number of unknown stresses is greater than the number of equations obtained from static considerations.

We will suppose then that we have a girder like that considered in § 2, but in which there are vertical bars or posts as shown in Fig. 1, Pl. II.

Taking the same forces acting at the end AB as in the first case and considering a section at the line IL, we have applied to the bars cut through forces $T_1$, $T_2$, $T_3$, $T_4$, equal to the stresses which they carry. For equilibrium of the portion of the girder to the right of IL, we shall have the three expressions (1) of the first case, from which we deduce, using the value (3) of the bending moment B,

$$\left.\begin{aligned} T_1 &= -T_4\cos a + \frac{P}{\tan a} + \frac{B}{h} - P'' \\ T_2 &= -T_4\cos a + \frac{P}{\tan a} - \frac{B}{h} - P'' \\ T_3 &= T_4 - \frac{2P}{\sin a} \end{aligned}\right\} \quad . \quad . \quad (7)$$

Now if the posts were absolutely rigid bars, i.e. incapable of extension or compression under the action of the forces to which they are subjected, it is clear that their internal work would be zero. Therefore, taking A = cross-sectional area, $l$ = length, and E the elastic modulus of the horizontal bars carrying stresses $T_1$, $T_2$; $A_1$, $l_1$, $E_1$ the corresponding quantities for the diagonal bars carrying stresses $T_3$, $T_4$, we shall have for the internal work of the whole structure

$$W_i = \frac{1}{2}\sum\left\{\left(\frac{T_1^2 + T_2^2}{\epsilon}\right) + \left(\frac{T_3^2 + T_4^2}{\epsilon_1}\right)\right\},$$

where

$$\epsilon = \frac{AE}{l}, \quad \epsilon_1 = \frac{A_1E_1}{l_1}.$$

By means of the three formulæ (7) we can eliminate the stresses $T_1$, $T_2$, $T_3$ from the expression for the internal work, so that it becomes a function of the stresses in the bars parallel to DG only. To obtain these stresses we must, in accordance with the theorem of the differential coefficients of internal work, equate to zero the differential coefficients of the internal work with regard to these same stresses. But we see that the term

$$\frac{T_1^2 + T_2^2}{2\epsilon} + \frac{T_3^2 + T_4^2}{2\epsilon_1}$$

will, after elimination of $T_1$, $T_2$, $T_3$, contain the unknown $T_4$ only, and each similar term will reduce to a function containing the stress in one only of the bars parallel to DG. Thus, on equating to zero the differential coefficient with regard to $T_4$, we shall obtain the equation

$$\frac{T_1\frac{dT_1}{dT_4} + T_2\frac{dT_2}{dT_4}}{\epsilon} + \frac{T_3\frac{dT_3}{dT_4} + T_4}{\epsilon_1} = 0;$$

and noting that $\frac{dT_1}{dT_4} = \frac{dT_2}{dT_4} = -\cos a, \quad \frac{dT_3}{dT_4} = 1,$

this becomes $$-\frac{(T_1 + T_2)\cos a}{\epsilon} + \frac{T_3 + T_4}{\epsilon_1} = 0 \qquad (8)$$

This equation, combined with the three others given above, completes the number required to determine the four unknowns. We thus obtain

$$\left.\begin{aligned}
T_1 &= \frac{B}{h} - \frac{P''\frac{\epsilon}{\epsilon_1}}{\cos^2 a + \frac{\epsilon}{\epsilon_1}} \\
T_2 &= -\frac{B}{h} - \frac{P''\frac{\epsilon}{\epsilon_1}}{\cos^2 a + \frac{\epsilon}{\epsilon_1}} \\
T_3 &= -\frac{P}{\sin a} - \frac{P''\cos a}{\cos^2 a + \frac{\epsilon}{\epsilon_1}} \\
T_4 &= \frac{P}{\sin a} - \frac{P''\cos a}{\cos^2 a + \frac{\epsilon}{\epsilon_1}}
\end{aligned}\right\} \qquad (9)$$

We see then, on comparing these results with those which we obtained in the first case, that the provision of absolutely rigid posts diminishes by $\dfrac{P'' \cos^2 a}{\cos^2 a + \dfrac{\epsilon}{\epsilon_1}}$ the stress in all the horizontal bars, and by $\dfrac{P'' \cos a}{\cos^2 a + \dfrac{\epsilon}{\epsilon_1}}$ the stress in all the diagonals.

If we call $T_5$ the stress in a post, and if we express the equilibrium of the stresses meeting at any node, we shall have, among others, the equation

$$(T_3 + T_4) \sin a + T_5 = 0,$$

and therefore

$$T_5 = \frac{2P'' \sin a \cos a}{\cos^2 a + \dfrac{\epsilon}{\epsilon_1}} \quad . \quad . \quad . \quad (10)$$

This is the stress in all the posts.

We see that if the end of the girder were subjected only to the force 2P and to the two forces P′, P″ being zero, the stress in the posts would be zero, and the horizontal and diagonal bars would carry exactly the same stresses as if the posts were absent.

**5. Ordinary Lattice Girder.**—Now let us consider an ordinary lattice girder (Fig. 2, Pl. II), i.e. one with continuous flanges, the diagonals being attached to the flanges by rivets instead of by pin-joints.

If the flanges and diagonals are small, as is most usually the case, each of them will have a very small moment of inertia compared with that of the figure made up of the sections of the two flanges; so that we shall depart very slightly from the truth by assuming that the moment of inertia of each flange and diagonal is zero, this being equivalent to regarding the girder as a pin-jointed frame and bringing us back, therefore, to the case which we have considered above.

We shall thus assume that the two flanges are reduced to two series of bars arranged in two horizontal lines and then first calculate the distance apart that these two lines must be taken, i.e. the height of the theoretical frame which we substitute for the given girder.

Now by analogy with what we have found for solid beams we will assume that in the flanges the normal stress $t_x$ is variable from one point to another in accordance with a linear function of the distance of the point from the neutral axis; moreover, if for a transverse section we call

P = normal pressure,
B = bending moment,
S = shearing force,
A = area, and I = moment of inertia of the figure formed by the sections of the two flanges,
$z$ = distance of a point in the section from the neutral axis,

we will assume that the intensity of compressive stress

$$c_x = \frac{P}{A} + \frac{Bz}{I} \quad . \quad . \quad . \quad . \quad (11)$$

which is the same formula as we have employed for solid beams.

This formula is based upon the assumption that after strain transverse sections of the girder are still plane and perpendicular to the centre line, which cannot be far from true.

On this assumption we will consider the system of parallel stresses $c_x$ applied to all the points in the section of a flange; the resultant of this system will be at a distance from the neutral axis equal to

$$\frac{\Sigma\left(\frac{P}{A} + \frac{Bz}{I}\right)z\,dA}{\Sigma\left(\frac{P}{A} + \frac{Bz}{I}\right)dA}$$

the summations being extended over the whole area of a flange. But we have

$$\Sigma dA = \frac{A}{2}, \quad \Sigma z^2 d\,A = \frac{I}{2};$$

and if we call

$$\Sigma z\,dA = M_0 . \quad . \quad . \quad . \quad . \quad (12)$$

and multiply the above formula by 2, we have for the distance between the stress centres of the two flanges the formula

$$h = 2 \times \frac{\frac{PM_0}{A} + \frac{B}{2}}{\frac{P}{2} + \frac{BM_0}{I}} . \quad . \quad . \quad . \quad . \quad (13)$$

If the normal pressure P = 0, this formula will give

$$h = \frac{I}{M_0} \quad . \quad . \quad . \quad . \quad . \quad (14)$$

and if, on the other hand, the bending moment B = 0 we shall have

$$h = \frac{4M_0}{A} . \quad . \quad . \quad . \quad . \quad (15)$$

This last result is the distance between the centroids of the flanges. From another point of view, we can see that the first result $\frac{I}{M_1}$ is always greater than $\frac{4M_1}{A}$, but less than the total depth of the girder. Thus, since for flanges of T section the distance between their centroids differs very little from the total depth of the girder, it follows that the two extreme results given above for the value of $h$ may be regarded with a close approximation as equal, and that we may therefore take

$$h = \frac{I}{M_0},$$

whatever be the ratio $\frac{B}{P}$.

Now the lattice bars form lozenge-shaped figures of which we will call

$d$ the diagonal length perpendicular to the centre line of the girder,

$d'$ the diagonal length parallel to the centre line of the girder,

$a$ the angle which the bars make with the centre line.

We will assume that the stresses in the girder will be changed very little if we substitute for the system of diagonal bars another system with bars parallel to the first but slightly closer together or further apart, provided that their section is altered in the inverse ratio to the distances so that the volume of metal in the diagonals remains the same.

With these assumptions we will suppose that on dividing the theoretical height of the girder by the diagonal $d$ we obtain a whole number $k$ plus a fraction $i$; then if we take the new imaginary system of diagonals as having, corresponding to $d$ and $d'$, lengths $\frac{d(k+i)}{k}$ and $\frac{d'(k+i)}{k}$, it is clear that the new length $\frac{d(k+i)}{k}$ will be exactly $\frac{h}{k}$, so that in the second system the bars may be attached in pairs to the same point in the lines representing the flanges. We shall then have a framed girder similar to that considered in § 3, and therefore the stress in each diagonal will be

$$\frac{S}{2k \sin a}.$$

As the bars actually employed in the girder are closer than those in the imaginary structure in the ratio $\frac{k}{k+i}$, it follows that the stress in them will be

$$\frac{S}{2k \sin a} \times \frac{k}{k+i} = \frac{S}{2(k+i) \sin a}$$

but we have

$$k + i = \frac{h}{d} = \frac{I}{dM_0}$$

$$\sin a = \frac{d}{\sqrt{d^2 + d'^2}},$$

so that the stresses in the diagonal bars will be given by the expression

$$\frac{M_0}{I} \times \frac{S\sqrt{d^2 + d'^2}}{2} \quad . \quad . \quad . \quad . \quad (16)$$

and if $A_d$ is the cross-sectional area of the bars, their stress per square metre will be

$$\frac{M_0}{I} \times \frac{S\sqrt{d^2 + d'^2}}{2A_d} \quad . \quad . \quad . \quad . \quad (17)$$

**6. Internal Work of Lattice Girders.**—The internal work of lattice girders is composed of two parts, one for the flanges and the other for the diagonals.

For an infinitely thin slice of thickness $ds$, the first is given by

$$dW_{i_f} = \frac{ds}{2E}\iint C_X^2 dy dz,$$

the double integral being extended over the whole section of the two flanges. Substituting

$$C_X = \frac{P}{A} + \frac{Bz}{I},$$

we obtain, as for solid beams,

$$dW_{i_f} = \frac{ds}{2E}\left(\frac{P^2}{A} + \frac{B^2}{I}\right).$$

To express in general terms the internal work due to the diagonal bars, we will take for their length the theoretical length $\frac{h}{\sin a}$, so that, calling $T_a$ the stress in any diagonal bar, $A_a$ its cross-sectional area, and $E'$ its elastic modulus, its internal work will be given by the expression

$$\frac{T_d^2 \dfrac{h}{\sin a}}{2E'A};$$

and since in a length of the girder equal to the horizontal length $d'$ there are two lengths of diagonal bars, it follows that the internal work due to the diagonal bars for each unit length of the girder will be equal to

$$\frac{2}{d'} \times \frac{T_d^2 h}{2E'A_d \sin a}.$$

But we have found that

$$T_d = \frac{M_0}{I} \times \frac{S\sqrt{d^2 + d'^2}}{2}; \quad h = \frac{I}{M_0}; \quad \sin a = \frac{d}{\sqrt{d^2 + d'^2}}$$

so that the above expression becomes equal to

$$\frac{S^2}{2E'A_d} \times \frac{M_0}{I} \cdot \frac{(d^2 + d'^2)^{\frac{3}{2}}}{2dd'}.$$

Then putting

$$F = \frac{M_0}{I} \cdot \frac{(d^2 + d'^2)^{\frac{3}{2}}}{2dd'},$$

we shall have for the internal work due to the diagonals of a length $ds$ of the girder

$$dW_{i_d} = \frac{F \cdot S^2}{2E'A_d} \cdot ds,$$

and the total internal work of the element

$$= dW_i = dW_{i_f} + dW_{i_d} = \frac{ds}{2E}\left(\frac{P^2}{A} + \frac{B^2}{I}\right) + \frac{FS^2}{2E'A_d} \cdot ds \qquad (18)$$

When the diagonals are inclined at 45°, $d = d'$ and we have

$$F = \frac{M_0}{I} d\sqrt{2}.$$

We have obtained formulæ for lattice girders by considering a girder subjected to forces at its end only, the stresses in all the diagonals being the same. But, as in the case of solid beams, the formulæ obtained for this special case can be extended to all lattice girders, whatever be the distribution of the load.

We will only note that the stress in any diagonal bar cannot change between two points, if in this interval it is not connected to any other bar; in the same way the magnitude of the normal stresses C must be regarded as the same throughout a length of flange between two nodes. If, however, the distance between the diagonals is a small fraction of the height of the girder and this height is very small compared with the length, we may regard the stresses in the flanges as varying in a continuous manner. For the diagonal bars we suppose the girder to be divided into lengths equal to the horizontal diagonal of the lozenge-shaped figures by means of transverse sections through the nodes, and we consider the stresses in all the diagonals between these sections as constant. But in the application of the formula

$$\frac{FS^2}{2E'A_d} \cdot ds,$$

when we have to integrate for the whole length of the girder, we shall find almost exactly the same result by considering the shearing force as varying continuously.

We will add that formula (18) differs from formula (14) of the previous chapter for solid beams only by the changing of the glide or shear modulus G into the modulus E′ of the diagonal bars, and by the substitution of the area of a diagonal bar for the area of the section of the beam. In the following chapters, therefore, we shall always use the formulæ for solid beams, but it is to be understood that all the results which we shall obtain will be applicable also to lattice girders by making the changes indicated above.

## CHAPTER VI.

### FORMULÆ FOR THE INTERNAL WORK OF DIFFERENT SOLID BODIES.

**1. Internal Work of a Cable or of a Prismatic Bar Pin-Jointed at its Ends.**—If we call

$l$ = the length of cable or bar,
A = its cross-sectional area,
E = elastic modulus (Young's modulus),
T = tension,

the internal work is given by the formula

$$W_i = \frac{1}{2}\frac{T^2 l}{EA} \quad . \quad . \quad . \quad . \quad (1)$$

**2. Internal work of a prismatic body loaded uniformly along its whole length and loaded in any manner at its ends** (Fig. 3, Pl. II).

Let ABC be the centre line of the body and suppose it to be subjected to the following forces:—

(*a*) At the section C to normal and tangential forces giving a normal pressure $P_0$, a shear $S_0$ along one of the principal axes of the section, and a bending moment $B_0$ about the other principal axis.

(*b*) To a uniformly distributed load along its whole length in the plane through the centre line of the body parallel to the shearing force $S_0$, the load per unit length having a normal component $w_n$ and an axial component $w_a$ acting from C towards A.

(*c*) At the section A to a normal thrust $P_2$, shearing force $S_2$, and bending moment $B_2$.

At a section D at distance $x$ from the end C we shall have

$$\text{Bending moment} = B = B_0 + S_0 x + \frac{w_n x^2}{2}.$$

$$\text{Normal pressure} = P = P_0 + w_a x.$$

$$\text{Shearing force} = S = S_0 + w_n x.$$

And, if $2l$ = length of the body,
$A$ = area of cross-section,
$I$ = moment of inertia of section,
$E$ = elastic (Young's) modulus,
$G$ = Glide or shear modulus,
we shall have for the internal work of the prism

$$W_i = \frac{1}{2EI}\int_0^{2l} B^2dx + \frac{1}{2EA}\int_0^{2l} P^2dx + \frac{F}{2GA}\int_0^{2l} S^2dx.$$

We can substitute for B, P, S their values in terms of $x$ given above and perform the integrations. But we should first note that if in the expression for B we put $x = 2l$ we must obtain $B_2$; i.e.

$$B_2 = B_0 + 2lS_0 + 2w_nl^2,$$

so that we can eliminate $S_0$ from the expression for B, thus obtaining

$$B = B_0 + \left(\frac{B_2 - B_0}{2l} - w_nl\right)x + \frac{w_nx^2}{2} \quad . \quad . \quad (2)$$

Further, we see that we have

$$P_2 = P_0 + 2w_al$$
$$S_2 = S_0 + 2w_nl.$$

Making use of formula (2) and of these values of $P_2$, $S_2$ and integrating and simplifying, we have the following formula :—

$$\left.\begin{aligned} W_i = \frac{2l}{2EI}&\left\{\frac{(B_0^2 + B_0B_2 + B_2^2)}{3} - \frac{(B_0 + B_2)w_nl^2}{3} + \frac{2w_n^2l^4}{15}\right\} \\ &+ \frac{2l}{2EA}\left(\frac{P_0^2 + P_2^2}{2} - \frac{2}{3}w_a^2l^2\right) + \frac{2Fl}{2GA}\left(\frac{S_0^2 + S_2^2}{2} - \frac{2}{3}w_n^2l^2\right)\end{aligned}\right\} \quad (3)$$

We can express this in another form by introducing the bending moment $B_1$, pressure $P_1$, and shear $S_1$ at the centre of the beam; for if in the general expression for B, P, and S we put $x = l$, we get

$$B_1 = \frac{B_0 + B_2}{2} - \frac{w_nl^2}{2}$$
$$P_1 = P_0 + w_al$$
$$S_1 = S_0 + w_nl.$$

Thus $B_0^2 + 2B_0B_2 + B_2^2 - 2(B_0 + B_2)w_nl^2 + w_n^2l^4 = 4B_1^2$, so that, adding $B_0^2 + B_2^2 - \frac{1}{5}w_n^2l^4$ and dividing by 3, we get

$$2\left\{\frac{(B_0^2 + B_0B_2 + B_2^2)}{3} - \frac{(B_0 + B_2)w_nl^2}{3} + \frac{2w_n^2l^4}{15}\right\} = \frac{B_0^2 + 4B_1^2 + B_2^2}{3} - \frac{w_n^2l^4}{15}.$$

We also have
$$P_1^2 = P_0^2 + 2P_0 w_a l + w_a^2 l^2$$
$$P_2 = P_0 + 2w_a l,$$
$$P_2^2 = P_0^2 + 4P_0 w_a l + 4w_a^2 l^2,$$
whence it follows that $P_2^2 - 2P_1^2 = -P_0^2 + 2w_a^2 l^2$

i.e. rearranging and multiplying by 2,

$$2(P_0^2 + P_2^2) - 4w_a^2 l^2 = 4P_1^2.$$

By adding $P_0^2 + P_2^2$ to each side and dividing by 3 we get

$$2\left(\frac{P_0^2 + P_2^2}{2} - \frac{2}{3}w_a^2 l^2\right) = \frac{P_0^2 + 4P_1^2 + P_2^2}{3}$$

and similarly

$$2\left(\frac{S_0^2 + S_2^2}{2} - \frac{2}{3}w_n^2 l^2\right) = \frac{S_0^2 + 4S_1^2 + S_2^2}{3}.$$

Putting these results into equation (3), we get

$$\left.\begin{aligned} W_i = \frac{l}{2EI}\left(\frac{B_0^2 + 4B_1^2 + B_2^2}{3} - \frac{w_n^2 l^4}{15}\right) \\ + \frac{l}{2EA}\left(\frac{P_0^2 + 4P_1^2 + P_2^2}{3}\right) + \frac{Fl}{2GA}\left(\frac{S_0^2 + 4S_1^2 + S_2^2}{3}\right) \end{aligned}\right\} \quad . \qquad (4)$$

**3. General Note.**—Since, in the use of formulæ for the internal work of solid bodies, we shall always have to take differential coefficients with regard to forces applied at given points, or with regard to bending moments occurring at given sections, it is clear that these unknown forces or bending moments enter only into the expressions for $B_0$, $B_1$, $B_2$; $P_0$, $P_1$, $P_2$; $S_0$, $S_1$, $S_2$; consequently the terms contained in the expression for the internal work which are independent of these quantities will necessarily disappear on differentiating, and might, therefore, be omitted in expressions (3) and (4).

These expressions may therefore be written as follows:—
(3) becomes

$$\left.\begin{aligned} \frac{2l}{2EI}\left\{\frac{(B_0^2 + B_0B_2 + B_2^2)}{3} - \frac{(B_0 + B_2)w_n l^2}{3}\right\} + \frac{l}{2EA}(P_0^2 + P_2^2) \\ + \frac{Fl}{2GA}(S_0^2 + S_2^2) \end{aligned}\right\} \qquad (5)$$

and (4) becomes

$$\frac{l}{2EI}\left(\frac{B_0^2 + 4B_1^2 + B_2^2}{3}\right) + \frac{l}{2EA}\left(\frac{P_0^2 + 4P_1^2 + P_2^2}{3}\right)$$
$$+ \frac{Fl}{2GA}\left(\frac{S_0^2 + 4S_1^2 + S_2^2}{3}\right) \qquad (6)$$

**4. Internal work of a solid body of constant section, loaded in any manner at its ends and loaded uniformly along its length, when the centre line is bent to a circular arc having a very small rise in proportion to its chord** (Fig. 4, Pl. II).

Let ABC be the centre line of the body, contained in a vertical plane, $a$ being the angle which the chord makes with the horizontal, and let the forces applied to the body be as follows:—

(*a*) On the section A forces equivalent to resultants $Q_0$, $R_0$, one parallel to the chord and the other at right angles to it in the plane of the arc, and to a couple in this plane having a moment $B_0$.

(*b*) Along the whole centre line a uniformly distributed load of $w$ kilograms per metre run.

(*c*) On the section C forces equivalent to resultants $Q_2$, $R_2$ as in (*a*) and to a couple of moment $B_2$.

Between these forces there must be the three characteristic equations for the equilibrium of a system of forces in one plane applied to a free body.

Now let us find expressions for the bending moment, normal pressure, and shear at any section of the body.

If we take A as origin, the chord AC as the axis of $x$, and the perpendicular to it as the axis of $y$, the equation to the centre line of radius R will be

$$R - (r - y) = \sqrt{R^2 - (l' - x)^2},$$

$r$ being the rise of the arc and $2l'$ the length of the chord.

But since the rise is very small compared with the chord, it will be very small compared with the radius; we can therefore write, neglecting terms beyond the second,

$$\sqrt{R^2 - (l' - x)^2} = R - \frac{(l' - x)^2}{2R},$$

and therefore
$$y = r - \frac{(l' - x)^2}{2R}.$$

Now $2R = \dfrac{l'^2 + r^2}{r}$, and neglecting $r^2$ in comparison with $l'^2$ we get

$$y = \frac{r}{l'^2}(2l'x - x^2) \quad . \quad . \quad . \quad . \quad (7)$$

Now consider a section E, the arc AE being of length $l$ and $\beta$ being the angle which the normal to the curve at this section makes with that at the mid-point B.

An element of arc $nn' = dl$ will be loaded with $wdl$ which can be resolved into two forces, one $wdl \cos a$ perpendicular to the chord AC and the other $wdl \sin a$ parallel to it. We will write $w \cos a = w_1$ and $w \sin a = w_2$.

Thus if $x'$, $y'$ are the co-ordinates of the point $n$ we shall have

$$B = B_0 + Q_0 y - R_0 x + \int_0^x w_1(x - x')dl - \int_0^x w_2(y - y')dl$$
$$P = Q_0 \cos \beta + R_0 \sin \beta - ws \sin (a + \beta)$$
$$S = Q_0 \sin \beta - R_0 \cos \beta + ws \cos (a + \beta).$$

Now since $\beta$ is the angle which the normal at E makes with the axis of $y$, we have

$$\tan \beta = \frac{dy}{dx} = \frac{2r(l' - x)}{l'^2};$$

and since $r$ is very small compared with $l'$, $\tan \beta$ will always be so small that its square may be neglected in comparison with unity. Therefore in place of

$$dl = dx\sqrt{1 + \tan^2 \beta}, \sin \beta = \frac{\tan \beta}{\sqrt{1 + \tan^2 \beta}}, \cos \beta = \frac{1}{\sqrt{1 + \tan^2 \beta}},$$

we may take the approximations

$$dl = dx, \quad l = x, \quad \sin \beta = \frac{2r(l' - x)}{l'^2}, \quad \cos \beta = 1,$$

and therefore

$$\sin (a + \beta) = \sin a \cos \beta + \cos a \sin \beta = \sin a + \frac{2r(l' - x)}{l'^2} \cos a$$

$$\cos (a + \beta) = \cos a \cos \beta - \sin a \sin \beta = \cos a - \frac{2r(l' - x)}{l'^2} \sin a$$

$$\int_0^s w_1(x - x')dl = w_1 \int_0^x (x - x')dx' = \frac{w_1 x^2}{2}$$

$$\int_0^s w_2(y - y')dl = w_2 \int_0^x (y - y')dx' = \frac{w_2 r}{l'^2}\left(l'x^2 - \frac{2x^3}{3}\right).$$

The expressions for B, P, S then become

$$\left.\begin{aligned} B &= B_0 + \left(\frac{2r}{l'}Q_0 - R_0\right)x + \left(-\frac{Q_0 r}{l'^2} + \frac{w_1}{2} + \frac{w_2 r}{l'}\right)x^2 + \frac{2w_2 r x^3}{3l'^2} \\ P &= \left(Q_0 + \frac{2rR_0}{l'}\right) + \left(-\frac{2rR_0}{l'^2} - w_2 - \frac{2rw_1}{l'}\right)x + \frac{2rw_1 x^2}{l'^2} \\ S &= \left(\frac{2rQ_0}{l'} - R_0\right) + \left(-\frac{2rQ_0}{l'^2} + w_1 - \frac{2rw_2}{l'}\right)x + \frac{2rw_2 x^2}{l'^2}. \end{aligned}\right\} \quad (8)$$

We observe that the shearing force is the differential coefficient of the bending moment, as it should be in accordance with the general rule proved in § 2. Moreover, as we must have $P = P_0$, $S = S_0$ for $x = 0$, it follows that we shall have

$$P_0 = Q_0 + \frac{2rR_0}{l'}, \quad S_0 = \frac{2rQ_0}{l'} - R_0;$$

hence the three formulæ (8) may be put in the following form :—

$$\left.\begin{aligned} B - B_0 - \frac{2w_2 r x^3}{3l'^2} &= Cx + Dx^2 \\ P - P_0 &= C_1 x + D_1 x^2 \\ S - S_0 &= C_2 x + D_2 x^2 \end{aligned}\right\} \qquad (9)$$

where C, D; $C_1$, $D_1$; $C_2$, $D_2$ represent the coefficients of $x$ and $x^2$ in formulæ (8). If we call $B_1$, $P_1$, $S_1$ the bending moment, normal pressure and shear at the mid-section B, the above formulæ must give $B = B_1$, $P = P_1$, $S = S_1$ for $x = l'$ and $B = B_2$, $P = P_2$ and $S = S_2$ for $x = 2l'$; from each of the equations (9), therefore, and the two deduced from them by putting $x = l'$ and $x = 2l'$, we can eliminate the coefficients of $x$ and $x^2$, thus obtaining the three following results:—

$$B = B_0 + (B_1 - B_0)\left(\frac{2l'x - x^2}{l'^2}\right) - (B_2 - B_0)\left(\frac{l'x - x^2}{2l'^2}\right) + \frac{2w_2 r}{3l'^2}(x^3 - 3l'x^2 + 2l'^2 x),$$

$$P = P_0 + (P_1 - P_0)\left(\frac{2l'x - x^2}{l'^2}\right) - (P_2 - P_0)\left(\frac{l'x - x^2}{2l'^2}\right),$$

$$S = S_0 + (S_1 - S_0)\left(\frac{2l'x - x^2}{l'^2}\right) - (S_2 - S_0)\left(\frac{l'x - x^2}{2l'^2}\right).$$

Now the internal work of the whole body, taking $dl = dx$ and $l = x$ throughout, is given by

$$W_i = \frac{1}{2EI}\int_0^{2l'} B^2 dx + \frac{1}{2EA}\int_0^{2l'} P^2 dx + \frac{F}{2GA}\int_0^{2l'} S^2 dx.$$

Substituting for B, P, S their values given above and effecting the integrations, we obtain

$$\left.\begin{aligned} W_i = \frac{l'}{2EI}\Big[&\frac{B_0^2 + 4B_1^2 + B_2^2}{3} - \frac{(B_0 - 2B_1 + B_2)^2}{15} \\ &\frac{8w_2 r l'}{45}(B_2 - B_0) + \frac{64w_2^2 r^2 l'^2}{945}\Big] \\ + \frac{l'}{2EA}\Big[&\frac{P_0^2 + 4P_1^2 + P_2^2}{3} - \frac{(P_0 - 2P_1 + P_2)^2}{15}\Big] \\ + \frac{Fl'}{2GA}\Big[&\frac{S_0^2 + 4S_1^2 + S_2^2}{3} - \frac{(S_0 - 2S_1 + S_2)^2}{15}\Big] \end{aligned}\right\} \qquad (10)$$

This formula can be simplified, as stated in § 3; we may write

$$\begin{aligned} P_1 - P_0 &= C_1 l' + D_1 l'^2 \\ P_2 - P_0 &= 2C_1 l' + 4D_1 l'^2, \end{aligned}$$

and by subtracting from the second the first multiplied by 2, we get

$$P_0 - 2P_1 + P_2 = 2D_1 l'^2,$$

i.e. seeing that $D_1 = \dfrac{2rw_1}{l'^2}$,

$$P_0 - 2P_1 + P_2 = 4rw_1;$$

which shows that the quantity $P_0 - 2P_1 + P_2$ depends on the weight $w_1$ only.

In the same way we find

$$S_0 - 2S_1 + S_2 = 4rw_2.$$

Neglecting the terms independent of the bending moments, normal pressures, and shears, the expression for the internal work becomes

$$\left.\begin{aligned}\frac{l'}{2EI}\left[\frac{B_0^2+4B_1^2+B_2^2}{3}-\frac{(B_0-2B_1+B_2)^2}{15}-\frac{8w_2l'r(B_2-B_0)}{45}\right]\\ +\frac{l'}{2EA}\left(\frac{P_0^2+4P_1^2+P_2^2}{3}\right)+\frac{Fl'}{2GA}\left(\frac{S_0^2+4S_1^2+S_2^2}{3}\right)\end{aligned}\right\} \quad (11)$$

Since the rise has been assumed to be very small in comparison with the chord, the term $\frac{8w_2l'r}{45}(B_2 - B_0)$ will be very small in comparison with $\frac{B_0^2 + 4B_1^2 + B_2^2}{3}$.

The same holds true for the term $\frac{(B_0 - 2B_1+B_2)^2}{15}$, because we have

$$B_1 - B_0 - \frac{2}{3}w_2l'r = Cl' + Dl'^2$$

$$B_2 - B_0 - \frac{16}{3}w_2l'r = 2Cl' + 4Dl'^2,$$

and by subtracting from the second equation the first multiplied by 2, we get

$$B_0 - 2B_1 + B_2 - 4w_2l'r = 2Dl'^2,$$

i.e. putting for D its value $-\frac{rQ_0}{l'^2} + \frac{w_1}{2} - \frac{w_2r}{l^1}$,

$$B_0 - 2B_1 + B_2 = -2Q_0r + w_1l'^2 + 2w_2l'r,$$

and therefore

$$(B_0 - 2B_1 + B_2)^2 = 4Q_0^2r^2 - 4Q_0r(w_1l'^2 + 2w_2l'r) + (w_1l'^2 + 2w_2l'r)^2.$$

The last term on the right-hand side may be neglected, since it is independent of $Q_0$, and the two first terms contain $r$ as a factor and are therefore very small when the rise is small compared with the chord. We can thus often neglect the quantity

$$-\frac{(B_0 - 2B_1 + B_2)^2}{15} - \frac{8w_2l'r(B_2 - B_0)}{45}$$

in expression (11), which then becomes

$$\frac{l'}{2EI}\left(\frac{B_0^2 + 4B_1^2 + B_2^2}{3}\right) + \frac{l'}{2EA}\left(\frac{P_0^2 + 4P_1^2 + P_2^2}{3}\right) + \frac{Fl'}{2GA}\left(\frac{S_0^2 + 4S_1^2 + S_2^2}{3}\right) \quad (12)$$

This is the formula which we shall adopt for general use in practical applications.

**5. Internal work of a solid body of constant section, loaded continuously throughout its length and having its centre line following any plane curve.**

We will first divide the centre line of the body into an even number of equal small parts, so that each of the segments of the solid containing two consecutive small parts of the centre line will have a very small rise compared with its length and may be regarded as uniformly loaded.

If now we call 0, 1, 2, . . . $n$ the sections corresponding to the points of division, and if we call

$l$ = length of each portion 0, 1 ; 1 ; 2, etc., of the centre line

A = cross-sectional area,

I = moment of inertia,

E, G = the elastic and glide moduli,

$B_0$, $B_1$, etc. = the bending moments about the sections 0, 1, etc.,

$P_0$, $P_1$ . . . = the normal pressures,

$S_0$, $S_1$ . . . = the shearing forces,

we then have the following expression for the internal work of the body, obtained by successive applications of formula (12) to each double part 0, 2 ; 2, 4 ; etc.

$$\left.\begin{aligned} &\frac{l}{2EI}\left(\frac{B_0^2 + 4B_1^2 + 2B_2^2 + 4B_3^2 + \ldots 4B_{n-1}^2 + B_n^2}{3}\right) \\ &+ \frac{l}{2EA}\left(\frac{P_0^2 + 4P_1^2 + 2P_2^2 + 4P_3^2 + \ldots 4P_{n-1}^2 + P_n^2}{3}\right) \\ &+ \frac{Fl}{2GA}\left(\frac{S_0^2 + 4S_1^2 + 2S_2^2 + 4S_3^2 + \ldots 4S_{n-1}^2 + S_n^2}{3}\right) \end{aligned}\right\} \qquad (13)$$

**6. Internal work of a solid body of variable section loaded continuously throughout its length and having its centre line curved in any manner.**

The internal work of any solid body satisfying the conditions set out in § 11, ch. iv. is given by the formula

$$W_i = \frac{1}{2E}\int\frac{B^2}{I}dl + \frac{1}{2E}\int\frac{P^2}{A}dl + \frac{1}{2G}\int\frac{FS^2}{A}dl,$$

the integration being extended over the whole length of the body.

Now if we plot a curve of $\frac{B^2}{I}$ against $l$, the area of this curve will be $\int\frac{B^2}{I}dl$.

But by Simpson's rule this area can be expressed approximately as follows: divide the base of the curve into an even number $n$ of equal parts each of length $l$, and let $y_0, y_1 \dots y_n$ be the corresponding ordinates; then the area is given by the expression

$$\frac{l}{3}(y_0 + 4y_1 + 2y_2 + 4y_3 + \dots + 4y_{n-1} + y_n)$$

and since $y_0 = \frac{B_0^2}{I_0}$, $y_1 = \frac{B_1^2}{I_1}$, etc., we have

$$\int\frac{B^2dl}{I} = \frac{l}{3}\left(\frac{B_0^2}{I_0} + \frac{4B_1^2}{I_1} + \frac{2B_2^2}{I_2} + \frac{4B_3^2}{I_3} + \dots + \frac{4B_{n-1}^2}{I_{n-1}} + \frac{B_n^2}{I}\right).$$

The values of $\int\frac{P^2}{A}dl$ and $\int\frac{FS^2}{A}dl$ may be obtained similarly, so that the following expression gives the internal work of the whole body:—

$$\left.\begin{aligned} &\frac{1}{2E}\cdot\frac{l}{3}\left(\frac{B_0^2}{I_0} + \frac{4B_1^2}{I_1} + \frac{2B_2^2}{I_2} + \frac{4B_3^2}{I_3} + \dots + \frac{4B_{n-1}^2}{I_{n-1}} + \frac{B_n^2}{I_n}\right) \\ &+ \frac{1}{2E}\cdot\frac{l}{3}\left(\frac{P_0^2}{A_0} + \frac{4P_1^2}{A_1} + \frac{2P_2^2}{A_2} + \frac{4P_3^2}{A_3} + \dots + \frac{4P_{n-1}^2}{A_{n-1}} + \frac{P_n^2}{A_n}\right) \\ &+ \frac{1}{2G}\cdot\frac{l}{3}\left(\frac{F_0S_0^2}{A_0} + \frac{4F_1S_1^2}{A_1} + \frac{2F_2S_2^2}{A_2} + \frac{4F_3S_3^2}{A_3} + \dots \right. \\ &\qquad\qquad \left. + \frac{4F_{n-1}S_{n-1}^2}{A_{n-1}} + \frac{F_nS_n^2}{A_n}\right) \end{aligned}\right\} \quad (14)$$

It should be noted that Simpson's rule can only be applied when the curve is a continuous one; to be able, therefore, to apply the above formula, the quantities B, P, S, I, A, F must be continuous functions of the arc $l$. If some of these functions are discontinuous, i.e. if they have abrupt changes of value, the body must be divided into several portions, the divisions being taken at the points where the abrupt changes occur. We can then apply expression (14) to each portion and add together the results.

It will also be seen that when the body is of constant section (14) becomes the same as (13); this should of course be the case, but it is interesting to note that (13) was obtained by quite a different method from (14).

# CHAPTER VII.

## THEORY OF STRAIGHT BEAMS.

**1. Definitions.**—The straight beams that we shall consider in this chapter are assumed to satisfy the conditions set out in chap. iv., § 11, i.e.

(1) That the dimensions of the transverse of the section beam are very small compared with its length.

(2) That the principal axes of the beam sections pass through their centroids and are perpendicular to each other.

(3) That one principal axis is vertical.

(4) That all the loads applied on the right hand of any section and also the reactions at fixed points on that side pass through the centre line of the beam, or are so disposed on one side and the other of this line that in considering the loads and reactions together they are equivalent to a vertical resultant and a couple having for its arm the horizontal principal axis of the section under consideration.

We will call the *plane of bending* the vertical plane passing through the centre line of the beam, and the *neutral axis* of any section the horizontal principal axis.

Beams satisfying these conditions will have their centre lines always in the planes of bending, and the neutral axes will still be horizontal after deflection.

**2. Beam Fixed at One End (Cantilever),** (Fig. 5, Pl. II).—In the first place let us consider a beam fixed at one end and subjected to the following forces:—

(1) To a load distributed along its length according to any law, continuous or discontinuous.

(2) On the end B to normal forces distributed according to a linear law giving zero resultant at the centre of the section and a bending moment $B_1$ about the neutral axis.

(3) On the same end B to vertical shearing forces acting upwards and of resultant $S_1$.

We will now determine the equation to the centre line of the beam after bending and the angular deflection of the end B.

If $l$ is the length of the beam and $M_{1x'}$ the moment about the section D at distance $x'$ from A of all the loads on the length DB, it is clear that we shall have

$$\left.\begin{aligned} B_{x'} &= B_1 - S_1(l - x') + M_{1x'} \\ S_{x'} &= S_1 + \frac{dM_{1x'}}{dx'} \end{aligned}\right\} \quad . \quad . \quad . \quad (1)$$

in which it should be noted that $\dfrac{dM_{1x'}}{dx'}$ will be continuous or discontinuous according as $M_{1x'}$ is continuous or discontinuous.

These two formulæ are applicable for all values of $x'$; but if at a point C at distance $x$ from the fixed end we place a fresh load W, it is clear that for the length AC, i.e. between $x' = 0$ and $x' = x$, the following equations will hold :—

$$B = B_1 - S_1(l - x') + M_{1x'} + W(x - x')$$
$$S = S_1 + \frac{dM_{1x'}}{dx'} - W.$$

Now the internal work of the beam is given by

$$W_i = \frac{1}{2}\int_0^l \frac{B^2 dx}{EI} + \frac{1}{2}\int_0^l \frac{FS^2}{GA} dx,$$

and $\dfrac{dW_i}{dW}$ gives the deflection of the point of application of W; but we note that each integral should be divided into two parts, one from 0 to $x$ and the other from $x$ to $l$. The second does not contain W and therefore disappears on differentiating. We then have, calling $d_n$ the deflection at the point C of the beam,

$$d_n = \int_0^x \frac{B}{EI} \cdot \frac{dB}{dW} dx' + \int_0^x \frac{FS}{GA} \frac{dS}{dW} dx',$$

or

$$d_n = \int_0^x \frac{B(x - x')}{EI} dx' - \int_0^x \frac{FS}{GA} dx'$$

since

$$\frac{dB}{dW} = (x - x') \text{ and } \frac{dS}{dW} = -1.$$

This formula must hold for all values of W, and therefore also when $W = 0$; in this case B and S are given by formulæ (1) so that we get

$$\left.\begin{aligned} d_n = B_1\int_0^x \frac{(x - x')dx'}{EI} - S_1\int_0^x \frac{(l - x')(x - x')}{EI} dx' \\ + \int_0^x \frac{M_{1x'}(x - x')}{EI} dx' - S_1\int_0^x \frac{F}{GA} dx' - \int_0^x \frac{F}{GA} \frac{dM_{1x'}}{dx'} dx' \end{aligned}\right\} \quad . \quad (2)$$

This is the equation to the curve adopted by the centre line of the beam after deflection.

To obtain the angular deflection $\theta_1$ of the extreme end B we must differentiate the internal work with regard to $B_1$; this internal work is given by

$$W_i = \frac{1}{2}\int_0^l \frac{B^2}{EI}dx' + \frac{1}{2}\int_0^l \frac{FS^2}{GA}dx'$$

where B and S are given by formulæ (1). Since the second term does not depend upon $B_1$, it disappears upon differentiating, so that we have

$$\theta_1 = \int_0^l \frac{B}{EI}\frac{dB}{dB_1}dx'$$

$$= \int_0^l \frac{B}{EI}dx'.$$

Inserting the value of B and changing $x'$ to $x$ we have

$$\theta_1 = B_1\int_0^l \frac{dx}{EI} - S_1\int_0^l \frac{(l-x)dx}{EI} + \int_0^l \frac{M_{1x}dx}{EI} \quad . \quad . \quad (3)$$

Now consider the special case in which the beam is loaded with a weight W at the end B and with a uniformly distributed load of $w$ kg. per metre run throughout.

We shall then have the following results, noting that both the normal and shear forces on the end B are zero :—

$$B_1 = 0, \quad S_1 = 0$$

$$M_{1x'} = W(l-x') + \frac{1}{2}w(l-x')^2; \quad \frac{dM_{1x'}}{dx'} = -W - w(l-x'),$$

$$M_{1x} = W(l-x) - \frac{1}{2}w(l-x)^2; \quad \frac{dM_{1x}}{dx} = -W - w(l-x),$$

and therefore

$$\left.\begin{aligned} d_n &= W\int_0^x \frac{(l-x')(x-x')}{EI}dx' + \frac{w}{2}\int_0^x \frac{(l-x')^2(x-x')}{EI}dx' \\ &\qquad + W\int_0^x \frac{F dx'}{GA} + w\int_0^x \frac{F(l-x')}{GA}dx' \\ \theta_1 &= W\int_0^l \frac{(l-x)dx}{EI} + \frac{w}{2}\int_0^l \frac{(l-x)^2 dx}{EI} \end{aligned}\right\} \quad (4)$$

If the beam is homogeneous and of constant section, the products EI and GA will be constant and may be taken outside the integration sign, so that we shall have

$$d_n = \frac{x^2}{EI}\left[W\left(\frac{l}{2} - \frac{x}{6}\right) + \frac{w}{2}\left(\frac{l^2}{2} - \frac{lx}{3} + \frac{x^2}{12}\right)\right] + \frac{Fx}{GA}\left[W + w\left(l - \frac{x}{2}\right)\right]$$

$$\theta_1 = \frac{l^2}{EI}\left(\frac{W}{2} + \frac{wl}{6}\right).$$

For the deflection at the end we shall have, on putting $x = l$,

$$d_n = \frac{l^3}{EI}\left(\frac{W}{3} + \frac{wl}{8}\right) + \frac{Fl}{GA}\left(W + \frac{wl}{2}\right) \quad . \quad . \quad (5)$$

We can easily show that in this formula the term due to shear, i.e. that containing the factor F, is very small compared with the other, due to bending, if the transverse dimensions of the beam are very small compared with its length. Thus, if we take a rectangular section of breadth $b$ and depth $d$ we shall have

$$A = bd; \quad I = \frac{bd^3}{12},$$

and we may take $F = \frac{6}{5}$, provided that $d$ is equal to or greater than $b$.

If, moreover, the body is isotropic, we shall have

$$G = \tfrac{2}{5}E;$$

therefore formula (5) will become :—

$$d_n = \frac{12l^3}{Ebd^3}\left(\frac{W}{3} + \frac{wl}{8}\right) + \frac{3l}{Ebd}\left(W + \frac{wl}{2}\right)$$

$$= \frac{12l^3}{Ebd^3}\left[\left(\frac{W}{3} + \frac{wl}{8}\right) + \left(\frac{W}{4} + \frac{wl}{8}\right)\frac{d^2}{l^2}\right];$$

thus the second term inside the brackets, i.e. that due to shear, will be very small compared with the first due to bending, if the transverse dimensions of the beam are small compared with its length, for then $\frac{d^2}{l^2}$ will be a very small quantity.

If, for example, we have $d < \frac{l}{10}$, as is nearly always the case, we see that the deflection due to shear is not $\frac{1}{100}$ of that due to bending.

**3. Beam Fixed at One End and Simply Supported at the Other** (Fig. 6, Pl. II).—Let the horizontal beam AB be fixed at A and simply supported at B, so that the end B must always remain at the same level as A whatever be the load carried by the beam.

We will also suppose that no force is applied at the end B, so that the bending moment there will be zero and the shearing force will be equal to the vertical reaction $R_B$.

We may assume that the support at B will be relieved from load if we apply at this point a vertical force equal to $R_B$, or a system of vertical forces whose resultant is equal to $R_B$. We then have a beam fixed at one end as in the preceding case but subject to the condition that the deflection at the end B must be zero. For such a beam the deflection at any point and the angular deflection at the point B will be given by equations (2) and (3) of the previous paragraph by putting $B_1 = 0$ and changing $S_1$ to $R_B$; we thus have

$$\left.\begin{aligned} d_n &= -R_B\left[\int_0^x \frac{(l-x')(x-x')}{EI}dx' + \int_0^x \frac{F}{GA}dx'\right] \\ &\qquad + \int_0^x \frac{M_{x'}(x-x')}{EI}dx' - \int_0^x \frac{F}{GA}\frac{dM_{1x'}}{dx'}dx' \\ \theta_B &= -R_B\int_0^l \frac{(l-x)dx}{EI} + \int_0^l \frac{M_{1x}}{EI}dx. \end{aligned}\right\} \quad (6)$$

In these equations the reaction $R_B$ is unknown, but as we must have $d_n = 0$ for $x = l$, $R_B$ is given by the equation

$$\left.\begin{aligned} &-R_B\left[\int_0^l \frac{(l-x')^2}{EI}dx' + \int_0^l \frac{F}{GA}dx'\right] \\ &\qquad + \int_0^l M_{1x'}\frac{(l-x')}{EI}dx' - \int_0^l \frac{F}{GA}\frac{dM_{1x'}}{dx'}dx' = 0 \end{aligned}\right\} \quad (7)$$

If, for example, the load is a uniformly distributed one of intensity $w$ we shall have

$$M_{1x'} = \frac{w}{2}(l-x')^2; \quad \frac{dM_{1x'}}{dx'} = -w(l-x')$$

and therefore

$$\left.\begin{aligned} &-R_B\left[\int_0^l \frac{(l-x')^2}{EI}dx' + \int_0^l \frac{F}{GA}dx'\right] \\ &\qquad + w\left[\frac{1}{2}\int_0^l \frac{(l-x')^3}{EI}dx' + \int_0^l \frac{F(l-x')}{GA}dx'\right] = 0 \end{aligned}\right\} \quad (8)$$

If, moreover, the beam is prismatic and homogeneous, the quantities EI and $\frac{F}{GA}$ will be constant, and the integrations can be effected, thus giving, on dividing by $l$,

$$-R_B\left(\frac{l^2}{3EI} + \frac{F}{GA}\right) + w\left(\frac{l^3}{8EI} + \frac{Fl}{2GA}\right) = 0 \quad . \quad (9)$$

We can show, as in the previous case, that if the transverse dimensions of the beam are very small compared with its length, the term containing F in each bracket is very small compared with the other, and can be neglected. We then have

$$R_B = \frac{3}{8}wl.$$

**4. Beam Fixed at Both Ends** (Fig. 7, Pl. II.).—We will assume that the beam is horizontal before loading and fixed at both ends. After loading there will be at each end a bending moment and a shearing force; we can then imagine the beam freed at the end B by applying to it forces induced by the fixing. We will call $B_1$ the moment of these forces about the neutral axis and $S_1$ the sum of the upward shear forces, so that $B_1$ will be the bending moment and $S_1$ the shearing force for the section B.

Thus we come back again to the problem considered in § 2, with the sole difference that $B_1$ and $S_1$ are now unknown quantities which have to be determined by the condition that the deflection of the right-hand end must be zero and its angular deflection also zero.

We must then equate to zero the value of $d_n$ given by equation (2) on putting $x = l$, and the value of $\theta_1$ given by equation (3), thus obtaining the following two equations:—

$$\left.\begin{aligned} &B_1\int_0^l \frac{(l - x')}{EI}dx' - S_1\left[\int_0^l \frac{(l - x')^2}{EI}dx' + \int_0^l \frac{F}{GA}dx'\right] \\ &\quad + \int_0^l \frac{M_{1x'}(l - x')}{EI}dx' - \int_0^l \frac{F}{GA}\frac{dM_{1x'}}{dx'}dx' = 0 \\ &B_1\int_0^l \frac{dx}{EI} - S_1\int_0^l \frac{(l - x)}{EI}dx + \int_0^l \frac{M_{1x'}}{EI}dx = 0 \end{aligned}\right\} . \quad (11)$$

By means of these equations we determine the values of $B_1$, $S_1$, and then by substituting these values in equation (2) we obtain the equation to the curve adopted by the centre line after deflection.

*Single Isolated Load.*—Suppose for example that the beam is homogeneous and of constant section and that the load consists of a weight W at distance $a$ from the end A.

We shall have between $x = 0$ and $x = a$

$$M_{1x} = W(a - x), \quad \frac{dM_{1x}}{dx} = - W$$

and between $x = a$ and $x = l$

$$M_{1x} = 0, \quad \frac{dM_{1x}}{dx} = 0.$$

Equations (11) then become

$$\frac{B_1 l^2}{2EI} - S_1\left(\frac{l^3}{3EI} + \frac{Fl}{GA}\right) + W\left[\frac{a^2(3l - a)}{6EI} + \frac{Fa}{GA}\right] = 0$$

$$\frac{B_1 l}{EI} - \frac{S_1 l^2}{2EI} + \frac{Wa^2}{2EI} = 0.$$

From these two equations we deduce

$$\left.\begin{aligned} S_1\left(\frac{l^3}{12EI} + \frac{Fl}{GA}\right) - W\left[\frac{a^2(3l - 2a)}{12EI} + \frac{Fa}{GA}\right] = 0 \\ B_1\left(\frac{l^2}{6EI} + \frac{2F}{GA}\right) - \frac{W}{l}\left[\frac{a^2l(l - a)}{6EI} + \frac{Aa(l - a)}{GA}\right] = 0 \end{aligned}\right\} \quad (12)$$

Having thus obtained the values of the two unknowns, we can introduce them into equation (2) to obtain the equation to the elastic line; this line will be composed of two distinct portions, one to the left and one to the right of the point of application of the load W. For the latter, which is comprised between $x = a$ and $x = l$, we divide the integrals of equation (2) into two parts, one between $x' = 0$ and $x' = a$, for which we shall have

$$M_{1x'} = W(a - x'), \quad \frac{dM_{1x'}}{dx'} = -W,$$

and the other between $x' = a$ and $x' = x$, for which we shall have

$$M_{1x'} = 0, \quad \frac{dM_{1x'}}{dx'} = 0.$$

Then effecting the integrations, we shall obtain

$$d_n = \frac{B_1x^2}{2EI} - S_1\left[\frac{x^2(3l - x)}{6EI} + \frac{Fx}{GA}\right] + W\left[\frac{a^2(3x - a)}{6EI} + \frac{Fa}{GA}\right].$$

If we have $a < \frac{l}{2}$, this equation will hold for $x = \frac{l}{2}$, and the deflection at the centre of the beam will be

$$\text{centre } d_n = \frac{B_1l^2}{8EI} - S_1\left(\frac{5l^3}{48EI} + \frac{Fl}{2GA}\right) + W\left(\frac{a^2(3l - 2a)}{12EI} + \frac{Fa}{GA}\right) \quad (13)$$

in which we can substitute for $B_1$, $S_1$ the values found above.

In accordance with what we have proved at the end of § 2, if the transverse dimensions of the beam are very small compared with its length, we may neglect the terms due to shear; this gives

$$\left.\begin{aligned} S_1 &= \frac{Wa^2(3l - 2a)}{l^3} \\ B_1 &= \frac{Wa^2(l - a)}{l^2} \\ \text{Centre } d_n &= \frac{Wa^2(3l - 4a)}{48EI} \end{aligned}\right\} \quad . \quad . \quad . \quad (14)$$

*Two Equal Isolated Loads Equidistant from the Ends.*—If the beam carries two loads W each at distance $a$ from one end, we shall have on the principle of superposition

$$\left.\begin{aligned} S_1 &= \frac{Wa^2(3l - 2a)}{l^3} + \frac{W(l-a)^2(3l - 2l + 2a)}{l^3} = W \\ B_1 &= \frac{Wa^2(l - a)}{l^2} + \frac{W(l - a)^2 a}{l^2} = \frac{Wa(l - a)}{l} \\ \text{Centre } d_n &= \frac{2Wa^2(3l - 4a)}{48EI} = \frac{Wa^2(3l - 4a)}{24EI} \end{aligned}\right\} \qquad (15)$$

*Uniformly Distributed Load.*—If $w$ is the intensity of the load, we shall have in this case

$$M_{1x} = \frac{w}{2}(l - x)^2; \quad \frac{dM_{1x}}{dx} = -w(l - x),$$

so that substituting in (11) and integrating we shall have

$$\frac{B_1 l^2}{2EI} - S_1\left(\frac{l^3}{3EI} + \frac{Fl}{GA}\right) + w\left(\frac{l^4}{8EI} + \frac{Fl^2}{2GA}\right) = 0$$

$$\frac{B_1 l}{EI} - \frac{S_1 l^2}{2EI} + \frac{wl^3}{6EI} = 0,$$

from which we deduce

$$S_1\left(\frac{l^3}{12EI} + \frac{Fl}{GA}\right) - w\left(\frac{l^4}{24EI} + \frac{Fl^2}{2GA}\right) = 0$$

$$B_1\left(\frac{l^2}{6EI} + \frac{2F}{GA}\right) - w\left(\frac{l^4}{72EI} + \frac{Fl^2}{6GA}\right) = 0$$

or

$$\left.\begin{aligned} S_1 &= \frac{wl}{2} \\ B_1 &= \frac{wl^2}{12} \end{aligned}\right\} \quad . \quad . \quad . \quad . \quad (16)$$

To obtain the equation for the deflected form of the beam we will substitute in equation (2)

$$M_{1x'} = \frac{w}{2}(l - x')^2; \quad \frac{dM_{1x'}}{dx'} = -w(l - x'),$$

and, integrating on the assumption that $\frac{1}{EI}$ and $\frac{F}{GA}$ are constant, we get

$$d_n = \frac{B_1 x^2}{2EI} - S_1\left[\frac{x^2(3l - x)}{6EI} + \frac{Fx}{GA}\right] + w\left[\frac{x^2(6l^2 - 4lx + x^2)}{24EI} + \frac{Fx(2l - x)}{2GA}\right]$$

i.e. substituting for $B_1$, $S_1$ and reducing,

$$d_n = w\left(\frac{l^2x^2 - 2lx^3 + x^4}{24EI}\right) + \frac{wFx(l - x)}{2GA} \quad . \quad . \quad (17)$$

At the centre where $x = \frac{l}{2}$ this gives

$$\text{centre } d_n = \frac{wl^4}{384\text{EI}} + \frac{\text{F}wl^2}{8\text{GA}} \quad . \quad . \quad . \quad (18)$$

*Uniform Load and Two Isolated Loads Equidistant from the Ends.*—By the principle of superposition, and neglecting the terms due to shear, we have for this case

$$\left.\begin{aligned} \text{S}_1 &= \text{W} + \frac{wl}{2} \\ \text{B}_1 &= \frac{\text{W}a(l - a)}{l} + \frac{wl^2}{12} \\ \text{centre } d_n &= \frac{\text{W}a^2(3l - 4a)}{24\text{EI}} + \frac{wl^4}{384\text{EI}} \end{aligned}\right\} \quad . \quad . \quad (19)$$

These formulæ are necessary for the design of cross girders of tubular bridges.

**5. Beam Simply Supported at its Ends** (Fig. 8, Pl. II).—Let AB be a beam loaded in any manner and subjected at its ends to any conditions, i.e. simply supported, supported at one end and fixed at the other, or fixed at each end, or even prolonged beyond the supports.

We may suppose the beam AB to be cut at two sections infinitely close to A and B but within these two points, i.e. one to the right of A and the other to the left of B. The beam will continue to be in equilibrium provided that at the two sections we apply external forces equal to the resultant stresses, so that we can study the elastic equilibrium of the beam by regarding it as a free solid body acted upon by the following forces :—

1. By weights distributed along the beam in any manner.

2. By forces applied to the left-hand section at A and giving for this section a bending moment $\text{B}_0$, a shearing force $\text{S}_0$ acting upward, and zero normal pressure.

3. By forces applied to the right-hand section at B and giving a bending moment $\text{B}_1$, a shearing force $\text{S}_1$ acting upward, and zero normal pressure.

As it is clear that the normal pressure will be zero for all sections of the beam, we have only to find expressions for the bending moment and shearing force at any section.

Adopting our previous notation, we shall have for a section C at distance $x$ from A

$$\left.\begin{aligned} \text{B} &= \text{B}_1 - \text{S}_1(l - x) + \text{M}_{1x} \\ \text{S} &= \text{S}_1 + \frac{d\text{M}_{1x}}{dx} \end{aligned}\right\} \quad . \quad . \quad . \quad (20)$$

We will note that any section divides the beam into two portions and that the shearing force acts in opposite directions according as the section is considered as belonging to one or the other of these portions; thus as we assume $S_1$ to be acting upwards, and as $\frac{dM_{1x}}{dx}$ also represents an upward vertical force, it is clear that $S_1 + \frac{dM_{1x}}{dx}$ expresses the shearing force on the section C considered as the end of the portion AC. If on the other hand we wished to express the shear at C considered as part of BC, we should write $-\left(S_1 + \frac{dM_{1x}}{dx}\right)$.

At A, where $x = 0$, the first of the formulæ (20) must give $B = B_0$, and the second $S = -S_0$, so that we have

$$\left.\begin{aligned} B_0 &= B_1 - S_1 l + M_{1,0} \\ S_0 &= -\frac{B_1 - B_0}{l} - \frac{M_{1,0}}{l} - \frac{dM_{1,0}}{dx} \end{aligned}\right\} \quad . \quad . \quad (21)$$

Now the internal work of the whole beam will be given by

$$W_i = \tfrac{1}{2}\int_0^l \frac{B^2}{EI}dx + \tfrac{1}{2}\int_0^l \frac{FS^2}{GA}dx,$$

and the differential coefficients with regard to $B_0$, $B_1$ give the angular deflections $\theta_0$, $\theta_1$ at the extreme ends, while that with regard to any one of the loads upon the beam gives the deflection at the point of application of that load. These deflections will be with reference to the original centre line of the body if the centres of the end sections are regarded as fixed; the shear reactions $S_0$, $S_1$ must therefore be regarded as functions of the other forces applied to the structure and will be given by equations (21). We then have

$$\theta_1 = \int_0^l \frac{B}{EI}\frac{dB}{dB_1}dx + \int_0^l \frac{FS}{GA}\frac{dS}{dB_1}dx$$

$$\theta_0 = \int_0^l \frac{B}{EI}\frac{dB}{dB_0}dx + \int_0^l \frac{FS}{GA}\frac{dS}{dB_0}dx.$$

Now on substituting in formulæ (20) the value of $S_1$ given by the first of the equations (21) we obtain

$$\left.\begin{aligned} B &= \frac{B_1 x}{l} + \frac{B_0(l - x)}{l} - \frac{M_{1,0}(l - x)}{l} + M_{1x} \\ S &= \frac{B_1 - B_0}{l} + \frac{M_{1,0}}{l} + \frac{dM_{1x}}{dx} \end{aligned}\right\} \quad . \quad (22)$$

and therefore we have

$$\frac{dB}{dB_1} = \frac{x}{l}; \quad \frac{dB}{dB_0} = \frac{l - x}{l}$$
$$\frac{dS}{dB_1} = \frac{1}{l}, \quad \frac{dS}{dB_0} = -\frac{1}{l},$$

so that

$$\left.\begin{aligned} \theta_1 &= \frac{1}{l}\int_0^l \frac{Bx}{EI}dx + \frac{1}{l}\int_0^l \frac{FS}{GA}dx \\ \theta_1 &= \frac{1}{l}\int_0^l \frac{B(l-x)}{EI}dx - \frac{1}{l}\int_0^l \frac{FS}{GA}dx \end{aligned}\right\} \quad . \quad . \quad (23)$$

Adding these equations, we obtain

$$\theta_0 + \theta_1 = \int_0^l \frac{B}{EI}dx$$

and we can, therefore, substitute for equations (23) the following :—

$$\theta_1 = \frac{1}{l}\int_0^l \frac{Bx}{EI}dx + \frac{1}{l}\int_0^l \frac{FS}{GA}dx,$$

$$\theta_0 = \int_0^l \frac{B}{EI}dx - \theta_1.$$

Substituting for B its value (22) and for S the value given by the second equation (20), we obtain

$$\left.\begin{aligned} \theta_1 = \frac{B_0 - M_{1,0}}{l^2}\int_0^l \frac{(l-x)x}{EI}dx + \frac{B_1}{l^2}\int_0^l \frac{x^2}{EI}dx + \frac{1}{l}\int_0^l \frac{M_{1x} \cdot x}{EI}dx \\ + \frac{S_1}{l}\int_0^l \frac{F}{GA}dx + \frac{1}{l}\int_0^l \frac{F}{GA} \cdot \frac{dM_{1x}}{dx}dx \\ \theta_0 = -\theta_1 + \frac{B_0 - M_{1,0}}{l}\int_0^l \frac{(l-x)}{EI}dx + \frac{B_1}{l}\int_0^l \frac{x}{EI}dx + \int_0^l \frac{M_{1x}}{EI}dx \end{aligned}\right\} \quad (24)$$

In these equations the shearing force $S_1$ can be expressed as a function of $B_0$, $B_1$ by means of the first equation (21). Now let us find the deflection of a point on the centre line of the beam at distance $x$ from the end A supposing that a load W is applied at this point.

For any section at distance $x'$ from A we shall have

$$B = \frac{B_1 x'}{l} + \frac{B_0(l - x')}{l} - \frac{M_{1,0}(l - x')}{l} + M_{1x'}$$

$$S = S_1 + \frac{dM_{1x'}}{dx'}.$$

Since the load W enters into the function $M_{1x'}$ when $x' < x$ and in $M_{1,0}$, we can employ it in the expressions by writing

$$M_{1x'} = \psi(x') + W(x - x'), \quad \frac{dM_{1x'}}{dx'} = \frac{d\psi(x')}{dx'} - W$$

$$M_{1,0} = \psi(0) + Wx, \quad \frac{dM_{1,0}}{dx'} = \frac{d\psi(0)}{dx'} + W.$$

The first two equations hold only for $0 < x' < x$, since for values of $x'$ in which $x < x' < l$, $M_1x'$ will not contain W, and we shall have

$$M_{1x'} = \psi(x'), \quad \frac{dM_{1x'}}{dx'} = \frac{d\psi(x')}{dx'}.$$

Thus the expressions for B and S which we have just given can be put in the following form :—

(1) $0 < x' < x$

$$B = \frac{B_1 x'}{l} + \frac{B_0(l - x')}{l} + \frac{\psi(0)(l - x')}{l} + \psi(x') - \frac{W(l - x)x'}{l},$$

$$S = \frac{(B_1 - B_0)}{l} + \frac{\psi(0)}{l} + \frac{d\psi(x')}{dx'} - \frac{W(l - x)}{l},$$

(2) $x < x' < l$

$$B = \frac{B_1 x'}{l} + \frac{B_0(l - x')}{l} - \frac{\psi(0)(l - x')}{l} + \psi(x') - \frac{W(l - x')x}{l},$$

$$S = \frac{(B_1 - B_0)}{l} + \frac{\psi(0)}{l} + \frac{d\psi(x')}{dx'} + \frac{Wx}{l};$$

so that if for all values of $x'$ we write

$$B' = \frac{B_1 x'}{l} + \frac{B_0(l - x')}{l} - \frac{\psi(0)(l - x')}{l} + \psi(x'),$$

$$S' = \frac{(B_1 - B_0)}{l} + \frac{\psi(0)}{l} + \frac{d\psi(x')}{dx'},$$

we shall have :—

for $0 < x' < x$

$$B = B' - \frac{W(l - x)x'}{l},$$

$$S = S' - \frac{W(l - x)}{l};$$

for $x < x' < l$

$$B = B' - \frac{W(l - x')x}{l},$$

$$S = S' + \frac{Wx}{l}.$$

The internal work of the whole beam is given by the formula

$$W_i = \frac{1}{2}\int_0^x \frac{\left(B' - \frac{W(l - x)}{l} \cdot x'\right)^2 dx'}{EI} + \frac{1}{2}\int_x^l \frac{\left(B' - \frac{W(l - x')x}{l}\right)^2 dx'}{EI}$$

$$+ \frac{1}{2}\int_0^x \frac{F\left(S' - W\frac{(l - x)}{l}\right)^2 dx'}{GA} + \frac{1}{2}\int_x^l \frac{F\left(S' + \frac{Wx}{l}\right)^2 dx'}{GA},$$

of which the differential coefficient with regard to W will give the deflection of the point of application of this load.

We thus have

$$d_n = -\frac{l-x}{l}\left[\int_0^x \frac{\left(B' - \frac{W(l-x)x'}{l}\right)x'dx'}{EI} + \int_0^x \frac{F\left(S' - \frac{W(l-x)}{l}\right)dx'}{GA}\right]$$
$$-\frac{x}{l}\left[\int_x^l \frac{\left(B' - \frac{W(l-x')x}{l}\right)(l-x')dx'}{EI} - \int_x^l \frac{F\left(S' + \frac{Wx}{l}\right)dx'}{GA}\right]$$

or

$$\left.\begin{aligned} d_n = -\frac{l-x}{l}\left[\int_0^x \frac{Bx'}{EI}dx' + \int_0^x \frac{FS}{GA}dx'\right] \\ -\frac{x}{l}\left[\int_x^l \frac{B(l-x')}{EI}dx' - \int_x^l \frac{FS}{GA}dx'\right] \end{aligned}\right\} \quad (25)$$

It will be noted that the quantity in the second brackets will be equal to $l\theta_0$, as follows from the second equation (25), if the integration be extended from 0 to $l$; we may therefore in place of this quantity employ the expression

$$l\theta_0 - \left[\int_0^x \frac{B(l-x')}{EI}dx' - \int_0^x \frac{FS}{GA}dx'\right]$$

and thus have

$$d_n = -\theta_0 x + \left[\int_0^x \frac{B(x-x')dx'}{EI} - \int_0^x \frac{FS}{GA}dx'\right] \quad . \quad (26)$$

As the weight W may be as small as we please, and may even be zero, it is clear that equation (26) will give the deflection of any point in the beam.

Now substituting for B its value (22) after changing $x$ to $x'$, and for S its value given by the second equation (20) with a similar change, we shall have

$$\left.\begin{aligned} d_n = -\theta_0 x + \frac{(B_0 - M_{1,0})}{l}\int_0^x \frac{(l-x')(x-x')dx'}{EI} + \frac{B_1}{l}\int_0^x \frac{x'(x-x')}{EI}dx' \\ + \int_0^x \frac{M_{1x'}(x-x')}{EI}dx' - S_1\int_0^x \frac{F}{GA}dx' - \int_0^x \frac{F}{GA}\frac{dM_{1x'}}{dx'}dx' \end{aligned}\right\} \quad (27)$$

which is the curve to the centre line of the beam after bending.

It should be noted that $S_1$ can be expressed as a function of $B_0$, $B_1$ by means of the first equation (21).

**6. Simply Supported Beam the Ends of which are not Subjected to any Normal Force.**—In the last paragraph we have considered the general case, and from it we can deduce formulæ for the particular cases which arise most frequently in practice. If there is no normal force at either end and the beam is simply supported at both ends, we must put

$$B_0 = 0, \quad B_1 = 0,$$

in the formulæ obtained above; if the beam is fixed at the left-hand end and simply supported at the other end, we must put $B_1 = 0$ and express the condition that $\theta_0$ is zero in order to obtain $B_0$; and if, finally, the beam is fixed at both ends we determine $B_1$, $B_0$ by making the angular deflection at each end zero.

We have already considered the last two cases and so will now only consider the case of the simply supported beam.

Putting $B_1 = B_0 = 0$, formulæ (21) to (24) become

$$\left.\begin{aligned}
S_1 &= \frac{M_{1,0}}{l}\\
S_0 &= -\frac{M_{1,0}}{l} - \frac{dM_{1,0}}{dx}\\
B &= -\frac{M_{1,0}(l-x)}{l} + M_{1,x}\\
\theta_1 &= -\frac{M_{1,0}}{l^2}\left[\int_0^l \frac{(l-x)x\,dx}{EI} - \int_0^l \frac{F\,dx}{GA}\right] + \frac{1}{l}\left[\int_0^l \frac{M_{1x}x\,dx}{EI} + \int_0^l \frac{F\,dM_{1x}}{GA\,dx}dx\right]\\
\theta_0 &= -\theta_1 - \frac{M_{1,0}}{l}\int_0^l \frac{(l-x)}{EI}dx + \int_0^l \frac{M_{1x}}{EI}\,.\,dx\\
d_n &= -\theta_0 x - \frac{M_{1,0}}{l}\int_0^x \frac{(l-x')(x-x')}{EI}dx' + \int_0^x \frac{M_{1x'}(x-x')}{EI}dx'\\
&\quad - \frac{M_{1,0}}{l}\int_0^x \frac{F}{GA}dx' - \int_0^x \frac{F}{GA}\frac{dM_{1x'}}{dx'}\,.\,dx'
\end{aligned}\right\} \quad (28)$$

Now let us consider the following special cases:—

1. *Isolated load* W *at distance* $a$ *from left-hand end.*—For $0 < x < a$ we have

$$M_{1x} = W(a-x), \quad \frac{dM_{1x}}{dx} = -W.$$

$$\therefore M_{1,0} = Wa, \quad \frac{dM_{1,0}}{dx} = -W$$

for $a < x < l$

$$M_{1x} = 0, \quad \frac{dM_{1x}}{dx} = 0.$$

Thus the first, second, fourth, and fifth of equations (28) become

$$\left.\begin{aligned}
S_1 &= \frac{Wa}{l}, \quad S_0 = \frac{W(l-a)}{l} \\
\theta_1 &= -\frac{Wa}{l^2}\left[\int_0^l \frac{(l-x)x\,dx}{EI} - \int_0^l \frac{F\,dx}{GA}\right] \\
&\qquad + \frac{W}{l}\left[\int_0^a \frac{(a-x)x\,dx}{EI} - \int_0^a \frac{F\,dx}{GA}\right] \\
\theta_0 &= -\theta_1 - \frac{Wa}{l}\int_0^l \frac{(l-x)dx}{EI} + W\int_0^a \frac{(a-x)dx}{EI}
\end{aligned}\right\} \quad (29)$$

The third and sixth are different according as we apply them to the portion to the right or left of the load W; for the latter, i.e. from $x = a$ to $x = l$, they become

$$B = -\frac{Wa(l-x)}{l} \quad . \quad . \quad . \quad . \quad (30)$$

$$\left.\begin{aligned}
d_n = -\theta_0 x - \frac{Wa}{l}\int_0^x \frac{(l-x')(x-x')}{EI}dx' + W\int_0^a \frac{(a-x')(x-x')}{EI}dx' \\
- \frac{Wa}{l}\int_0^x \frac{F}{GA}dx' + W\int_0^a \frac{F}{GA}dx'
\end{aligned}\right\} \quad (31)$$

If the beam is homogeneous and of constant section, the quantities EI, $\frac{F}{GA}$ being constant, we can effect the integrations and obtain

$$\left.\begin{aligned}
\theta_1 &= -\frac{Wa}{l}\frac{(l^2-a^2)}{6EI} \\
\theta_0 &= -\frac{Wa}{l}\frac{(l-a)(2l-a)}{6EI} \\
d_n &= -\theta_0 x - \frac{Wa}{l}\left[\frac{x^2(3l-x)}{6EI} + \frac{Fx}{GA}\right] + W\left[\frac{a^2(3x-a)}{6EI} + \frac{Fa}{GA}\right]
\end{aligned}\right\} \quad (32)$$

If we assume that $a < \frac{l}{2}$, this expression for $d_n$, which must hold between $x = a$ and $x = l$, gives the deflection at the centre on putting $x = \frac{l}{2}$; we thus obtain

$$\text{Central } d_n = Wa\left[\frac{(3l^2 - 4a^2)}{48EI} + \frac{F}{2GA}\right] \quad . \quad . \quad (33)$$

2. *Two isolated loads* W *at distance a from the ends.*—If the beam is loaded with two loads, each at distance $a$ from one end, it is clear from the principle of superposition that the angular deflections $\theta_0$, $\theta_1$ must be equal to each other and each equal to the sum of the two values found above for the case of the single isolated load. In the same way, the central deflection

will be double that given by (33); we shall therefore have in this case

$$\left.\begin{aligned}\theta_0 = \theta_1 &= -\frac{Wa(l-a)}{2EI}\\ \text{Central } d_n &= Wa\left[\frac{(3l^2-4a^2)}{24EI}+\frac{F}{GA}\right]\end{aligned}\right\} \quad . \quad . \quad (34)$$

3. *Uniformly distributed load.*—If $w$ is the intensity of the load we shall have

$$M_{1x} = \frac{w}{2}(l-x)^2, \quad \frac{dM_{1x}}{dx} = -w(l-x)$$

and $\qquad \therefore M_{1,0} = \frac{wl^2}{2}, \quad \frac{dM_{1,0}}{dx} = -wl.$

If, moreover, we assume that the beam is symmetrical about the centre, we shall have $\theta_0 = \theta_1$ and obtain the following results:—

$$\left.\begin{aligned}S_0 = S_1 &= \frac{wl}{2}\\ B &= -\frac{wx}{z}(l-x)\\ \theta_0 = \theta_1 = \frac{\theta_0+\theta_1}{2} &= -\frac{w}{4}\left[l\int_0^l\frac{(l-x)dx}{EI}-\int_0^l\frac{(l-x)^2dx}{EI}\right]\\ &= -\frac{w}{4}\int_0^l\frac{x(l-x)dx}{EI}\\ d_n = -\theta_0 x &- \frac{wl}{2}\left[\int_0^x\frac{(l-x')(x-x')}{EI}dx'+\int_0^x\frac{Fdx'}{GA}\right]\\ &+ w\left[\int_0^x\frac{(l-x')^2(x-x')}{EI}dx'+2\int_0^x\frac{F(l-x')}{GA}dx'\right]\end{aligned}\right\} \quad (35)$$

Since the beam is symmetrical about its centre, it follows that $\frac{x(l-x)}{EI}$ has equal values for equal distances from the centre, so that

$$\int_0^l\frac{x(l-x)}{EI}dx = 2\int_0^{\frac{l}{2}}\frac{x(l-x)}{EI}dx.$$

Moreover, we may collect similar terms in the expression for $d_n$, so that the last two formulæ may be put in the following form:—

$$\theta_0 = \theta_1 = -\frac{w}{2}\int_0^{\frac{l}{2}}\frac{x(l-x)}{EI}dx \quad . \quad . \quad . \quad . \quad (36)$$

$$d_n = -\theta_0 x - \frac{w}{2}\int_0^x\frac{x'(l-x')(x-x')dx'}{EI}+\frac{w}{2}\int_0^x\frac{F(l-2x')}{GA}dx' \quad (37)$$

To determine the central deflection we put $x = \frac{l}{2}$ in (37); changing $x'$ to $x$ and substituting for $\theta_0$ its value (36) and effecting the necessary reductions, we have

$$\text{Central } d_n = \frac{w}{2}\int_0^{\frac{l}{2}} \frac{x^2(l-x)dx}{\text{EI}} + \frac{w}{2}\int_0^{\frac{l}{2}} \frac{\text{F}(l-2x)}{\text{GA}}dx \quad . \quad (38)$$

In practice we often have to apply this formula to girder bridges; but in this case the beam will be homogeneous and divided into several sections in each of which A, I, and F will be constant. The application of formula (38) then becomes very simple. For lattice girders it will happen that the different lengths for which I is constant will not be the same as those for which $\frac{\text{F}}{\text{A}}$ is constant, but that will not make the application of formula (38) difficult.

If the beam is of constant section throughout, we can effect the integrations, and obtain

$$\left.\begin{aligned} \theta_0 = \theta_1 &= -\frac{wl^3}{24\text{EI}} \\ \text{Central } d_n &= \frac{5wl^4}{384\text{EI}} + \frac{\text{F}wl^2}{8\text{GA}} = \frac{wl^2}{8}\left(\frac{5l^2}{48\text{EI}} + \frac{\text{F}}{\text{GA}}\right) \end{aligned}\right\} \quad . \quad (39)$$

If the beam is loaded simultaneously with a uniform load and the two loads W considered in the second case, we shall obtain the following results by addition :—

$$\left.\begin{aligned} \theta_0 = \theta_1 &= -\frac{\text{W}a(l-a)}{2\text{EI}} - \frac{wl^3}{24\text{EI}} \\ \text{Central } d_n &= \text{W}a\left(\frac{3l^2 - 4a^2}{24\text{EI}} + \frac{\text{F}}{\text{GA}}\right) + \frac{wl^2}{8}\left(\frac{5l^2}{48\text{EI}} + \frac{\text{F}}{\text{GA}}\right) \end{aligned}\right\} \quad (40)$$

**7. Beams Continuous over Several Supports.**—Consider a beam which is continuous over a number of supports and is loaded in any manner. It is clear that each portion of the beam comprised between two consecutive supports could be regarded as a separate beam, supported at its ends, carrying its load, and subjected at its ends to normal forces giving bending moments equal to those which occur at the supports in the continuous beam.

Now formulæ (21), (22), (24), (27) of § 5 show that, when we know the loading and the end bending moments of a beam supported at its ends, we can find the bending moment and shearing force at any section, as well as the slope or angular deflection at the ends and the equation to the elastic line of the beam.

We thus see that in order to determine the conditions of equilibrium of a continuous beam we have only to determine the bending moment at each of the supports.

We will now proceed to do this, noting in the first place that in order to simplify the discussion we will use the term "span" to denote each portion of the beam between two consecutive supports.

To analyse the most general case we will assume that the supports are capable of lowering under the action of the pressure which the beam exerts upon them; or rather, as amounts to the same thing, that these supports are not initially at exactly the same level.

Let A, B, C (Fig. 9, Pl. II) be three consecutive supports which were initially at the same level, and let $\alpha, \beta, \gamma$ be the extremely small drop of these points from their original position.

The angles which the straight lines AB, BC make with the horizontal after loading will be

$$\frac{\beta - \alpha}{l}, \quad \frac{\gamma - \beta}{l_1}$$

and therefore $\angle$JBC between BC and AB produced will be

$$\frac{\gamma - \beta}{l_1} - \frac{\beta - \alpha}{l}.$$

Now let BH be the final position of the section, initially vertical, corresponding to the support B. In accordance with what we have said above, if $B_0$, $B_1$, $B_2$ are the support bending moments, we can regard the span AB as a separate loaded beam supported at its ends and having normal forces at its ends giving moments $B_0$, $B_1$, the section B having before deformation the direction BH′ at right angles to AB; in the same way BC may be regarded as having end moments $B_1$, $B_2$, the section at B having initially the direction BH″ at right angles to BC.

We see from the figure that between the three angles HBH′, HBH″, H′BH″ we have the relation

$$\angle H'BH'' = \angle HBH'' - \angle HBH',$$

and as $\angle$H′BH″ is equal to $\angle$JBC which BC makes with AB produced, it follows that

$$\frac{\gamma - \beta}{l_1} - \frac{\beta - \alpha}{l} = \angle HBH'' - \angle HBH'.$$

If we call $W_i$ the internal work of the span AB, and $W_i'$ that of the span BC, we know by the theorem of the differential coefficients of internal work that

$$\angle \mathrm{HBH}' = -\frac{d\mathrm{W}_i}{d\mathrm{B}_1}$$

$$\angle \mathrm{HBH}'' = -\frac{d\mathrm{W}_i'}{d\mathrm{B}_1}.$$

The negative sign in the first expression is due to the fact that the angular deflection of the section, from BH′ to BH, occurs in the opposite direction to that which we have assumed for the bending moment $B_1$ considered as applied to the end of the span AB. Substituting in the previous equation we have

$$\frac{\gamma - \beta}{l_1} - \frac{\beta - a}{l} = \frac{d\mathrm{W}_i}{d\mathrm{B}_1} - \frac{d\mathrm{W}_i'}{d\mathrm{B}_1} . \qquad . \qquad . \qquad (41)$$

Now let us consider in the span AB a section G at distance $x$ from the end A, and let us call $M_{1x}$ the moment about G of all the loads on GB, the sum of these loads being $\frac{dM_{1x}}{dx}$. By formulæ (22) we have for the section G

$$\left.\begin{aligned} \mathrm{B}_\mathrm{G} &= \frac{\mathrm{B}_0(l - x)}{l} + \frac{\mathrm{B}_1 x}{l} - \frac{\mathrm{M}_{1,\,0}(l - x)}{l} + \mathrm{M}_{1x} \\ \mathrm{S}_\mathrm{G} &= \frac{\mathrm{B}_1 - \mathrm{B}_0}{l} + \frac{\mathrm{M}_{1,\,0}}{l} + \frac{d\mathrm{M}_{1x}}{dx} \end{aligned}\right\} \quad . \qquad (42)$$

For the internal work of the span AB we have

$$\mathrm{W}_i = \frac{1}{2}\int_0^l \frac{\mathrm{B}_\mathrm{G}^2 dx}{\mathrm{EI}} + \frac{1}{2}\int_0^l \frac{\mathrm{F}\mathrm{S}_\mathrm{G}^2 dx}{\mathrm{GA}}.$$

$$\frac{d\mathrm{W}_i}{d\mathrm{B}_1} = \frac{1}{l}\left[\int_0^l \frac{\mathrm{B}_\mathrm{G} x dx}{\mathrm{EI}} + \int_0^l \frac{\mathrm{F}\mathrm{S}_\mathrm{G} dx}{\mathrm{GA}}\right].$$

In the same way if we take a section K in the span BC at distance $y$ from C, and adopt similar notation, we shall have

$$\left.\begin{aligned} \mathrm{B}_\mathrm{K} &= \frac{\mathrm{B}_2(l_1 - y)}{l_1} + \frac{\mathrm{B}_1 y}{l_1} - \frac{\mathrm{M}'_{1,\,0}(l_1 - y)}{l_1} + \mathrm{M}_{1y} \\ \mathrm{S}_\mathrm{K} &= \frac{\mathrm{B}_1 - \mathrm{B}_2}{l_1} + \frac{\mathrm{M}'_{1,\,0}}{l_1} + \frac{d\mathrm{M}'_{1y}}{dy} \end{aligned}\right\} \qquad (43)$$

The internal work for the span BC will be

$$\mathrm{W}_i' = \frac{1}{2}\int_0^{l_1} \frac{\mathrm{B}_\mathrm{K}^2 dy}{\mathrm{EI}} + \frac{1}{2}\int_0^{l_1} \frac{\mathrm{F}\mathrm{S}_\mathrm{K}^2 dy}{\mathrm{GA}};$$

and therefore we shall have

$$\frac{d\mathrm{W}_i'}{d\mathrm{B}_1} = \frac{1}{l_1}\left[\int_0^{l_1} \frac{\mathrm{B}_\mathrm{K} y dy}{\mathrm{EI}} + \int_0^{l_1} \frac{\mathrm{F}\mathrm{S}_\mathrm{K} dy}{\mathrm{GA}}\right].$$

Substituting in equation (41) we obtain

$$\left.\begin{aligned}\frac{\gamma-\beta}{l_1}-\frac{\beta-a}{l}=\frac{1}{l}\left[\int_0^l\frac{\mathrm{B_G}x\,dx}{\mathrm{EI}}+\int_0^l\frac{\mathrm{FS_G}dx}{\mathrm{GA}}\right]&\\+\frac{1}{l_1}\left[\int_0^{l_1}\frac{\mathrm{B_K}y\,dy}{\mathrm{EI}}+\int_0^{l_1}\frac{\mathrm{FS_K}dy}{\mathrm{GA}}\right]&\end{aligned}\right\}\qquad(44)$$

This equation between the three bending moments $B_0$, $B_1$, $B_2$ can be easily developed by substituting for $B_G$, $S_G$, $B_K$, $S_K$ their values given by equations (42) and (43).

The above is the generalised form of Clapeyron's well-known "Theorem of Three Moments".

Suppose that the beam contains $n$ spans, i.e. $(n + 1)$ supports; if the beam is simply supported at the ends, i.e. if the bending moments at the extreme ends are zero, we shall require to find only $(n - 1)$ support bending moments, and therefore $(n - 1)$ unknowns.

Now by applying equation (44) to the first and second spans from the left, then to the second and third, and so on, we shall obtain exactly $(n - 1)$ equations each containing three support bending moments, except the first and last, which contain only two unknowns, since the moments at the extreme ends are zero.

If the beam is fixed at its ends, the extreme support bending moments will no longer be zero, and we shall have $(n + 1)$ unknowns, which will all be contained in the $(n - 1)$ equations obtained by successive applications of formula (44); thus this set of equations will not suffice for the determination of all the unknowns, and we must complete it by expressing the condition that the extreme sections remain vertical.

Now the slope of the section A of the span AB is given by

$$\theta_0=\frac{d\mathrm{W}_i}{d\mathrm{B}_0}=\frac{1}{l}\left[\int_0^l\frac{\mathrm{B_G}(l-x)dx}{\mathrm{EI}}-\int_0^l\frac{\mathrm{FS_G}dx}{\mathrm{GA}}\right]\qquad .\quad(45)$$

and that of the section C of the span BC by

$$\theta_2=\frac{d\mathrm{W}_i'}{d\mathrm{B}_2}=\frac{1}{l_1}\left[\int_0^l\frac{\mathrm{B_K}(l_1-y)dy}{\mathrm{EI}}-\int_0^{l_1}\frac{\mathrm{FS_K}dy}{\mathrm{GA}}\right]\qquad .\quad(46)$$

But as the axis of the span AB is inclined at an angle $\frac{(\beta-a)}{l}$ to the horizontal, and that of the span BC at an angle $\frac{\gamma-\beta}{l_1}$, it follows that if the sections at A and C are to remain vertical we must have

$$\frac{\beta-a}{l}-\theta_0=0,\quad\frac{\gamma-\beta}{l_1}+\theta_2=0,$$

and on substituting for $\theta_0$, $\theta_2$ in terms of (45), (46) after having multiplied the first equation by $l$ and the second by $l_1$, we get

$$(\beta - a) = \int_0^l \frac{B_G(l - x)dx}{EI} - \int_0^l \frac{FS_G dx}{GA} \quad . \quad . \quad (47)$$

$$(\gamma - \beta) = -\int_0^{l_1} \frac{B_K(l_1 - y)dy}{EI} + \int_0^{l_1} \frac{FS_K dy}{GA} \quad . \quad . \quad (48)$$

Applying equation (47) to the extreme left-hand span and equation (48) to the extreme right-hand span, we shall have the two additional formulæ required for the determination of all the bending moments.

In practice, the most common case of continuous beams occurs in iron bridges, where the extreme ends are never fixed and where the drop of the supports, i.e. of piles or piers, is zero or insignificant; so that we shall only have to apply formula (44) by putting $a = \beta = \gamma = 0$, thus reducing it to the following form

$$\frac{1}{l}\left[\int_0^l \frac{B_G x dx}{EI} + \int_0^l \frac{FS_G dx}{GA}\right] + \frac{1}{l_1}\left[\int_0^{l_1} \frac{B_K y dy}{EI} + \int_0^{l_1} \frac{FS_K dy}{GA}\right] = 0$$

Substituting in this equation for the bending moments and shears their values given by formulæ (42), (43), we obtain an equation between the three moments $B_0$, $B_1$, $B_2$, and, adopting the following abbreviations,

$$\left.\begin{aligned}
&\frac{1}{l^2}\left(\int_0^l \frac{(l - x)x dx}{EI} - \int_0^l \frac{F dx}{GA}\right) = L; \\
&\qquad\qquad \frac{1}{l_1^2}\left(\int_0^{l_1} \frac{(l_1 - y)y dy}{EI} - \int_0^{l_1} \frac{F dy}{GA}\right) = L_1 \\
&\frac{1}{l^2}\left(\int_0^l \frac{x^2 dx}{EI} + \int_0^l \frac{F dx}{GA}\right) = N; \quad \frac{1}{l_1^2}\left(\int_0^{l_1} \frac{y^2 dy}{EI} + \int_0^{l_1} \frac{F dy}{GA}\right) = N_1 \\
&\frac{1}{l}\left(\int_0^l \frac{M_{1x} x dx}{EI} + \int_0^l \frac{F}{GA} \frac{dM_{1x} dx}{dx}\right) = Q; \\
&\frac{1}{l_1}\left(\int_0^{l_1} \frac{M'_{1y} y dy}{EI} + \int_0^{l_1} \frac{F}{GA} \frac{dM'_{1y} dy}{dy}\right) = Q_1
\end{aligned}\right\} \quad (49)$$

this equation then becomes

$$B_0 L + B_1(N + N_1) + B_2 L_1 = M_{1,0} . L - Q + M'_{1,0} L_1 - Q_1 \quad (50)$$

It is important to note that the quantities L, N, $L_1$, $N_1$ are independent of the loading upon the beam, so that when we wish to study different systems of loading upon a given beam, these coefficients have to be calculated only once; the coefficients Q, $Q_1$, on the other hand, depend upon the loading and must be calculated for each case.

Generally, in the main beams of an iron bridge, the moment of inertia remains constant for one portion and then changes abruptly and remains constant for the succeeding portion and so on; the same is true for the quantity $\frac{F}{A}$, but, in the case of lattice girders, the points where the moment of inertia changes are not the same as those where $\frac{F}{A}$ changes. Thus for ordinary cases which arise in practice the calculation of L, N, Q, $L_1$, $N_1$, $Q_1$ will be quite simple.

We will now examine certain particular cases of loading when the continuous beam is homogenous and of the same section throughout.

The quantities EI, $\frac{F}{GA}$ being constant, we can effect the integrations, obtaining the following results:—

$$\left.\begin{aligned} L &= \frac{1}{l}\left(\frac{l^2}{6EI} - \frac{F}{GA}\right) \quad L_1 = \frac{1}{l_1}\left(\frac{l_1^2}{6EI} - \frac{F}{GA}\right) \\ N &= \frac{1}{l}\left(\frac{l^2}{3EI} + \frac{F}{GA}\right) \quad N_1 = \frac{1}{l_1}\left(\frac{l_1^2}{3EI} + \frac{F}{GA}\right) \end{aligned}\right\} \quad . \quad (51)$$

$$Q = \frac{1}{l}\left(\frac{1}{EI}\int_0^l M_{1x}x\,dx + \frac{F}{GA}\int_0^l \frac{dM_{1x}}{dx}dx\right)$$

$$Q_1 = \frac{1}{l_1}\left(\frac{1}{EI}\int_0^{l_1} M'_{1y}y\,dy + \frac{F}{GA}\int_0^{l_1} \frac{dM'_{1y}}{dy}dy\right).$$

Now we have

$$\int_0^l \frac{dM_{1x}}{dx}dx = M_{1l} - M_{1,0}; \quad \int_0^{l_1} \frac{dM'_{1y}}{dy}dy = M'_{1l_1} - M'_{1,0}$$

and since $M_{1l} = 0$ and $M'_{1l_1} = 0$ we deduce

$$Q = \frac{1}{l}\left(\frac{1}{EI}\int_0^l M_{1x}x\,dx - \frac{FM_{1,0}}{GA}\right); \quad Q_1 = \frac{1}{l_1}\left(\frac{1}{EI}\int_0^{l_1} M'_{1y}y\,dy - \frac{FM'_{1,0}}{GA}\right)$$

and therefore

$$\left.\begin{aligned} M_{1,0}L - Q &= \frac{1}{EIl}\left(\frac{l^2M_{1,0}}{6} - \int_0^l M_{1x}x\,dx\right) \\ M'_{1,0}L_1 - Q_1 &= \frac{1}{EIl_1}\left(\frac{l_1^2M'_{1,0}}{6} - \int_0^{l_1} M'_{1y}y\,dy\right) \end{aligned}\right\} \quad . \quad (52)$$

**Isolated Loads.**—Suppose, for example, that the span AB carries a load W at distances $a$, $b$ from A and B, and the span BC a load $W_1$ at distances $a_1$, $b_1$ from B and C (Fig. 10, Pl. II).

For the span AB we have for $0 < x < a$

$$M_{1x} = W(a - x)$$

and

$$M_{1,0} = Wa;$$

for $a < x < l$ we have

$$M_{1x} = 0.$$

$$\therefore \int_0^l M_{1x} x dx = W\int_0^a (a - x)x dx = \frac{Wa^3}{6}$$

and $$M_{1,0}L - Q = \frac{1}{l}\frac{Wa(l^2 - a^2)}{6EI}.$$

Similarly

$$M'_{1,0}L_1 - Q_1 = \frac{1}{l_1}\frac{W_1 b_1(l_1^2 - b_1^2)}{6EI}$$

and equation (50) becomes

$$\left.\begin{aligned}\frac{B_0}{l}\left(\frac{l^2}{6EI} - \frac{F}{GA}\right) + \frac{B_1(l + l_1)}{ll_1}\left(\frac{ll_1}{3EI} + \frac{F}{GA}\right) + \frac{B_2}{l_1}\left(\frac{l_1^2}{6EI} - \frac{F}{GA}\right)\\ = \frac{Wa(l^2 - a^2)}{6EIl} + \frac{W_1 b_1(l_1^2 - b_1^2)}{6EIl_1}\end{aligned}\right\} \quad (53)$$

If we neglect the terms due to shear, i.e. those containing $\frac{F}{GA}$, this formula simplifies considerably, and on multiplying throughout by 6EI becomes

$$B_0 l + 2B_1(l + l_1) + B_2 l_1 = \frac{Wa(l^2 - a^2)}{l} + \frac{W_1 b_1(l_1^2 - b_1^2)}{l_1}. \quad (54)$$

**Uniformly Distributed Load.**—Now suppose that the spans AB, BC are loaded with uniformly distributed loads, the first with intensity $w$ and the second with intensity $w_1$; we shall then have

$$M_{1x} = \frac{w}{2}(l - x)^2, \quad M'_{1y} = \frac{w_1}{2}(l_1 - y)^2$$

$$M_{1,0} = \frac{wl^2}{2}, \qquad M'_{1,0} = \frac{w_1 l_1^2}{2}$$

and therefore

$$\int_0^l M_{1x} x dx = \frac{w}{2}\int_0^l (l - x)^2 x dx = \frac{wl^4}{24}$$

$$\int_0^{l_1} M'_{1y} y dy = \frac{w_1}{2}\int_0^{l_1} (l_1 - y)^2 y dy = \frac{w_1 l_1^4}{24}$$

$$M_{1,0}L - Q = \frac{wl^3}{24EI}; \quad M'_{1,0}L_1 - Q_1 = \frac{w_1 l_1^3}{24EI}$$

Substituting these and the results (51) in equation (50) we get

$$\left.\begin{aligned}\frac{B_0}{l}\left(\frac{l^2}{6EI} - \frac{F}{GA}\right) + \frac{B_1(l + l_1)}{ll_1}\left(\frac{ll_1}{3EI} + \frac{F}{GA}\right) + \frac{B_2}{l_1}\left(\frac{l_1^2}{6EI} - \frac{F}{GA}\right)\\ = \frac{wl^3 - w_1 l_1^3}{24EI}\end{aligned}\right\} \quad (55)$$

If we neglect the terms due to shear containing $\frac{F}{GA}$, which are generally relatively very small, the equation simplifies and becomes after multiplying throughout by 6EI

$$B_0 l + 2B_1(l + l_1) + B_2 l_1 = \frac{wl^3 + w_1 l_1^3}{4} \quad . \quad . \quad (56)$$

This is the very important equation given by Clapeyron which has made possible the introduction into ordinary practice of calculations for continuous iron bridges.

**8. Practical formulæ for the design of single-span bridges —Diagrams of bending moment and shear for a uniform load rolling across the bridge.**

For the design of single-span bridges we usually assume a load uniformly distributed over a portion of the span.

Let ABB′A′ (Fig. 11, Pl. II) be a straight horizontal beam simply supported at its ends and loaded over a portion C′D′ with a uniform load of intensity $w$.

If we call

$$\begin{aligned} l &= \text{the span,} \\ a &= \text{the length A}'\text{C}', \\ b &= \text{the length A}'\text{D}', \end{aligned}$$

we find that the reactions are given by

$$R_A = w(b - a)\frac{2l - a - b}{2l}, \quad R_B = w(b - a)\frac{(a + b)}{2l}.$$

The bending moment at distance $x$ from A will be

$$\begin{aligned} B &= -R_A x \text{ if } 0 < x < a \\ &= -R_A x + \frac{w}{2}(x - a)^2 \text{ if } a < x < b \\ &= -R_B(l - x) \text{ if } b < x < l. \end{aligned}$$

Therefore if we take the straight line A′B′ as the axis of $x$ and A′ the origin, and draw the *bending moment diagram*, i.e. a curve whose ordinates are proportional to the bending moments, we see that this diagram will be composed of two straight lines A′E, FB′ and of a curve EGF, which will be a parabola with its vertex on GG′, as can be seen from its equation

$$B = -R_A x + \frac{w}{2}(x - a)^2 \quad . \quad . \quad . \quad (57)$$

This parabola is tangential to the straight lines A′E, B′F at E and F; in fact for $x = a$, equation (57) gives

$$B = -R_A a, \quad \frac{dB}{dx} = -R_A,$$

which shows that the parabola passes through E and that

it is tangential to A′E. In the same way by putting $x = b$ we see that the parabola also passes through F and is tangential to FB′.

It will be noted that the parameter of the parabola is $\frac{2}{w}$ whatever be the lengths $a$ and $b$, from which it follows that the parabola will be the higher the smaller the length $a$ and the greater the loaded portion C′D′ become. Therefore the bending moment at any section will reach its maximum value when the load covers the whole span.

For practical purposes therefore we will take for *the bending moment diagram a parabola passing through the ends A and B, the axis being vertical and the parameter* $\frac{2}{w}$.

Now let us determine the *shear diagram*, i.e. the curve whose abscissæ are the distances $x$ and whose ordinates give at each section the maximum shearing force that can occur as the lengths $a$ and $b$ vary.

As the shearing forces are the differential coefficients of the bending moments, it is clear that the expression for the shearing force will be as follows :—

$$\begin{aligned} S &= -R_A \text{ for the length AC} \\ &= -R_A + w(x - a) \text{ for the length CD} \\ &= R_B \text{ for the length DB ;} \end{aligned}$$

i.e. when the portion CD only is loaded, the shear diagram is made up of a straight line HI parallel to the axis of $x$ and having an ordinate proportional to $-R_A$, of another straight line IL with slope proportional to $w$, and finally of a third straight line LP parallel to the axis of $x$ and having an ordinate proportional to $R_B$. The point G′ where the shear diagram cuts the base is that at which the differential coefficient of the bending moment is zero, i.e. a point of maximum bending moment.

In practice we generally draw below the base the portion of the shear diagram which would come above, as is shown at G′L′P′.

We must now determine the loci of the points I, L′ as $a$ and $b$ vary. Now we see in the first place that if we vary the quantities $a$, $b$ in such a manner that the difference $(b - a)$ remains constant, i.e. if we move the load bodily along the span, the loci of I, L′ are two exactly similar curves symmetrical with regard to the centre of the beam.

If we diminish $a$ while maintaining $b$ constant we see that the ordinate of the point L′ increases, because it is proportional to

$$R_B = \frac{w(b^2 - a^2)}{2l}.$$

Thus the greatest value of the ordinate DL′ will correspond to $a = 0$ and will be equal to $S = \frac{wb^2}{2l}$.

Taking values of $b$ as abscissæ and of S as ordinates, it follows from this equation that the curve of maximum shear will be a parabola AQT with its vertex at A and its axis vertical, the parameter being $\frac{2l}{w}$.

As we must consider another parabola equal to the first and symmetrical about the centre of the beam, we see that the curve of maximum shears due to the rolling or live load will consist of two parabolic arcs QU and QT.

**9. Diagrams of bending moment and shear for combined dead and rolling loads.**—The dead load on bridges is usually regarded as uniformly distributed along the length.

We will suppose that the dead load is $w_d$ kg. per metre run and the live, rolling or super-load $w_s$ kg. per metre run. It is clear that if we consider the dead load alone, the bending moment at a distance $x$ from the support A will be

$$B_d = -\frac{1}{2}w_d x(l - x);$$

and we have seen that for the super-load alone the maximum bending moment at any section occurs when the whole span is covered, so that we shall have

$$B_s = -\frac{1}{2}w_s x(l - x).$$

When therefore we consider the dead load and super-load acting together, the maximum bending moment on a section at distance $x$ from the support will be

$$B = -\frac{1}{2}(w_d + w_s)x(l - x),$$

*i.e. the bending moment diagram will be a parabola passing through the ends of the beam, its axis being vertical and its parameter* $\frac{2}{w_d + w_s}$.

Now let us consider the shearing force.

For a section at distance $x$ from the support A, the dead load alone gives a shearing force

$$S_d = -w_d\left(\frac{l}{2} - x\right).$$

But for the rolling or super-load we have seen that the maximum shearing force that can occur at this section will be

$$S_s = \frac{w_s x^2}{2l}$$

if the section is in the right-hand half of the beam, i.e. if $x > \frac{l}{2}$; thus for all values of $x$ satisfying this condition the maximum shearing force due to the combined loads will be

$$S = -w_d\left(\frac{l}{2} - x\right) + \frac{w_s x^2}{2l}$$

*therefore the maximum shear diagram for the right-hand half of the beam will be a parabolic arc having a vertical axis and a parameter equal to* $\frac{2l}{w_s}$, *and that for the left-hand half will be a similar symmetrical parabolic arc.*

To facilitate the drawing of this diagram we note that its mid-ordinate is $\frac{w_s l}{8}$, and its end ordinates $\frac{(w_d + w_s)l}{2}$, as will be seen by putting $x = \frac{l}{2}$ and $x = l$ in the above formula.

**10. Bending Moment and Shear Diagrams for an Isolated Rolling Load.**—When the load W is at a distance $x$ from the end A (Fig. 12, Pl. II) we have below the load a bending moment

$$B = -\frac{Wx(l - x)}{l}, \quad . \quad . \quad . \quad (58)$$

and the bending moment diagram for this position of the load consists of two straight lines AD, DB, the ordinate CD representing the bending moment at the section C. The diagram of maximum bending moments when the load moves across the span is the locus of the point D; now this locus is clearly given by equation (58), so that the diagram required is a parabola passing through the ends A and B, the axis being vertical and the parameter being $\frac{l}{W}$.

In the case of a uniformly distributed load of intensity $w$, we have seen that the bending moment diagram is also a parabola, its parameter being $\frac{2}{w}$, so that the two parabolas will coincide if $\frac{l}{W} = \frac{2}{w}$ or $w = \frac{2W}{l}$. We see therefore that *a load W rolling across a beam produces the same bending moment as a load of twice the amount uniformly distributed over the whole span.*

Next let us consider the shearing forces.

When the load W is at distance $x$ from the end A, the shearing force between A and C is equal to $R_A$, i.e. to $\frac{W(l-x)}{l}$, and between C and B it is equal to $R_B$, i.e. to $\frac{Wx}{l}$, so that the shear diagram is made up of two horizontal straight lines EF, GH.

To find the maximum shear diagram when the load crosses, we must find the locus of the point F for the left-hand half of the beam and then repeat this locus symmetrically for the other half.

Now $CF = S = \frac{W(l-x)}{l}$, and therefore the locus of F is a straight line, and since $S = W$ for $x = 0$ and $S = 0$ for $x = l$, we see that this straight line passes through B and that its ordinate AI is equal to W; at the centre, where $x = \frac{l}{2}$, the ordinate will be $\frac{W}{2}$. Therefore *the diagram of maximum shear consists of two straight lines IL, LN passing respectively through B and A.*

**11. Bending Moment and Shear Diagrams for Iron Bridges.** —We have seen that the uniform load $w$, equivalent as regards bending moment to an isolated rolling load W, is given by the relation $wl = 2W$.

Multiplying by $\frac{l}{8}$ we have

$$\frac{wl^2}{8} = \frac{Wl}{4};$$

now $\frac{wl^2}{8}$ is the maximum bending moment due to the uniform load and $\frac{Wl}{4}$ is the maximum bending moment which the load W can induce in traversing the span; it follows, therefore, that the equivalent uniform load can be found by the condition that the maximum bending moment shall be the same in the two cases.

This relation has been used in practice to simplify the calculations for railway bridges. As the trains which have to traverse the bridge have not uniformly distributed loads, *we find the maximum bending moment which can arise and then design the bridge as if it had a uniformly distributed load which would produce this bending moment.*

If this bending moment is B, we shall find the equivalent uniform load $w$ by the equation

$$B = \frac{wl^2}{8}$$

or
$$w = \frac{8B}{l^2},$$

$w$ being expressed in weight units per metre run.

But we see from our previous analysis that this reasoning for bending moments cannot be correct for shearing forces; for in the case of an isolated rolling load we have seen that the maximum shear diagram consists of two straight lines IL, LN giving at the centre a shearing force which is one-half that at the ends, whereas with a uniformly distributed load the shear diagram consists of two straight lines IL′, L′N giving zero shear at the centre, and therefore smaller shears than those due to the rolling load at all other sections.

We believe then that for shearing forces it is necessary in each particular case to take the heaviest train load for which the bridge is intended and find the maximum shear at every point as this load crosses the span.

**12. Deflection at the centre of a single span bridge carrying a uniformly distributed load.**—For this case we have, by equation (38), § 6,

$$\text{Central } d_n = \frac{w}{2}\int_0^{\frac{l}{2}} \frac{x^2(l-x)dx}{EI} + \frac{w}{2}\int_0^{\frac{l}{2}} \frac{F(l-2x)dx}{GA};$$

this shows that the deflection consists of two parts, one $= \frac{w}{2}\int_0^{\frac{l}{2}} \frac{x^2(l-x)dx}{EI}$ due to bending moment, and the other $= \frac{w}{2}\int_0^{\frac{l}{2}} \frac{F(l-2x)dx}{GA}$ due to shear.

These formulæ apply only to beams which are symmetrical about their centre.

If we suppose the beam to be divided into a number of elements symmetrical in pairs with regard to the centre C, and such that the end elements $a_1$, $a_1$ (Fig. 13, Pl. II) have a constant moment of inertia $I_1$, the next two a constant moment of inertia $I_2$, and so on, we can perform the integrations and we shall have

Central deflection due to bending moment

$$= \frac{w}{2E}\left[\frac{l}{3}\left(\frac{a_1^3}{I_1} + \frac{a_2^3 - a_1^3}{I_2} + \frac{a_3^3 - a_2^3}{I_3} + \frac{\left(\frac{l}{2}\right)^3 - a_3^3}{I_4}\right) - \frac{1}{4}\left(\frac{a_1^4}{I_1} + \frac{a_2^4 - a_1^4}{I_2} + \frac{a_3^4 - a_2^4}{I_3} + \frac{\left(\frac{l}{2}\right)^4 - a_3^4}{I_4}\right)\right]$$

To calculate the central deflection due to shear we will suppose that the beam is divided into symmetrical elements (Fig. 14, Pl. II) such that $\frac{F}{A}$ is constant for each pair and equal to $\frac{F_1}{A_1}$ for the end two and for the next two equal to $\frac{F_2}{A_2}$, and so on, and on performing the integrations we shall have

Deflection due to shear

$$= \frac{w}{2G}\left[ l\left( \frac{F_1 b_1}{A_1} + \frac{F_2(b_2 - b_1)}{A_2} + \frac{F_3(b_3 - b_2)}{A_3} + \frac{F_4\left(\frac{l}{2} - b_3\right)}{A_4} \right) - \left( \frac{F_1 b_1^2}{A_1} + \frac{F_2(b_2^2 - b_1^2)}{A_2} + \frac{F_3(b_3^2 - b_2^2)}{A_3} + \frac{F_4\left(\frac{l^2}{4} - b_3^2\right)}{A_4} \right) \right]$$

In general it happens that solid web beams are divided into the same elements for bending moments as for shears, so that we change $b$ into $a$, but this does not give much advantage in the calculations. In lattice girders, on the other hand, this hardly ever happens, so that we have kept separate the two expressions making up the central deflection.

It should be understood that in the case of a lattice girder formula (38) should be modified by substituting the direct elastic modulus E for the glide modulus G and the area $A_d$ of the diagonal bracing for the area A of the section of the beam.

**13. Iron Bridges Composed of Continuous Beams.**—In single span iron bridges, the bending moment and shearing force at any section depend only upon the load and not upon the transverse dimensions of the beam. On the other hand, in bridges in which the main beams extend from one end to the other without interruption at the intermediate piers, it follows from formulæ (44) and 50 of § 7 that the bending moments at the piers, and therefore the bending moment and shearing force at any other section, depend also upon the dimensions of the beam.

Now as these dimensions are the very unknown quantities which we have to calculate, we see that considerable difficulty would arise if we wished to solve the problem in a direct manner. It is for this reason that in practice we have been obliged up to the present to calculate continuous beams by means of formula (56), which is deduced from (50), on the assumptions that each span carries a uniformly distributed load, that deformations due to shear are negligible, and that the moment of inertia of the beam section is constant throughout.

This procedure is undoubtedly the only rational one for preliminary calculations, because it enables us to calculate approximately the bending moment and shearing force at any section without knowing the dimensions of the beam. We can then, by the results thus obtained, divide each span into a number of lengths, giving to each a constant section such that it can resist the maximum stresses to which it will be subjected.

Generally we restrict ourselves to this operation, but it is clear that to determine the exact stresses which will occur in the beam we must make fresh calculations and determine a fresh bending moment and shear diagram, taking into account the variation both of the moment of inertia and of the quantity $\frac{F}{A}$. This second calculation does not present any difficulty, because the beam is now divided into lengths for each of which I and $\frac{F}{A}$ are constant, so that the integrations of formula (44) can be readily performed.

We will also note that generally continuous bridges do not have their spans equal, the two end spans being usually equal in length to each other but a little shorter than the rest.

Finally, we will add that when we have to perform the calculations for a continuous girder bridge, it is sufficient in practice to consider the cases of the spans loaded one at a time and two consecutive spans at a time.

To calculate the deflection $d_n$ at the centre of any span of a continuous beam we employ formula (25) of § 5 by putting $x = \frac{l}{2}$, which gives, on omitting the dash in $x'$

$$\text{Central } d_n = -\frac{1}{2}\left[\int_0^{\frac{l}{2}} \frac{Bx\,dx}{EI} + \int_{\frac{l}{2}}^{l} \frac{B(l-x)\,dx}{EI}\right] - \frac{1}{2}\left[\int_0^{\frac{l}{2}} \frac{FS\,dx}{GA} - \int_{\frac{l}{2}}^{l} \frac{FS\,dx}{GA}\right].$$

This deflection is made up, as we can see, of two parts,

one $$-\frac{1}{2}\left[\int_0^{\frac{l}{2}} \frac{Bx\,dx}{EI} + \int_{\frac{l}{2}}^{l} \frac{B(l-x)\,dx}{EI}\right]$$

due to bending moment,

and the other $$= -\frac{1}{2}\left[\int_0^{\frac{l}{2}} \frac{FS\,dx}{GA} - \int_{\frac{l}{2}}^{l} \frac{FS\,dx}{GA}\right]$$

due to shear.

Since we assume that the span under consideration carries a uniformly distributed load, we must put in (22) of § 5

$$M_{1,0} = \frac{wl^2}{2}, \quad M_{1x} = \frac{w}{2}(l - x)^2$$

which gives

$$B = \frac{B_0(l - x)}{l} + \frac{B_1 x}{l} - \frac{w}{2}(l - x)x$$

and therefore

$$S = \frac{(B_1 - B_0)}{l} - w\left(\frac{l}{2} - x\right).$$

These last two formulæ enable us to determine the bending moment and shearing force at any section when we have determined the bending moments at the interior supports.

Substituting in the above formulæ we have

Central deflection due to bending moment

$$= -\frac{B_0}{2l}\left[\int_0^{\frac{l}{2}}\frac{(l - x)x\,dx}{EI} + \int_{\frac{l}{2}}^{l}\frac{(l - x)^2dx}{EI}\right]$$

$$-\frac{B_1}{2l}\left[\int_0^{\frac{l}{2}}\frac{x^2dx}{EI} + \int_{\frac{l}{2}}^{l}\frac{(l - x)x\,dx}{EI}\right]$$

$$+\frac{w}{4}\left[\int_0^{\frac{l}{2}}\frac{(l - x)x^2dx}{EI} + \int_{\frac{l}{2}}^{l}\frac{(l - x)^2x\,dx}{EI}\right],$$

Central deflection due to shear

$$= -\frac{(B_1 - B_0)}{2l}\left[\int_0^{\frac{l}{2}}\frac{F\,dx}{GA} - \int_{\frac{l}{2}}^{l}\frac{F\,dx}{GA}\right]$$

$$+\frac{w}{2}\left[\int_0^{\frac{l}{2}}\frac{F(l - 2x)dx}{2GA} - \int_{\frac{l}{2}}^{l}\frac{F(l - 2x)dx}{2GA}\right].$$

In most cases in practice these formulæ become simplified whether we apply them to end or to intermediate spans, as the following considerations show.

1. *End Spans.*—If we take, for instance, the left-hand support we have $B_0 = 0$ and therefore

Central deflection due to bending moment

$$= -\frac{B_1}{2l}\left[\int_0^{\frac{l}{2}}\frac{x^2dx}{EI} + \int_{\frac{l}{2}}^{l}\frac{(l - x)x\,dx}{EI}\right]$$

$$+\frac{w}{4}\left[\int_0^{\frac{l}{2}}\frac{(l - x)x^2dx}{EI} + \int_{\frac{l}{2}}^{l}\frac{(l - x)^2x\,dx}{EI}\right],$$

Central deflection due to shear

$$= -\frac{B_1}{2l}\left[\int_0^{\frac{l}{2}}\frac{Fdx}{GA} - \int_{\frac{l}{2}}^{l}\frac{Fdx}{GA}\right]$$

$$+\frac{w}{2}\left[\int_0^{\frac{l}{2}}\frac{F(l-2x)dx}{2GA} - \int_{\frac{l}{2}}^{l}\frac{F(l-2x)dx}{2GA}\right].$$

2. *Intermediate Spans.*—These spans, as we have noted already, are usually equal in length to each other and symmetrical about their centre lines; in this case we can simplify considerably, because if we put $(l - x) = y$ in the integrals from $\frac{l}{2}$ to $l$, i.e. if for these integrals we take the right-hand end for origin, we have

$$\int_{\frac{l}{2}}^{l}\frac{(l-x)^2dx}{EI} = \int_0^{\frac{l}{2}}\frac{y^2dy}{EI};\quad \int_{\frac{l}{2}}^{l}\frac{(l-x)xdx}{EI} = \int_0^{\frac{l}{2}}\frac{(l-y)ydy}{EI};$$

$$\int_{\frac{l}{2}}^{l}\frac{(l-x)^2xdx}{EI} = \int_0^{\frac{l}{2}}\frac{y^2(l-y)dy}{EI},\quad \int_{\frac{l}{2}}^{l}\frac{Fdx}{GA} = \int_0^{\frac{l}{2}}\frac{Fdy}{GA}$$

$$\int_{\frac{l}{2}}^{l}\frac{F(l-2x)dx}{2GA} = -\int_0^{\frac{l}{2}}\frac{F(l-2y)dy}{2GA}$$

and as, by symmetry, these integrals in $y$ are equal to those which would be obtained by exchanging $x$ for $y$, we have

Central deflection due to bending moment

$$= -\frac{(B_0 + B_1)}{2l}\left[\int_0^{\frac{l}{2}}\frac{(l-x)xdx}{EI} + \int_0^{\frac{l}{2}}\frac{x^2dx}{EI}\right] + \frac{w}{2}\int_0^{\frac{l}{2}}\frac{(l-x)x^2dx}{EI},$$

$$= -\frac{(B_0 + B_1)}{2}\int_0^{\frac{l}{2}}\frac{xdx}{EI} + \frac{w}{2}\int_0^{\frac{l}{2}}\frac{(l-x)x^2dx}{EI}.$$

Central deflection due to shear

$$= \frac{w}{2}\int_0^{\frac{l}{2}}\frac{F(l-2x)dx}{GA}.$$

The use of these formulæ is very simple in practical cases because each span will generally be divided into several segments each of which will have a constant section.

We will repeat here the observation made with regard to beams of single span, that when lattice girders are employed it will be necessary to substitute, in the expression for the deflection due to shear, E′ for the bracing for G, and for A the area $A_d$ of the diagonal bars.

With regard to the coefficient F, we have given formulæ for its calculation, both for solid web and lattice-braced beams.

**14. Observations on Formula (25).**—We will now show that we can pass at once from formula (25) to equation (44).

Now the first term of formula (25), i.e.

$$-\frac{(l-x)}{l}\left[\int_0^x \frac{\mathrm{B}x'dx'}{\mathrm{EI}} + \int_0^x \frac{\mathrm{FS}dx'}{\mathrm{GA}}\right]$$

relates to the segment AC (Fig. 8, Pl. II) whilst the second term

$$-\frac{x}{l}\left[\int_x^l \frac{\mathrm{B}(l-x')dx'}{\mathrm{EI}} - \int_x^l \frac{\mathrm{FS}}{\mathrm{GA}}dx'\right]$$

relates to the segment BC; we can write this second term in the following form:—

$$-\frac{x}{l}\left[-\int_{l-x}^0 \frac{\mathrm{B}(l-x')}{\mathrm{EI}}d(l-x') + \int_{l-x}^0 \frac{\mathrm{FS}}{\mathrm{GA}}d(l-x')\right],$$

and therefore by putting $(l-x') = y'$, $(l-x) = y$ and reversing the limits of integration, which necessitates alteration of the sign of the terms, we obtain

$$-\frac{x}{l}\left[\int_0^y \frac{\mathrm{B}y'dy'}{\mathrm{EI}} - \int_0^y \frac{\mathrm{FS}}{\mathrm{GA}}dy'\right].$$

We can, for greater clearness, write B′ for B and S′ for S, to distinguish the bending moment and shearing force on the segment CB from the similar quantities for the segment AC. Thus formula (25) can be written in the form

$$d_n = -\frac{y}{l}\left[\int_0^x \frac{\mathrm{B}'x'dx'}{\mathrm{EI}} + \int_0^x \frac{\mathrm{FS}}{\mathrm{GA}}dx'\right]$$
$$-\frac{x}{l}\left[\int_0^y \frac{\mathrm{B}'y'dy'}{\mathrm{EI}} - \int_0^y \frac{\mathrm{FS}'}{\mathrm{GA}}dy'\right].$$

We know that we can express the bending moment B and shearing force S, for any section on the length AC, as a function of the loads which it carries, and of the bending moments at the sections A, C; and in the same way the bending moment B′ and shearing force S′ for any section on the length BC can be expressed as a function of the loads which it carries and of the bending moments at the sections B, C. It follows, therefore, that we could express the deflection of the point C as a function of the three bending moments, i.e. of those at the ends of the beam and that at the section C.

It should be noted that formula (25) gives the deflection at the point C on the assumption that the ends of the beam cannot drop; in other words, it gives the lowering of the point C with reference to the straight line joining the extreme ends of the beam after deformation; thus if the ends A, B drop respectively by amounts $a$, $\gamma$ during the loading, and the point C by an amount $\beta$, it is clear that the deflection of this point, with reference to the straight line joining the extreme ends after loading, will be (noting that $x + y = l$)

$$\beta - \left(a + \frac{\gamma - a}{l}x\right) = \frac{(\beta - \gamma)x + (\beta - a)y}{l}.$$

Substituting this value for $d_n$ in the formula obtained above and then dividing by $-\frac{xy}{l}$, we obtain

$$\frac{(\gamma - \beta)}{y} - \frac{(\beta - a)}{x} = \frac{1}{x}\left[\int_0^x \frac{\mathrm{B}x'dx'}{\mathrm{EI}} + \int_0^x \frac{\mathrm{FS}dx'}{\mathrm{GA}}\right]$$
$$+ \frac{1}{y}\left[\int_0^y \frac{\mathrm{B}'y'dy'}{\mathrm{EI}} - \int_0^y \frac{\mathrm{FS}'dy'}{\mathrm{GA}}\right].$$

If we wish to apply this formula to continuous beams, we regard two consecutive spans AB, BC as forming a single span, by assuming that we have removed the support B and substituted an upward vertical force. Then to obtain a relation between the bending moments at the three supports A, B, C we have only to change $x$ to $l$ and $y$ to $l_1$ in the last formula, and we shall have equation (44) at once.

## CHAPTER VIII.

### THEORY OF COLUMNS.

**1. Curve assumed by a vertical beam when loaded at the top** (Fig. 15, Pl. II).

Let AB be a straight vertical beam fixed at its lower end, the principal axes of the sections of the beam being contained in two planes at right angles to each other cutting a horizontal plane in XX′, YY′.

Let us suppose that on the upper end there are applied :—

1. Downwardly acting vertical forces or loads distributed uniformly on each line parallel to XX′ and in a linear relation parallel to YY′.

2. Tangential forces parallel to YY′ having a resultant $S_1$. We will call $P_1$ the resultant pressure or sum of all the forces or loads applied on the end B, and $B_1$ the sum of their moments about the principal axis parallel to XX′.

We have now to find the angular rotation of the upper end and the curved form of the centre line after bending.

Now if we suppose that after deformation the centre line of the body takes the curved form AB′, and if we call $d_n$ the deflection, i.e. the horizontal displacement BB′ of the centre of the uppermost section, and $y$ the ordinate of a point C′ on the curve, i.e. the displacement CC′ of the centroid of the section C, it is clear that, for the section corresponding to a point C at distance $x$ from the base, the bending moment, normal pressure, and shearing force will be given by the following formulæ :—

$$\left.\begin{aligned} B &= B_1 + P_1(d_n - y) + S_1(l - x) \\ P &= P_1 \\ S &= S_1 \end{aligned}\right\} \quad . \quad . \quad (1)$$

We take for the shearing forces the constant value $S_1$, neglecting the component of the pressure P due to the small inclination of the sections to the horizontal after bending.

Adopting our usual notation, the internal work will be expressed by the formula

$$W_i = \frac{1}{2}\int_0^l \frac{B^2 dx}{EI} + \frac{1}{2}\int_0^l \frac{P^2 dx}{EA} + \frac{1}{2}\int_0^l \frac{FS^2}{GA},$$

the differential coefficient of which with regard to $B_1$ must give the angular deflection $\theta_1$ of the top section. The second and third integrals do not contain $B_1$ and therefore disappear on differentiating; we therefore have, noting that $\frac{dB}{dB_1} = 1$,

$$\theta_1 = \int_0^l \frac{B}{EI}\frac{dB}{dB_1}dx = \int_0^l \frac{Bdx}{EI} \quad . \quad . \quad . \quad (2)$$

whence, substituting for B its value (1)

$$\theta_1 = B_1\int_0^l \frac{dx}{EI} + P_1\int_0^l \frac{(d_n - y)dx}{EI} + S_1\int_0^l \frac{(l - x)dx}{EI} \quad . \quad (2A)$$

We see that we can only make use of this formula after having found $y$ as a function of $x$, i.e. after having determined the curve followed by the centre line of the column after deformation.

To find the equation to this curve, let us suppose that a horizontal force $F_h$ and a vertical force $F_v$, as small as desired, are applied at the centre of the section C; then the bending moment, normal pressure and shear for the section D at a distance $x' < x$ from the base will be given by the following equations:—

$$\left.\begin{aligned} B &= B_1 + P_1(d_n - y') + S_1(l - x') + F_h(x - x') + F_v(y - y') \\ P &= P_1 + F_v \\ S &= S_1 + F_h \end{aligned}\right\} \quad (3)$$

For sections above C, the bending moment, normal pressure, and shearing force will not contain F , $F_v$, i.e. the quantities will be given by equations (1). Thus, as the horizontal and vertical displacements of the point C are the differential coefficients with regard to $F_h$ and $F_v$ of the internal work of the column, and as the portion of this work due to the part CB does not contain $F_h$ or $F_v$, it is sufficient to consider the internal work of the portion AC, i.e. the expression

$$\frac{1}{2}\int_0^x \frac{B^2dx'}{EI} + \frac{1}{2}\int_0^x \frac{P^2dx'}{EA} + \frac{1}{2}\int_0^x \frac{FS^2}{GA}dx',$$

the values of B, P, and S being given by equations (3).

Taking the differential coefficients of this formula with regard to $F_h$ and $F_v$ we obtain the horizontal and vertical displacements $y$, $z$ of the point C. Thus, since formulæ (3) give

$$\frac{dB}{dF_h} = x - x'; \quad \frac{dP}{dF_h} = 0; \quad \frac{dS}{dF_h} = 1$$
$$\frac{dB}{dF_v} = y - y'; \quad \frac{dP}{dF_v} = 1; \quad \frac{dS}{dF_v} = 0$$

we obtain

$$\left.\begin{aligned} y &= \int_0^x \frac{B(x - x')dx'}{EI} + \int_0^x \frac{FS}{GA}dx' \\ z &= \int_0^x \frac{B(y - y')dx'}{EI} + \int_0^x \frac{P}{EA}dx' \end{aligned}\right\} \quad . \quad . \quad (4)$$

To obtain the displacements $y$, $z$ when the forces $F_h$, $F_v$ are non-existent, we must substitute in these equations the expressions for B, P, S in (3) after having made $F_h = F_v = 0$, i.e.

$$\left.\begin{aligned} B &= B_1 + P_1(d_n - y') + S_1(l - x') \\ P &= P_1 \\ S &= S_1 \end{aligned}\right\} \quad . \quad . \quad (5)$$

It should be noted that under the integration sign in formulæ (4) we ought to take for $x, x'$ the abscissæ after deformation of the points under consideration, i.e. the original abscissæ diminished by the vertical displacements given by the second of formulæ (4). If we wished to proceed with a rigid analysis, the solution of the problem would be very difficult; but as the vertical displacements are always extremely small, it is clear that we may with a more than sufficient degree of accuracy take the original abscissæ for $x$ and $x'$ in formulæ (4). Then the first of these formulæ will suffice to obtain the ordinates of the curved formation of the centre-line of the column under load, and we can use the second to calculate the diminution in the original abscissæ. Moreover, we see that since the ordinates $y$ are extremely small with reference to the abscissæ $x$, the integral

$$\int_0^x \frac{B(y - y')dx'}{EI}$$

will be very small compared with the integral

$$\int_0^x \frac{B(x - x')dx'}{EI},$$

so that we can neglect the first term in the expression for $z$ and take simply

$$z = \int_0^x \frac{P_1 dx'}{EA} \quad . \quad . \quad . \quad . \quad (6)$$

Now if we differentiate the first of the formulæ (4) with regard to $x$, noting that $x$ is regarded as constant in the integration, we have, by the rule for differentiating under the integration sign,

$$\frac{dy}{dx} = \int_0^x \frac{B dx'}{EI} + \frac{FS_1}{GA} \quad . \quad . \quad . \quad . \quad (7)$$

The abscissa $x$ does not at present enter into the quantity under the integration sign,* but only in the superior limit of the integral; therefore on differentiating a second time with regard to $x$ we obtain

$$\frac{d^2y}{dx^2}=\frac{B}{EI}+\frac{S_1 d\left(\frac{F}{GA}\right)}{dx}=\frac{B_1+P_1(d_n-y)+S_1(l-x)}{EI}+S_1\frac{d\left(\frac{F}{GA}\right)}{dx} \quad (8)$$

**2. Integration of Equation (8) for Homogeneous Columns of Constant Section.**—If the column is homogeneous and prismatic (i.e. of constant section), the quantities EI and $\frac{F}{GA}$ will be constant and the second term of equation (8) will disappear, and we can easily integrate the equation by writing

$$B_1 + P_1(d_n - y) + S_1(l - x) = (B_1 + P_1 d_n)u,$$

thus obtaining

$$-P_1\frac{dy}{dx} - S_1 = (B_1 + P_1 d_n)\frac{du}{dx}$$

$$-P_1\frac{d^2y}{dx^2} = (B_1 + P_1 d_n)\frac{d^2u}{dx^2},$$

so that equation (8) becomes

$$\frac{d^2u}{dx^2}+\frac{P_1 u}{EI}=0.$$

The general solution of this equation is

$$u=\frac{B_1+P_1(d_n-y)+S_1(l-x)}{B_1+P_1 d_n}=C_1\cos\left(x\sqrt{\frac{P_1}{EI}}\right)+C_2\sin\left(x\sqrt{\frac{P_1}{EI}}\right),$$

$C_1$ and $C_2$ being arbitrary constants.

To determine these constants we will note that for $x = 0$ we must have $y = 0$, $\frac{dy}{dx}=\frac{FS_1}{GA}$, as result from the first equation (4) and equation (7); from this we get

$$C_1=\frac{B_1+P_1 d_n+S_1 l}{B_1+P_1 d_n},\quad C_2=-\sqrt{\frac{EI}{P_1}}\cdot\frac{\frac{FP_1}{GA}+1}{B_1+P_1 d_n}\cdot S_1.$$

* We can also obtain formula (7) as follows by writing the value of $y$ as

$$y=x\int_0^x\frac{B\,dx'}{EI}-\int_0^x\frac{Bx'\,dx'}{EI}+\int_0^x\frac{FS\,dx'}{GA};$$

since the variable $x$ with regard to which we have to differentiate no longer comes under the integration sign, we have at once

$$\frac{dy}{dx}=\int_0^x\frac{B\,dx'}{EI}+\frac{xB}{EI}-\frac{Bx}{EI}+\frac{FS}{GA}=\int_0^x\frac{B\,dx'}{EI}+\frac{FS_1}{GA}.$$

We can neglect in the value of $C_2$ the quantity $\frac{FP_1}{GA}$ in comparison with unity, thus obtaining

$$C_2 = -\sqrt{\frac{EI}{P_1}} \cdot \frac{S_1}{B_1 + P_1 d_n}.$$

For a homogeneous rectangular section we shall have $G = \frac{2E}{5}$, $F = \frac{6}{5}$, and therefore $\frac{FP_1}{GA} = \frac{3P_1}{EA}$; now since the quantity $\frac{P_1}{EA}$ represents the shortening of a unit length of the column and is extremely small, we can neglect it absolutely in comparison with unity, so that the general solution becomes in this case

$$B_1 + P_1(d_n - y) + S_1(l - x) = (B_1 + P_1 d_n + S_1 l) \cos\left(x\sqrt{\frac{P_1}{EI}}\right) - S_1\sqrt{\frac{EI}{P_1}} \sin\left(x\sqrt{\frac{P_1}{EI}}\right) \quad (9)$$

**3. Discussion of Equation (9) when $S_1$ is Zero.**—If $S_1 = 0$, equation (9) becomes

$$\frac{B_1 + P_1(d_n - y)}{B_1 + P_1 d_n} = \cos\left(x\sqrt{\frac{P_1}{EI}}\right) \quad . \quad . \quad . \quad (10)$$

As this equation contains the deflection $d_n$, which is unknown, we determine it by the condition that $y = d_n$ for $x = l$, so that

$$\frac{B_1}{B_1 + P_1 d_n} = \cos\left(l\sqrt{\frac{P_1}{EI}}\right)$$

which gives

$$d_n = \frac{B_1\left\{1 - \cos\left(l\sqrt{\frac{P_1}{EI}}\right)\right\}}{P_1 \cos\left(l\sqrt{\frac{P_1}{EI}}\right)} \quad . \quad . \quad . \quad (11)$$

If the load is uniformly distributed on the top, i.e. if $B_1 = 0$, it follows from formula (11) that in general the deflection $d_n$ will also be zero. Nevertheless, since this formula can be written in the form

$$d_n \cos\left(l\sqrt{\frac{P_1}{EI}}\right) = \frac{B_1}{P_1}\left\{1 - \cos\left(l\sqrt{\frac{P_1}{EI}}\right)\right\}$$

we see that when $B_1 = 0$,

$$d_n \cos\left(l\sqrt{\frac{P_1}{EI}}\right) = 0;$$

this equation will be satisfied either if $d_n = 0$, or if $\cos\left(l\sqrt{\frac{P_1}{EI}}\right) = 0$, i.e. if

$$l\sqrt{\frac{P_1}{EI}} = (2i + 1)\frac{\pi}{2} \quad . \quad . \quad . \quad (12)$$

$i$ being any integer.

From equation (12) we deduce

$$P_1 = (2i + 1)^2\frac{\pi^2}{4} \cdot \frac{EI}{l^2} \quad . \quad . \quad . \quad (13)$$

The smallest value of this expression corresponds to $i = 0$ and gives

$$P_1 = \frac{\pi^2 EI}{4l^2} \quad . \quad . \quad . \quad . \quad (14)$$

We see therefore that if on the top of the column we apply a pressure $P_1$ and a bending moment $B_1$, the column will bend whatever be the value of $P_1$ and $B_1$, and the resulting curve will be given by equation (10). If after bending we remove the bending moment $B_1$, the column will return to its straight vertical form for all values of $P_1$ less than that given by formula (14), because then the equation $d_n \cos\left(l\sqrt{\frac{P_1}{EI}}\right) = 0$ can be satisfied only if $d_n = 0$. But if the load $P_1$ has exactly this value we see that the beam will remain bent, because the above equation will be satisfied without making $d_n = 0$, and the remarkable fact is that this will be true whatever be the value of this deflection, so long as it is very small, for this is the condition assumed in order that our equations may be applicable.

The equation to the curve which the centre line of the column assumes after bending (Fig. 16, Pl. II), is obtained from equation (10) by putting $B_1 = 0$ and substituting for $P_1$ its value (14), so that we have

$$y = d_n\left(1 - \cos\frac{\pi x}{2l}\right) \quad . \quad . \quad . \quad (15)$$

We see that the ordinates of this curve are proportional to the deflection $d_n$ which, as we have noted above, remains arbitrary.

If we have a prismatic column subjected to the load given by equation (14), it is clear that it will not bend if it is absolutely homogeneous; but if owing to some external cause the upper end is displaced, it will bend to the curve determined by equation (15) and will remain in the bent condition. From this it follows that by increasing the load even by an extremely small amount the deflection will be greatly increased, and failure of the column may result.

We see then that in practice the load on columns must be kept considerably below that given by formula (14).

We will add another consideration of the problem. This formula gives the least value of $P_1$ that will be capable of maintaining the column bent; but formula (13) gives a large number of other values, for we can put $i = 1, 2, 3$, etc. To appreciate the meaning of these different values of $P_1$, we will note that if in equation (9) we put $B_1 = 0$ and then substitute for $P_1$ its general value (13), we obtain for the equation to the curve

$$y = d_n\left\{1 - \cos\frac{(2i+1)\pi x}{2l}\right\}.$$

Take for example the case where $i = 1$; we then have

$$y = d_n\left(1 - \cos\frac{3\pi x}{2l}\right)$$

and we can easily see that it represents a curve (Fig. 17, Pl. II) which gives the following values for ordinates of the points B, C, D

$$d_n,\ 2d_n,\ d_n.$$

Thus the value

$$P_1 = \frac{3^2\pi^2}{4}\cdot\frac{EI}{l^2} = \frac{9\pi^2}{4}\frac{EI}{l^2}$$

is the thrust which is capable of maintaining the beam ABCD bent to the form shown in Fig 17, Pl. II.

We will notice that the value of $P_1$ corresponding to $i = 1$ can be obtained by observing that the portions AB, BC, CD are exactly equal and that each may be regarded as fixed at one of its ends and loaded with the pressure $P_1$ at the other; for example, the portion CD may be regarded as fixed at C because the tangent is vertical at this point. Thus the value of the load capable of maintaining it bent will be given by formula (14), taking $\frac{l}{3}$ as the length, *i.e.* we shall have

$$P_1 = \frac{\pi^2}{4}\cdot\frac{EI}{\left(\frac{l}{3}\right)^2} = \frac{9\pi^2}{4}\cdot\frac{EI}{l^2}$$

as we obtain from formula (13) by putting $i = 1$. A similar discussion will serve for $i = 2$, $i = 3$, etc.

**4. Application to another particular case, the horizontal force $S_1$ being still zero** (Fig. 18, Pl. II).

If the column is fixed at the top as well as at the bottom, or rather, if the top of the column is forced to remain horizontal and is prevented from lateral displacement, the angular deflection and the lateral deflection at the top must both be zero. Then, deducing from the equation to the curve the value of $y$ in terms of $x$ after having put $d_n = 0$, and putting this value in formula (2) and integrating, we should obtain $\theta_1 = 0$, and the resulting equation would enable us to determine the bending moment $B_1$.

Suppose, for example, that the column is prismatic and homogeneous; then the curve of its deflected form is given by putting $d_n = 0$ in equation (9), i.e.

$$y = \frac{B_1}{P_1}\left\{1 - \cos\left(x\sqrt{\frac{P_1}{EI}}\right)\right\} \quad . \quad . \quad . \quad (16)$$

This equation must give $y = 0$ for $x = 0$ and $x = l$, which necessitates that

$$\cos\left(l\sqrt{\frac{P_1}{EI}}\right) = 1$$

or

$$l\sqrt{\frac{P_1}{EI}} = 2i\pi \quad . \quad . \quad . \quad . \quad (17)$$

$i$ being any integer.

Introducing in formula (2) the expression for $y$ given by equation (16) and effecting the integrations, which present no difficulties in this case since EI is constant, we obtain

$$\theta_1 = \frac{B_1 \sin\left(l\sqrt{\frac{P_1}{EI}}\right)}{\sqrt{P_1 EI}}.$$

Now the right-hand side of this equation clearly becomes zero on substituting for $l\sqrt{\frac{P_1}{EI}}$ the values $2i\pi$ given by equation (17), so that the upper end of the column will remain horizontal in this case.

From equation (17) we deduce

$$P_1 = \frac{4i^2\pi^2 EI}{l^2} \quad . \quad . \quad . \quad . \quad (18)$$

and for $i = 1$

$$P_1 = \frac{4\pi^2 EI}{l^2} \quad . \quad . \quad . \quad . \quad (19)$$

This is the least value of the load which will be capable of maintaining bent the column AB (Fig. 18, Pl. II) when the top end must remain horizontal and cannot move laterally.

We have said that on the upper end we must then have a bending moment $B_1$; to determine the value of this moment we notice that if in equation (16) for the curve of the column we put $x = \frac{l}{2}$, we should obtain the deflection at the centre. This gives, seeing that $\frac{l}{2}\sqrt{\frac{P_1}{EI}} = \pi$,

$$\text{Central } d_n = \frac{2B_1}{P_1}$$

or

$$B_1 = \frac{P_1}{2} \times \text{central } d_n.$$

Now in order to have a pressure $P_1$ and bending moment $B_1$ at the top of the column, it will suffice if the load $P_1$ is distributed according to a linear law in such a manner that the resultant acts at distance $\frac{B_1}{P_1}$ or half the central deflection from the centre of the section. Thus in the case in which the upper end of the column is forced to descend vertically while remaining horizontal, it is clear that to keep the deflection at a given value, the value of $P_1$ must be the same whatever this deflection may be, but its eccentricity must be equal to half the deflection.

We have only examined in the above treatment the case in which $i = 1$; the case in which $i = 2$ will hold if the centre of the column is forced to remain on the vertical AB (Fig. 19, Pl. II).

We should take $i = 3$ when the two points of the column which divide it into three equal parts are forced to remain on the vertical AB.

As the discussion of these cases presents no difficulties we will not consider them further.

**5. Third Special Case.**—Suppose that the column, assumed to be homogeneous and prismatic, is hinged at the centre of the bottom and is loaded at the top with a uniformly distributed, i.e. a central, load, the centre of the top being forced to remain on the vertical line through the centre of the lower end.

This case can be deduced from that first considered, for it is clear that after bending the curve of the centre line of the column will be made up of two exactly equal portions AB, BC (Fig. 20, Pl. II), and that the central section will remain horizontal; thus each half of the column, for example BC, may be regarded as a column fixed at its lower end and axially loaded at the other, as we considered in the first special case.

Thus if $l$ is the total length of the column we can find the critical load $P_1$ sufficient to maintain it in its deflected form by putting $\frac{l}{2}$ for $l$ in formula (14), thus obtaining

$$P_1 = \frac{\pi^2 EI}{l^2} \quad . \quad . \quad . \quad . \quad (20)$$

If the centre point of the column is held so that it cannot deflect, then the column will bend as shown in Fig. 21, i.e. it will consist of two parts each of which satisfies the above conditions, so that the value of $P_1$ in this case necessary to maintain the deflected form is obtained by putting $\frac{l}{2}$ for $l$ in formula (20), thus giving

$$P_1 = \frac{4\pi^2 EI}{l^2} \quad . \quad . \quad . \quad . \quad . \quad (21)$$

**6. Discussion of Equation (9) when the Bending Moment $B_1$ is Zero.**—When the bending moment $B_1$ is zero, and the shear force $S_1$ has a real value, equation (9) becomes

$$P_1(d_n - y) + S_1(l - x) = (P_1 d_n + S_1 l) \cos\left(x\sqrt{\frac{P_1}{EI}}\right) - S_1\sqrt{\frac{EI}{P_1}} \sin\left(x\sqrt{\frac{P_1}{EI}}\right) \quad (22)$$

This equation contains the deflection $d_n$, which is unknown; to determine it we note that we must have $y = d_n$ for $x = l$, which gives

$$(P_1 d_n + S_1 l) \cos\left(l\sqrt{\frac{P_1}{EI}}\right) - S_1\sqrt{\frac{EI}{P_1}} \sin\left(l\sqrt{\frac{P_1}{EI}}\right) = 0 \quad (23)$$

from which we can deduce the value of $d_n$ in terms of $P_1$ and $S_1$, or else the value of one of the latter when the other and $d_n$ are given.

Let us find, for example, the condition necessary in order that the centre of the upper end shall not deflect laterally, i.e. that $d_n = 0$. In this case formula (23) becomes

$$S_1\left[l \cos\left(l\sqrt{\frac{P_1}{EI}}\right) - \sqrt{\frac{EI}{P_1}} \sin\left(l\sqrt{\frac{P_1}{EI}}\right)\right] = 0,$$

and in order that this may be satisfied we must have

$$l \cos\left(l\sqrt{\frac{P_1}{EI}}\right) - \sqrt{\frac{EI}{P_1}} \sin\left(l\sqrt{\frac{P_1}{EI}}\right) = 0,$$

or, dividing by

$$\sqrt{\frac{EI}{P_1}} \cos\left(l\sqrt{\frac{P_1}{EI}}\right), \quad l\sqrt{\frac{P_1}{EI}} - \tan\left(l\sqrt{\frac{P_1}{EI}}\right) = 0 \quad . \quad (24)$$

This transcendental equation gives an infinite number of values of $l\sqrt{\frac{P_1}{EI}}$ all comprised in the expression $(2i + 1)\pi + \alpha$, $i$ being an integer and $\alpha$ a quantity which is different for the different values but always less than $\frac{\pi}{2}$. The smallest root occurs for $i = 0$, and as we find that $\alpha$ in this case is very nearly equal to 78°, we have

$$l\sqrt{\frac{P_1}{EI}} = \pi + \frac{78\pi}{180} = \frac{129\pi}{90}$$

and therefore

$$P_1 = \left(\frac{129}{90}\right)^2 \pi^2 \frac{EI}{l^2} = 2{\cdot}05 \frac{\pi^2 EI}{l^2} \quad . \quad . \quad . \quad (25)$$

Such then is the least value of the pressure $P_1$ capable of keeping the column bent when the centre of the upper section cannot deflect laterally (Fig. 22, Pl. II).

It will be seen that for the case $d_{,,} = 0$ the equation to the curve taken up by the centre line becomes

$$-P_1 y = S_1 \left\{ -(l-x) + l \cos\left(x\sqrt{\frac{P_1}{EI}}\right) - \sqrt{\frac{EI}{P_1}} \sin\left(x\sqrt{\frac{P_1}{EI}}\right)\right\} \quad (26)$$

As the load $P_1$ has the value found above, we see clearly that the value of $S_1$ will remain indeterminate, and that we could alter all the ordinates $y$ in any ratio whatever, so long as they remain small, provided that $S_1$ changed in the same ratio.

This peculiar property is similar to that which we have proved in § 4 with reference to the bending moment $B_1$.

# CHAPTER IX.

## THEORY OF CURVED RIBS AND ARCHES.

**1. Strains in Structures Curved in One Plane.**—The structures which we shall consider here are assumed to satisfy the conditions set out in Chapter IV, § 11, i.e. :—

1. The centre line of the structure will be a plane curve.
2. Any section will have one of its principal axes in the plane of the curve, and the other, which we will call the *axis of bending*, perpendicular to this plane.
3. All forces applied to the structure will be distributed in such a manner that those on the right of any section will be statically equivalent to a couple of moment B about the axis of bending and to two forces, P, S, applied at the centroid of the section, the first normal to it and the second in the direction of the principal axis contained in the plane of the curve.
4. The transverse dimensions of the structure will be very small in comparison with its length.

If these conditions are satisfied it is clear that under strain the centre line of the structure will not depart from its original plane. Thus, as we propose to determine only the angular deflections of the sections of the structure and the displacements of points upon its centre line, we can refer the structure to two axes of co-ordinates in its plane.

Let AB (Fig. 1, Pl. III) be the centre line of a curved structure referred to two retangular axes OX, OY, and let the member be securely fixed at the end A; we will endeavour to find the displacements parallel to the axes of any point C on the centre line and also the angular rotation of the section at this point.

Let us call

$x_1$, $y_1$ the co-ordinates of the point C.
$\Delta x_1$, $\Delta y_1$ the displacements of this point parallel to the axes.
$\theta_1$ the angular deflection or rotation of the section at C.
$l_1$ the length of the centre line AC.
$x$, $y$ the co-ordinates of any point D on the curve between A and C.
$l$ the length of the centre line AD.

$B_1$, $P_1$, $S_1$ the bending moment, normal pressure, and shearing force for the section D.

A, I, F the area, moment of inertia, and shear coefficient for this section.

E the elastic modulus of the material.

If we suppose that on the section C, besides the external forces applied, there act additional forces statically equivalent to forces $F_x$, $F_y$ parallel to the axes and to a couple of moment $B_D$ about the axis of bending of the section D; and if, moreover, we assume that all the forces on the portion BC, including the reactions of the fixed points, if there are any, may be regarded as known external forces, it is clear that the bending moment, normal pressure, and shearing force at the section D will become

$$B = B_1 + B_D - F_x(y_1 - y) + F_y(x_1 - x)$$

$$P = P_1 + F_x\frac{dx}{dl} + F_y\frac{dy}{dl}$$

$$S = S_1 + F_x\frac{dy}{dl} - F_y\frac{dx}{dl}.$$

Now, the internal work of the portion AC is expressed by the formula

$$W_i = \frac{1}{2}\int_0^{l_1}\frac{B^2}{EI}dl + \frac{1}{2}\int_0^{l_1}\frac{P^2}{EA}dl + \frac{1}{2}\int_0^{l_1}\frac{FS^2}{GA}dl \quad . \quad (1)$$

and that for the portion CB cannot contain either the forces $F_x$, $F_y$ or the couple $B_D$. Therefore the displacements $\Delta x_1$, $\Delta y_1$ of the point C and the angular rotation $\theta_1$ of the corresponding section will be obtained by taking the differential coefficients of expression (1) with regard to $F_x$, $F_y$, and $B_D$ and then putting in the results $F_x = 0$, $F_y = 0$, $B_D = 0$.

This gives

$$\left.\begin{aligned} \Delta x_1 &= -\int_0^{l_1}\frac{B(y_1 - y)}{EI}dl + \int_0^{l_1}\frac{P}{EA}\frac{dx}{dl}dl + \int_0^{l_1}\frac{FS}{GA}\frac{dy}{dl}dl \\ \Delta y_1 &= \int_0^{l_1}\frac{B(x_1 - x)}{EI}dl + \int_0^{l_1}\frac{P}{EA}\frac{dy}{dl}dl - \int_0^{l_1}\frac{FS}{GA}\frac{dx}{dl}dl \\ \theta_1 &= \int_0^{l_1}\frac{Bdl}{EI}. \end{aligned}\right\} \quad (2)$$

Let us suppose that the integrations contained in these expressions are all effected, so that we have the displacements $\Delta x_1$, $\Delta y_1$, in terms of $x_1$ and $y_1$. Then if we call $X_1$, $Y_1$ the co-ordinates of the point C on the centre line after strain, we shall have

$$\left.\begin{aligned} X_1 &= x_1 - \Delta x_1 \\ Y_1 &= y_1 - \Delta y_1 \end{aligned}\right\} \quad . \quad . \quad . \quad . \quad (3)$$

and as $\Delta x_1$ and $\Delta y_1$ are known functions of $x_1$, $y_1$, we can eliminate the original co-ordinates between the two equations (3) and the equation of the centre line of the member before strain; this will give an equation between $X_1$ and $Y_1$ which will be the equation of the centre line of the member after strain.

We could then by differentiation obtain the general expression for $\frac{dY_1}{dX_1}$, i.e. the direction of the tangent to the centre line. But the value of $\frac{dY_1}{dX_1}$ can be obtained directly by differentiating equations (3) with regard to the length $l_1$ of the curve. This gives

$$\frac{dX_1}{dl_1} = \frac{dx_1}{dl_1} - \frac{d\Delta x_1}{dl_1}$$

$$\frac{dY_1}{dl_1} = \frac{dy_1}{dl_1} - \frac{d\Delta y_1}{dl_1}$$

and therefore

$$\frac{dY_1}{dX_1} = \frac{\frac{dy_1}{dl_1} - \frac{d\Delta y_1}{dl_1}}{\frac{dx_1}{dl_1} - \frac{d\Delta x_1}{dl_1}} \quad . \quad . \quad . \quad . \quad (4)$$

The differentiation of $\Delta x_1$, $\Delta y_1$ with regard to $l_1$ does not present any difficulty for the last two terms of each expression, for we have only to omit the integration sign by referring the differentials to the point C, this affecting all the quantities with subscript 1. But for the first term of each expression, i.e. for the term containing the moment B, we proceed in another way as follows:—

We have

$$\int_0^{l_1} \frac{B(y_1 - y)dl}{EI} = y_1 \int_0^{l_1} \frac{Bdl}{EI} - \int_0^{l_1} \frac{Bydl}{EI};$$

therefore on differentiating with regard to $l_1$ we have

$$\left(\frac{dy_1}{dl_1}\int_0^{l_1} \frac{Bdl}{EI} + y_1 \frac{B_1}{E_1 I_1}\right) - \frac{B_1 y_1}{E_1 I_1} = \frac{dy_1}{dl_1}\int_0^{l_1} \frac{Bdl}{EI}.$$

In the same way we find

$$\frac{d}{dl_1}\int_0^{l_1} \frac{B(x_1 - x)dl}{EI} = \frac{dx_1}{dl_1}\int_0^{l_1} \frac{Bdl}{EI}.$$

Then substituting in expression (4) we shall have

$$\frac{dY_1}{dX_1} = \frac{\frac{dy_1}{dl_1}\left(1 - \frac{P_1}{E_1 A_1}\right) - \frac{dx_1}{dl_1}\left(\int_0^{l_1} \frac{Bdl}{EI} - \frac{F_1 S_1}{G_1 A_1}\right)}{\frac{dx_1}{dl_1}\left(1 - \frac{P_1}{E_1 A_1}\right) + \frac{dy_1}{dl_1}\left(\int_0^{l_1} \frac{Bdl}{EI} - \frac{F_1 S_1}{G_1 A_1}\right)} \qquad (5)$$

**2. Cases in which the above Formulæ are Inapplicable.**—In order that the above formulæ may be applicable directly, one end of the member must be securely fixed, i.e. it cannot move parallel to its plane or turn about its axis of bending; because it is from one end that we must take the integrals involved in the formulæ given in the preceding paragraph.

Now it may and often does happen that we have to consider a structure in which neither of the ends is fixed; this is the case in arched roof members and in iron arches with pivot bearings.

We will give the formulæ applicable to these cases.

Let AB (Fig. 2, Pl. III) be a curved structure having end portions AI, BH of such forms that the structure has point bearings O and V; suppose that the first of these points is fixed and that the other is forced to remain in the horizontal plane of the bearing, so that it can be displaced only by sliding on this plane.

Then the structure can exert only vertical forces upon its supports, so that we may consider the right-hand support removed provided that we apply at the point V a reaction force $R_v$ acting vertically upwards and equal to the pressure on the support.

We will refer the structure to two axes OX, OY passing through the fixed point O, the axis OX being horizontal and the axis OY vertical.

To find the displacements of the point C on the centre line of the structure, the co-ordinates of which are $x_1$, $y_1$, we will imagine that in addition to the forces actually acting on the section at C there are applied other forces statically equivalent to a couple of moment $B_c$ and to horizontal and vertical forces $F_h$, $F_v$.

Thus the reaction $R_v$ will change, and to find its new value $R_v'$ we must express the condition that the moment of the difference $R_v' - R_v$ about the point O is equal to the sum of the moments about this point of the additional forces. This gives

$$(R_v' - R_v)l' = B_c - F_h y_1 + F_v x_1$$

from which we deduce

$$R_v' = R_v + \frac{B_c - F_h y_1 + F_v x_1}{l'} \quad . \quad . \quad . \quad (6)$$

Now if we consider a section D between A and C having co-ordinates $x$, $y$, and if we call B, P, S the bending moment, normal pressure, and shearing force before the application of these additional forces, it is clear that after these forces are applied we shall have

$$\left.\begin{aligned} B' &= B - (R_v' - R_v)(l' - x) + B_c - F_h(x_1 - y) + F_v(x_1 - y) \\ P' &= P - (R_v' - R_v)\frac{dy}{dl} + F_h\frac{dx}{dl} + F_v\frac{dy}{dl} \\ S' &= S + (R_v' - R_v)\frac{dx}{dl} + F_h\frac{dy}{dl} - F_v\frac{dx}{dl} \end{aligned}\right\} \quad (7)$$

For a section E between C and B we shall have, if $x$, $y$ are its co-ordinates,

$$\left.\begin{aligned} B'' &= B - (R_v' - R_v)(l' - x) \\ P'' &= P - (R_v' - R_v)\frac{dy}{dl} \\ S'' &= S + (R_v' - R_v)\frac{dx}{dl} \end{aligned}\right\} \quad . \quad . \quad . \quad (8)$$

Now if $l_1$ is the length of the arc AC and $l_2$ the length of the whole arc AB, we know that the internal work of the whole body will be given by

$$W_i = \frac{1}{2}\left[\int_0^{l_1}\frac{B'^2 dl}{EI} + \int_{l_1}^{l_2}\frac{B''^2 dl}{EI}\right] + \frac{1}{2}\left[\int_0^{l_1}\frac{P'^2 dl}{EA} + \int_{l_1}^{l_2}\frac{P''^2 dl}{EA}\right] + \frac{1}{2}\left[\int_0^{l_1}\frac{FS'^2 dl}{GA} + \int_{l_1}^{l_2}\frac{FS''^2 dl}{GA}\right].$$

We obtain the displacements $\Delta x$, $\Delta y$ of the centre of the section C and its angular rotation $\theta$ by taking the differential coefficients of this expression with regard to $F_h$, $F_v$, $B_c$, having regard to equations (6), (7), and (8), which give the following results:—

$$\begin{cases} \dfrac{dB'}{dF_h} = \dfrac{y_1}{l'}(l' - x) - (y_1 - y), & \dfrac{dP'}{dF_h} = \dfrac{y_1}{l'}\dfrac{dy}{dl} + \dfrac{dx}{dl}, \quad \dfrac{dS'}{dF_h} = -\dfrac{y_1}{l'}\dfrac{dx}{dl} + \dfrac{dy}{dl} \\ \dfrac{dB''}{dF_h} = \dfrac{y_1}{l'}(l' - x), & \dfrac{dP''}{dF_h} = \dfrac{y_1}{l'}\dfrac{dy}{dl}, \quad \dfrac{dS''}{dF_h} = -\dfrac{y_1}{l'}\dfrac{dx}{dl} \end{cases}$$

$$\begin{cases} \dfrac{dB'}{dF_v} = -\dfrac{x_1}{l'}(l' - x) + (x_1 - x), & \dfrac{dP'}{dF_v} = -\dfrac{x_1}{l'}\dfrac{dy}{dl} + \dfrac{dy}{dl}, \quad \dfrac{dS'}{dF_v} = \dfrac{x_1}{l'}\dfrac{dx}{dl} - \dfrac{dx}{dl} \\ \dfrac{dB''}{dF_v} = -\dfrac{x_1}{l'}(l' - x), & \dfrac{dP''}{dF_v} = -\dfrac{x_1}{l'}\dfrac{dy}{dl}, \quad \dfrac{dS''}{dF_v} = \dfrac{x_1}{l'}\dfrac{dx}{dl} \end{cases}$$

$$\begin{cases} \dfrac{dB'}{dB_c} = -\dfrac{(l' - x)}{l'} + 1, & \dfrac{dP'}{dB_c} = -\dfrac{1}{l'}\dfrac{dy}{dl}, \quad \dfrac{dS'}{dB_c} = \dfrac{1}{l'}\dfrac{dx}{dl} \\ \dfrac{dB''}{dB_c} = -\dfrac{(l' - x)}{l'}, & \dfrac{dP''}{dB_c} = -\dfrac{1}{l'}\dfrac{dy}{dl}, \quad \dfrac{dS''}{dB_c} = \dfrac{1}{l'}\dfrac{dx}{dl} \end{cases}$$

Then, taking the differential coefficient of the internal work with regard to $F_h$, we shall have

$$\Delta x_1 = \int_0^{l_1}\frac{B'}{EI}\left[\frac{y_1}{l'}(l'-x)-(y_1-y)\right]dl + \int_{l_1}^{l_2}\frac{B''}{EI}\frac{y_1}{l'}(l'-x)\,dl$$
$$+\int_0^{l_1}\frac{P'}{EA}\left[\frac{y_1}{l'}\frac{dy}{dl}+\frac{dx}{dl}\right]dl + \int_{l_1}^{l_2}\frac{P''}{EA}\frac{y_1}{l'}\frac{dy}{dl}\,dl$$
$$+\int_0^{l_1}\frac{FS'}{GA}\left[-\frac{y_1}{l'}\frac{dx}{dl}+\frac{dy}{dl}\right]dl + \int_{l_1}^{l_2}\frac{FS''}{GA}\left[-\frac{y_1}{l'}\frac{dx}{dl}\right]dl.$$

We must now put $F_h = 0$, $F_v = 0$, $B_c = 0$, which reduces B′, P′, S′ and B″, P″, S″ to B, P, S. We can thus combine in this expression the terms containing the same quantity under the integration sign, and then obtain the first of the following results, the other two being deduced in a similar manner:—

$$\left.\begin{aligned}
\Delta x_1 &= \frac{y_1}{l'}\left[\int_0^{l_2}\frac{B(l'-x)dl}{EI}+\int_0^{l_2}\frac{P}{EA}\frac{dy}{dl}\,dl-\int_0^{l_2}\frac{FS}{GA}\frac{dx}{dl}\,dl\right]\\
&\quad+\left[-\int_0^{l_1}\frac{B(y_1-y)dl}{EI}+\int_0^{l_1}\frac{P}{EA}\frac{dx}{dl}\,dl+\int_0^{l_1}\frac{FS}{GA}\frac{dy}{dl}\,dl\right]\\
\Delta y_1 &= -\frac{x_1}{l'}\left[\int_0^{l_2}\frac{B(l'-x)dl}{EI}+\int_0^{l_2}\frac{P}{EA}\frac{dy}{dl}\,dl-\int_0^{l_2}\frac{FS}{GA}\frac{dx}{dl}\,dl\right]\\
&\quad+\left[\int_0^{l_1}\frac{B(x_1-x)dl}{EI}+\int_0^{l_1}\frac{P}{EA}\frac{dy}{dl}\,dl-\int_0^{l_1}\frac{FS}{GA}\frac{dx}{dl}\,dl\right]\\
\theta_1 &= -\frac{1}{l'}\left[\int_0^{l_2}\frac{B(l'-x)dl}{EI}+\int_0^{l_2}\frac{P}{EA}\frac{dy}{dl}\,dl-\int_0^{l_2}\frac{FS}{GA}\frac{dx}{dl}\,dl\right]\\
&\quad+\int_0^{l_1}\frac{B\,dl}{EI}
\end{aligned}\right\}\quad(9)$$

We can simplify these formulæ, for if we call $\theta_0$ the rotation of the section B corresponding to $l_1 = 0$, the third equation becomes

$$\theta_0 = -\frac{1}{l'}\left[\int_0^{l_2}\frac{B(l'-x)dl}{EI}+\int_0^{l_2}\frac{P}{EA}\frac{dy}{dl}\,dl-\int_0^{l_2}\frac{FS}{GA}\frac{dx}{dl}\,dl\right]\qquad(10)$$

and therefore we have

$$\left.\begin{aligned}
\Delta x_1 &= -\theta_0 y_1+\left[-\int_0^{l_1}\frac{B(y_1-y)dl}{EI}+\int_0^{l_1}\frac{P}{EA}\frac{dx}{dl}\,dl+\int_0^{l_1}\frac{FS}{GA}\frac{dy}{dl}\,dl\right]\\
\Delta y_1 &= \theta_0 x_1+\left[\int_0^{l_1}\frac{B(x_1-x)dl}{EI}+\int_0^{l_1}\frac{P}{EA}\frac{dy}{dl}\,dl-\int_0^{l_1}\frac{FS}{GA}\frac{dx}{dl}\,dl\right]\\
\theta_1 &= \theta_0+\int_0^{l_1}\frac{B\,dl}{EI}
\end{aligned}\right\}\quad(11)$$

If we compare these three equations with group (2), we find the striking relation that the right-hand sides of equations (11) differ from the right-hand sides of equations (2) only in the addition of the linear terms in $\theta_0$. We could in fact now prove that we might pass from equations (2) to equations (11) in a simple manner; but the treatment which we have followed to arrive at equations (11) is not less simple, and it has the advantage of being more direct and of showing better the application of the theorem of the differential coefficients of internal work.

**3. Application to an Arch with Rounded Ends.**—Consider a structure of which the centre line AB (Fig. 3, Pl. III) is a circular arc with a horizontal chord, and suppose that this structure has rounded ends whose centres are on the tangents to the ends of the centre line and that the planes of the bearings are perpendicular to the centre line.

In this case it is clear that on account of compressibility the contact will not be upon a line only but will extend a little on each side of the line of contact of the ends A, B, in such a manner that the pressure per square metre will not in general be sufficient to cause crushing.

We see, moreover, that the contact surface on each bearing will be very narrow and that the centre of pressure will correspond almost exactly with the end of the centre line.

We will call the vertical central section the crown of the arch and we will assume that the arch is symmetrical about this section, so that any two sections equidistant from the crown will be exactly similar.

If the ends of the arch are so held that they cannot slide on their bearings, the determination of the deformations comes within the case studied in the previous paragraph, because as the point B is fixed we can regard it as compelled to remain on a horizontal plane.

We must here suppose that we have removed the right-hand bearing and that we have applied at the point B an upward vertical force $V_B$ and a horizontal thrust H acting from B to A, these two forces being equal and opposite to the components of the pressure of the arch upon its bearing.

The vertical reaction $V_B$ can be determined by expressing the condition that its moment about the point A must be equal to the moment of all the loads upon the arch about the same point.

The horizontal thrust must be determined by expressing the condition that the horizontal displacement of the point B is zero. Now, if we assume that we have obtained the expression for the internal work of the whole arch in terms of the loads upon the arch and of $V_B$ and H, we can substitute the known value for $V_B$ and leave the expression with H as the only unknown. In accordance with the theorem of the differential coefficients of internal work, the differential coefficient of this expression with regard to H will give the horizontal displacement of the point B, and since this displacement must be zero, it follows that we must equate to zero the differential coefficient with regard to H of the internal work of the whole arch; or, in general, we have the following rule:—

*The value of the horizontal thrust will be given by the value of H which will make the internal work of the arch a minimum.*

We can, moreover, obtain the value of H by making use of the first of formulæ (11).

When we have found the value of H, we can find by means of formulæ (11) the displacements of any point of the arch and the angular deflection of any section.

To give an example of this kind of calculation, we will assume that the arch is homogeneous and of constant section, and that it carries a load uniformly distributed along the horizontal. As in this case everything is symmetrical about the crown, the internal work for the whole arch will be twice that for each half. Now if we call

- R the radius of the centre line of the arch,
- $\phi_1$ the angle subtended by half the arch at the centre,
- $\phi$ the angle which any section makes with the vertical,
- B the bending moment for this section,
- P the normal pressure ,, ,, ,,
- S the shearing force ,, ,, ,,
- $w$ the load per unit length of the horizontal,

we shall have

$$V_A = V_B = wR \sin \phi_1$$

$$\left.\begin{aligned} B &= HR(\cos \phi - \cos \phi_1) - \frac{wR^2}{2}(\sin^2 \phi_1 - \sin^2 \phi) \\ P &= H \cos \phi + wR \sin^2 \phi \\ S &= -H \sin \phi + wR \sin \phi \cos \phi \end{aligned}\right\} \quad (12)$$

As in this case the arch element $dl = Rd\phi$, it is clear that the internal work of the half-arch is given by

$$\frac{R}{2}\int_0^{\phi_1}\frac{B^2 d\phi}{EI} + \frac{R}{2}\int_0^{\phi_1}\frac{P^2 d\phi}{EA} + \frac{R}{2}\int_0^{\phi_1}\frac{FS^2 d\phi}{GA},$$

so that for the whole arch, since E, G, A, I, F are constant, we shall have

$$W_i = \frac{R}{EI}\int_0^{\phi_1} B^2 d\phi + \frac{R}{EA}\int_0^{\phi_1} P^2 d\phi + \frac{RF}{GA}\int_0^{\phi_1} S^2 d\phi.$$

In accordance with what we have said previously, we must equate to zero the differential coefficient of this expression with regard to H to determine this quantity; thus, having regard to equations (12) we obtain, on dividing by 2R,

$$\frac{R}{EI}\int_0^{\phi_1} B(\cos\phi - \cos\phi_1)d\phi + \frac{1}{EA}\int_0^{\phi_1} P\cos\phi d\phi - \frac{F}{GA}\int_0^{\phi_1} S\sin\phi d\phi = 0,$$

whence, after substituting for B, P, and S their values (12) and effecting the integrations,

$$\left.\begin{aligned} &H\left\{\frac{R^2}{EI}\left(\frac{\phi_1 - 3\sin\phi_1\cos\phi_1}{2} + \phi_1\cos^2\phi_1\right) + \frac{1}{EA}\left(\frac{\phi_1 + \sin\phi_1\cos\phi_1}{2}\right)\right.\\ &\qquad\qquad\left. + \frac{F}{GA}\left(\frac{\phi_1 - \sin\phi_1\cos\phi_1}{2}\right)\right\}\\ &= wR\left[\frac{R^2}{2EI}\left\{\frac{2}{3}\sin^3\phi_1 - \phi_1\sin^2\phi_1\cos\phi_1 + \cos\phi_1\left(\frac{\phi_1 - \sin\phi_1\cos\phi_1}{2}\right)\right\}\right.\\ &\qquad\qquad\left. + \left(-\frac{1}{EA} + \frac{F}{GA}\right)\frac{\sin^3\phi_1}{3}\right] \end{aligned}\right\} \quad (13)$$

from which we can calculate the value of H.

When the rise of the arch is very small in comparison with the chord, this formula can be simplified considerably, but it will be better to treat the question directly instead of expanding the trigonometrical functions in series.

If we call $2l'$ the length of the chord and $r$ the rise of the arch, and $x$, $y$ the co-ordinates of C referred to the chord AB and its perpendicular AY (Fig. 3, Pl. III), we can take for the equation of the arc AB, as we saw in Chap. VI, § 4,

$$y = \frac{r}{l'^2}(2l'x - x^2)$$

from which we deduce

$$\frac{dy}{dx} = \frac{2r}{l'^2}(l' - x).$$

Since the quantity $\frac{dy}{dx}$ is very small, we may neglect its square in comparison with unity, which gives

$$\frac{dl}{dx} = \sqrt{1 + \left(\frac{dy}{dx}\right)^2} = 1, \text{ or } dl = dx,$$

and therefore

$$\frac{dx}{dl} = 1; \ \frac{dy}{dl} = \frac{2r(l' - x)}{l'^2}.$$

Thus the bending moment, normal pressure, and shearing force for the section C will be expressed by the following formulæ :—

$$\left.\begin{aligned}
B &= Hy - wl'x + \frac{wx^2}{2} = \left(\frac{Hr}{l'^2} - \frac{w}{2}\right)(2l'x - x^2),\\
P &= H\frac{dx}{dl} + wl'\frac{dy}{dl} - wx\frac{dy}{dl} = H + \frac{2wr(l'-x)^2}{l'^2},\\
S &= H\frac{dy}{dl} - wl'\frac{dx}{dl} + wx\frac{dx}{dl} = 2\left(\frac{Hr}{l'^2} - \frac{w}{2}\right)(l'-x).
\end{aligned}\right\} \quad (14)$$

The internal work of the whole arch is given by

$$W_i = \frac{1}{2EI}\int_0^{2l'} B^2 dl + \frac{1}{2EA}\int_0^{2l'} P^2 dl + \frac{F}{2GA}\int_0^{2l'} S^2 dl,$$

from which we get, by putting in the values of B, P, and S in (14), taking $dl = dx$, and effecting the integrations,

$$W_i = \frac{1}{2EI}\left(\frac{Hr}{l'^2} - \frac{w}{2}\right)^2\frac{16l'^5}{15} + \frac{2l'}{2EA}\left(H^2 + \frac{4Hwr}{3} + \frac{4}{5}w^2r^2\right) + \frac{F}{2GA}\left(\frac{Hr}{l'^2} - \frac{w}{2}\right)^2\frac{8l'^3}{3}.$$

On differentiating this with regard to H and equating to zero, we get

$$\left(\frac{8}{15}\frac{r^2}{EI} + \frac{F}{GA}\frac{4r^2}{3l'^2}\right)\left(H - \frac{wl'^2}{2r}\right) + \frac{1}{EA}\left(H + \frac{2}{3}wr\right) = 0,$$

which may be written as follows :—

$$\left(\frac{8}{15}\frac{r^2}{EI} + \frac{4r^2}{3l'^2}\frac{F}{GA}\right)\left(H - \frac{wl'^2}{2r}\right) + \frac{1}{EA}\left(H - \frac{wl'^2}{2r} + \frac{wl'^2}{2r} + \frac{2wr}{3}\right) = 0$$

i.e.

$$\left(\frac{8r^2}{15EI} + \frac{1}{EA} + \frac{4r^2}{3l'^2}\frac{F}{GA}\right)\left(H - \frac{wl'^2}{2r}\right) + \frac{1}{EA}\left(\frac{wl'^2}{2r} + \frac{2wr}{3}\right) = 0, \quad (15)$$

from which the value of H can easily be found.

If we neglect the last term on the left-hand side, we shall obtain the equation

$$H - \frac{wl'^2}{2r} = 0$$

and it will follow from equations (14) that the bending moment and shearing force will be zero for all sections ; but this omission cannot be justified and the value of H should be derived from the complete equation (15). What we can always do, to a sufficiently accurate degree of approximation, to simplify the equation, is to neglect the term $\frac{4r^2}{3l'^2}\cdot\frac{F}{GA}$ which is generally small in comparison with $\frac{1}{EA}$, since $\frac{r^2}{l'^2}$ is very small, and also to neglect $\frac{2}{3}wr$ in comparison with $\frac{wl'^2}{2r}$; equation (15) then becomes

$$\left(\frac{8r^2}{15\text{EI}} + \frac{1}{\text{EA}}\right)\left(\text{H} - \frac{wl'^2}{2r}\right) + \frac{wl'^2}{2r\text{EA}} = 0,$$

from which we deduce

$$\text{H} = \frac{wl'^2}{2r}\left(1 - \frac{\frac{\text{I}}{\text{A}}}{\frac{8r^2}{15} + \frac{\text{I}}{\text{A}}}\right) \quad . \quad . \quad . \quad (16)$$

We can now substitute this value in equations (14) to calculate the bending moment, normal thrust and shearing force at any point.

If, for example, we consider the section at the crown, i.e. if we put $x = l'$, we have

$$\text{B} = -\frac{15}{8}\frac{\text{I}}{\text{A}r}\,.\,\text{H}$$

$$\text{P} = \text{H}$$

$$\text{S} = 0.$$

We can now derive a very simple formula for calculating at once the maximum stress at the crown section, for if we call $d$ the depth of the section and $n$ the distance from the centre line to the most compressed edge (the extrados), we know that the maximum compressive stress at extrados

$$= c_e = \frac{\text{P}}{\text{A}} - \frac{\text{B}n}{\text{I}} = \frac{\text{H}}{\text{A}}\left(1 + \frac{15n}{8r}\right) \quad . \quad . \quad (17)$$

and at intrados

$$c_i = \frac{\text{P}}{\text{A}} + \frac{\text{B}(d - n)}{\text{I}} = \frac{\text{H}}{\text{A}}\left\{1 - \frac{15(d - n)}{8r}\right\}.$$

We can easily see by examination of formula (16) that in general we can take in practical applications, to a sufficient degree of approximation,

$$\text{H} = \frac{wl'^2}{2r}$$

so that the formula for the maximum stress becomes

$$c_e = \frac{wl'^2}{2r\text{A}}\left(1 + \frac{15n}{8r}\right) \quad . \quad . \quad . \quad (17\text{A})$$

Up to the present it has been the practice to regard arches of small rise loaded uniformly along the horizontal as arches in which the bending moment is zero throughout, i.e. in which the pressure on any section is uniformly distributed, and H has been taken equal to $\frac{wl'^2}{2r}$, so that the compressive stress at the crown is

$$c = \frac{wl'^2}{2r\text{A}},$$

and this is less than the more exact stress given by formula (17A). If, for example, we have $\frac{n}{r} = \frac{1}{8}$, we shall have by formula (17A)

$$c_e = \left(1 + \frac{15}{64}\right)\frac{wl'^2}{2rA}$$

$$= 1{\cdot}23\frac{wl'^2}{2rA},$$

so that the old theory will give an error of about 20 per cent. This error will be greater still if $\frac{n}{r} > \frac{1}{8}$.

**4. Application to Arches with Flat Ends** (Fig. 4, Pl. III).—We will now consider the case of an arch with flat ends, i.e. resting upon flat bearings perpendicular to its centre line. If the ends of the arch are held so that they cannot slide on their bearings, the end sections cannot turn, and the arch may be regarded as fixed at each end; to determine the deformation, therefore, we may make use of the formulæ given in § 1. In each case we must imagine that one of the supports is removed, for example, the right-hand one, and that there are applied at that end normal and tangential forces giving a bending moment $B_1$, normal pressure $P_1$, and shearing force $S_1$, these three quantities being determined by the condition that the section can neither turn about its axis of bending nor slide parallel to or perpendicular to its plane.

But if the arch has its supports at the same level and is symmetrical about a vertical plane through the crown so that any two sections equidistant from this plane are exactly similar, we can simplify the calculation by following the procedure which we will explain later.

Let $B_0$, $P_0$, $S_0$ be the bending moment, normal pressure, and shearing force at the crown. We can imagine the arch to be cut through at this point and consider separately the two half-arches, provided that we apply to the crown ends of each forces giving a bending moment $B_0$, a normal pressure $P_0$, and a shearing force $S_0$. Naturally this moment and these forces will be considered as acting in opposite directions for the two half-arches.

We will express the internal work $W_{il}$ for the left-hand half-arch and that $W_{ir}$ for the right-hand half-arch as functions of the forces applied to each half and of the three quantities $B_0$, $P_0$, $S_0$; then the differential coefficients

$$\frac{dW_{il}}{dB_0}, \quad \frac{dW_{il}}{dP_0}, \quad \frac{dW_{il}}{dS_0}$$

will express the rotation of the crown section of the left-hand half-arch and the horizontal and vertical displacements of its centre; whilst the differential coefficients

$$\frac{dW_{ir}}{dB_0}, \quad \frac{dW_{ir}}{dP_0}, \quad \frac{dW_{ir}}{dS_0}$$

will express the corresponding quantities for the right-hand half-arch Now these last three differential coefficients must give results respectively numerically equal to the first but reversed in sign; for in the first place the crown section must be displaced in only one way, no matter in which half-arch we consider it, and the quantities $B_0$, $P_0$, $S_0$ are taken as of opposite sign for the two half-arches. We therefore have

$$\frac{dW_{il}}{dB_0} + \frac{dW_{ir}}{dB_0} = 0, \quad \text{or} \quad \frac{d(W_{il} + W_{ir})}{dB_0} = 0,$$

$$\frac{dW_{il}}{dP_0} + \frac{dW_{ir}}{dP_0} = 0, \quad \text{or} \quad \frac{d(W_{il} + W_{ir})}{dP_0} = 0,$$

$$\frac{dW_{il}}{dS_0} + \frac{dW_{ir}}{dS_0} = 0, \quad \text{or} \quad \frac{d(W_{il} + W_{ir})}{dS_0} = 0;$$

from which it follows that to determine the unknowns $B_0$, $P_0$, $S_0$ we must equate to zero the differential coefficients with regard to each of these quantities of the internal work of the whole arch, i.e. *express the condition that the internal work is a minimum.*

When the values of the three unknown quantities have been determined, we calculate the displacements and rotation of any section by means of equations (2).

Let us examine, for instance, the particular case in which the arch has a constant section and carries a load uniformly distributed along the horizontal.

We will suppose that the centre line of the arch is a circular arc (Fig. 4, Pl. III), so that as everything is symmetrical about the crown the shearing force at this section must be zero. We shall therefore have only two unknowns, i.e. $B_0$ and $P_0$, the bending moment and normal pressure at the crown. Moreover, since the internal work of each half-arch is the same we need only consider that of one of them.

Now if we call $\phi$ the angle which a section C makes with the crown section, and $\phi_1$ the angle subtended at the centre by the half-arch, we shall have the following formulæ for the bending moment, normal pressure, and shearing force at the section C :—

$$\left.\begin{aligned} B &= B_0 - P_0R(1 - \cos\phi) + \frac{wR^2\sin^2\phi}{2} \\ P &= P_0\cos\phi + wR\sin^2\phi \\ S &= -P_0\sin\phi + wR\sin\phi\cos\phi. \end{aligned}\right\} \quad . \quad (18)$$

The internal work of each half-arch is given by the formula

$$W_i = \frac{R}{2EI}\int_0^{\phi_1} B^2 d\phi + \frac{R}{2EA}\int_0^{\phi_1} P^2 d\phi + \frac{FR}{2GA}\int_0^{\phi_1} S^2 d\phi,$$

and it is clear that on equating to zero the differential coefficients of this expression with regard to $B_0$ and $P_0$ we shall have the two equations

$$\frac{1}{EI}\int_0^{\phi_1} Bd\phi = 0 \quad . \quad . \quad . \quad . \quad (19)$$

$$-\frac{R}{EI}\int_0^{\phi_1} B(1 - \cos\phi)d\phi + \frac{1}{EA}\int_0^{\phi_1} P\cos\phi d\phi - \frac{F}{GA}\int_0^{\phi_1} S\sin\phi d\phi = 0 \quad (20)$$

In view of (19) we can omit the term

$$\frac{R}{EI}\int_0^{\phi_1} Bd\phi$$

from (20) which then becomes

$$\frac{R}{EI}\int_0^{\phi_1} B\cos\phi d\phi + \frac{1}{EA}\int_0^{\phi_1} P\cos\phi d\phi - \frac{F}{GA}\int_0^{\phi_1} S\sin\phi d\phi = 0 \quad (21)$$

Then substituting for B, P, S their values given in equations (18) and effecting the integrations, equations (19) and (21) become respectively

$$B_0\phi_1 - P_0R(\phi_1 - \sin\phi_1) + \frac{wR^2}{2}\left(\frac{\phi_1 - \sin\phi_1\cos\phi_1}{2}\right) = 0 \quad (22)$$

and

$$\left.\begin{aligned} &\frac{R}{EI}\left[B_0\sin\phi_1 - P_0R\left(\sin\phi_1 - \frac{\phi_1 + \sin\phi_1\cos\phi_1}{2}\right) + \frac{wR^2\sin^3\phi_1}{6}\right] \\ &\quad + \frac{1}{EA}\left[P_0\left(\frac{\phi_1 + \sin\phi_1\cos\phi_1}{2}\right) + \frac{wR\sin^3\phi_1}{3}\right] \\ &\quad + \frac{F}{GA}\left[P_0\left(\frac{\phi_1 - \sin\phi_1\cos\phi_1}{2}\right) - \frac{wR\sin^3\phi_1}{3}\right] \end{aligned}\right\} = 0 \quad (23)$$

from which we can deduce the values of $B_0$, $P_0$.

When the rise is very small in comparison with the chord, we can simplify these formulæ by expanding the sines and cosines in terms of the angles, taking account only of the first terms, but we will treat this case directly as it is very important.

Taking in this case the centre of the crown section as the

origin, the axis of $x$ horizontal, and of $y$ vertical, the equation to the centre line of the arch will be

$$x^2 = (2R - y)y,$$

or approximately

$$x^2 = 2Ry,$$

neglecting $y$ in comparison with 2R, since it is smaller than the rise, and this is very small compared with the radius.

If we call $l'$ the half-chord and $r$ the rise, this equation should give $y = r$ for $x = l'$, so that we have

$$l'^2 = 2Rr,$$

and we can therefore eliminate R from the previous equation, thus obtaining

$$y = \frac{rx^2}{l'^2} \quad . \quad . \quad . \quad . \quad (24)$$

from which we deduce

$$\frac{dy}{dx} = \frac{2rx}{l'^2}.$$

It follows therefore that the quantity $\frac{dy}{dx}$ will be very small, and that its square will be negligible compared with unity; this gives

$$\frac{dl}{dx} = \sqrt{1 + \left(\frac{dy}{dx}\right)^2} = 1, \text{ or } dl = dx,$$

and

$$\frac{dy}{dl} = \frac{dy}{dx} = \frac{2rx}{l'^2}.$$

We therefore have for any section whose abscissa is $x$,

$$\left.\begin{aligned} B &= B_0 - P_0 y + \frac{wx^2}{2} = B_0 - \left(P_0 - \frac{wl'^2}{2r}\right)\frac{rx^2}{l'^2} \\ P &= P_0\frac{dx}{dl} + wx\frac{dy}{dl} = P_0 + \frac{2wrx^2}{l'^2} \\ S &= -P_0\frac{dy}{dl} + wx\frac{dx}{dl} = -\left(P_0 - \frac{wl'^2}{2r}\right)\frac{2rx}{l'^2} \end{aligned}\right\}; \qquad (25)$$

seeing that for the degree of accuracy to which we are working, we may take $dl = dx$, i.e. the abscissa instead of the arc, the formula for the internal work for the half-arch may be put into the form

$$W_i = \frac{1}{2EI}\int_0^{l'} B^2 dx + \frac{1}{2EA}\int_0^{l'} P^2 dx + \frac{F}{2GA}\int_0^{l'} S^2 dx,$$

of which the differential coefficients with regard to $B_0$ and $P_0$, when equated to zero, give

$$\left.\begin{array}{l}\displaystyle\int_0^{l'} \mathrm{B}dx = 0 \\ \displaystyle -\frac{1}{\mathrm{EI}}\frac{r}{l'^2}\int_0^{l'} \mathrm{B}x^2dx + \frac{r}{\mathrm{EA}}\int_0^{l'} \mathrm{P}dx - \frac{\mathrm{F}}{\mathrm{GA}}\cdot\frac{2r}{l'^2}\int_0^{l'} \mathrm{S}xdx = 0\end{array}\right\} \quad (26)$$

Substituting in these equations the values for B, P, and S given in (25) and effecting the integrations, we obtain

$$\left.\begin{array}{l}\mathrm{B}_0 - \left(\mathrm{P}_0 - \dfrac{wl'^2}{2r}\right)\dfrac{r}{3} = 0 \\ -\dfrac{rl'}{\mathrm{EI}}\left[\dfrac{\mathrm{B}_0}{3} - \left(\mathrm{P}_0 - \dfrac{wl'^2}{2r}\right)\dfrac{r}{5}\right] + \dfrac{l'}{\mathrm{EA}}\left(\mathrm{P}_0 + \dfrac{2wr}{3}\right) + \dfrac{\mathrm{F}l'4r^2}{\mathrm{GA}3l'^2}\left(\mathrm{P}_0 - \dfrac{wl'^2}{2r}\right) = 0\end{array}\right\} \quad (27)$$

From the first we deduce

$$\mathrm{B}_0 = \left(\mathrm{P}_0 - \frac{wl'^2}{2r}\right)\frac{r}{3}. \quad . \quad . \quad . \quad (28)$$

and the second becomes, on inserting this value of $\mathrm{B}_0$,

$$\left(\frac{4r^2}{45\mathrm{EI}} + \frac{4\mathrm{F}r}{3\mathrm{GA}l'^2}\right)\left(\mathrm{P}_0 - \frac{wl'^2}{2r}\right) + \frac{1}{\mathrm{EA}}\left(\mathrm{P}_0 + \frac{2wr}{3}\right) = 0,$$

which we may write in the form

$$\left(\frac{4r^2}{45\mathrm{EI}} + \frac{1}{\mathrm{EA}} + \frac{4\mathrm{F}r^2}{3\mathrm{GA}l'^2}\right)\left(\mathrm{P}_0 - \frac{wl'^2}{2r}\right) + \frac{1}{\mathrm{EA}}\left(\frac{wl'^2}{2r} + \frac{2wr}{3}\right) = 0 \quad (29)$$

from which the value of $\mathrm{P}_0$ can easily be computed. We then substitute this value in the expression for $\mathrm{B}_0$, and we have the bending moment at the crown in terms of the weight $w$ only.

To obtain the bending moment, normal pressure, and shearing force at the supports we must put $x = l'$ in formulæ (25), thus obtaining

$$\left.\begin{array}{l}\mathrm{B}_1 = \mathrm{B}_0 - \left(\mathrm{P}_0 - \dfrac{wl'^2}{2r}\right)r = \dfrac{2}{3}\left(\mathrm{P}_0 - \dfrac{wl'^2}{2r}\right)r \\ \mathrm{P}_1 = \mathrm{P}_0 + 2wr \\ \mathrm{S}_1 = -\left(\mathrm{P}_0 - \dfrac{wl'^2}{2r}\right)\dfrac{2r}{l'}\end{array}\right\} \quad . \quad (30)$$

By means of these formulæ we can calculate the maximum normal and shear stresses at the ends.

**5. Application to Arches of any Form Carrying a Continuous Load.**—The two special cases which we have studied in the two previous paragraphs, although very important from the theoretical standpoint, are not of much value in practice, because it seldom happens that the cross-section of the arch is constant, and still more seldom that the load is uniformly distributed over the centre line of the arch.

The most frequent and most general case is that in which

the cross-section varies and in which the load is distributed according to any law on the extrados instead of on the centre line.

The study of the elastic equilibrium of an arch under these conditions is not more difficult than in the two particular cases considered above; we might even say that it becomes more simple, by expressing the internal work by means of the approximate formula (14) given in Chap. VI, § 6. To prove the truth of this assertion, suppose that we have an arch with cylindrical ends symmetrical about the crown section, the section varying in a continuous manner; we will take the load as continuous on each half of the arch but of equal or different intensities on the two sides of the crown.

In this case there will be only one unknown, i.e. the horizontal thrust H.

We will divide the centre line of each half-arch into an even number of equal parts, and will consider the cross-sections through the points of division; for each of these sections we can express the bending moment, normal pressure, and shearing force in terms of the load upon the arch and of the unknown thrust H. Now, if we calculate the area A, moment of inertia I, and shear coefficient F for each section, we can express by means of formula (14) of Chapter VI the internal work of the whole arch. As this expression cannot contain any unknowns except H and is of the second degree, it will clearly be of the form

$$\frac{1}{2}(\alpha H^2 + 2\beta H + \gamma),$$

$\alpha$, $\beta$, $\gamma$ being numerical coefficients.

The differential coefficient of this expression with regard to H will give the amount by which the abutments approach each other, considering the abutments as free and acted upon by the force H. As this amount of approach must be zero, we have

$$\alpha H + \beta = 0$$

or

$$H = -\frac{\beta}{\alpha}.$$

The horizontal thrust being now known, we introduce it into the expressions for the bending moment, normal pressure, and shearing force on the sections considered, and thus obtain the numerical values of these quantities.

The accuracy of this approximate method will increase as we increase the number of elements into which we divide the centre line of the arch, i.e. with the decrease in size of these elements.

The calculations will not be longer than if we could effect the integrations to obtain at once the final value of the horizontal thrust, for, even in the cases in which this integration is possible, it is clear that after having obtained the value of H we must find the bending moment, normal pressure, and shearing force for a fairly large number of sections, which leads again to the operations necessary for the application of the above method. There is, however, this difference in favour of the latter, that it requires only very simple arithmetical operations, whilst the method of direct integration is possible only in very simple cases of loading and form of arch, and even in these cases it generally requires fairly difficult calculations.

We will not add any other explanations, because we shall give later in the numerical applications several complete examples of calculation by the method indicated above.

**6. Line of Pressure.**—Let B be the bending moment and P the normal pressure at any section.

We know that the stresses over this section are distributed according to a linear law and are such that their sum is equal to P and their moment about the axis of bending is equal to B.

If we divide this moment by the resultant P, the result gives the distance from the axis of bending at which this resultant must be applied to give the moment B, i.e. the distance from the centroid line of the section to the *load point* or point at which the resultant of the normal stresses on the section acts. We will call this distance the *eccentricity of the load*.

We will call the *line of pressure* of a structure the line joining the load points.

If in addition to the normal stresses we consider the shear stresses whose resultant is equal to the shearing force S, we will call the resultant of this force and the normal pressure the *absolute pressure* at the section considered.

This pressure will usually be oblique to the section ; if we consider two sections infinitely close, the straight lines representing their absolute pressures will intersect at a point ; the locus of this point is a curve tangential to the absolute pressure at any section, i.e. the curve is the envelope of the absolute pressures at all the sections.

This envelope is called by some authors the *curve of pressure*. In general it differs very little from the line of pressure, at least for structures with transverse dimensions very small in comparison with their length.

We see then that the straight line representing the absolute pressure on a section is not in general absolutely tangential to the line of pressure; but it is approximately so and will cut the line of pressure at an extremely small angle.

It follows that if the line of pressure of a structure is everywhere very slightly inclined to the centre line of the structure, the absolute pressure on any section will make a very small angle with the normal to the section, so that the shearing force will also be very small compared with the normal pressure.

# CHAPTER X.

## THEORY OF COMPOSITE STRUCTURES.

**1. Determination of Unknown Stresses.**—We have proved in §§ 4, 5, 6, and 8 of Chapter II, the theorem of the differential coefficients of internal work and that of least work for structures of any type; and we have seen that these theorems, the second of which is only a corollary of the first, are sufficient to determine all the unknown stresses occurring in the structures.

We will now deal with a number of points to make more clear the value of these theorems and the manner of their use in practical applications.

**2. Change of the Unknowns.**—Suppose that the internal work of an elastic structure is expressed in terms of certain unknown forces P, Q, R . . . which, according to the theorem of the differential coefficients of internal work, must be found by equating to zero the differential coefficients of the internal work with regard to these same unknowns.

It often happens that these forces can be expressed in terms of an equal number of unknowns $p, q, r$ . . . connected to the first by linear relations in such a manner that we have

$$\left.\begin{aligned} P &= a_1 p + b_1 q + c_1 r + \dots \\ Q &= a_2 p + b_2 q + c_2 r + \dots \\ R &= a_3 p + b_3 q + c_3 r + \dots \\ &\dots\dots \end{aligned}\right\} \quad . \quad . \quad . \quad (1)$$

$a_1, b_1, c_1$ . . . ; $a_2, b_2, c_2$ . . . ; $a_3, b_3, c_3$ . . . being coefficients independent of the unknowns.

We will now prove a very important property, i.e. *If we express the internal work of a structure in terms of the unknowns $p, q, r$ . . ., the latter can be determined directly by equating to zero the differential coefficients of the internal work with regard to themselves.*

If we call $W_i$ the internal work of the structure we can express $W_i$ in terms of the original unknowns P, Q, R . . ., and the latter in terms of the new unknowns $p, q, r$ . . ., by equations (1). We then have, in accordance with the rules of the differential calculus,

$$\frac{dW_i}{dp} = \frac{dW_i}{dP} \cdot \frac{dP}{dp} + \frac{dW_i}{dQ} \cdot \frac{dQ}{dp} + \frac{dW_i}{dR} \cdot \frac{dR}{dp} + \ldots;$$

substituting here for $\frac{dP}{dp}, \frac{dQ}{dp}, \frac{dR}{dp} \ldots$ their values $a_1, a_2, a_3 \ldots$ we obtain the first of the following relations, and the others are obtained in the same manner :—

$$\frac{dW_i}{dp} = a_1\frac{dW_i}{dP} + a_2\frac{dW_i}{dQ} + a_3\frac{dW_i}{dR} + \ldots$$
$$\frac{dW_i}{dq} = b_1\frac{dW_i}{dP} + b_2\frac{dW_i}{dQ} + b_3\frac{dW_i}{dR} + \ldots$$
$$\frac{dW_i}{dr} = c_1\frac{dW_i}{dP} + c_2\frac{dW_i}{dQ} + c_3\frac{dW_i}{dR} + \ldots$$
$$\cdot \quad \cdot \quad \cdot \quad \cdot \quad \cdot \quad \cdot \quad \cdot \quad \cdot$$

Now we have by hypothesis

$$\frac{dW_i}{dP} = 0, \quad \frac{dW_i}{dQ} = 0, \quad \frac{dW_i}{dR} = 0 \ldots;$$

then we shall also have

$$\frac{dW_i}{dp} = 0, \quad \frac{dW_i}{dq} = 0, \quad \frac{dW_i}{dr} = 0 \ldots;$$

which proves the theorem enunciated above.

We could deduce this theorem at once from a consideration of least internal work; for if the internal work must be a minimum when expressed in terms of the original unknowns P, Q, R, it must also be a minimum in terms of the new unknowns $p, q, r$ which depend upon the first. We believe, however, that the very simple direct proof which we have given will be found to be the more satisfactory.

**3. Calculations for Symmetrical Structures.**—We have already seen that in nearly every case the elastic structures employed in practice are plane, i.e. they satisfy the conditions set out in Chap. IV, § 11.

Now it happens also that these structures are nearly always composed of two parts which are exactly equal and symmetrically arranged.

In this case the calculation of the unknown stresses is capable of considerable simplification, and this holds equally if the external forces are equal and symmetrically arranged on the two parts or if there is no such symmetrical relation.

To show how these simplifications are attained we will examine a particular case, for what we shall say for this case can be applied without difficulty to all other cases.

Consider then a structure composed of an arch ACB (Fig. 5,

Pl. III) in one piece, and tie-rods AD, DE, etc., connected to each other and to the arch by bolts or rivets. We will suppose that this structure is symmetrical about the vertical CE and has rounded ends by means of which it has point contact on each support.

We will also assume that the end B rests upon a flat plate supported on rollers so that it can slide freely, and is incapable of resisting any horizontal thrust.

We will call $T_{L1} \dots T_{L5}$ the tensions in the left-hand tie bars and $T_{R1} \dots T_{R5}$ the corresponding tensions on the right.

It will be observed in the first place that as the tensions in the four bars meeting at D must be in equilibrium we shall have two equations by which we can deduce $T_{L3}$ and $T_{L4}$ in terms of $T_{L1}$ and $T_{L2}$; the same reasoning holds for the point F, and as the inclinations of the bars meeting at F are the same as of those meeting at D, the two pairs of equations will have the same coefficients, so that we shall have

$$\left.\begin{aligned} T_{L3} &= \alpha T_{L1} + \beta T_{L2} \\ T_{L4} &= \gamma T_{L1} + \delta T_{L2} \end{aligned}\right\} \quad . \quad . \quad . \quad (2)$$

$$\left.\begin{aligned} T_{R3} &= \alpha T_{R1} + \beta T_{R2} \\ T_{R4} &= \gamma T_{R1} + \delta T_{R2} \end{aligned}\right\} \quad . \quad . \quad . \quad (3)$$

$\alpha$, $\beta$, $\gamma$, and $\delta$ being numerical coefficients.

If we then express the condition that the bars meeting at the node E must be in equilibrium, we shall obtain two equations from which we can deduce $T_{L5}$, $T_{R5}$ in terms of $T_{L2}$, $T_{R2}$, and we see from symmetry that these equations must have the same coefficients, so that if $\lambda$ and $\mu$ are these coefficients we shall have

$$\left.\begin{aligned} T_{R5} &= \lambda T_{R2} + \mu T_{L2} \\ T_{L5} &= \lambda T_{L2} + \mu T_{R2} \end{aligned}\right\} \quad . \quad . \quad . \quad . \quad (4)$$

We see then that by means of equations (2) to (4) the tensions in the ten tie-bars can be expressed in terms of four of them only, viz. $T_{L1}$, $T_{L2}$, $T_{R1}$, $T_{R2}$.

Now if we call

$l_1 \dots l_5$ the lengths of the five pairs of tie-bars,
$A_1 \dots A_5$ their cross-sectional areas,
$E_1 \dots E_5$ their direct elastic moduli,

it is clear that the internal work of the tie-bars will be expressed by the formula

$$\frac{1}{2}\left\{\frac{(T_{L1}{}^2 + T_{R1}{}^2)l_1}{E_1A_1} + \frac{(T_{L2}{}^2 + T^2{}_{R2})l_2}{E_2A_2} + \dots \frac{(T_{L5}{}^2 + T^2{}_{R5})l_5}{E_5A_5}\right\};$$

eliminating $T_{L3}$, $T_{L4}$, $T_{L5}$, $T_{R3}$, $T_{R4}$, $T_{R5}$ by means of equations (2) to (4), we obtain an expression of the form

$$\frac{1}{2}[C(T_{L1}^2 + T_{R1}^2) + D(T_{L2}^2 + T_{R2}^2) + 2H(T_{L1}T_{L2} + T_{R1}T_{R2}) + 2JT_{L2}T_{R2}] \quad (5)$$

C, D, H, J being numerical coefficients; this expression is symmetrical with reference to $T_L$ and $T_R$.

If we now imagine the tie-bar AD to be removed and replaced by equal and opposite forces $T_{L1}$ applied at A and D, we see readily that the bending moment, normal pressure, and shearing force on any section of the portion AG of the arch can be expressed in terms of the vertical reaction at the support A, the load upon AG, and the tension $T_{L1}$; thus the internal work of the portion AG will be a function of the second degree in $T_{L1}$ in which the coefficient of $T_{L1}^2$ will depend only upon the lever-arms of the forces and their direction, whilst the coefficient of $T_{L1}$ will depend also upon the load upon the arch and will be a linear function of it.

If we now suppose the tie-bars DE, DH removed and the tie-bar AD replaced, and imagine forces to be applied at D, E, H equal to the tensions in the bars removed and in the same directions, we can express the bending moment, normal pressure, and shearing force for any section in the portion GH in terms of the reaction at A, the load between A and H and the tensions $T_{L2}$, $T_{L4}$.

In the expressions thus obtained we can eliminate $T_{L4}$ by means of the second equation (2), and then the internal work of the portion GH can be expressed by a function of the second degree in $T_{L1}$, $T_{L2}$, in which the squares and product of $T_{L1}$, $T_{L2}$ will have coefficients independent of the loads on the arch whilst the terms of the first degree in $T_{L1}$, $T_{L2}$ will have coefficients which are linear functions of these loads.

If, finally, we imagine the two tie-rods DH, DE replaced and the other two EI, EF removed, and forces equivalent to the tensions substituted at E, I, F, we shall be able to express the bending moment, normal pressure, and shearing force for any section in the portion HC in terms of the reaction at A, the loads on the half-arch AC and the tensions $T_{R2}$, $T_{R5}$; in the expressions thus obtained we eliminate $T_{R5}$ by the second equation (4), and we are thus able to express the internal work by means of a function of the second degree in $T_{L2}$, $T_{R2}$.

Adding the expressions obtained for the internal work of the portions AG, GH, HC of the left-hand half-arch, we obtain for the internal work of the half-arch an expression of the form

$$\frac{1}{2}[LT_{L1}^2 + MT_{L2}^2 + KT_{R2}^2 + 2PT_{L1}T_{L2} + 2QT_{L2}T_{R2} + 2RT_{L1} + 2ST_{L2} + 2UT_{R2} + V]$$

where L. M, K, P, Q are coefficients independent of the tensions in the tie-bars and of the loads on the arch and dependent only on the lever-arms of the tie-bars and their directions; whilst R, S, U are linear functions of the loads upon the arch.

It is clear, moreover, that the internal work of the right-hand half-arch will be expressed by a formula quite analogous to the above from which it can be deduced by reversing $T_R$ and $T_L$ and substituting suitable coefficients R′, S′, U′ for R, S, U, which will also be linear functions of the loads.

Thus the internal work of the whole arch will be expressed by the formula

$$\frac{1}{2}\left\{\begin{array}{l} L(T_{L1}^2 + T^2_{R1}) + (M+K)(T_{L2}^2 + T_{R2}^2) + 2P(T_{L1}T_{L2} \\ + T_{R1}T_{R2}) + 4QT_{L2}T_{R2} + 2RT_{L1} + 2R'T_{R1} + 2(S+U')T_{L2} \\ \qquad + 2(S'+U)T_{R2} + V + V' \end{array}\right\} \quad (6)$$

So that by adding (5) and (6) we shall have the following value for the internal work of the whole structure:—

$$\frac{1}{2}\left\{\begin{array}{l} (C+L)(T_{L1}^2 + T_{R1}^2) + (D+M+K)(T_{L2}^2 + T_{R2}^2) \\ \quad + 2(H+P)(T_{L1}T_{L2} + T_{R1}T_{R2}) + 2(J+2Q)T_{L2}T_{R2} + \\ 2RT_{L1} + 2R'T_{R1} + 2(S+U')T_{L2} + 2(S'+U)T_{R2} + V + V' \end{array}\right\}$$

or in more simplified form,

$$\frac{1}{2}\left\{\begin{array}{l} a(T_{L1}^2 + T_{R1}^2) + b(T_{L2}^2 + T_{R2}^2) + 2c(T_{L1}T_{L2} + T_{R1}T_{R2}) \\ \quad + 2dT_{L2}T_{R2} + 2RT_{L1} + 2R'T_{R1} + 2fT_{L2} + 2gT_{R2} + h \end{array}\right\} \quad (7)$$

$a$, $b$, $c$, etc., being numerical coefficients.

To determine the four unknown tensions $T_{L1}$, $T_{L2}$, $T_{R1}$, $T_{R2}$, we must equate to zero the differential coefficients of the above expression with regard to these quantities. Now it can easily be seen that the differential coefficient with regard to $T_{L1}$ will contain only $T_{L1}$, $T_{L2}$, and that with regard to $T_{R1}$ only $T_{R1}$, $T_{R2}$, while each of the differential coefficients with regard to $T_{L2}$, $T_{R2}$ will contain three unknowns.

But we can simplify the calculation by putting

$$\left.\begin{array}{ll} T_{R1} + T_{L1} = X_1, & T_{R2} + T_{L2} = X_2 \\ T_{L1} - T_{R1} = Y_1, & T_{L2} - T_{R2} = Y_2 \end{array}\right\} \quad . \quad . \quad (8)$$

from which we deduce the relations

$$\left.\begin{array}{ll} T_{L1} = \dfrac{X_1 + Y_1}{2}, & T_{L2} = \dfrac{X_2 + Y_2}{2} \\ T_{R1} = \dfrac{X_1 - Y_1}{2}, & T_{R2} = \dfrac{X_2 - Y_2}{2} \end{array}\right\} \quad . \quad . \quad (9)$$

and therefore

$$\left.\begin{array}{ll} T_{L1}^2 + T_{R1}^2 = \dfrac{X_1^2 + Y_1^2}{2}, & T_{L2}^2 + T_{R2}^2 = \dfrac{X_2^2 + Y_2^2}{2} \\ T_{L1}T_{L2} + T_{R1}T_{R2} = \dfrac{X_1X_2 + Y_1Y_2}{2}, & T_{L2}T_{R2} = \dfrac{X_2^2 - Y_2^2}{4} \end{array}\right\} \quad (10)$$

Substituting in formula (7) we see that it separates into two parts, of which one contains only $X_1$, $X_2$ and the other only $Y_1$, $Y_2$. We thus obtain :—

$$\frac{1}{4}\left\{\begin{array}{l}[aX_1^2 + (b+d)X_2^2 + 2cX_1X_2 + 2(R+R')X_1 + 2(f+g)X_2] \\ + [aY_1^2 + (b-d)Y_2^2 + 2cY_1Y_2 + 2(R-R')Y_1 \\ \qquad\qquad + 2(f-g)Y_2] + h\end{array}\right\} \quad (11)$$

In accordance with the theorem of the change of the unknowns proved in § 2, we can equate to zero the differential coefficients of this expression with regard to $X_1$, $X_2$, $Y_1$, $Y_2$, instead of the differential coefficients of expression (7) with regard to $T_{L1}$, $T_{L2}$, $T_{R1}$, $T_{R2}$, so that we shall obtain the following groups of equations for the two unknowns :—

$$\left.\begin{array}{r} aX_1 + cX_2 + (R + R') = 0 \\ cX_1 + (b+d)X_2 + (f+g) = 0 \end{array}\right\} \quad . \quad . \quad (12)$$

$$\left.\begin{array}{r} aY_1 + cY_2 + (R - R') = 0 \\ cY_1 + (b-d)Y_2 + (f-g) = 0 \end{array}\right\} \quad . \quad . \quad (13)$$

From these equations we can deduce without difficulty the values of $X_1$, $X_2$, $Y_1$, $Y_2$, and then find the values of $T_{L1}$, $T_{L2}$, $T_{R1}$, $T_{R2}$ from equations (8).

In the majority of the applications which we shall give later we shall make use of the simplification indicated above, since all the structures we shall have to consider will be symmetrical.

**4. Use of the Principle of Superposition for the Study of Composite Structures.**—We have shown in Chap. I, § 19, and in Chap. II, § 2, that in all elastic structures, whether pin-jointed or not, the internal stresses, the displacements of the different points, and the angular deflections of the sections are linear functions of the external forces, and that therefore the principle of superposition can be applied; for example, the tension in a tie-bar when the structure is loaded with different loads P, Q, R, etc., is the sum of the tensions to which the tie-bar would be subjected if the loads were separately applied. This greatly facilitates calculations.

Suppose, for example, that for the calculation of the unknowns we have to take the square of the function $w$ of the external forces. As we have

$$w = \alpha P + \beta Q + \gamma R \ldots \quad . \quad . \quad . \quad (14)$$

it is clear that if we consider these external forces one by one we shall only have to square the single terms $\alpha P$, $\beta Q$, $\gamma R$ . . .; but if the forces have to be considered simultaneously we have to square the general expression for $w$, and, therefore, besides the squares of these quantities we shall have all the possible combinations of the products of these quantities taken two at a time.

Another advantage of the principle of superposition is shown by the following consideration: suppose that in order to study the conditions of equilibrium of a truss symmetrical about its centre we have first to determine the unknown stresses for the following two cases:—

1. Taking only the dead load which is assumed to be symmetrically distributed with regard to the centre.
2. Leaving the dead load out of account and considering only the superload of $w$ kilogram per sq. metre of the surface on the left-hand half of the truss.

It is clear that we can by this principle determine the unknown stresses in the different cases that have to be considered in practice and which we are going to examine later.

To make the matter more clear we will apply our reasoning to two symmetrical tie-bars which we will assume to exist in the structure, and we will call $T'$ the tension in these two tie-bars obtained in the first calculation, in which we considered only the dead load, and $T_L$, $T_R$ the tensions obtained for the left and right-hand tie-bars respectively in the second calculation, i.e. for a superload of $w$ kg. per sq. metre on the left-hand half of the truss, leaving the dead load out of account.

It follows from the symmetry of the structure that if the superload were assumed to be only on the right-hand half of the truss, the left-hand tie-bar would have a tension $T_R$ and the right-hand tie-bar a tension $T_L$; it follows from this by the principle of superposition that if the superload were on both sides of the truss, each tie-bar would have a tension equal to

$$T_L + T_R \, . \qquad (15)$$

If we consider the dead load and the superload on each side acting together, the tension in these tie-bars will be

$$T = T' + T_L + T_R \, . \qquad (16)$$

If we consider the dead load and the superload on the left-hand side only, the tension in the left-hand tie-bar will be

$$T' + T_L \, . \qquad (17)$$

and that in the right-hand tie-bar will be

$$T' + T_R \, . \qquad (18)$$

We can consider the further case in which the superload on the truss is $w'$ on the left-hand half and $w''$ on the right-hand half. Then the dead load alone gives a tension $T'$ in each tie-bar, and the superload $w'$ on the left gives tensions in the left and right-hand tie-bars equal respectively to

$$\frac{w'}{w}T_L, \quad \frac{w'}{w}T_R, \quad . \quad . \quad . \quad . \quad . \quad (19)$$

and the superload $w''$ on the right-hand side gives tensions in these bars equal respectively to

$$\frac{w''T_R}{w}, \quad \frac{w''T_L}{w}. \quad . \quad . \quad . \quad . \quad . \quad (20)$$

Therefore the final tension in the left-hand tie-bar will be

$$T' + \frac{w'}{w}T_L + \frac{w''}{w}T_R, \quad . \quad . \quad . \quad . \quad (21)$$

and in the right-hand tie-bar

$$T' + \frac{w'}{w}T_R + \frac{w''}{w}T_L. \quad . \quad . \quad . \quad . \quad (22)$$

If in a symmetrical truss there is a horizontal tie-bar with its centre vertically below the apex of the truss, it will have the same tension for the superload on the left-hand half only as for that on the right-hand half only, so that its tension for the superload on both halves (leaving the dead load out of account) will be twice that to which it is subjected for the superload on one side only.

From several of the practical applications to be given later in this book, where we shall make calculations considering at first the full load, i.e. dead load and superload on both sides, and then the superload only on the left-hand side (leaving the dead load out of account), we shall see that we can deduce from these two cases the third case in which we take account of the dead load and the superload on one half only.

Now we have obtained above the formula

$$T = T' + T_L + T_R$$

which gives the tension in the two symmetrical tie-bars for the dead load and superload over the whole truss. From this formula we get

$$T' + T_L = T - T_R. \quad . \quad . \quad . \quad . \quad (23)$$

$$T' + T_R = T - T_L. \quad . \quad . \quad . \quad . \quad (24)$$

and as $T' + T_L$ and $T' + T_R$ aré the tensions in the two tie-bars for the case of the dead load and the superload on the left-hand half only, it follows that

*the tension in one of the tie-bars in the case of dead load and superload on the left-hand half is obtained by subtracting from the tension which the same tie-bar would have for full load that which the other symmetrical tie-bar would have for left-hand superload only.*

All that we have said in the present treatment for the tensions in the two symmetrical tie-bars holds equally well for the bending moments, normal pressures, and shearing forces for symmetrical sections.

If, for example, in an arch symmetrical about the crown we call B and P the bending moment and normal pressure for the crown section in the case of a superload on the left-hand half of the arch only, omitting the dead load, it is clear that in the case in which there is an equal superload on the right-hand half also, the bending moment and normal pressure at the crown will be 2B and 2P, but their ratio will be $\frac{B}{P}$ as in the first case. Thus *the line of pressure passes through the same point at the crown both for the superload on one side only and for the superload on both sides.*

**5. Proof of the small influence on the stresses in elastic structures of the terms due to shear.**—All the structures considered in the later part of this book are made up of a solid member subjected to compression, bending, or shear, and of straight bars attached to this member and to themselves by pin-joints; these bars can therefore be subjected only to axial compressive or tensile stresses.

Sometimes the part subjected to bending will form the whole structure, as in the case of hinged and fixed arches considered in Chap. IX, §§ 3 and 4, in which the abutments prevent displacement of the ends of the arch.

It is clear that in the structures defined above we have only to determine the line of pressure for the member subjected to bending, because for the articulated bars this line will coincide with their axis.

Now we shall realise on examining all the numerical applications which we give later that the line of pressure of the member subjected to bending passes as much on one side of the centre line as on the other, remaining always very near to this centre line and at a very small inclination to it, i.e. the tangent to the line of pressure at any point makes a very small angle with the tangent to the centre line at the corresponding point.

But it follows from what we have said in § 6 of the previous chapter that the resultant pressure acting on any section, i.e. the resultant of the normal pressure and the shearing force, is very nearly tangential to the line of pressure; from which it follows that the normal pressure will be everywhere nearly equal to the resultant pressure, while the shearing force will be very small compared with the latter.

Moreover, as the line of pressure passes as much on one side of the centre line as on the other, we see that not only will the shearing force be everywhere very small compared with the normal pressure, but that it will be positive for some segments and negative for others, whilst the normal pressure will be positive on each section.

For the member subjected to bending, therefore, *the deformation due to the shearing force will be very small compared with that due to normal pressure ;* and, since each is in general very small compared with that due to bending, as we have proved in Chap. IV, § 8, and, moreover, as the unknown stresses occurring in elastic structures depend on their deformations, we may conclude that it will be nearly always useless to take account of the effect of the shearing stresses when determining the unknown stresses or reactions.

As this point is very important in curtailing calculations, and as we shall derive benefit from it in all numerical applications, we think it useful to prove its truth by means of some particular examples, which will serve to show that the influence of the terms due to shear is very small compared with that of the terms due to compression, although the former taken separately in each member of the equations might be greater than the latter.

Take first the case of the arch with hinged ends which we have considered in Chap. IX, § 3.

The only unknown in the problem, which is the horizontal thrust, is given by formula (13), Chap. IX ; if the angle $\phi$ is equal to $\frac{\pi}{2}$, i.e. if the centre line of the arch is a semicircle, this formula gives

$$H = \frac{4w\mathrm{R}}{3\pi}\left(\frac{\frac{\mathrm{R}^2}{\mathrm{EI}} - \frac{1}{\mathrm{EA}} + \frac{\mathrm{F}}{\mathrm{GA}}}{\frac{\mathrm{R}^2}{\mathrm{EI}} + \frac{1}{\mathrm{EA}} + \frac{\mathrm{F}}{\mathrm{GA}}}\right)$$

or

$$H = \frac{4w\mathrm{R}}{3\pi}\frac{(a - x)}{(a + y)}$$

where

$$a = \frac{\mathrm{R}^2}{\mathrm{EI}},\ x = \frac{1}{\mathrm{EA}} - \frac{\mathrm{F}}{\mathrm{GA}},\ y = \frac{1}{\mathrm{EA}} + \frac{\mathrm{F}}{\mathrm{GA}}.$$

The quantities $x$ and $y$ will be very small compared with $a$, as we have already shown several times ; thus, neglecting powers of $\frac{x}{a}$ and $\frac{y}{a}$ above the first, we shall have

$$\frac{a - x}{a + y} = \frac{a - x}{a}\left(1 + \frac{y}{a}\right)^{-1} = \frac{(a - x)\left(1 - \frac{y}{a}\right)}{a} = \frac{a - x - y}{a},$$

and the expression for H will become on putting back the values of $a$, $x$, and $y$

$$\mathrm{H} = \frac{4w\mathrm{R}}{3\pi}\left(\frac{\frac{\mathrm{R}^2}{\mathrm{EI}} - \frac{2}{\mathrm{EA}}}{\frac{\mathrm{R}^2}{\mathrm{EI}}}\right)$$

We should evidently have obtained this result if we had started with the formula

$$\mathrm{H} = \frac{4w\mathrm{R}}{3\pi}\left(\frac{\frac{\mathrm{R}^2}{\mathrm{EI}} - \frac{1}{\mathrm{EA}}}{\frac{\mathrm{R}^2}{\mathrm{EI}} + \frac{1}{\mathrm{EA}}}\right)$$

in which the term $\frac{\mathrm{F}}{\mathrm{GA}}$ due to shear has been omitted from the numerator and the denominator, which shows that this term has a very small effect upon the value of H.

It should be noted that the term $\frac{\mathrm{F}}{\mathrm{GA}}$ is not very small compared with the term $\frac{1}{\mathrm{EA}}$, but that on the contrary it will usually be greater; if nevertheless it has practically no influence upon the value of H, this is due to the fact that it has the same sign as the term $\frac{\mathrm{R}^2}{\mathrm{EI}}$ in the numerator and denominator, whilst the term $\frac{1}{\mathrm{EA}}$ has opposite signs.

When the rise of the arch is very small compared with the chord, the horizontal thrust is determined by equation (15), Chap. IX, i.e.

$$\left(\frac{8r^2}{15\mathrm{EI}} + \frac{1}{\mathrm{EA}} + \frac{4r^2}{3l'^2}\cdot\frac{\mathrm{F}}{\mathrm{GA}}\right)\left(\mathrm{H} - \frac{wl'^2}{2r}\right) + \frac{1}{\mathrm{EA}}\left(\frac{wl'^2}{2r} + \frac{2wr}{3}\right) = 0 \quad (26)$$

As the quantity $\frac{\mathrm{F}}{\mathrm{GA}}$ is of the same order of magnitude as the quantity $\frac{1}{\mathrm{EA}}$, and $\frac{r^2}{l'^2}$ is very small, we can neglect the term $\frac{4r^2}{3l'^2}\cdot\frac{\mathrm{F}}{\mathrm{GA}}$ in comparison with $\frac{1}{\mathrm{EA}}$, so that the preceding equation becomes

$$\left(\frac{8r^2}{15\mathrm{EI}} + \frac{1}{\mathrm{EA}}\right)\left(\mathrm{H} - \frac{wl'^2}{2r}\right) + \frac{1}{\mathrm{EA}}\left(\frac{wl'^2}{2r} + \frac{2}{3}wr\right) = 0,$$

i.e. it does not contain any term due to shear."

Now let us take the case of the arch with fixed ends that we have studied in § 4 of the previous chapter.

We have here the two unknowns $B_0$, $P_0$ which are determined by the two equations (22) and (23), Chap. IX.

If the angle $\phi$ is equal to $\frac{\pi}{2}$, i.e. if the centre line of the arch is a semicircle, these two equations become

$$\left.\begin{aligned} &\frac{B_0\pi}{2} - P_0R\left(\frac{\pi}{2} - 1\right) + \frac{\pi}{8}wR^2 = 0, \\ &\frac{R}{EI}\left[B_0 - P_0R\left(1 - \frac{\pi}{4}\right) + \frac{wR^2}{6}\right] + \frac{1}{EA}\left(P_0\frac{\pi}{4} + \frac{wR}{3}\right) \\ &\qquad + \frac{F}{GA}\left(\frac{P_0\pi}{4} - \frac{wR}{3}\right) = 0 \end{aligned}\right\} \quad (27)$$

and on substituting in the second the value of $B_0$ deduced from the first and collecting like terms we obtain

$$\left[\frac{R^2}{EI}\left(\frac{\pi}{4} - \frac{2}{\pi}\right) + \frac{1}{EA}\cdot\frac{\pi}{4} + \frac{F}{GA}\cdot\frac{\pi}{4}\right]P_0 = \left(\frac{R^2}{12EI} - \frac{1}{3EA} + \frac{F}{3GA}\right)wR\,;$$

seeing that we have approximately

$$\frac{\pi}{4} - \frac{2}{\pi} = \frac{10}{53}\cdot\frac{\pi}{4},$$

this formula gives

$$P_0 = \frac{4wR}{3\pi}\cdot\frac{\left(\frac{R^2}{4EI} - \frac{1}{EA} + \frac{F}{GA}\right)}{\left(\frac{10R^2}{53EI} + \frac{1}{EA} + \frac{F}{GA}\right)} \quad . \quad . \quad (28)$$

or

$$P_0 = \frac{4wR}{3\pi}\frac{(a - x)}{(b + y)},$$

where

$$a = \frac{R^2}{4EI}\,; \quad b = \frac{10R^2}{53EI}\,; \quad x = \frac{1}{EA} - \frac{F}{GA}\,; \quad y = \frac{1}{EA} + \frac{F}{GA}.$$

As the quantities $x$ and $y$ are very small compared with $a$ and $b$, we may take

$$\frac{a - x}{b + y} = \frac{a - x}{b}\left(1 + \frac{y}{b}\right)^{-1} = \frac{(a - x)\left(1 - \frac{y}{b}\right)}{b} = \frac{a - x - \frac{ay}{b}}{b},$$

neglecting powers of $\frac{x}{a}$ and $\frac{y}{b}$ above the first; substituting the values of $a$, $b$, $x$, $y$, we have

$$P_0 = \frac{4wR}{3\pi}\frac{\left(\frac{R^2}{4EI} - \frac{93}{40}\cdot\frac{1}{EA} - \frac{13}{40}\cdot\frac{F}{GA}\right)}{\frac{10R^2}{53EI}},$$

We need only consider the numerator of $P_0$, or rather the quantity

$$\frac{R^2}{4EI} - \frac{93}{40}\cdot\frac{1}{EA} - \frac{13}{40}\cdot\frac{F}{GA},$$

and to compare with each other the last two terms which we have already shown to be small compared with the first.

Now if we assume that the arch is homogeneous and isotropic and that its section is rectangular, we shall have

$$F = \frac{6}{5}, \quad G = \frac{2}{5}E.$$

and

$$\frac{13}{40}\cdot\frac{F}{GA} = \frac{39}{40}\cdot\frac{1}{EA},$$

so that the above expression will become

$$\frac{R^2}{4EI} - \frac{93}{40}\cdot\frac{1}{EA} - \frac{39}{40}\cdot\frac{1}{EA} = \frac{R^2}{4EI} - \frac{132}{40}\cdot\frac{1}{EA}.$$

As the term $\frac{13}{40}\cdot\frac{F}{GA} = \frac{39}{40}\cdot\frac{1}{EA}$ is almost exactly $\frac{2}{7}$ of the term $\frac{132}{40}\cdot\frac{1}{EA}$, we see that the term due to shear is only $\frac{2}{7}$ of the sum of this term and that due to the normal thrust; as, moreover, this sum is very small compared with the term $\frac{R^2}{4EI}$ due to bending, it follows that we can neglect the $\frac{2}{7}$ without making too serious an error.

Further, we should bear in mind that a semicircular arch with fixed ends presents one of the most unfavourable cases from the point of view of shear; for the line of pressure is considerably inclined for a large portion of its length to the centre line of the arch, i.e. the shearing forces are not very small compared with the normal pressures.

We can easily see by means of equations (22) and (23), Chap. IX, that when the centre line of the arch is a circular arc less than a semicircle, the effect of shear will be negligible compared with that of normal pressure to a degree of approximation which becomes closer as the rise becomes smaller compared with the chord.

Let us consider, for example, the case in which the rise is very small compared with the chord. The value of the thrust $P_0$ is then given by equation (29), Chap. IX, which differs from equation (15) analysed above only in the coefficient of the term due to bending; we are thus still led to the conclusion that the effect of shear upon the value of $P_0$ is insignificant and that consequently it is not necessary to take it into account.

Summing up, we see that among the four cases considered, there are three in which the effect of shear is almost nil, and only one in which this effect is comparable with that due to the normal pressure, but still much smaller. Thus in practical applications we can first obtain the unknowns by neglecting shear; we can then draw the line of pressure. If it has everywhere only a small inclination to the centre line of the structure the results may be regarded as absolutely exact; but if this inclination is appreciable over a large portion of the line of pressure we ought to make a fresh calculation, taking into account also the internal work due to shear.

This latter condition will not arise in any of the numerical applications which we shall give later.

**6. Theorem of the Effects of Temperature Variation.**—When an elastic structure has been erected at a certain temperature it usually happens that, on a change of temperature, stresses are induced which add to those produced by the external forces. For example, when the temperature of a metal arch varies, the chord tends to increase or decrease in length; but if it is fixed at the ends this change in length cannot take place, and so additional stresses are induced in the arch.

Now let us suppose that an elastic structure has been erected at a certain temperature $t°$ at which the lengths of the various members are exactly those geometrically necessary for the composition of the structure; let us suppose also that one end is made free of each of the redundant bars, which cannot freely change in length themselves and which would prevent the change in length of the others.

Let, for example, ABCD (Fig. 6, Pl. III) be a structure composed of members AB, BC, DH, rigidly fixed to each other, and of a tie-bar AC hinged at its ends.

We see readily that when the temperature varies, these members cannot freely change in length, because each is fixed at both ends to other members. But if we imagine the member BC to be cut through at the section EF, and the end C of the tie-bar AC to be free, we see that each member will be attached to the others only at one end and therefore that it can conform to the changes due to temperature.

Now in accordance with our hypothesis, at the initial temperature $t°$ the two structures DBFE, CFE will join exactly at the section EF and the tie-bar will have a length exactly equal to the distance between the points A, C which it has to connect

But at another temperature $t_1°$ the section EF will become E′F′, E″F″ (Fig. 7, Pl. III); moreover, the distance AC will become AC′, while the length of the tie-bar will become AC″.

We will call

$\lambda$ the difference between the lengths AC′ and AC″;

$\theta$ the very small angle between E′F′, E″F″;

$\xi$, $\eta$ the projections on the normal to the initial position of the section EF, and on that section itself, of the straight line joining the centres of the sections E′F′ and E″F″.

If we now suppose that we apply to the structure ABCD (Fig. 6, Pl. III), in which all the original connections are assumed to be remade, any external forces, the structure will be strained and the strains as well as the stresses will be functions of the temperature; but it is clear that for any given temperature we could express the stresses at any point if we knew the tension T of the tie-bar AC and the bending moment B, normal pressure P, and shearing force S on the section EF.

To determine the values which these four unknowns will have at the temperature $t_1°$, we will suppose that the structure is in the first place in the condition represented by Fig. 7 and that we apply the following external forces:—

At the point C′ a force T acting from C′ towards A.

At the end C″ of the tie-bar another force T acting from C″ towards C′.

At the section E′F′ normal and tangential forces giving a bending moment B, normal pressure P, and shearing force S.

Finally, at the section E″F″, normal and tangential forces equal and opposite to those on the section E′F′.

As regards the direction of the forces applied on E′F′, E″F″ we will assume

1. That the moments B tend to turn these sections so as to diminish the angle between them.

2. That the forces P are true normal pressures, i.e. that they tend to increase the projection on the normal to the section EF of the distance between the centres of the two sections E′F′, E″F″.

3. Finally, that the shearing forces S have such direction that they tend to diminish the projection of this distance on the section EF.

The four quantities T, B, P, S must be such that under their action and that of the given forces the points C′, C″, and the sections E′F′, E″F″ are brought into coincidence.

For greater clearness we will call T′, T″ the two forces applied at C′, C″ and B′, P′, S′ and B″, P″, S″ the bending moment, normal pressure, and shearing force on the sections E′F′, E″F″.

In accordance with what we have said already, we can express the bending moment, normal pressure, and shearing force at any section of the members composing the structure, and therefore the internal work, as a function of the given external forces and of the quantities T′, T″; B′, B″; P′, P″; S′, S″.

Let $W_i$ be this internal work, then in the first place we note that the sums

$$\frac{dW_i}{dT'}+\frac{dW_i}{dT''};\quad \frac{dW_i}{dB'}+\frac{dW_i}{dB''};\quad \frac{dW_i}{dP'}+\frac{dW_i}{dP''};\quad \frac{dW_i}{dS'}+\frac{dW_i}{dS''}$$

are only the differential coefficients of $W_i$ with regard to the quantities T, B, P, S; in fact, if we consider T′, T″, for example, as functions of T given by the equations

$$T''=T;\quad T'=T$$

we obtain

$$\frac{dW_i}{dT}=\frac{dW_i}{dT'}\cdot\frac{dT'}{dT}+\frac{dW_i}{dT''}\cdot\frac{dT''}{dT}$$

and since

$$\frac{dT'}{dT}=\frac{dT''}{dT}=1,$$

it follows that

$$\frac{dW}{dT}=\frac{dW_i}{dT'}+\frac{dW_i}{dT''}.$$

Now this expression gives the sum of the displacements of the points C′, C″ towards each other, i.e. the diminution of the distance C′C″, so that in order that these two points shall coincide after deformation, we must have

$$\frac{dW_i}{dT'}+\frac{dW_i}{dT''}=\lambda,$$

i.e.

$$\frac{dW_i}{dT}=\lambda \quad . \quad . \quad . \quad (29)$$

The formula $\frac{dW_i}{dB'}+\frac{dW_i}{dB''}$ gives the sum of the angular deflections of the two sections E′F′, E″F″, i.e. the diminution of the angle between these two sections, and in order that they shall coincide after deformation, we must have

$$\frac{dW_i}{dB'}+\frac{dW_i}{dB''}=\theta,$$

i.e.

$$\frac{dW_i}{dB}=\theta \quad . \quad . \quad . \quad . \quad . \quad (30)$$

Similarly we must have

$$\frac{dW_i}{dP}=-\xi \quad . \quad . \quad . \quad . \quad . \quad (31)$$

$$\frac{dW_i}{dS}=\eta \quad . \quad . \quad . \quad . \quad . \quad (32)$$

The four equations (29) to (32) enable the four unknowns T, B, P, S to be found.

We have, up to the present, applied our reasonings to an elastic structure of a particular form; but we can easily see that they apply to every case. If, for example, a structure contains several redundant tie-bars and several fixed members that can be cut through at one section without making the structure deficient, we shall have an unknown tension T for each such bar, and three unknowns B, P, S for each such member. To determine all the unknowns we shall have for each tie-bar an equation similar to (29), and for each member three equations similar to (30) to (32).

It should be observed that equations (29) to (32) can be obtained by equating to zero the differential coefficients with regard to T, B, P, and S of the function

$$W_i - T\lambda - (B\theta - P\xi + S\eta).$$

In the general case in which there are several redundant tie-bars, and several fixed members that could be cut through at one section without making the structure deficient, we obtain all the equations for the determination of the unknowns by equating to zero the differential coefficients with regard to these unknowns of the function

$$W_i - \Sigma T\lambda - \Sigma(B\theta - P\xi + S\eta).$$

We then arrive at the following theorem: *The stresses occurring in the structure under consideration at the temperature* $t_1^\circ$ *are those which make a minimum the internal work of the structure diminished by as many terms similar to* $T\lambda$ *as there are redundant bars and by as many expressions similar to* $(B\theta - P\xi + S\eta)$ *as there are fixed members which can be cut through without rendering the structure deficient.*

It will be seen that this theorem contains as a particular case the theorem of least work, because at the original temperature the quantities $\lambda$, $\theta$, $\xi$, $\eta$ are all zero, and therefore it is only the internal work of the structure which has to be a minimum.

Finally we note that the quantities $\lambda$, $\theta$, $\xi$, $\eta$ are functions of the difference of temperature $(t_1^\circ - t^\circ)$ as well as of the form and nature of the component parts of the structure, but are absolutely independent of the external forces.

We can even express them in general by the formula $a(t_1^\circ - t^\circ)$ $a$ being a numerical coefficient different for each of these quantities but independent of $(t_1^\circ - t^\circ)$.

**7. Simple Graphical Constructions to Replace Numerical Calculations.**—In the numerical investigations on the equilibrium of elastic structures we have always to derive expressions for the bending moments, normal pressures, and shearing forces for the members making up the structure; it often happens that there are four concurrent tie-bars and we have to express the tensions of two of them in terms of the tensions in the other two.

To effect these operations rigidly we could calculate all the necessary quantities, i.e. the lever-arms of the forces, the angles which they make with different sections, the angles between the tie-bars, etc., by means of geometrical or trigonometrical methods. In practice, however, it is generally sufficient to measure all these quantities from a diagram carefully drawn to a sufficiently large scale.

We will give here some notes on the graphical processes which we shall apply in the second portion of this book.

*Bending Moments.*—To obtain the expression for the bending moment on a given section of a body we must know the lever-arm of each of the forces applied on one side of the given section, including in these forces not only the applied loads but also the tensions in the tie-bars, reactions on the supports, etc. Therefore on a diagram representing the structure, we mark the sections about which we require to find the bending moments, and from the centroid of each section we draw perpendiculars on to all the forces known and unknown which combine to produce the bending moment at such section. By the side of each perpendicular we write its length measured on the diagram.

*Normal Pressures and Shearing Forces.*—To obtain the expression for the normal pressure or that for the shearing force at any section we have to consider the same forces which combine to produce the bending moment; we have to know in addition their projections on the normal to the section to obtain the normal pressure and their projections parallel to the section to obtain the shearing force. The values of these projections are obtained by multiplying each force by the sine or cosine of the angle which it makes with the section.

We can obtain by a single graphical construction the sines and cosines of the angles which any force makes with different sections. On a line parallel to the direction of the force take a length AB (Fig. 9, Pl. III) to represent unity to any convenient scale, and on this line as diameter describe a semicircle; then draw from the point B lines $BC_0$, $BC_1$, etc., parallel to the given sections and join $C_0$, $C_1$, etc., to A, these lines being perpendicular to $BC_0$, $BC_1$, etc., respectively. If we measure $AC_0$, $AC_1$, etc., to our given scale they will give the sines of the angles which the direction AB makes with the sections considered, the lengths $BC_0$, $BC_1$, etc., giving the cosines.

In all the practical examples which we give later we shall make use of this graphical construction.

*Resolution of Forces.*—Let O (Fig. 10, Pl. III) be a node at which meet four pin-jointed bars belonging to an elastic structure in equilibrium under the action of external forces, and let $T_1$, $T_2$, $T_3$, $T_4$ be the tensions in these bars.

As the four forces are in equilibrium, it is clear that two of them can be expressed in terms of the other two, so that when the latter are known the former can easily be determined.

Suppose, for example, that we have to express $T_2$ and $T_3$ in terms of $T_1$ and $T_4$.

Take any point A on a line in the direction of $T_3$ and drop perpendiculars $AA_1$, $AA_2$, $AA_4$ on to the directions of the tensions $T_1$, $T_2$, $T_4$. As the algebraic sum of the moments of the four tensions about any point must be zero, it is clear that if we take moments about the point A, the moment of $T_3$ being zero, we shall have

$$T_2 \,.\, AA_2 - T_1 \,.\, AA_1 - T_4 \,.\, AA_4 = 0.$$

If therefore we take the length $AA_2$ as unity and call the other perpendiculars ($AA_1$, $AA_4$) $a_1$, $a_4$ respectively, this equation will become

$$T_2 = T_1 a_1 + T_4 a_4.$$

Now take on the direction of the tension $T_2$ a length $OB = OA$, and drop perpendiculars $BB_1$, $BB_3$, $BB_4$ on to the directions of the other tensions. If we express the condition that the sum of the moments about B must be zero we shall have

$$T_3 \,.\, BB_3 - T_1 \,.\, BB_1 + T_4 \,.\, BB_4 = 0,$$

but since clearly $BB_3 = AA_2 = 1$, we shall have, if we write $BB_1$, $BB_4$ as $b_1$, $b_4$ respectively,

$$T_3 - T_1 b_1 - T_4 b_4.$$

We see therefore that after having made the graphical construction indicated above, we express the tensions $T_2$, $T_3$ in terms of $T_1$, $T_4$ by means of the two equations

$$T_2 = T_1 a_1 + T_4 a_4$$
$$T_3 = T_1 b_1 - T_4 b_4.$$

# CHAPTER XI.

## IMPERFECTLY ELASTIC STRUCTURES SUCH AS MASONRY ARCHES.

**1. Some Notes on Imperfectly Elastic Structures.**—We call imperfectly elastic those bodies which after being compressed do not return exactly to their original form when the external forces are removed. Mortar is such a body when it has not set hard.

Now as structures composed of these bodies, although not perfectly elastic, are compressible and therefore liable to deformation, and as, moreover, their deformations are greater when the forces producing them are greater, it follows that the deformations are functions of the external forces.

Therefore if we have an imperfectly elastic structure subject to certain geometrical conditions, we determine the unknown quantities, i.e. the reactions or the bending moment, normal pressure, and shearing force on the fixed ends, by expressing the geometrical conditions which the structure must satisfy.

If, moreover, the compressibility of the imperfectly elastic bodies composing the structures is proportional to the pressures, as is true within certain limits, it is easy to see that the distribution of the internal stresses will be found in exactly the same way as in perfectly elastic bodies. From this it follows that the formulæ and theorems that we have derived for perfectly elastic structures and bodies will hold also for imperfectly elastic ones, the most common example of which in practice is that of the masonry arch when the mortar has not fully set.

Mortar may be regarded as quite incapable of tensile resistance, so that at joints at which tension would occur, if it were possible, there will be cracks. But we can see easily that all the formulæ and theorems which we have previously given will be rigorously applicable; the calculations will, however, be somewhat longer, because we do not know to start with the dimensions of all the sections.

The same holds for ashlar masonry arches constructed without mortar.

For example, if an arch has a small rise compared with the span it nearly always happens that at all the sections of a portion ABED near to the springings (Fig. 8, Pl. III) tensile stresses tend to be produced near to the extrados, but from the section ED to the crown each section is compressed throughout.

It is for this reason that on removal of the centering of very shallow elliptic arches cracks nearly always develop at the springings which start at the extrados and extend more or less into the arch.

In this case if we consider any section of the portion ABED we see that it must be compressed on the part contained between the intrados and a certain curve CE, and that there is neither compression nor tension between this curve and the extrados.

Thus, when we have to study the conditions of equilibrium of such an arch we can apply to it the formulæ and theorems previously found, but for the portion ABED we must consider the resisting section as limited by the curve CE and not by the extrados.

The portion BCE need only be regarded as a load.

The difficulty consists then in the determination of the curve CE which is unknown to start with; but even this difficulty can be overcome, without having recourse to very long calculations, by the method of successive approximations, of which we will give an example in the study of an ashlar masonry bridge over the Doire at Turin, constructed in 1827 by Charles Mosca. The arch of this bridge was constructed partly without mortar and partly by interposing a layer of mortar between the blocks of granite, the thickness of each layer being variable.

We do not think it necessary to add further considerations to prove that the calculations for imperfectly elastic structures can be made in accordance with the same principles as those adopted for perfectly elastic structures; for these principles reduce to this, that in all solid materials the stresses are functions of the strains produced by the external forces.

The objection might be raised that if the external forces applied to an imperfectly elastic body increase from zero to certain values and then decrease to zero again and increase further later on, the condition of the body with regard to the strains is not exactly the same in the first period as in successive periods for the same system of external forces, as would occur in perfectly elastic bodies.

This objection is doubtless valid, but it cannot detract in general from what we have said above.

Moreover, it should be noted that this variation in external forces never occurs in practice except within very confined limits. For example, in a masonry bridge it is only the superload which can vary, and this is always a very small fraction of the whole load.

# PART II.

## APPLICATION TO PRACTICAL EXAMPLES

# CHAPTER XII.

## STUDY OF A BEAM STRENGTHENED BY TWO TIE-RODS AND A CAST-IRON STRUT.*

(Plate IV, Figs. 1, 2.)

**1. General Data.—**

| | | |
|---|---|---|
| Clear span between supports . | | = 8·00 m. |
| Distance from the intersection of the tie-rods to the axis of the beam . . . . . | | = ·60 m. |
| Sectional area of the beam . . | ·20 m. × ·20 m. | = ·04 sq. m. |
| Diameter of tie-rods . . . | | = ·026 m. |

The strut is of cast iron and of cruciform section, and we will regard it as absolutely rigid on account of its short length and of the small pressure to which it is subjected.

The load on the beam, including its own weight = 180 kg. per metre run.

**2. Calculation of the Tension in the Tie-rods.**—We will call

$T$ = the tension in the tie-rods,

$P$ = the pressure on the strut,

$\beta$ = the angle which the tie-rods make with the axis of the beam;

then

$$\sin \beta = \frac{\cdot 60}{\sqrt{4^2 + \cdot 60^2}} = \cdot 1482$$

$$\cos \beta = \frac{4}{\sqrt{4^2 + \cdot 60^2}} = \cdot 9889,$$

and therefore

$$P = 2T \sin \beta = \cdot 2964T \quad . \quad . \quad . \quad (1)$$

Throughout the beam there is a constant thrust of value $T \cos \beta = \cdot 9889T$.

Each reaction = 180 × 4 = 720 kg.

* The majority of the applications which we give here are taken from a publication by the Art Bureau of the Northern Italian Railways entitled "Practical Applications of the Theory of the Equilibrium of Elastic Systems". We have modified them slightly to bring them into closer agreement with the theory that we have expounded in the preceding chapters.

The bending moment at the ends of the beam is zero, and at the centre section C it is equal to

$$B_c = -R \times 4 + \frac{w}{2} \times 4^2 + T \times 4 \sin \beta,$$

i.e. $B_c = -1440 + {\cdot}5928T$.

Now the area and the moment of inertia of the beam section have values $A = {\cdot}04$ and $I = \frac{{\cdot}20 \times {\cdot}20^3}{12} = {\cdot}000133$ respectively, and if we call E the elastic modulus of the timber, the internal work or strain energy of the beam due to compression will be expressed by the formula

$$W_{i_1} = \frac{8}{2EA} \times (T \cos \beta)^2 = \frac{1}{2E} \times 195{\cdot}58T^2 \quad . \quad (2)$$

and that due to bending by

$$W_{i_2} = 2 \times \frac{4}{2EI}\left\{\frac{B_c^2}{3} - \frac{B_c w \times 4^2}{12}\right\} = \frac{1}{2E}(7046T^2 - 21395000 \times 2T) \quad (3)$$

Finally, if E′ is the elastic modulus of the iron, since the length of each tie-rod is 4·045 m., and its cross-sectional area $A' = {\cdot}000531$ sq. m., the internal work of the tie-rods is given by

$$W_{i_3} = 2 \times \frac{4{\cdot}045}{2E'A'} \cdot T^2 = \frac{15235T^2}{2E'},$$

i.e. assuming $E' = 12E$, we have

$$W_{i_3} = \frac{1270T^2}{2E} \quad . \quad . \quad . \quad . \quad . \quad (4)$$

The internal work of the whole structure is obtained by adding the three formulæ (2), (3), and (4), thus giving the result

$$\text{Total internal work} = \frac{1}{2E}(8512T^2 - 21395000 \times 2T) \quad (5)$$

To obtain the value of T we must equate to zero the differential coefficient of this expression with regard to T, which gives

$$8512T - 21395000 = 0,$$

$$\text{i.e. } T = 2514 \text{ kg.}$$

**3. Maximum Stress per sq. metre in the Tie-rods and Beam.**—The stress per sq. metre in the tie-rods will be

$$t = \frac{2514}{{\cdot}000531} = 4{,}730{,}000.$$

The bending moment about a section at distance $x$ from the nearest end is

$$B = -Rx + T \sin \beta \, . \, x + \frac{wx^2}{2} = -347x + 90x^2.$$

This has a maximum value for $x = \frac{347}{180} = 1{\cdot}93$, giving $B = -335$ kg. m.; from this it follows that the maximum stress per sq. metre in the beam is given by

$$c = \frac{T\cos\beta}{A} - \frac{Bn}{I} = \frac{2486}{\cdot 04} + \frac{335 \times \cdot 1}{\cdot 000133} = 313{,}000,$$

or 31·30 kg. per sq. cm.

**4. Effect of Temperature Variation.**—The previous results hold good if the temperature remains the same as that at the time of erection, which we will call the *mean temperature*.

If the temperature increases or decreases by $t_1^\circ$ C., each tie-rod, if free, will elongate or contract by an amount $\frac{4{\cdot}045t_1^\circ}{82500} = \cdot 000049t_1^\circ$, taking $\frac{1}{82500}$ as the coefficient of expansion for iron.

As regards the beam we may assume that even if it were free to expand the variation in length due to change of temperature would be negligible. Finally, we will neglect the expansion of the strut, partly on account of its small length and partly on account of the manner in which it acts in the structure.

From the above it is clear that if the erection of the beam had to be effected at a temperature $t_1^\circ$ C. above or below the mean, each tie-rod would have a length of $\cdot 000049t_1^\circ$ in excess or short of that which is geometrically necessary.

Thus to find from the internal work the tension in the tie-rods at this temperature, assuming the beam to have been erected and loaded at the mean temperature, we should have first to add to formula (5) the term

$$\pm 2 \times \cdot 000049t_1^\circ T = \pm \cdot 000098t_1^\circ T, \text{ giving}$$

$$\text{Total internal work} = \frac{1}{2E}(8512T^2 - 21395000 \times 2T) \pm \cdot 000098t_1^\circ T;$$

equating to zero the differential coefficient of this expression with regard to T, we have

$$\frac{1}{E}(8512T - 21395000) \pm \cdot 000098t_1^\circ = 0,$$

$$\text{i.e. } T = 2514 \mp 17{\cdot}27t_1^\circ. \qquad (6)$$

taking $E = 1{,}500{,}000{,}000$.

The − sign relates to increases and the + sign to decreases in temperature.

If we suppose that the mean temperature is 15° C., the maximum and minimum being 40° C. and − 10° C. respectively, the maximum value of $t_1°$ will be 25, and it will be necessary to examine separately the conditions of resistance of the beam at the two extreme temperatures.

1. *Temperature* 40° C.—Putting $t_1° = 25$ and taking the − sign in (6), we have

$$T = 2514 - 432 = 2082 \text{ kg.},$$

and the tensile stress per sq. metre of the tie-rod will be

$$t = \frac{2082}{\cdot 000531} = 3{,}920{,}000 \text{ kg. per. sq. metre.}$$

The normal pressure or thrust through the beam will be $T \cos \beta = 2082 \times \cdot 9889 = 2059$ kg., and the bending moment about a section at distance $x$ from the nearest support will be

$$B = -Rx + T \sin \beta \,.\, x + \frac{wx^2}{2} = -411x + 90x^2;$$

the maximum value occurs for $x = \frac{411}{180}$ and equals − 469 kg. m.

Thus at a temperature of 40° C., the maximum compressive stress in the timber is given by

$$c = \frac{2059}{\cdot 04} + \frac{469 \times \cdot 10}{\cdot 000133} = 405{,}000 \text{ kg. per sq. metre.}$$

$$\text{i.e. } c = 40{\cdot}5 \text{ kg. per sq. cm.}$$

2. *Temperature* − 10° C.—In this case we must put $t_1° = 25$ in (6) and take the plus sign; we then have

$$T = 2514 + 432 = 2946 \text{ kg.}$$

and the tensile stress in the tie-rod

$$t = \frac{2946}{\cdot 000531} = 5{,}548{,}000 \text{ kg. per sq. metre.}$$

The thrust throughout the beam is

$$T \cos \beta = 2946 \times \cdot 9889 = 2913 \text{ kg.},$$

and the bending moment about a section at distance $x$ from the nearest end is

$$B = -Rx + T \sin \beta \,.\, x + \frac{wx^2}{2} = -283x + 90x^2,$$

the maximum numerical value of which occurs at the centre of the beam, where $x = 4{\cdot}00$, and is given by

$$B_c = +308 \text{ kg. m.}$$

It follows that the maximum compressive stress in the beam is

$$c = \frac{2913}{\cdot 04} + \frac{308 \times \cdot 10}{\cdot 000133} = 304{,}400 \text{ kg. per sq. metre.}$$

**5. Conclusion.**—Reviewing the foregoing results, we see

that if the temperature varies between − 10° C. and + 40° C., the maximum tensile stress in the tie-rod varies between 5,548,000 and 3,920,000 kg. per sq. metre, and the maximum compressive stress in the timber varies between 30·44 and 40·5 kg. per sq. cm.

It therefore follows that in calculations of structures composed of timber and iron it is essential to take account of temperature variation.*

* Fig. 2, Pl. IV, shows the bending moment curves for the beam at mean temperature, and at the two extreme temperatures. The curves are parabolas, of which the parameter is half the load per metre run, and give a graphic representation of the laws of distribution of the bending stresses in the different sections.

## CHAPTER XIII.

### STUDY OF A BEAM STRENGTHENED BY THREE TIE-BARS AND TWO CAST-IRON STRUTS.

(Plate IV, Figs. 3 and 4.)

**1. General Data.**—

| | |
|---|---|
| Clear span between supports . . . . . | = 8·00 m. |
| Distance of each strut from the nearest end . | = 2·75 m. |
| Distance between struts . . . . . . | = 2·50 m. |
| Distance from horizontal tie-bar to axis of the beam . . . . . . . . . | = ·60 m. |
| Cross-sectional area of beam = ·20 × ·20 . . | = ·04 sq. m. |
| Diameter of tie-bars . . . . . . . | = ·024 m. |

The struts are of cast iron and cruciform in section, and we will assume them to be perfectly rigid, partly on account of their short length and partly on account of the small stresses to which they are subjected.

The load on the beam, including its own weight, is 180 kg. per metre run.

**2. Calculation of the Tension in the Tie-bars.**—We will call

$T$ = the tension in the side tie-bars.
$T_1$ = the tension in the horizontal tie-bar.
$P$ = the pressure on the strut.
$\beta$ = the angle which the side tie-bars make with the axis of the beam.

We then have

$$\sin\beta = \frac{\cdot 60}{\sqrt{2\cdot 75^2 + \cdot 60^2}} = \cdot 2131$$

$$\cos\beta = \frac{2\cdot 75}{\sqrt{2\cdot 75^2 + \cdot 60^2}} = \cdot 9734,$$

and therefore

$$\left.\begin{aligned} T_1 &= \cdot 9734T \\ P &= \cdot 2131T \end{aligned}\right\} \quad . \quad . \quad . \quad . \quad (1)$$

The thrust is constant throughout the beam, and is equal to $T_1$.

The vertical reaction at each support

$$= R = 180 \times 4{\cdot}00 = 720 \text{ kg.}$$

The bending moments at the ends of the beam and at the sections above the struts are respectively

$$B_1 = 0,$$

$$B_2 = -2{\cdot}75R + {\cdot}60T\cos\beta + \frac{w}{2}2{\cdot}75^2 = -1299{\cdot}4 + {\cdot}584T.$$

Then if we call

$E$ = the elastic modulus of the timber,

$A = {\cdot}20 \times {\cdot}20 = {\cdot}04$ the cross-sectional area of the beam,

$I = \dfrac{{\cdot}20 \times {\cdot}20^3}{12} = {\cdot}000133$ its moment of inertia,

we can express the internal work as follows:—

due to compression alone

$$= \frac{8}{2EA}\cdot T_1^2 = \frac{1}{2E}\cdot\frac{8 \times {\cdot}9734^2}{{\cdot}04}T^2 = \frac{189{\cdot}48}{2E}T^2; \quad . \quad (2)$$

due to bending of the two end portions

$$= 2 \times \frac{2{\cdot}75}{2EI}\left(\frac{B_2^2}{3} - \frac{B_2}{12}\cdot 2{\cdot}75^2 w\right);$$

due to bending of the centre portion

$$= \frac{2{\cdot}5}{2EI}\left(B_2^2 - \frac{B_2}{6}\cdot 2{\cdot}5^2 w\right).$$

Adding these last two formulæ, substituting for $B_2$ its value in terms of T, and giving to I and $w$ their numerical values, we have the following value for the internal work of the whole beam due to bending, viz. :—

$$\frac{1}{2EI}\left(\frac{13B_2^2}{3} - \frac{2{\cdot}75^3 + 2{\cdot}50^3}{6}wB_2\right) = \frac{1}{2E}(11084T^2 - 27055054 \times 2T) \ (3)$$

Finally, the internal work of the tie-rods is given by

$$\frac{1}{2E'A'}(2 \times 2{\cdot}815T^2 + 2{\cdot}50T_1^2);$$

if we substitute in this formula the value for $T_1$ given by formula (1) and note that $A' = {\cdot}000452$ and assume that $E' = 12E$, we get

$$\frac{1}{2E}\cdot 1475T^2 \quad . \quad . \quad . \quad . \quad (4)$$

Adding results (2), (3), and (4) we find

$$\text{Total internal work} = \frac{1}{2E}(12748T^2 - 27055054 \times 2T) \quad . \quad (5)$$

To find the value of T we equate to zero the differential coefficient of this expression with regard to T and obtain

$$12748T - 27055054 = 0$$

or

$$T = \frac{27055054}{12748} = 2122{\cdot}2.$$

**3. Maximum Stresses per Sq. Metre in the Tie-rods and in the Beam.**—If in equations (1) we substitute for T the above value, we obtain

$$T_1 = 2065{\cdot}7\,;\ P = 452{\cdot}2\,;$$

thus the stress per sq. metre in the tie-bars will be:—

for the inclined bars $\frac{2122{\cdot}2}{{\cdot}000452} = 4{,}700{,}000$ kg. per. sq. m.

for the horizontal bar $\frac{2065{\cdot}7}{{\cdot}000452} = 4{,}570{,}000$ kg. per sq. m.

As regards the struts, it is unnecessary to consider them further, because it is obvious that their stress is so small that they will always possess more than sufficient strength.

To find the maximum compressive stress in the timber, we first note that the normal pressure is equal to $T_1$ throughout the beam. The bending moment varies, and its value for a section at a distance $x < 2{\cdot}75$ m. from one of the supports, is given by the formula

$$B = -Rx + T\sin\beta\,.\,x + \frac{wx^2}{2} = -267{\cdot}8x + 90x^2.$$

The maximum numerical value between $x = 0$ and $x = 2{\cdot}75$ m. occurs for $x = \frac{267{\cdot}8}{180} = 1{\cdot}49$ m. and equals $-199$ kg. m.

For the portion between the struts, we find that the maximum bending moment occurs at the mid section, and that its value is

$$B = -4R + \frac{w}{2}\,.\,4^2 + {\cdot}6T_1 = -200{\cdot}58.$$

We see then that the maximum bending moment for the whole beam occurs at the centre; the maximum stress in the timber will therefore also occur at the centre, and will be given by

$$c = \frac{T_1}{A} - \frac{Bn}{I} = \frac{2065{\cdot}7}{{\cdot}04} + \frac{200{\cdot}58 \times {\cdot}1}{{\cdot}000133} = 201{,}702 \text{ kg. per sq. m.},$$

i.e. 20·17 kg. per sq. cm.

**4. Temperature Effects.**—Since variation of temperature has practically no effect upon wood, we will assume that the timber beam alone, if free, would neither expand nor contract with change of temperature, while the tie-rods if free would expand or contract. We will neglect the effect of temperature on the struts, both on account of their short length and of the manner in which they act in the structure.

We will call the temperature at which the different parts of the beam were assembled the mean temperature, and we will take $\frac{1}{82500}$ as the coefficient of expansion for iron.

With these assumptions, it is clear that if the temperature increased or decreased by $t_1^\circ$ C. from the mean temperature, the centre tie-bar, if free, would expand or contract by an amount $\frac{2{\cdot}5t_1^\circ}{82500}$, and each of the side tie-bars by an amount $\frac{2{\cdot}815t_1^\circ}{82500}$.

If then we wished to assemble the different parts of the structure at the new temperature, we see readily that, after having placed the horizontal tie-bar in position, each of the side tie-bars would have an excess or deficiency of length given by the expression

$$\frac{1}{2} \cdot \frac{2{\cdot}5t_1^\circ}{82500} \cos\beta + \frac{2{\cdot}815t_1^\circ}{82500} = {\cdot}000049t_1^\circ.$$

Thus, in accordance with a theorem proved in the theoretical portion of this book (the structure having been erected at the mean temperature), in order to determine the tension T in the side tie-rods at a temperature of $t_1^\circ$ C. above or below the mean, under the load previously taken, we must add to formula (5), which gives the internal work of the whole system, the term

$$\pm 2 \times {\cdot}000049t_1^\circ\mathrm{T} = \pm {\cdot}000098t_1^\circ\mathrm{T},$$

thus obtaining

$$\frac{1}{2\mathrm{E}}(12748\mathrm{T}^2 - 27055054 \times 2\mathrm{T}) \pm {\cdot}000098t_1^\circ\mathrm{T};$$

then equating to zero the differential coefficient of this expression with regard to T, we have

$$\frac{1}{\mathrm{E}}(12748\mathrm{T} - 27055054) \pm {\cdot}000098t_1^\circ = 0,$$

from which, by taking $\mathrm{E} = 1{\cdot}5 \times 10^9$ we get

$$\mathrm{T} = 2122{\cdot}2 \mp 11{\cdot}53t_1^\circ \qquad . \quad . \quad . \quad (6)$$

In this equation we must take the upper sign for increase and the lower sign for decrease of temperature.

Let us suppose that the erection or mean temperature is 15° C., and that the temperature may vary between − 10° C. and + 40° C., and let us examine the two extreme cases.

*Case* 1. *Temperature* 40° *C.*—We must put $t_1^\circ = 25$ in equation (6) and take the upper sign; we then have

$$T = 2122{\cdot}2 - 11{\cdot}53 \times 25 = 1833{\cdot}9 \text{ kg.};$$

this gives a stress in the side tie-rods

$$\frac{1833{\cdot}9}{{\cdot}000452} = 4{,}057{,}000 \text{ kg. per sq. metre.}$$

To find the maximum stress in the timber, we see in the first place that the normal pressure P, which is constant throughout, is equal to T cos $\beta = 1785$ kg.; the bending moment varies and its maximum numerical value occurs at the centre of the beam where

$$B = -4R + \frac{w}{2} \cdot 4^2 + {\cdot}6P = -369 \text{ kg. m.}$$

It follows that the maximum stress in the timber is

$$c = \frac{1785}{{\cdot}04} + \frac{369 \times {\cdot}1}{{\cdot}000133} = 322{,}069 \text{ kg. per sq. metre,}$$

i.e. $c = 32{\cdot}21$ kg. per sq. cm.

*Case* 2. *Temperature* $-$ 10° C.—We still make $t_1^\circ = 25$, but we must take the plus sign, giving

$$T = 2122{\cdot}2 + 11{\cdot}53 \times 25 = 2410{\cdot}4;$$

thus the stress in the side tie-bars

$$t^\circ = \frac{2410{\cdot}4}{{\cdot}000452} = 5{,}332{,}000 \text{ kg. per sq. metre.}$$

In this case P = T cos $\beta = 2410{\cdot}4 \times {\cdot}9734 = 2346$ kg.; the bending moment for a section at distance $x$ from the nearest support, $x$ being $< 2{\cdot}75$, is given by

$$B = -Rx + T \sin \beta \,.\, x + \frac{wx^2}{2} = -206{\cdot}4x + 90x^2.$$

The maximum value of this corresponds to $x = \dfrac{206{\cdot}4}{180} = 1{\cdot}15$, and is $B = -118{\cdot}3$ kg. m.

For the centre section we find readily that $B = -42{\cdot}4$, from which it follows that the maximum stress in the timber occurs at the end portions, and that its value is

$$c = \frac{2346}{{\cdot}04} + \frac{118{\cdot}3 \times {\cdot}10}{{\cdot}000133} = 147{,}600 \text{ kg. per sq. metre,}$$

i.e. $c = 14{\cdot}76$ kg. per sq. cm.*

* Fig. 4, Pl. IV, shows the bending moment curves for the mean and the extreme temperatures; these curves are parabolic.

**5. Conclusion.**—The preceding results show what a large effect temperature variation has upon the stresses in a structure composed of timber and iron.

In the trussed beam considered above, it follows that if the temperature varies from $-10°$ C. to $+40°$ C., i.e. between limits that are commonly exceeded, at all events in Northern Italy, the stress per sq. metre varies between following limits:—

For side tie-bars between 4,057,000 and 5,332,000.
For the timber beam between 322,069 and 147,600.

# CHAPTER XIV.

## STUDY OF A BEAM STRENGHTENED BY THREE IRON TIE-BARS AND TWO CAST-IRON STRUTS.

(Plate IV, Figs. 5 and 6.)

**1. General Data.**—

| | |
|---|---|
| Distance between centres of supports . . . . | = 7 m. |
| Distance between edges of supports, i.e. clear span | = 6·7 m. |
| Distance of each strut from the nearest support . | = 2·18 m. |
| Distance between the two struts . . . . . | = 2·34 m. |
| Distance from horizontal tie-bar to axis of beam . | = ·60 m. |
| Vertical projection of tie-bars, ·60 − ·13 m. . . | = ·47 m. |
| Cross sectional area of beam = ·20 × ·20 m. . . | = ·04 m. |
| Diameter of tie-bars . . . . . . . | = ·025 m |

The struts are of cast iron and are cruciform in section; we will regard them as perfectly rigid.

The load upon the beam, including its own weight,
per metre run = $w$ . . . . . . . . = 250 kg.

**2. Calculation of the Tension in the Tie-bars.**—We will call
T the tension in the two side tie-bars.
$T_1$ the tension in the horizontal tie-bar.
P the pressure on the struts.
$\beta$ the angle which the side tie-bars make with the horizontal.

Then
$$\sin \beta = \frac{\cdot 47}{\sqrt{2{\cdot}07^2 + {\cdot}47^2}} = \cdot 2214;$$
$$\cos \beta = \frac{2{\cdot}07}{\sqrt{2{\cdot}07^2 + {\cdot}47^2}} = \cdot 9752;$$
and therefore
$$\left.\begin{array}{l} T_1 = \cdot 9752T \\ P\ = \cdot 2214T \end{array}\right\} \quad . \quad . \quad . \quad . \quad (1)$$

For all sections of the beam between the extreme ends of the tie-bars, the normal thrust is constant and equal to $T_1$.

For the two short lengths between the ends of the tie-bars and the edges of the bearings of the beam (a length of 11 m. at each end) the normal thrust is zero.

The bending moments at the points which divide up the beam into different parts are given by the following formulæ :—

At the edges of the bearings

$$B_0 = \frac{w \times \cdot 15^2}{2} = 2 \cdot 81 \text{ kg. m.}$$

At sections at distance ·11 m. from the above, i.e. corresponding to the points of connection of the ends of the tie-rods,

$$B_1 = - \cdot 11R + \frac{w}{2} \times \cdot 26^2 + \cdot 13 \,.\, T \cos \beta = - 87 \cdot 80 + \cdot 1268T.$$

At the sections above the struts

$$B_2 = - 2 \cdot 18R + \frac{w}{2} \times 2 \cdot 33^2 + \cdot 60T \cos \beta = - 1228 \cdot 89 + \cdot 5851T.$$

Now if E is the elastic modulus of the timber, the area A and moment of inertia I being given by

$$A = \cdot 20 \times \cdot 20 = \cdot 04 \text{ sq. m.}$$

$$I = \frac{\cdot 20 \times \cdot 20^3}{12} = \cdot 000133 \text{ m.}^4,$$

we can easily find an expression for the internal work.

Due to the normal thrust alone we have

$$\text{Internal work} = \frac{(2 \cdot 07 \times 2 + 2 \cdot 34)T_1^2}{2EA} = \frac{6 \cdot 48 \times \cdot 9752^2 T^2}{2E \times \cdot 04}$$

$$= \frac{154 \cdot 06T^2}{2E} \quad . \quad . \quad . \quad . \quad . \quad (2)$$

Due to bending alone we have :—

For the two short lengths between the bearings and the points of connection of the tie-bars :—

$$\text{Internal work} = 2 \,.\, \frac{\cdot 11}{2EI}\left[\frac{B_0^2 + B_0B_1 + B_1^2}{3} - \frac{\cdot 11^2 w}{12}(B_0 + B_1)\right]$$

$$= \frac{1}{2E}(9T^2 - 607 \times 2T) \quad . \quad . \quad . \quad . \quad (3)$$

For the two lengths between the ends of the tie-rods and the struts :—

$$\text{Internal work} = 2 \,.\, \frac{2 \cdot 07}{2EI}\left[\frac{B_1^2 + B_1B_2 + B_2^2}{3} - \frac{2 \cdot 07^2 w}{12}(B_1 + B_2)\right]$$

$$= \frac{1}{2E}(4488T^2 - 9640068 \times 2T) \quad . \quad . \quad . \quad . \quad (4)$$

For the length between the two struts :—

$$\text{Internal work} = \frac{2{\cdot}34}{2\text{EI}}\left[\text{B}_2{}^2 - \frac{w}{6} \times 2{\cdot}34^2 \times \text{B}_2\right]$$

$$= \frac{1}{2\text{E}}(6021\text{T}^2 - 13842407\,.\,2\text{T}) \quad . \quad . \quad . \quad (5)$$

Finally, since the area A′ of the tie-bars = ·000491, we have for them

$$\text{Internal work} = \frac{1}{2\text{E}'\text{A}'}(2 \times 2{\cdot}123\text{T}^2 + 2{\cdot}34\text{T}^2) = \frac{13179\text{T}^2}{2\text{E}'},$$

and taking E′ = 12E this becomes

$$\frac{1098\text{T}^2}{2\text{E}} \quad . \quad . \quad . \quad . \quad . \quad (6)$$

If we add formulæ (2) to (6) we get

$$\text{Total internal work} = \frac{1}{2\text{E}}(11770\text{T}^2 - 23483082 \times 2\text{T}) \quad . \quad (7)$$

To find the value of the tension T we must equate to zero the differential coefficient of the above expression with regard to T, thus obtaining the equation

$$11770\text{T} - 23483082 = 0$$

or

$$\text{T} = 1995 \text{ kg.}$$

**3. Maximum Stress per Sq. Metre in the Tie-bars and Beam.**—If we insert the above value of T in equations (1), we have $\text{T}_1 = 1945{\cdot}5$ kg.; P = 441·7 kg.

The side tie-bars are most heavily stressed, their stress being

$$t = \frac{1995}{{\cdot}000491} = 4{,}063{,}000 \text{ kg. per sq. metre.}$$

To obtain the maximum compressive stress in the beam we must first know the maximum bending moment. Now, putting T = 1995 in the expressions for $\text{B}_1$ and $\text{B}_2$ given in the previous paragraph, we have

$$\text{B}_1 = 170{\cdot}2 \text{ kg. m.}; \ \text{B}_2 = -\ 61\ 6 \text{ kg. m.}$$

Then remembering that the bending moment diagrams for the three portions of the beam are parabolas having the same parameter, viz. half the load per metre run, i.e. $\frac{w}{2} = 125$, we see that the maximum bending moment occurs at the centre of the beam. It is given by

$$\text{B} = -\ 3{\cdot}35\text{R} + \frac{250}{2} \times 3{\cdot}5^2 + {\cdot}60\text{T}_1 = -\ 232 \text{ kg. m.}$$

As the normal pressure is equal to $T_1 = 1946$, the maximum compressive stress is given by

$$c = \frac{1946}{\cdot 04} + \frac{232 \times \cdot 10}{\cdot 000133} = 223{,}490 \text{ kg. per sq. metre,}$$

i.e. $c = 22{\cdot}35$ kg. per sq. cm.

**4. Temperature Effects.**—The previous results hold good at the temperature at which the structure is assembled, and which we will call the *mean temperature.*

At a temperature of $t_1^\circ$ C. above or below the mean temperature, the lengths of the different portions would no longer be proportional to their previous values, and so erection could not be effected without inducing stresses even before the beam was loaded.

As the variations in length of the timber due to changes in temperature may be regarded as nil, we will assume that the length of the timber beam remains constant when the temperature varies.

We will neglect also the effect of temperature upon the struts, because they are short compared with the tie-bars, and their arrangement is such that variations in length would have only a slight influence upon the calculation, even if they were comparable with those of the tie-bars.

Now if the temperature increases or diminishes by $t_1^\circ$ C. from the mean, the tie-bars will elongate or contract by the following amounts :—

$$\text{For the centre tie-bar } \frac{2{\cdot}34t_1^\circ}{82500},$$

$$\text{For each side tie-bar } \frac{2{\cdot}123t_1^\circ}{82500},$$

$\frac{1}{82500}$ being the coefficient of expansion of iron.

If erection has to be effected at a temperature of $t_1^\circ$ C. above or below the mean temperature, it is clear that after having placed the centre tie-bar, each of the side tie-bars would have an excess or shortage of length given by

$$\frac{1}{2} \cdot \frac{2{\cdot}34t_1^\circ}{82500} \cos\beta + \frac{2{\cdot}123t_1^\circ}{82500} = \cdot 000040t_1^\circ.$$

If the erection has been effected at the mean temperature, to find the value of the tension T at a temperature of $t_1^\circ$ C. above or below the mean, we shall have to add to formula (7), which expresses the internal work of the whole system, the term

$$\pm 2 \times \cdot 000040t_1^\circ\text{T} = \pm \cdot 000080t_1^\circ\text{T},$$

thus giving

$$\frac{1}{2E}(11770T^2 - 23483082 \times 2T) \pm \cdot 000080 t_1^\circ T,$$

and then equate to zero the differential coefficient of this expression with regard to T. We thus obtain the equation

$$\frac{1}{E}(11770T - 23483082) \pm \cdot 000080 t_1^\circ = 0,$$

whence $$T = 1995 \mp 10 \cdot 20 t_1^\circ, \qquad (8)$$

taking $E = 1 \cdot 5 \times 10^9$, and taking the upper sign for increases in temperature and the lower for diminutions.

We will suppose that the mean temperature is 15° C., and that the possible limits of variation are 40° C and − 10° C. We will examine these two limiting cases.

*Case I. Temperature* 40° *C.*—Taking $t_1^\circ = 25$ in formula (8), and adopting the upper sign, we have

$$T = 1995 - 255 = 1740 \text{ kg.},$$

and therefore $$T_1 = T \cos \beta = 1697.$$

These forces are less than occur at the mean temperature.

To obtain the value of the maximum stress in the timber, we first note that in this structure the maximum bending moment occurs at the middle of the beam, and that its value is

$$B = -3 \cdot 35R + \frac{250}{2} \times 3 \cdot 5^2 + \cdot 6T_1 = -382 \text{ kg. m.};$$

as the normal pressure is equal to $T_1 = 1697$, it follows that the maximum stress in the timber is given by

$$c = \frac{1697}{\cdot 04} + \frac{382 \times \cdot 10}{\cdot 000133} = 329{,}643 \text{ kg. per sq. metre,}$$

i.e. 32·96 kg. per sq. cm.

*Case II. Temperature* − 10° *C.*—In this case we still have $t_1^\circ = 25$, but we take the lower sign, thus obtaining

$$T = 1995 + 255 = 2250,$$

and therefore $$T_1 = 2250 \cos \beta = 2194.$$

The maximum stress in the iron tie-bars occurs in the side bars, and is given by

$$t = \frac{2250}{\cdot 000491} = 4{,}583{,}000 \text{ kg. per sq. metre.}$$

As for the timber beam, it is unnecessary to find the maximum stress, because it can easily be seen that it must be less than in the previous case.*

* In Fig. 6, Pl. 4, the bending moment curves are shown for the mean and the two extreme temperatures. These curves are parabolic, and show that at − 10° C. the maximum bending moments are less than at 40° C.

**5. Conclusion.**—We see from the results obtained in the different cases considered that the maximum tension in the tie-bars occurs at the temperature of $\dot{-}$ 10° C., and that its value is 4·58 kg. per sq. mm.; whilst the maximum stress in the timber occurs at the temperature of 40° C. and its value is 32·96 kg. per sq. cm.

We may thus conclude that the conditions of stability of the trussed beam are quite satisfactory.

# CHAPTER XV.

## STUDY OF A ROOF WITH IRON TRUSSES WITHOUT TIE-RODS.*

(Plate V.)

**1. General Data.**—

| | |
|---|---|
| Clear span . . . . . | = 48 m. |
| Rise . . . . . . . | = 24 m. |
| Radius of intrados . . . | = 24 m. |
| Angle subtended at centre . . | = 180°. |
| Length of development of intrados | = 75·398 m. |
| Radial depth of section . . . | = 1·816 m. |
| Distance between principals . . | = 7 m. |

The ends of each truss are of cast iron for a height of 5·48 m. measured on the intrados, and serve as bearings fixed to the masonry. The portion above the bearings is of wrought iron, and is the only part that we need consider in the calculations; it must be regarded as fixed at the ends.

The angle subtended at the centre by the wrought-iron portion of the arch is 153° 16′.

**2. Loads.**—It should be noted that the covering of the roof is not uniform, for at the sides it is formed of zinc sheets placed upon boards ·03 m. thick, while near the top it is of glass.

As, however, these two forms of covering have almost the same weight per sq. metre, we will assume that the dead weight of the roof is uniformly distributed over its whole surface.

We therefore get the following loading per sq. metre:—

| | |
|---|---|
| Superload due to snow or wind . | = 60 kg. |
| Covering of zinc or glass . . . | = 29 ,, |
| Purlins . . . . . . . | = 18 ,, |
| Total for calculation of purlins . | 107 kg. |
| Weight of roof truss . . . | 19 ,, |
| Total for calculation of truss . | 126 kg. |

* The trusses of the main roof of the railway station at Turin have the same general form, the same chord, and the same rise as those which we are investigating here; they differ only in the height, which is 2 m. instead of 1·816, in the dimensions of the iron plates, and in the method of supporting the ends. Comparison of our trusses with those at Turin shows that the latter have sufficient strength, if not a little too much.

Load per metre run on the extrados of arch :—

| | | |
|---|---|---|
| Dead load = (126 − 60) × 7 | . . | = 462 kg. |
| Superload = 60 × 7 . | . . . . | = 420 ,, |
| Total . | . . . . . . . | 882 kg. |

**3. Principles and general formulæ for the case of the superload on the left-hand half of the truss, neglecting the dead load.**

As indicated in Fig. 1, one half of the wrought-iron portion of the arch has been divided into six equal segments,* the length of each of which is 5·565 m. measured on the centre line and 5·769 m. measured on the extrados.

We may as an approximation assume that the centre of gravity of the load carried by each of these six segments acts through the centre of the extrados of such segment.

We may also assume that only the upper five segments are loaded, and that the segment 0, 1, the extrados of which is almost entirely covered by the masonry, is not. In fact, for this segment we need only take account of the weight of the truss itself, which is so small compared with the weight of the covering and with the superload that it may be neglected, especially as this segment is near to the springings, so that its weight, even if it were not very small, would have very little effect upon the results of the calculation.

We will take as the unknowns of our problem

$$\left.\begin{array}{l} B = \text{bending moment} \\ H = \text{horizontal thrust} \\ S = \text{shearing force} \end{array}\right\} \text{at the crown,}$$

as functions of which we can express the bending moment, normal pressure, and shear for any section to the right or left. In fact, if we consider the arch to be divided into two equal parts, each will remain in equilibrium provided that at the section 6 we apply two forces H and S at the centre and a couple whose moment is equal to B. For the section 4 on the left, for example, all the forces acting from the right are: H, S, the moment B acting on the section 6, and the load of 420 kg. per metre run on the extrados of the two segments 4, 5, and 5, 6. It follows that to express the bending moment $B_4$, normal

* It is essential to divide the half-arch into an even number of segments, because, according to the formulæ given in § 4, we may divide the half-arch into any number of parts, but for each of these we must consider the mid-section, i.e. we must divide each part into two; this always gives an even number of segments.

pressure $P_4$, and shear $S_4$ at the section 4 on the left, we have the following formulæ :—*

$$B_4 = B - 2{\cdot}45H - 10{\cdot}76S + 2423 \times 7{\cdot}84 + 2423 \times 2{\cdot}25$$
$$P_4 = {\cdot}902H - {\cdot}432(S - 2423 \times 2)$$
$$S_4 = {\cdot}432H + {\cdot}902(S - 2423 \times 2).$$

For the section 4 on the right the bending moment, normal pressure and shear are given by the same expressions with the signs of the terms in S reversed and with the purely numerical terms omitted, because we have assumed the right-hand half to be unloaded. We thus have

$$B'_4 = B - 2{\cdot}45H + 10{\cdot}76S$$
$$P'_4 = {\cdot}902H + {\cdot}432S$$
$$S'_4 = {\cdot}432H - {\cdot}902S.$$

Writing similar formulæ for sections 0, 1, 2, . . . both to right and left, and simplifying the numerical terms, we then compile the following table (see opposite page).

**4. Determination of the Value of the Unknowns.**—The two half-arches considered separately deflect under the action of the forces applied to them, but their deflections must be such that the section 6 of the one follows that of the other in such a way that these two sections keep in contact. We therefore have the three following geometrical conditions :—

(1) *and* (2) *The horizontal and vertical displacements of the centre of the section at the crown of the left-hand half-arch must be respectively equal to and opposite in sign to those of the centre of the corresponding section of the right-hand half-arch.*

(3) *The angular tilt of the crown section of the left-hand half-arch must be equal and opposite to that of the right-hand half-arch.*

Wo find the values of B, H, and S which satisfy these conditions by expressing the internal work of the whole arch in terms of these unknowns, and then equating to zero the differential coefficients of it with regard to each of them. We will call

E = elastic modulus of the iron,
A = area of section of arch,
I = moment of inertia of the section

and we will take

* The lever-arms of the forces for deriving the expressions for the bending moments are written in Fig. 1. For deriving the expressions for P and S we must know the sines and cosines of the angles which the sections make with the vertical. Now, all these sines and cosines can be found at once by the simple construction shown in Fig. 2.

| Sections. | Bending Moments. | Normal Pressures. | Shearing Forces. |
|---|---|---|---|
| | I.—LEFT-HAND HALF-ARCH. | | |
| 0 | $B_0 = B - 19{\cdot}22H - 24{\cdot}25S + 135688$ | $P_0 = {\cdot}211H - {\cdot}977S + 11836$ | $S_0 = {\cdot}977H + {\cdot}211S - 2556$ |
| 1 | $B_1 = B - 13{\cdot}99H - 22{\cdot}39S + 113614$ | $P_1 = {\cdot}438H - {\cdot}899S + 10891$ | $S_1 = {\cdot}899H + {\cdot}438S - 5306$ |
| 2 | $B_2 = B - 9{\cdot}30H - 19{\cdot}41S + 83279$ | $P_2 = {\cdot}627H - {\cdot}779S + 7550$ | $S_2 = {\cdot}779H + {\cdot}627S - 6077$ |
| 3 | $B_3 = B - 5{\cdot}39H - 15{\cdot}47S + 51949$ | $P_3 = {\cdot}784H - {\cdot}621S + 4514$ | $S_3 = {\cdot}621H + {\cdot}784S - 5699$ |
| 4 | $B_4 = B - 2{\cdot}45H - 10{\cdot}76S + 24472$ | $P_4 = {\cdot}902H - {\cdot}432S + 2093$ | $S_4 = {\cdot}432H + {\cdot}902S - 4371$ |
| 5 | $B_5 = B - {\cdot}62H - 5{\cdot}52S + 6300$ | $P_5 = {\cdot}975H - {\cdot}222S + 538$ | $S_5 = {\cdot}222H + {\cdot}975S - 2362$ |
| 6 | $B_6 = B$ | $P_6 = 1{\cdot}000H$ | $S_6 = + 1{\cdot}000S$ |
| | II.—RIGHT-HAND HALF-ARCH. | | |
| 0 | $B_0' = B - 19{\cdot}22H + 24{\cdot}25S$ | $P_0' = {\cdot}211H + {\cdot}977S$ | $S_0' = {\cdot}977H - {\cdot}211S$ |
| 1 | $B_1' = B - 13{\cdot}99H + 22{\cdot}39S$ | $P_1' = {\cdot}438H + {\cdot}899S$ | $S_1' = {\cdot}899H - {\cdot}438S$ |
| 2 | $B_2' = B - 9{\cdot}30H + 19{\cdot}41S$ | $P_2' = {\cdot}627H + {\cdot}779S$ | $S_2' = {\cdot}779H - {\cdot}627S$ |
| 3 | $B_3' = B - 5{\cdot}39H + 15{\cdot}47S$ | $P_3' = {\cdot}784H + {\cdot}621S$ | $S_3' = {\cdot}621H - {\cdot}784S$ |
| 4 | $B_4' = B - 2{\cdot}45H + 10{\cdot}76S$ | $P_4' = {\cdot}902H + {\cdot}432S$ | $S_4' = {\cdot}432H - {\cdot}902S$ |
| 5 | $B_5' = B - {\cdot}62H + 5{\cdot}52S$ | $P_5' = {\cdot}975H + {\cdot}222S$ | $S_5' = {\cdot}222H - {\cdot}975S$ |
| 6 | $B_6' = B$ | $P_6' = 1{\cdot}000H$ | $S_6' = - 1{\cdot}000S$ |

$$\Sigma(B^2 + B'^2) = \frac{1}{3}\left\{\begin{array}{l}(B_0^2 + B_0'^2) + 4(B_1^2 + B_1'^2) \\ \quad + 2(B_2^2 + B_2'^2) + \ldots + (B_6^2 + B_6'^2)\end{array}\right\}$$

$$\Sigma(P^2 + P'^2) = \frac{1}{3}\left\{\begin{array}{l}(P_0^2 + P_0'^2) + 4(P_1^2 + P_1'^2) \\ \quad + 2(P_2^2 + P_2'^2) + \ldots + (P_6^2 + P_6'^2)\end{array}\right\}$$

If we remember that the length of each segment measured along the centre line is 5·564 m., we have as an expression for the internal work of the whole arch (neglecting as very small the portion due to shear),

$$\frac{5{\cdot}564}{2EI}\Sigma(B^2 + B'^2) + \frac{5{\cdot}564}{2EA}\Sigma(P^2 + P'^2);$$

i.e. omitting the numerical factor $\frac{5{\cdot}564}{2E}$, which will not affect the equations obtained by equating to zero the differential coefficients with regard to B, H, S,

$$\Sigma(B^2 + B'^2) + \frac{I}{A}\Sigma(P^2 + P'^2) \quad . \quad . \quad . \quad (1)$$

The section of the arch being that shown in Fig. 3, we have

$$A = ({\cdot}22 \times 1{\cdot}816) - ({\cdot}072 \times 1{\cdot}8) - ({\cdot}122 \times 1{\cdot}782) - ({\cdot}018 \times 1{\cdot}66) - ({\cdot}008 \times 1{\cdot}36) = {\cdot}011756 \text{ sq. m.}$$

$$I = \frac{1}{12}\left\{({\cdot}22 \times 1{\cdot}816^3) - ({\cdot}072 \times 1{\cdot}8^3) - ({\cdot}122 \times 1{\cdot}782^3) - ({\cdot}018 \times 1{\cdot}66^3) - ({\cdot}008 \times 1{\cdot}36^3)\right\} = {\cdot}008735 \text{ m.}^4$$

$$\text{and } \frac{I}{A} = \frac{{\cdot}008735}{{\cdot}011756} = {\cdot}743.$$

If we now substitute for $B_0$, $B_0'$ . . ., $P_0$, $P_0'$ . . ., their expressions in terms of B, H, S given in the previous tables, we obtain

$$\Sigma(B^2 + B'^2) = 12B^2 - 81{\cdot}82 \times 2BH + 346214 \times 2B + 970H^2 - 3923436 \times 2H + 3105{\cdot}02S^2 - 6859647 \times 2S;^*$$

$$\frac{I}{A}\Sigma(P^2 + P'^2) = 5{\cdot}18H^2 + 12040 \,.\, 2H + 4{\cdot}40S^2 - 18784 \,.\, 2S;$$

and formula (1) becomes

$$12B^2 - 81{\cdot}82 \times 2BH + 346214 \times 2B + 975{\cdot}18H^2 - 3911306 \times 2H + 3109{\cdot}42S^2 - 6878431 \times 2S \quad . \quad . \quad . \quad (2)$$

Equating to zero the differential coefficients of this expression with regard to B, H, S, we obtain the following equations :—

* It will be seen from these expressions that the second part of formula (1), i.e. that containing the pressures P, has a very small value compared with the first part containing the bending moments. We could as a first approximation take account only of the first part of formula (1) which involves neither the area nor the moment of inertia of the arch section. We should thus find the approximate values of the three unknowns B, H, S in the case of a load on a half-arch only; then by the principle of superposition it would be easy to obtain the values of B, H for the case of the dead load and the superload over the whole arch, and afterwards to calculate the bending moment, normal thrust, and shear for the sections 0, 1, 2 . . . 6.

These results would serve to determine the dimensions of the section; after that we could remake the calculation in the manner given in the text to determine more exactly the stresses developed in the different cases of loading.

In nearly every case in practice the first approximation will be sufficiently accurate.

$$12B - 81{\cdot}82H + 346214 = 0$$
$$- 81{\cdot}82B + 975{\cdot}18H - 3911306 = 0$$
$$3109{\cdot}42S - 6878431 = 0,$$

in which we see that the first two contain only B and H, and the third only S.

From these equations we get

$$B = - 3510 \text{ kg. m.}$$
$$H = 3715{\cdot}7 \text{ kg.}$$
$$S = 2212{\cdot}1 \text{ kg.}$$

**5. Line of pressure * and maximum stress per sq. metre in the iron, taking account of the dead load and of the superload both on the right and left of the crown.**

In this case, as the load is symmetrical, we have $S = 0$. Moreover, if we omit the dead load and take the superload on both sides, the values of B and H will be twice those which we have found above, i.e.

$$B = - 3510 \times 2 = - 7020; \; H = 3715{\cdot}7 \times 2 = 7431{\cdot}4.$$

Now the superload alone is 420 kg. per metre run, while the total load is 882 kg.; we have therefore for the total load

$$B = - 7020 \times \frac{882}{420} = - 14742 \text{ kg. m.}$$

$$H = 7431{\cdot}4 \times \frac{882}{420} = 15605{\cdot}9 \text{ kg.}$$

Knowing thus the bending moment and horizontal thrust at the crown, we can easily calculate the bending moment, normal pressure, and shear for each of the sections 0, 1, . . . 6. Dividing the bending moments by the corresponding normal pressures we obtain the eccentricities of the load. By means of the expression

$$\frac{P}{A} \pm \frac{Bn}{I},$$

in which P and B are the normal pressure and bending moment at any section, and A, I, $n$ the area, moment of inertia, and half-depth of that section, we obtain the stress per sq. metre at the intrados and extrados of the sections 0, 1, . . . 6.

Finally, if we call

* In this example, as in all following, we will give numerical tables in which one of the columns will give the distances between the line of pressure and the centre line of the arch. We will call these distances the eccentricities of the load.

$A_d = (\cdot 1 + 2 \times \cdot 031) \times \cdot 008 = \cdot 001296$, the area of the section of the diagonal bracing (Fig. 4, Pl. V),

$d$ = diagonal length of the squares formed by this bracing,

$$M_1 = \frac{1}{8}(\cdot 22 \times 1\cdot 816^2 - \cdot 072 \times 1\cdot 8^2 - \cdot 122 \times 1\cdot 782^2 - \cdot 018 \times 1\cdot 66^2 - \cdot 08 \times 1\cdot 36^2) = \cdot 005055,$$

the first moment of the half-section about the neutral axis,

$I = \cdot 008735$, the moment of inertia of the whole section,

$S$ = shear at any section,

we obtain the tensile and compressive stress in any section by the expression

$$\frac{M_1 S d\sqrt{2}}{2IA_d} = 476\cdot 7S.$$

All these results are collected in the following table:—

| Section. | Bending Moment. | Normal Pressure. | Shear. | Eccentricity of Load. | Compressive Stress per sq. m. At Intrados. | Compressive Stress per sq. m. At Extrados. | Stress per sq. m. in Diagonal Bracing. |
|---|---|---|---|---|---|---|---|
| 0 | – 29,742 | 28,148 | + 9878 | – 1·056 | – 702,000 | 5,461,000 | +4,709,000 |
| 1 | + 5,519 | 29,707 | + 2887 | + ·185 | +3,101,000 | 1,953,000 | +1,376,000 |
| 2 | + 15,009 | 25,641 | – 603 | + ·585 | +3,741,000 | 621,000 | – 287,000 |
| 3 | + 10,233 | 21,714 | – 2276 | + ·471 | +2,911,000 | 783,000 | –1,085,000 |
| 4 | – 1,585 | 18,472 | – 2436 | – ·085 | +1,406,000 | 1,736,000 | –1,161,000 |
| 5 | – 11,189 | 16,346 | – 1495 | – ·684 | + 227,000 | 2,553,000 | – 713,000 |
| 6 | – 14,742 | 15,606 | 0 | – ·944 | – 205,000 | 2,859,000 | 0 |

The deflection of the centre-line at the crown in the case of full load is given by the following formula:—

$$d_n = \frac{5\cdot 564}{3E}\left[\left(\frac{B_0x_0 + 4B_1x_1 + 2B_2x_2 + \ldots 4B_5x_5}{I}\right) + \left(\frac{P_0 \sin\phi_0 + 4P_1\sin\phi_1 + 2P_2\sin\phi_2 + \ldots + 4P_5\sin\phi_5}{A}\right)\right]$$

in which $B_0, B_1 \ldots P_0, P_1 \ldots$ represent the bending moments and normal prrssures in the sections 0, 1, 2 . . . 6; $I$ and $A$ the moment of inertia and area of the arch section; $x_0, x_1, x_2, \ldots$ the distances of the centres of the sections 0, 1, 2 . . . 5 from the vertical plane through the crown; and $\phi_0, \phi_1$, etc., the angles which these sections make with the vertical.*

* To prove this formula, suppose that the arch is cut through at the crown and that the right-hand half-truss is removed. We will consider the left-hand half-truss subjected to the following forces:—

1. The exterior forces to which it is subjected.
2. The horizontal forces acting at the crown, giving a resultant thrust H at the centre and a bending moment B.
3. A weight W′ applied at the crown.

We will call $B_0, B_1$, etc., the bending moments and $P_0, P_1$, etc., the normal

Substituting in the above formula the numerical values given in the preceding table and in Plate V, we obtain

$$\frac{B_0x_0 + 4B_1x_1 + 2B_2x_2 + \ldots + 4B_5x_5}{3I} = 27{,}001{,}764,$$

$$\frac{P_0 \sin\phi_0 + 4P_1 \sin\phi_1 + 2P_2 \sin\phi_2 + \ldots + 4P_5 \sin\phi_5}{3A}$$
$$= 7{,}274{,}977,$$

and therefore $$d_n = \frac{190{,}715{,}787}{E};$$

taking $E = 18 \times 10^9$ we get

$$d_n = \cdot 0106 \text{ m.}$$

**6. Line of pressure and maximum stress in the iron per sq. metre, taking the dead load into account and assuming that the superload is only on the left-hand half of the arch.**

We obtain the values of B, H, and S for this case by adding those which would arise for the dead load alone to those which would arise for the superload on the left side only.

Now for the dead load we shall have

$$B = -14742 \times \frac{462}{882} = -7722 \text{ kg. m.}$$

$$H = 15606 \times \frac{462}{882} = 8174 \text{ kg.}$$

$$S = 0,$$

and for the superload on the left side we shall have the values obtained in § 4; thus for the combined dead load and superload on the left we shall have

pressures on the sections 0, 1, etc., when the weight W′ is zero; $x_0$ $x_1$, etc., the horizontal distances of the sections 0, 1, etc., from the vertical through the crown; $\phi_0$, $\phi_1$, etc., the angles which the sections make with the vertical.

After the application of the weight W′ the bending moments and normal pressures have the following values:—

$$B_0 + W'x_0,\ B_1 + W'x_1,\ \ldots\ B_6;\ P_0 + W' \sin\phi_0,\ P_1 + W'\sin\phi_1,\ \ldots\ P_6,$$

and the internal work for the half-arch will be

$$\frac{5\cdot 564}{6E}\left[\left\{\frac{(B_0 + W'x_0)^2 + 4(B_1 + W'x_1)^2 + \ldots + B_6^2}{I}\right\} + \left\{\frac{(P_0 + W'\sin\phi_0)^2 + 4(P_1 + W'\sin\phi_1)^2 + \ldots + P_6^2}{A}\right\}\right].$$

The differential coefficient of this expression with regard to W′ gives the deflection $d_n$ at the crown for any value of W′. If this load is zero we put W′ = 0 and we shall then get

$$d_n = \frac{5\cdot 564}{3E}\left[\left(\frac{B_0x_0 + 4B_1x_1 + 2B_2x_2 + \ldots + 4B_5x_5}{I}\right) + \left(\frac{P_0 \sin\phi_0 + 4P_1\sin\phi_1 + 2P_2\sin\phi_2 + \ldots + 4P_5\sin\phi_5}{A}\right)\right].$$

$$B = -7722 - 3510 = -11{,}232 \text{ kg. m.}$$
$$H = 8174 + 3715{\cdot}7 = 11{,}890 \text{ kg.}$$
$$S = 2212 \text{ kg.}$$

Knowing the bending moment, normal pressure, and shear at the crown, we can easily calculate these three quantities for the sections 0, 1, 2 . . . both for the left and for the right; then, dividing the bending moments by the corresponding normal pressures, we shall obtain the eccentricities of the load, which will enable us to draw the line of pressure.* We thus obtain the following results:—

| Section. | Left-hand Half-arch. | | | | Right-hand Half-arch. | | | |
|---|---|---|---|---|---|---|---|---|
| | Bending Moment. | Normal Pressure. | Shear. | Eccen-tricity. | Bending Moment. | Normal Pressure. | Shear. | Eccen-tricity. |
| 0 | − 8,460 | 25,243 | + 6715 | − ·335 | − 36,861 | 17,689 | + 8337 | − 2·083 |
| 1 | + 11,483 | 26,091 | + 515 | + ·440 | − 3,073 | 19,177 | + 3884 | − ·160 |
| 2 | + 10,138 | 21,588 | − 2111 | + ·469 | + 12,733 | 17,484 | + 1192 | + ·728 |
| 3 | − 450 | 17,427 | − 2849 | − ·025 | + 16,043 | 15,661 | − 619 | + 1·024 |
| 4 | − 12,774 | 14,165 | − 2046 | − ·901 | + 10,359 | 13,983 | − 1666 | + ·741 |
| 5 | − 17,586 | 12,232 | − 163 | − 1·437 | + 536 | 12,676 | − 2115 | + ·042 |
| 6 | − 11,232 | 11,890 | + 2212 | − ·944 | − 11,232 | 11,890 | − 2212 | − ·944 |

If we compare the results of this table with those obtained in § 5 for the case of the superload on both sides, we see that the maximum bending moment occurs at section 0 of the unloaded half when the other half is loaded. But as the normal pressure is less in this case than when the arch is fully loaded, we cannot judge *a priori* in which of these two cases the maximum stress will occur.

Now when the superload is only on the left half of the arch, the maximum stress, at the right-hand section 0, is

$$c = \frac{17689}{{\cdot}011756} + \frac{36861 \times {\cdot}908}{{\cdot}008735} = 5{,}335{,}000 \text{ kg. per sq. metre,}$$

whilst that when the whole arch is loaded we have found to be 5,461,000 kg. per sq. metre.

With regard to shears, it will be seen that the maximum value occurs in the case of the arch fully loaded.

It follows that for trusses of the kind studied here it will usually be sufficient to examine the fully-loaded case.

* In Fig. 5, Pl. V, we have shown by the continuous line the line of pressure for the superload on the whole truss, and by the broken line that for the superload on the left half only. To save space we have folded back the left-hand half on the right hand.

We will add here that the same observation applies to the lines of pressure in Plates VI, VII, VIII, and XIII.

**7. Temperature Effects.**—Up to the present we have studied the equilibrium of the arch only with reference to the loading; we must now investigate the effect of change of temperature upon the stresses.

In accordance with the principle of superposition, we can study the effect of temperature, leaving entirely out of account the dead load and superload, for we have only to add to the temperature stresses those due to the loading to obtain those which will occur for the two causes combined.

We will call $t^\circ$ the *mean temperature* at which the arch was erected.

Now it is clear that if the ends of the arch were quite free instead of being fixed to their abutments, when the temperature increases from $t^\circ$ to $t_1^\circ$ all the dimensions would increase in exactly the same proportion, and therefore the final form of the arch would be geometrically similar to its initial form. Thus the chord, measured between the sections 0, which is 48·5 m. at the temperature $t^\circ$, will at the temperature $t_1^\circ$ be $48{\cdot}5\left(1 + \frac{t_1^\circ - t^\circ}{82500}\right)$, $\frac{1}{82500}$ being the coefficient of expansion of iron.

Hence if the truss had to be erected at the temperature $t_1^\circ$ the chord would exceed the necessary length by an amount $\frac{48{\cdot}5}{82500}(t_1^\circ - t^\circ)$. As for the angles of the extreme sections, they will not change in a free arch when the temperature changes, so that they will always remain exactly equal to the angles of the abutments.

To determine the stresses which occur at the temperature $t_1^\circ$ independently of the loading (erection taking place at temperature $t^\circ$), we must add to the internal work of the arch expressed in terms of B, H, and S the term $-\frac{48{\cdot}5(t_1^\circ - t^\circ)\mathrm{H}}{82500}$, and equate to zero the differential coefficients of the resulting expression with regard to B, H, and S.

Now the internal work of the whole arch under the action of the couple B and the forces H and S applied at the crown, neglecting the dead load and superload, is expressed by formula (2), § 4, multiplied by $\frac{5{\cdot}564}{2\mathrm{EI}}$, omitting the terms of the first degree in B, H, and S, since these terms depend on the loads and disappear when the load is nil. We have

$$\text{Internal work} = \frac{5{\cdot}564}{2\mathrm{EI}}(12\mathrm{B}^2 - 81{\cdot}82 \times 2\mathrm{BH} + 975{\cdot}18\mathrm{H}^2 + 3109{\cdot}42\mathrm{S}^2).$$

Adding to it the term $-\frac{48{\cdot}5(t_1^\circ - t^\circ)H}{82500} = -{\cdot}000588H(t_1^\circ - t^\circ)$ and then equating to zero the differential coefficients with regard to B, H, S, we obtain three equations which, multiplied by $\frac{EI}{5{\cdot}564}$, become

$$12B - 81{\cdot}82H = 0,$$
$$-81{\cdot}82B + 975{\cdot}18H = 16616(t_1^\circ - t^\circ)$$
$$3109{\cdot}42S = 0.$$

The last gives $S = 0$, and shows that the shear at the crown is zero, as would have been expected, because the truss is symmetrical about the crown.

The other two equations give us

$$B = 271{\cdot}57(t_1^\circ - t^\circ),$$
$$H = 39{\cdot}82(t_1^\circ - t^\circ).$$

Having thus determined the bending moment, normal pressure, and shear at the crown, we obtain the analogous quantities for the other sections 0, 1, 2, . . . 5 by substituting for B and H in the table of § 3 for the right-hand half-arch (i.e. for the unloaded portion) the values found above and putting $S = 0$. We thus obtain the following table:—

| Section. | Bending Moment. | Normal Pressure. | Shear. |
|---|---|---|---|
| 0 | $-493{\cdot}8(t_1^\circ - t^\circ)$ | $8{\cdot}4(t_1^\circ - t^\circ)$ | $38{\cdot}90(t_1^\circ - t^\circ)$ |
| 1 | $-285{\cdot}5(t_1^\circ - t^\circ)$ | $17{\cdot}4(t_1^\circ - t^\circ)$ | $35{\cdot}80(t_1^\circ - t^\circ)$ |
| 2 | $-98{\cdot}8(t_1^\circ - t^\circ)$ | $25{\cdot}0(t_1^\circ - t^\circ)$ | $31{\cdot}02(t_1^\circ - t^\circ)$ |
| 3 | $+56{\cdot}9(t_1^\circ - t^\circ)$ | $31{\cdot}2(t_1^\circ - t^\circ)$ | $24{\cdot}73(t_1^\circ - t^\circ)$ |
| 4 | $+174{\cdot}0(t_1^\circ - t^\circ)$ | $35{\cdot}9(t_1^\circ - t^\circ)$ | $17{\cdot}20(t_1^\circ - t^\circ)$ |
| 5 | $+246{\cdot}8(t_1^\circ - t^\circ)$ | $38{\cdot}8(t_1^\circ - t^\circ)$ | $8{\cdot}84(t_1^\circ - t^\circ)$ |
| 6 | $+271{\cdot}6(t_1^\circ - t^\circ)$ | $39{\cdot}8(t_1^\circ - t^\circ)$ | 0 |

Now let us assume that the temperature of erection is 15° C. and that the limits of temperature variation are − 10° C. and 40° C. We will examine the two limiting cases, taking the dead load and superload into account.

*Temperature* 40° C.—We put $t_1^\circ - t^\circ = 40 - 15 = 25$ in the preceding table, and the results thus obtained are added to the corresponding results in columns 2, 3, and 4 in the table given in § 5 for the case of the complete loading of the arch.

*Temperature* − 10° C.—We put $t_1^\circ - t^\circ = -10 - 15 = -25$ in the preceding table and add the results (taking account of signs) to those of § 5 exactly as for $t_1^\circ = 40^\circ$ C. We thus compile the following table:—

| Section. | Temperature 40° C. | | | | Temperature − 10° C. | | | |
|---|---|---|---|---|---|---|---|---|
| | Bending Moment. | Normal Pressure. | Shear. | Eccentricity of Load. | Bending Moment. | Normal Pressure. | Shear. | Eccentricity of Load. |
| 0 | − 42,084 | 28,358 | +10,850 | − 1·484 | − 17,399 | 27,938 | + 8,905 | − ·622 |
| 1 | − 1,618 | 30,142 | + 3,782 | − ·053 | − 12,656 | 29,272 | + 1,992 | − ·432 |
| 2 | + 12,539 | 20,266 | + 152 | + ·618 | + 17,479 | 25,016 | + 1,378 | + ·698 |
| 3 | + 11,655 | 22,494 | − 1,582 | + ·518 | + 8,811 | 20,934 | − 2,969 | + ·420 |
| 4 | + 2,765 | 19,369 | − 2,006 | + ·142 | + 5,935 | 17,574 | − 2,866 | + ·337 |
| 5 | − 5,019 | 17,316 | − 1,274 | − ·289 | − 17,359 | 15,376 | − 1,716 | − 1·128 |
| 6 | − 14,470 | 16,602 | 0 | − ·871 | − 15,014 | 14,612 | 0 | − 1·027 |

We see from these results how large may be the effects of temperature variation upon trusses whose ends are fixed.

At a temperature of 40° C. the maximum stress at the springings has a value

$$c = \frac{28358}{\cdot 011756} + \frac{42084 \times \cdot 908}{\cdot 008735} = 6{,}786{,}000 \text{ kg. per sq. metre,}$$

whilst at a temperature of − 10° C. the maximum stress is only

$$c = \frac{27938}{\cdot 011756} + \frac{17399 \times \cdot 908}{\cdot 008735} = 4{,}184{,}000 \text{ kg. per sq. metre.}$$

If, then, the truss were always loaded, we see that, under a temperature variation from − 10° C. to 40° C., the maximum stress would vary between 4,184,000 and 6,786,000 kg. per sq. metre.

As iron can safely carry a stress of 8,000,000 kg. per sq. metre, it follows that the truss considered has a sufficient factor of safety.

It is necessary to observe that the temperature of 40° C. could only occur when there was no snow load, so that the stress of 6,786,000 could never be reached. Nevertheless, as we have also to take account of wind loads, which may be quite large, it is wise to assume that even in summer the truss may be loaded as heavily as in winter, and to make sure that on this hypothesis the maximum stress remains below the limit adopted in practice.*

* The examination of the influence of temperature upon the stress distribution in trusses is specially necessary because the maximum stresses are often increased by the low temperatures which occur in winter, when the trusses may be fully loaded.

To understand this fully, it must be noted that increase of temperature causes increase of stress at the extrados, while diminution of temperature causes increase of stress at the intrados.

Now, for arches which have a large rise compared with the span, the maximum stress at the springings occurs at the extrados even at the mean temperature, and therefore it is rise of temperature that increases the maximum stress. On the other hand, for flat arches the maximum stress at the springings at the mean temperature occurs at the intrados, and therefore it is diminution in temperature that causes the increase in stress.

# CHAPTER XVI.

## STUDY OF AN ARCHED ROOF-TRUSS WITH A SINGLE TIE-ROD.*

(Plate VI.)

**1. General Data.**—

| | |
|---|---|
| Chord of the intrados . . . . | = 47·7 m. |
| Rise ,, ,, . . . . | = 12·1 m. |
| Radius ,, ,, . . . . | = 29·555 m. |
| Angle subtended at centre by intrados | = 107° 36′ 6″ |
| Length of development of intrados . | = 55·502 m. |
| Depth of arch section . . . . | = 1·2 m. |
| Distance between principals . . | = 7 m. |
| Length of each bearing . . . | = ·6 m. |

**2. Loads.**—From the bearings up to a certain height we will assume that the roof covering consists of wood planking ·03 m. thick faced with zinc plates of 14 gauge. From the end of this covering to the top, we will assume a covering of glass 6 mm. thick.

As the weight per sq. metre is almost identical for the two forms of covering, we will take it to be exactly the same over the whole roof, and will estimate the load as follows :—

| | |
|---|---|
| Superload due to snow or wind | 50 kg. per sq. metre. |
| Zinc or glass covering . | 27 ,, ,, |
| Purlins . . . . | 18 ,, ,, |
| | — |
| Total for calculation of purlins | 95 kg. per sq. metre. |
| Weight of truss . . . | 16 ,, ,, |
| | —— |
| Total for calculation of truss | 111 kg. per sq. metre. |

* Trusses of this form were employed for the main roof of the railway station at Genoa. They have the same span and rise as we have taken in the calculation, and differ only in the depth of the arch section, in the dimensions of the tie-rods, and the form of the lattice bracing. A comparison between the truss analysed in this chapter and those employed as above shows that in the latter a considerable saving in material might have been effected, and a sufficient factor of safety still obtained.

*Load per Metre Run on the Extrados of the Arch*:—

| | |
|---|---|
| Dead load (111 − 50) × 7 | = 427 kg. |
| Superload, 50 × 7 | = 350 ,, |
| Total | 777 kg. |

**3. Notation and General Formulæ for the Case of the Superload over the Whole Arch.**

As we see from Plate VI, the half-arch, except for the walled-in portion at the abutment, is divided into six equal elements, the length of each of which measured on the centre line is 4·61 m., and measured on the extrados 4·70 m.

As an approximation we may assume that the vertical line through the centre of gravity of the load carried by each element passes through the mid-point of the extrados of each.

Moreover, the first element near the springings is partly sunk into the wall, and for this portion there is neither the weight of covering nor the superload, but only the weight of the truss to consider; and for simplification we will neglect this small load, especially as, being near to the springings, it can have only a very small influence on the results.

Finally, the area of the bearing being sufficiently small, we may assume that each reaction acts through the centre of the bearing.

Having made these assumptions, we will call

$W$ = the total weight carried by each element 1, 2; 2, 3; . . .
$W_1$ = ,, ,, ,, ,, ,, the element 0, 1.
$R$ = the vertical reaction at each bearing, which is equal to half the total load carried by the truss.
$T$ = the tension in the tie-rod, which we have to determine.

We shall then have

$$W = 777 \times 4{\cdot}7 = 3651{\cdot}9 \text{ kg.}$$
$$W_1 = 777 \times 3{\cdot}82 = 2968{\cdot}14 \text{ kg.}$$
$$R = 3651{\cdot}9 \times 5 + 2968{\cdot}14 = 21227{\cdot}64 \text{ kg.}$$

With these data and the dimensions given on the drawing, we can now calculate in terms of $T$ the bending moment, normal pressure, and shear for each of the sections 0, 1, . . . 6; for example, for the section 2 we have

$$B_2 = 6{\cdot}88T - 6{\cdot}84R + 5{\cdot}3W_1 + 2{\cdot}2W$$
$$P_2 = {\cdot}82T + {\cdot}572(R - W_1 - W)$$
$$S_2 = {\cdot}572T - {\cdot}82(R - W_1 - W).$$

Substituting for $R$, $W_1$, and $W$ their numerical values, and simplifying, we compile the following table, in which T is the only unknown :—

| Section. | Bending Moment. | Normal Pressure. | Shear. |
|---|---|---|---|
| 0 | $B_0 = \cdot52T - 6{,}368$ | $P_0 = \cdot615T + 16{,}812$ | $S_0 = \cdot792T - 13{,}055$ |
| 1 | $B_1 = 3\cdot95T - 64{,}645$ | $P_1 = \cdot725T + 12{,}617$ | $S_1 = \cdot691T - 13{,}238$ |
| 2 | $B_2 = 6\cdot88T - 121{,}432$ | $P_2 = \cdot820T + 8{,}356$ | $S_2 = \cdot572T - 11{,}978$ |
| 3 | $B_3 = 9\cdot20T - 170{,}444$ | $P_3 = \cdot897T + 4{,}831$ | $S_3 = \cdot441T - 9{,}827$ |
| 4 | $B_4 = 10\cdot95T - 208{,}347$ | $P_4 = \cdot953T + 2{,}198$ | $S_4 = \cdot301T - 6{,}961$ |
| 5 | $B_5 = 11\cdot95T - 232{,}121$ | $P_5 = \cdot986T + 555$ | $S_5 = \cdot152T - 3{,}601$ |
| 6 | $B_6 = 12\cdot30T - 240{,}338$ | $P_6 = 1\cdot000T$ | $S_6 = 0$ |

**4. Determination of the Value of the Unknown T.**—We will call

$E$ = elastic modulus of the material.
$A$ = area of cross-section of arch.
$I$ = moment of inertia of cross-section.
$A_T$ = area of cross-section of tie-rod.

We will take

$$\Sigma B^2 = \frac{1}{3}(B_0^2 + 4B_1^2 + 2B_2^2 + \ldots + 4B_5^2 + B_6^2)$$

$$\Sigma P^2 = \frac{1}{3}(P_0^2 + 4P_1^2 + 2P_2^2 + \ldots + 4P_5^2 + P_6^2)$$

If we remember that the length of each element is 4·61 m., measured on the centre line, and that the length of the tie-rod is 48·1 m., we have the following expression for the internal work of the truss (neglecting the portion due to shear, which has very little effect) :—

$$\text{internal work} = 2\left(\frac{4\cdot61}{2EI}\Sigma B^2 + \frac{4\cdot61}{2EA}\Sigma P^2\right) + \frac{48\cdot1T^2}{2EA_T},$$

and dividing through by $\frac{2 \times 4\cdot61}{2EI}$, which will have no effect on the result obtained by equating to zero the differential coefficient with regard to T, we get

$$\Sigma B^2 + \frac{I}{A}\Sigma P^2 + \frac{I}{A_T}\cdot\frac{48\cdot1}{2 \times 4\cdot61}T^2 \quad . \quad . \quad (1)$$

Now from Fig. 3, Plate VI, we have

$$A = \cdot15 \times 1\cdot2 - \cdot1375 \times 1\cdot175 - \cdot0125 \times 1 = \cdot005937 \text{ sq. m.},$$

$$I = \frac{1}{12}\{(\cdot15 \times 1\cdot2^3) - (\cdot1375 \times 1\cdot175^3) - (\cdot0125 \times 1^3)\} = \cdot001970 \text{ m.}^4$$

Further, $A_T = \cdot 003318$ sq. m., since the diameter of the rod is ·065 m.

$$\therefore \frac{I}{A} = \cdot 332 ; \frac{I}{A} = \cdot 594.$$

Substituting for $B_0$, $B_1$, etc., $P_0$, $P_1$, etc., their values in terms of T given in the preceding table, we obtain

$$\Sigma B^2 = 486{\cdot}06T^2 - 9194095 \times 2T,^*$$

$$\frac{I}{A} \Sigma P^2 = 1{\cdot}52T^2 + 9344 \times 2T,$$

$$\frac{I}{A_T} \cdot \frac{48{\cdot}1}{2 \times 4{\cdot}61} \cdot T^2 = 3{\cdot}1T^2.\dagger$$

Adding together these results we have the following expression for formula (1):—

$$490{\cdot}68T^2 - 9184751 \times 2T \quad . \quad . \quad . \quad (2)$$

To obtain the value of the tension T we must express the condition that the relative horizontal deflection of the ends of the arch must be equal to the extension of the tie-rod; this is done by equating to zero the differential coefficient with regard to T of the internal work of the whole system, i.e. of expression (2). We thus obtain

$$490{\cdot}68T - 9184751 = 0,$$

$$\text{i.e. } T = 18718 \text{ kg.}$$

**5. Line of Pressure and Maximum Stresses per Sq. Metre in the Tie-rod and Arch.**

The stress in the tie-rod $= t = \frac{T}{A_T} = \frac{18718}{\cdot 003318} = 5{,}640{,}000$ kg. per sq. metre.

* It should be noted that we have omitted the numerical terms independent of T because they will disappear on differentiating with regard to T.

† We see that the quantities $\frac{I}{A} \Sigma P^2$ and $\frac{I}{A_T} \cdot \frac{48{\cdot}1T^2}{2 \times 4{\cdot}61}$ are very small compared with $\Sigma B^2$; as a first approximation, therefore, we could neglect these two quantities. We should thus obtain the following equation to determine T:—

$$486{\cdot}06T - 9194095 = 0,$$

$$\text{i.e. } T = 18915,$$

a value very little different from that which we have obtained above.

Substituting this value of T in the table in § 3 we obtain the bending moment, normal pressure, and shear for each of the sections 0, 1, . . . 6.

In this method of calculation it is not necessary to know the dimensions of the arch or tie-rod: thus the results obtained, although only approximate, will serve to determine these dimensions so that the maximum stresses are within the limits usually adopted in practice.

The approximation to the bending moments obtained by this preliminary calculation will not usually be sufficiently close, so that after having fixed the dimensions of the arch and tie-rod, it will always be necessary to make a fresh calculation according to the manner given in the text to determine more accurately the tension in the tie-rod and the stresses in the arch.

Now if we substitute the value of T in the table of § 3, we shall obtain the results given in the following table.

Dividing each bending moment by the corresponding normal pressure, we obtain the tabulated results for the eccentricity of loading.

By means of the formula

$$c = \frac{P}{A} \pm \frac{Bn}{I},$$

we obtain the tabulated values of the stresses at intrados and extrados.

Finally, we note that the diagonal bracing is of ⊔ section, (Fig. 4, Pl. VI), of which the area is

$$A_B = (\cdot 064 + 2 \times \cdot 031)\,\cdot 008 = \cdot 001008 \text{ sq. m.},$$

and that the diagonal length is $l = 1{\cdot}01$ m.; then if we calculate for the arch section the quantity

$$M_1 = \frac{1}{8}\{(\cdot 15 \times 1{\cdot}2^2) - (\cdot 1375 \times 1{\cdot}175^2) - (\cdot 0125 \times 1^2)\} = \cdot 001708,$$

and remember that I for the arch section = ·001970, we obtain for the stresses in the bracing

$$c_B = t_B = \frac{M_1 S l \sqrt{2}}{2IA_B} = 614S,$$

S being the shearing force at the section under consideration. The results of this calculation are given in the last column of the accompanying table.

| Section. | Bending Moment. | Normal Pressure. | Shear. | Eccentricity of Load. | Stresses per Sq. Metre at | | |
|---|---|---|---|---|---|---|---|
| | | | | | Intrados. | Extrados. | Bracing. |
| 0 | + 3,365 | 28,324 | + 1,770 | + ·117 | 5,781,500 | 3,759,900 | 1,086,780 |
| 1 | + 9,291 | 26,188 | − 304 | + ·355 | 7,240,700 | 1,581,300 | 186,656 |
| 2 | + 7,348 | 23,705 | − 1,271 | + ·310 | 6,230,600 | 1,754,800 | 780,394 |
| 3 | + 1,762 | 21,621 | − 1,572 | + ·081 | 4,178,340 | 3,105,060 | 965,208 |
| 4 | − 3,385 | 20,036 | − 1,327 | − ·169 | 3,344,100 | 4,405,300 | 814,778 |
| 5 | − 8,441 | 19,011 | − 756 | − ·444 | 689,800 | 5,714,400 | 464,184 |
| 6 | − 10,107 | 18,718 | 0 | − ·535 | 74,550 | 6,230,950 | 0 |

We thus see that

(1) The maximum stress at the intrados occurs at section 1, and its value is 7,240,700 kg. per sq. metre; whilst the maximum stress at the extrados occurs at the crown, and its value is 6,230,950 kg. per sq. metre.

(2) The maximum stress in tension or compression in the diagonal bracing occurs at the abutments, and is equal to 1,086,780 kg. per sq. metre.

**6. Examination of the Case in which only the Left-hand half of the Truss is Loaded.**

In this case, the total load on the truss per metre run of the extrados of the arch is made up of the dead load of 427 kg. and of the superload of 350 kg. on the left-hand half of the truss.

By the principle of superposition we can easily deduce the tension of the tie-bar in the present case from that which we found for the superload on the whole span.

If we call

$T_d$ the tension due to the dead load alone,
$T_s$ ,, ,, ,, ,, superload alone on the whole span,
$T'$ ,, ,, ,, ,, dead load + superload on half span,

we have
$$T' = T_d + \frac{T_s}{2}.$$

Now,
$$T_d = \frac{18718 \times 427}{777} = 10{,}287 \text{ kg.}$$
$$T_s = \frac{18718 \times 350}{777} = 8{,}431 \text{ kg.}$$

hence
$$T' = 14{,}502 \text{ kg.}$$

The reaction at the bearings is also made up partly by the dead load and partly by the superload distributed on the left-hand half of the truss. The first part is the same at each end, and is equal to the dead load on half the truss, i.e. $427 \times 3{\cdot}82 + 5 \times 427 \times 4{\cdot}70 = 11{,}666$.

The second part is greater for the left-hand bearing than for the right, and is easily determined, since we know the line of action of the superload on each of the left-hand elements. We thus find that the superload on the half-arch gives the following reactions:—

Left-hand reaction = 7316 kg.
Right-hand ,, = 2246 ,,

Therefore the total reactions are

Left-hand reaction = 7316 + 11666 = 18,982 kg.
Right-hand ,, = 2246 + 11666 = 13,912 ,,

Knowing thus the tension in the tie-bar and the reactions when only the left-hand half is loaded, we can calculate the bending moment, normal pressure, and shear at the sections 0, 1, 2, . . . 6 for the right and left-hand halves, and thus compile the following table:—

| Section. | Left-hand Half-arch. | | | | Right-hand Half-arch. | | | |
|---|---|---|---|---|---|---|---|---|
| | Bending Moment. | Normal Pressure. | Shear. | Eccentricity of Load. | Bending Moment. | Normal Pressure. | Shear. | Eccentricity of Load. |
| 0 | + 1,846 | 23,953 | − 188 | + ·077 | + 3,367 | 19,937 | + 2930 | + ·169 |
| 1 | + 26 | 21,579 | − 1589 | + ·002 | + 14,367 | 19,000 | + 1118 | + ·757 |
| 2 | − 6,298 | 18,963 | − 1842 | − ·330 | + 17,676 | 17,768 | − 130 | + ·997 |
| 3 | − 12,814 | 16,849 | − 1418 | − ·756 | + 15,537 | 16,654 | − 1020 | + ·937 |
| 4 | − 15,776 | 15,343 | − 456 | − 1·022 | + 10,515 | 15,704 | − 1601 | + ·675 |
| 5 | − 15,032 | 14,513 | + 816 | − 1·027 | + 1,933 | 14,946 | − 1990 | + ·138 |
| 6 | − 7,843 | 14,502 | + 2246 | − ·535 | − 7,840 | 14,502 | − 2246 | − ·535 |

From the results contained in the above table we see that in the left-hand half-arch, which is loaded, the maximum stress occurs at the extrados of section 4, and is given by

$$c = \frac{15341}{\cdot 005937} + \frac{15685 \times \cdot 6}{\cdot 00197} = 7{,}360{,}000 \text{ kg. per sq. metre,}$$

whilst in the right-hand half-arch, which is not loaded, the maximum stress occurs at the intrados of section 2, and is given by

$$c = \frac{17765}{\cdot 005937} + \frac{17717 \times \cdot 6}{\cdot 00197} = 8{,}388{,}000 \text{ kg. per sq. metre.}$$

Thus, in the truss with a single tie-rod that we have considered here the maximum stresses for one half loaded are greater than for the load over the whole span; this is due to the facility with which this form of structure lends itself to deformation.

**7. Temperature Effects.**—As one of the ends of the truss rests on rollers, it is clear that the truss can yield freely to the effects of temperature variation. Moreover, if we consider the arch and tie-rod separately it is clear that for a given increase in temperature the increase in length of the tie-rod will be equal to that of the chord of the arch. At any erection temperature, therefore, the tie-rod will be of the length geometrically necessary to join the two ends of the arch.

It follows that variations of temperature have no effect upon the stresses induced in the structure under consideration, at least when the variations are the same throughout the truss.

## CHAPTER XVII.

### STUDY OF A ROOF-TRUSS OF THE POLONCEAU TYPE.

(Plate VII.)

**1. General Data.**—The trusses are supported upon one side on the side wall of an adjacent building, and on the other side by cast-iron columns.

| | | |
|---|---|---|
| Distance from the wall to the axis of the columns | = | 26 m. |
| Length of the wall bearing . . . . | = | ·2 ,, |
| ,, ,, column bearing . . . . | = | ·2 ,, |
| Rise of the intrados . . . . . . | = | 4·9 ,, |
| Distance between principals . . . . | = | 3·75 ,, |

Lattice bracing is employed for the rafters.

The covering is of corrugated iron 1 mm. thick, with glass lights about 6 mm. thick towards the ridge.

**2. Loads.**—As the corrugated iron covering is lighter than the glass, we will assume conditions less favourable than actually obtain, by calculating for iron covering from the bearings to the struts and for glass from the struts to the ridge.

We will take for the maximum superload due to snow and wind 60 kg. per sq. metre; we thus compile the following tables :—

*Loads per sq. metre from bearings to struts*—

| | |
|---|---|
| Superload . . . . . . . | 60 kg. |
| Corrugated iron . . . . . | 11 ,, |
| Purlins . . . . . . . | 18 ,, |
| Total for calculation of purlins | 89 kg. |
| Weight of truss . . . . | 12 ,, |
| Total for calculation of truss | 101 kg. per sq. metre. |

*Loads per sq. metre from struts to ridge*—

| | |
|---|---|
| Superload . . . . . . . | 60 kg. |
| Glass cover, including iron glazing bars . . . . . . . | 24 ,, |
| Purlins . . . . . . . | 18 ,, |
| Total for calculation of purlins | 102 kg. |
| Weight of truss . . . . | 12 ,, |
| Total for calculation of truss | 114 kg. per sq. metre. |

As the distance between the principals is 3·75 m., the load per metre run of the rafter will be

(1) *From bearing to strut*—

Dead load (101 − 60) 3·75 = 154
Superload 60 × 3·75 = 225

Total . . 379 kg. per metre run.

(2) *From strut to ridge*—

Dead load (114 − 60) 3·75 = 203
Superload 60 × 3·75 = 225

Total . . 428 kg. per metre run.

**3. Notation and General Formulæ for the Superload on the Whole Truss.**—We will call $T_1$, $T_2$, $T_3$ the tensions in the three tie-rods and C the compression in the strut.

Let $w_1$, $w_2$ be the weights per metre run of the rafters between the bearings and struts, and those between struts and ridge respectively. Then the reactions at the bearings will be

$$R = 6{\cdot}79w_1 + 6{\cdot}93w_2.$$

For each of the sections 0, 1, 2, 3, 4 we now calculate the bending moment, normal pressure, and shear in terms of the unknown forces $T_1$ and C, and thus compile the following table:—

| Section. | Bending Moment. | Normal Pressure. | Shear. |
|---|---|---|---|
| 0 | $B_0 = {\cdot}1T_1 - (2{\cdot}436w_1 + 2{\cdot}564w_2)$ | $P_0 = {\cdot}96T_1 + 2{\cdot}13w_1 + 2{\cdot}29w_2$ | $S_0 = {\cdot}274T_1 - (6{\cdot}07w_1 + 6{\cdot}51w_2)$ |
| 1 | $B_1 = {\cdot}98T_1 - (17{\cdot}012w_1 + 23{\cdot}770w_2)$ | $P_1 = {\cdot}96T_1 + 1{\cdot}07w_1 + 2{\cdot}29w_2$ | $S_1 = {\cdot}274T_1 - (3{\cdot}03w_1 + 6{\cdot}51w_2)$ |
| 2 | $B_2 = 1{\cdot}87T_1 - (21{\cdot}655w_1 + 44{\cdot}906w_2)$ | $P_2 = {\cdot}96T_1 + 2{\cdot}29w_2$ | $S_2 = \begin{cases} {\cdot}274T_1 - 6{\cdot}51w_2 \\ {\cdot}274T_1 - C - 6{\cdot}51w_2 \end{cases}$ |
| 3 | $B_3 = 2{\cdot}76T_1 - 3{\cdot}23C - (21{\cdot}055w_1 + 60{\cdot}843w_2)$ | $P_3 = {\cdot}96T_1 + 1{\cdot}22w_2$ | $S_3 = {\cdot}274T_1 - C - 3{\cdot}47w_2$ |
| 4 | $B_4 = 3{\cdot}65T_1 - 6{\cdot}46C - (21{\cdot}655w_1 + 66{\cdot}927w_2)$ | $P_4 = {\cdot}96T_1 + {\cdot}16w_2$ | $S_4 = {\cdot}274T_1 - C - {\cdot}44w_2$ |

**4. Calculation of the Forces $T_1$ and C.**—

Let A and I be the area and moment of inertia of the rafter, $A_1$, $A_2$, $A_3$, $A_c$ the areas and $l_1 = 7{\cdot}1$, $l_2 = 2 \times 5{\cdot}82$, $l_3 = 7{\cdot}15$, $l_c = 1{\cdot}70$ the lengths of the tie-bars and strut.

E = elastic modulus of the iron.

We will assume that

$$\Sigma B^2 = \frac{B_0^2 + 4B_1^2 + 2B_2^2 + 4B_3^2 + B_4^2}{3}$$

$$\Sigma P^2 = \frac{P_0^2 + 4P_1^2 + 2P_2^2 + 4P_3^2 + P_4^2}{3}$$

$$\Sigma\frac{lT^2}{A} = \frac{7{\cdot}10T_1^2}{A_1} + \frac{5{\cdot}82T_2^2}{A_2} + \frac{7{\cdot}15T_3^2}{A_3} + \frac{1{\cdot}70C^2}{A_c}.$$

The length of the elements 0, 1; 1, 2; etc., being 3·23 m., we have the following expression for the internal work of the whole structure, neglecting only the portion due to shear:—

$$\text{Internal work} = 2\left[\frac{3{\cdot}23}{2EI}\Sigma B^2 + \frac{3{\cdot}23}{2EA}\Sigma P^2 + \frac{1}{2E}\Sigma\frac{lT^2}{A}\right].$$

Dividing through by $\frac{2 \times 3{\cdot}23}{2EI}$, which will not affect the equations which we shall obtain by equating to zero the differential coefficients of this expression with regard to $T_1$ and C, we get

$$\Sigma B^2 + \frac{I}{A}\Sigma P^2 + \frac{I}{3{\cdot}23}\Sigma\frac{lT^2}{A} \quad . \quad . \quad . \quad (1)$$

Taking the section in Fig. 4, Pl. VII, for the rafter, and $d_1 = 56$, $d_2 = 44$, $d_3 = 38$ mm. for the diameters of the tie-bars, and for the strut a cruciform section 60 × 60 × 10 mm., we have

$$A = {\cdot}1 \times {\cdot}5 - {\cdot}091 \times {\cdot}48 - {\cdot}009 \times {\cdot}35 = {\cdot}003170 \text{ sq. m.}$$

$$I = \frac{1}{12}({\cdot}1 \times {\cdot}5^3 - {\cdot}091 \times {\cdot}48^3 - {\cdot}009 \times {\cdot}35^3) = {\cdot}000170 \text{ m.}^4$$

$$A_1 = {\cdot}002463 \text{ sq. m.}$$
$$A_2 = {\cdot}001520 \quad ,,$$
$$A_3 = {\cdot}001134 \quad ,,$$
$$A_c = {\cdot}001100 \quad ,,$$

We therefore have

$$\frac{I}{A} = {\cdot}054;\ \frac{I}{A_1} = {\cdot}0693;\ \frac{I}{A_2} = {\cdot}1125;\ \frac{I}{A_3} = {\cdot}1510;\ \frac{I}{A_c} = {\cdot}1555.$$

$$\text{and } \frac{I}{3{\cdot}23}\Sigma\frac{lT^2}{A} = {\cdot}160T_1^2 + {\cdot}203T_2^2 + {\cdot}335T_3^2 + {\cdot}082C^2 \quad . \quad (2)$$

Now from the vector diagram of Fig. 3 we obtain the following expressions for $T_2$ and $T_3$ in terms of $T_1$ and C,

$$\left.\begin{aligned} T_2 &= {\cdot}915T_1 - 1{\cdot}695C \\ T_3 &= {\cdot}111T_1 + 1{\cdot}608C \end{aligned}\right\} \quad . \quad . \quad . \quad (3)$$

We can therefore eliminate $T_2$ and $T_3$ from equation (2), thus getting

$$\frac{I}{3{\cdot}23}\Sigma\frac{lT^2}{A} = {\cdot}3341T_1^2 - {\cdot}2557 \times 2T_1C + 1{\cdot}5211C^2.$$

Substituting in $\Sigma B^2$ and $\Sigma P^2$ for $B_0$, $B_1$, etc., $P_0$, $P_1$, etc., their values given in the table in § 3 we get

$$\Sigma B^2 = 18{\cdot}2128T_1^2 - 19{\cdot}7394 \times 2T_1C + 27{\cdot}8211C^2 - (155{\cdot}33w_1 + 392{\cdot}45w_2)\,.\,2T_1 + (139{\cdot}88w_1 + 406{\cdot}14w_2)\,.\,2C,^*$$

$$\frac{I}{A}\Sigma P^2 = {\cdot}1991T_1^2 + ({\cdot}11w_1 + {\cdot}36w_2)2T_1$$

and expression (1) becomes

$$18{\cdot}746T_1^2 - 19{\cdot}995T_1C + 29{\cdot}342C^2 - (155{\cdot}22w_1 + 392{\cdot}09w_2)2T_1 + (139{\cdot}88w_1 + 406{\cdot}14w_2)2C.$$

The differential coefficients of this expression must be equated to zero to determine the unknowns, thus giving

$$18{\cdot}746T_1 - 19{\cdot}995C - (155{\cdot}22w_1 + 392{\cdot}09w_2) = 0,$$
$$-19{\cdot}995T_1 + 29{\cdot}342C + (139{\cdot}88w_1 + 406{\cdot}14w_2) = 0;$$

from which we derive

$$\left.\begin{aligned} T_1 &= 11{\cdot}69w_1 + 22{\cdot}53w_2 \\ C &= 3{\cdot}20w_1 + 1{\cdot}51w_2 \end{aligned}\right\} \quad . \quad . \quad . \quad (4)$$

**5. Stresses per Sq. Metre in Tie-rods and Strut.**—If we now put $w_1 = 379$ and $w_2 = 428$ in equations (4) and then substitute the resulting values in equations (3), we obtain the following results :—

$$T_1 = 14{,}073 \text{ kg.}$$
$$T_2 = 9{,}726 \text{ ,,}$$
$$T_3 = 4{,}551 \text{ ,,}$$
$$C = 1{,}859 \text{ ,,}$$

Thus the stresses in these four members will be :—

$$t_1 = \frac{14073}{{\cdot}002463} = 5{,}700{,}000 \text{ kg. per sq. metre.}$$

$$t_2 = \frac{9726}{{\cdot}001520} = 6{,}370{,}000 \text{ ,, ,, ,,}$$

$$t_3 = \frac{4551}{{\cdot}001134} = 4{,}100{,}000 \text{ ,, ,, ,,}$$

$$c = \frac{1859}{{\cdot}0011} = 1{,}700{,}000 \text{ ,, ,, ,,}$$

* We see also for this structure that the quantities $\frac{I}{A}\Sigma P^2$ and $\frac{I}{3{\cdot}23}\Sigma\frac{lT^2}{A}$ are very small compared with the quantity $\Sigma B^2$. But it is easy to prove that if we neglect the two smaller terms, we shall obtain very different results from those given by the more accurate calculation, particularly as regards the bending moment in the rafter.

Therefore in Polonceau trusses we should not neglect the internal work due to the normal pressures in the rafter and to the forces in the tie-bars and struts.

**6. Line of Pressure and Maximum Stresses per Sq. Metre in the Rafters.**—If in the table in § 3 we put $w_1 = 379$, $w_2 = 428$ and substitute for $T_1$ and C the values found above, we obtain the values tabulated below in columns 2 to 4. Dividing the bending moments by the normal pressures we get the eccentricities of load as tabulated :—

| Section. | Bending Moment. | Normal Pressure. | Shear. | Eccentricity of Load. | Stresses per Sq. Metre. | | |
|---|---|---|---|---|---|---|---|
| | | | | | Intrados. | Extrados. | Diagonal Bracing. |
| 0 | − 613 | 15,271 | − 1238 | − ·040 | 3,911,420 | 5,723,180 | 3,073,950 |
| 1 | − 2830 | 14,869 | − 86 | − ·192 | 489,000 | 8,892,000 | 213,540 |
| 2 | − 1110 | 14,463 | {+ 1062<br>{− 800 | − ·080 | 2,852,200 | 6,272,800 | {2,636,950<br>{1,986,400 |
| 3 | − 1411 | 14,006 | + 501 | − ·106 | 2,215,300 | 6,621,300 | 1,243,980 |
| 4 | + 2505 | 13,552 | + 1798 | + ·175 | 7,779,400 | 770,600 | 4,464,430 |

By means of the formula

$$c = \frac{P}{A} \pm \frac{Bn}{I},$$

in which $c$ is the stress at intrados or extrados, $n$ is the neutral axis depth of the section, and B, A, I, P are as before, we calculate the stresses given in columns 6 and 7.

Finally, noting that the diagonal bracing is composed of flat iron bars ·05 m. × ·006 m., and calling

$A_D$ = ·0003 = the area of section of these bars,
$d$ = ·55 = the diagonal parallel to the rafter of the diamond-shaped figure formed by the bracing,
$d_1$ = ·42 = the other diagonal,
I = ·00017 = the moment of inertia of the rafter,
$M_1 = \frac{1}{8}(\cdot 1 \times \cdot 5^2 - \cdot 091 \times \cdot 48^2 - \cdot 009 \times \cdot 35^2) = \cdot 000366$ = the first moment of the half-section of the rafter about the neutral axis,
S = the shearing force at any section,

we calculate the stress per sq. metre in the diagonal bracing by means of the formula

$$\text{stress} = \frac{M_1 S \sqrt{d^2 + d_1^2}}{2IA_D} = 2483S,$$

thus obtaining the results in the last column of the table.

**7. Dimensions which should be Adopted.**—From the previous results we see that the tie-rod ($T_2$) is subjected to a stress of more than 6 kg. per sq. mm. which is higher than that usually allowed in practice; its diameter should therefore be increased from 44 mm. to 46 mm. We shall then have $A_2 = \cdot 001662$ and $t_2 = \frac{9726}{\cdot 001662} = 5{,}852{,}000$ kg. per sq. metre or say 5·8 kg. per sq. mm.

For the rafter, which is composed of bars of small section usually free from defects, we can allow a stress of 8 kg. per sq. mm. in tension or compression. Now as this stress is exceeded only at the extrados at the section between the bearing and strut, we can strengthen this part of the section by riveting to the T a plate of the same width and 8 mm. thick. We then shall have

$$A = \cdot 003970\,;\ I = \frac{1}{3}[\cdot 1(\cdot 213^3 + \cdot 295^3) - \cdot 091(\cdot 195^3 + \cdot 285^3) - \cdot 009(\cdot 13^3 + \cdot 022^3)] = \cdot 000212,$$

and the resulting stresses in section (2) become

$$\text{Intrados}\ \frac{14869}{\cdot 003970} - \frac{2830 \times \cdot 295}{\cdot 000212} = -\cdot 224{,}000 \text{ kg. per sq metre.}$$

$$\text{Extrados}\ \frac{14869}{\cdot 003970} + \frac{2830 \times \cdot 213}{\cdot 000212} = 6{,}615{,}000 \text{ kg. per sq. metre.}$$

**8. Examination of the case in which the superload is on one side only of the truss and the dead load is left out of account.**—Let $w$ be the load per metre run distributed over one of the rafters (say the left), and let us assume that on the other side there is no load, not even the weight of the truss itself.

Then the total load will be $(6\cdot 79 + 6\cdot 93)w = 13\cdot 72w$; and the reactions will be

$$\text{on left-hand side} = R_L = \frac{3}{4} \times 13\cdot 72w = 10\cdot 29w,$$

$$\text{,, right- ,, ,,} = R_R = \frac{1}{4} \times 13\cdot 72w = 3\cdot 43w.$$

Then if we call $T_{L1}$, $T_{L2}$, $T_{L3}$, $C_L$ the forces in the tie-rods and strut on the left and $T_{R1}$, $T_{R2}$, $T_{R3}$, $C_R$ the corresponding quantities on the right, and notice that $T_{L2}$ and $T_{R2}$ must be equal, since they refer to the same tie-rod, we can compile the following table (see next page), which is similar to that in § 3.

Substituting in expression (1) the values for B, P, T and C for each side, and putting also,

$$\left.\begin{aligned} T_{L2} &= \cdot 915T_{L1} - 1\cdot 695C_L \\ T_{L3} &= \cdot 111T_{L1} + 1\cdot 608C_L \end{aligned}\right\} \quad . \quad . \quad . \quad (5)$$

$$\left.\begin{aligned} T_{R2} &= \cdot 915T_{R1} - 1\cdot 695C_R \\ T_{R3} &= \cdot 111T_{R1} + 1\cdot 608C_R \end{aligned}\right\} \quad . \quad . \quad . \quad (6)$$

| Section. | Bending Moment. | Normal Pressure. | Shear. |
|---|---|---|---|
| | | 1. *Left-hand Rafter.* | |
| 0 | $B_0 = \cdot 1T_{L1} - 3\cdot 731w$ | $P_0 = \cdot 96T_{L1} + 3\cdot 29w$ | $S_0 = \cdot 274T_{L1} - 9\cdot 36w$ |
| 1 | $B_1 = \cdot 98T_{L1} - 29\cdot 009w$ | $P_1 = .96T_{L1} + 2\cdot 22w$ | $S_1 = \cdot 274T_{L1} - 6\cdot 33w$ |
| 2 | $B_2 = 1\cdot 87T_{L1} - 44\cdot 339w$ | $P_2 = \cdot 96T_{L1} + 1\cdot 15w$ | $S_2 = \begin{cases} \cdot 274T_{L1} - 3\cdot 29w \\ \cdot 274T_{L1} - C_L - 3\cdot 29w \end{cases}$ |
| 3 | $B_3 = 2\cdot 76T_{L1} - 3\cdot 23C_L - 49\cdot 81w$ | $P_3 = \cdot 96T_{L1} + \cdot 09w$ | $S_3 = \cdot 274T_{L1} - C_L - \cdot 25w$ |
| 4 | $B_4 = 3\cdot 65T_{L1} - 6\cdot 46C_L - 45\cdot 433w$ | $P_4 = \cdot 96T_{L1} - \cdot 98w$ | $S_4 = \cdot 274T_{L1} - C_L + 2\cdot 78w$ |
| | | 2. *Right-hand Rafter.* | |
| 0 | $B_0' = \cdot 1T_{R1} - 1\cdot 269w$ | $P_0' = \cdot 96T_{R1} + 1\cdot 13w$ | $S_0' = \cdot 274T_{R1} - 3\cdot 22w$ |
| 1 | $B_1' = \cdot 98T_{R1} - 11\cdot 765w$ | $P_1' = \cdot 96T_{R1} + 1\cdot 13w$ | $S_1' = \cdot 274T_{R1} - 3\cdot 22w$ |
| 2 | $B_2' = 1\cdot 87T_{R1} - 22\cdot 226w$ | $P_2' = \cdot 96T_{R1} + 1\cdot 13w$ | $S_2' = \begin{cases} \cdot 274T_{R1} - 3\cdot 22w \\ \cdot 274T_{R1} - C_R - 3\cdot 22w \end{cases}$ |
| 3 | $B_3' = 2\cdot 76T_{R1} - 3\cdot 23C_R - 32\cdot 685w$ | $P_3' = \cdot 96T_{R1} + 1\cdot 13w$ | $S_3' = \cdot 274T_{R1} - C_R - 3\cdot 22w$ |
| 4 | $B_4' = 3\cdot 65T_{R1} - 6\cdot 46C_R - 43\cdot 149w$ | $P_4' = \cdot 96T_{R1} + 1\cdot 13w$ | $S_4' = \cdot 274T_{R1} - C_R - 3\cdot 22w$ |

we get the expression

$$\left.\begin{aligned} &18\cdot 746\,(T_{L1}^2 + T_{R1}^2) - 19\cdot 992(2T_{L1}C_L + 2T_{R1}C_R) \\ &\qquad\qquad + 29\cdot 342(C_L^2 + C_R^2) \\ &- (331\cdot 625w \times 2T_{L1} + 215\cdot 685w \times 2T_{R1}) \\ &\qquad\qquad + (312\cdot 345w \times 2C_L + 233\cdot 675w \times 2C_R) \end{aligned}\right\} \quad (7)$$

Now since $T_{L2} = T_{R2}$ we get from equations (5) and (6)

$$\cdot 915T_{L1} - 1\cdot 695C_L = \cdot 915T_{R1} - 1\cdot 695C_R$$

i.e.
$$C_L - C_R = \cdot 54(T_{L1} - T_{R1}) \quad . \quad . \quad . \quad (8)$$

Having this relation between the four unknowns in expression (7) we must eliminate one and then equate to zero the differential coefficients of the expression with regard to the others. But we can simplify the calculation by changing the unknowns; for if we put

$$T_{L1} + T_{R1} = X_1; \quad C_L + C_R = X_2$$
$$T_{L1} - T_{R1} = Y_1; \quad C_L - C_R = Y_2$$

we shall have
$$\left.\begin{aligned} T_{L1} &= \frac{X_1 + Y_1}{2}; \quad C_L = \frac{X_2 + Y_2}{2} \\ T_{R1} &= \frac{X_1 - Y_1}{2}; \quad C_R = \frac{X_2 - Y_2}{2} \end{aligned}\right\} \quad . \quad . \quad (9)$$

Then putting in these values, expression (7) becomes

$$\left.\begin{aligned} &\frac{1}{2}[18\cdot 746X_1^2 - 19\cdot 992 \times 2X_1X_2 + 29\cdot 342X_2^2 \\ &\qquad\qquad - 547\cdot 31w \times 2X_1 + 546\cdot 02w \times 2X_2 \\ &+ 18\cdot 746Y_1^2 - 19\cdot 992 \times 2Y_1Y_2 + 29\cdot 409Y_2^2 \\ &\qquad\qquad - 115\cdot 94w \times 2Y_1 + 78\cdot 67w \times 2Y_2] \end{aligned}\right\} \quad (10)$$

and equation (8) becomes

$$Y_2 = \cdot 54Y_1 \quad . \quad . \quad . \quad . \quad . \quad (11)$$

We must now eliminate $Y_2$ from (10) and then equate to zero the differential coefficients with regard to $X_1$, $X_2$, and $Y_1$; but we see that the portion of expression (10) which contains $X_1$ and $X_2$ coincides exactly with the last formula for the internal work obtained in § 4 by putting $w_1 = w_2 = w$. Thus the values of $X_1$, $X_2$ which we should obtain from expression (10) by equating to zero the differential coefficients with regard to $X_1$, $X_2$, would be obtained from equations (4) by putting $w_1 = w_2 = w$.

We therefore have

$$X_1 = 34{\cdot}22w$$
$$X_2 = 4{\cdot}71w.$$

The last part of expression (10), i.e. that which contains $Y_1$ and $Y_2$, becomes after elimination of $Y_2$

$$\frac{1}{2}(5{\cdot}73Y_1^2 - 73{\cdot}46w \times 2Y_1);$$

the differential coefficient of which, with regard to $Y_1$, equated to zero, gives

$$5{\cdot}73Y_1 - 73{\cdot}46w = 0,$$

i.e.

$$Y_1 - 6{\cdot}43w.$$

Equation (11) then gives on substitution of this value

$$Y_2 = 3{\cdot}47w.$$

Equations (9) thus give:—

$$T_{L1} = 23{\cdot}54w; \quad C_L = 5{\cdot}82w$$
$$T_{R1} = 10{\cdot}68w; \quad C_R = -1{\cdot}12w$$

Putting $w = 225$ (see § 2) and then substituting the results obtained in formulæ (5), (6) and in the table on p. 280, we obtain the following results in kg. and kg.-m:—

$$T_{L1} = 5296, \quad T_{R1} = 2403$$
$$T_{L2} = T_{R2} = 2627$$
$$T_{L3} = 2693, \quad T_{R3} = -138$$
$$C_L = 1309, \quad C_R = -252$$

| Section. | Left-hand Rafter. | | | Right-hand Rafter. | | |
|---|---|---|---|---|---|---|
| | Bending Moment. | Normal Pressure. | Shear. | Bending Moment. | Normal Pressure. | Shear. |
| 0 | − 310 | 5800 | − 660 | − 45 | 2560 | − 68 |
| 1 | − 1337 | 5580 | + 23 | − 292 | 2560 | − 68 |
| 2 | − 73 | 5310 | { + 706<br>− 600 | − 507 | 2560 | { − 68<br>+ 178 |
| 3 | − 818 | 5100 | + 83 | + 92 | 2560 | + 178 |
| 4 | + 652 | 4850 | + 765 | + 690 | 2560 | + 178 |

**9. The case in which account is taken of the dead load on the whole truss and the superload on the left-hand side only.**

We know that this case is deduced from the two preceding ones by simple additions and subtractions. We thus obtain the following results:—

| Left-hand Side. | Right-hand Side. |
|---|---|
| $T_1 - T_{R1} = 11670$ | $T_1 - T_{L1} = 8777$ |
| $T_2 - T_{R2} = 7099$ | $T_2 - T_{L2} = 7099$ |
| $T_3 - T_{R3} = 4689$ | $T_3 - T_{L3} = 1858$ |
| $C - C_R = 2111$ | $C - C_L = 550$ |

| Section. | Left-hand Rafter. | | | | Right-hand Rafter. | | | |
|---|---|---|---|---|---|---|---|---|
| | Bending Moment. | Normal Pressure. | Shear. | Eccentricity of Load. | Bending Moment. | Normal Pressure. | Shear. | Eccentricity of Load. |
| 0 | − 568 | 12,711 | − 1170 | − ·044 | − 303 | 9471 | − 578 | − ·032 |
| 1 | − 2538 | 12,309 | − 18 | − ·208 | − 1493 | 9289 | − 63 | − ·162 |
| 2 | − 603 | 11,903 | { + 1130<br>− 978 | − ·054 | − 1037 | 9153 | { + 356<br>− 200 | − ·116 |
| 3 | − 1319 | 11,746 | + 323 | − ·136 | − 593 | 8906 | + 418 | − ·072 |
| 4 | + 1815 | 10,992 | + 1620 | + ·158 | + 1853 | 8702 | + 1033 | + ·204 |

Comparing these results with those obtained in § 6, we see that the only bending moment which is less for both sides loaded than for one side loaded is that at section 3; but as in the second case the normal pressure at this section is much less than in the first, the stress per sq. metre will be the greater when the load is on both sides.

As for the shearing forces, it is quite true that for some sections they are less for both sides loaded than for one side; but in the first place we note that the differences are small and secondly the greatest shear of all occurs when both sides are loaded, so that this case gives the maximum stress in the diagonal bracing.

We thus see that with the dimensions given there will be an ample factor of safety for all the cases considered.

**10. Temperature Stresses.**—As the structure is homogeneous and the ends of the rafters can expand or contract freely, one end resting upon cast-iron columns, it follows that temperature variations which are the same throughout the structure will have no effect upon the stresses.

We might examine the case of different temperature variations in different members, but we do not think these calculations necessary, because the temperature differences cannot be very great, so that their influence on the stresses is negligible.

For trusses of this type in which timber rafters are employed we ought to take account of temperature variations even if the variations are the same in all the members.

# CHAPTER XVIII.

## STUDY OF AN ARCHED ROOF-TRUSS WITH SEVERAL TIE-RODS.

(Plate VIII.)

**1. General Data.**—These are the same as for the previous example, except that the rise of the intrados is 6·12 m. instead of 4·9 m.

**2. Loads.**—We will also assume the same loading as in Chapter XVII.

**3. Notation and General Formulæ for the Case in which the Arch is Fully Loaded.**—Let $T_1$, $T_2$, $T_3$, $T_4$ be the tensions in the four tie-rods, and $w_1$, $w_2$ the load per metre run on the lower and upper portions of each arc.

We will divide the length of each of the two arcs into four equal parts, leaving out of account the plates near to the crown and springings.

The length of each portion measured on the centre line is 3·58 m., and 3·61 m. measured on the extrados.

The reaction at the ends will be

$$R = 3{\cdot}61 \times 2w_1 + (3{\cdot}61 \times 2 + {\cdot}22)w_2 = 7{\cdot}22w_1 + 7{\cdot}44w_2.$$

For each of the sections 0, 1, 2, 3, 4 we calculate the bending moment, normal pressure, and shear; e.g. for the § 1 we have the following expressions:—

$$\begin{aligned} B_1 &= 1{\cdot}85T_1 - 2{\cdot}89R + 3{\cdot}61 \times 1{\cdot}57w_1, \\ P_1 &= {\cdot}904T_1 + {\cdot}575(R - 3{\cdot}61w_1), \\ S_1 &= {\cdot}421T_1 - {\cdot}818(R - 3{\cdot}61w_1). \end{aligned}$$

For the other sections we have similar expressions, except that for sections 2, 3, and 4 the tension $T_4$ must also be considered.

All these quantities simplified, and with the value of R substituted, are tabulated below, the only unknowns being $T_1$ and $T_4$:—

| Section. | Bending Moment. | Normal Pressure. | Shear. |
|---|---|---|---|
| 0 | $B_0 = \cdot 06T_1 - (1\cdot 08w_1 + 1\cdot 12w_2)$ | $P_0 = \cdot 818T_1 + 5\cdot 10w_1 + 5\cdot 26w_2$ | $S_0 = \cdot 576T_1 - (5\cdot 09w_1 + 5\cdot 25w_2)$ |
| 1 | $B_1 = 1\cdot 85T_1 - (15\cdot 2w_1 + 21\cdot 5w_2)$ | $P_1 = \cdot 904T_1 + 2\cdot 08w_1 + 4\cdot 28w_2$ | $S_1 = \cdot 421T_1 - (2\cdot 95w_1 + 6\cdot 09w_2)$ |
| 2 | $B_2 = 3\cdot 09T_1 - (20\cdot 18w_1 + 44\ 42w_2)$ | $P_2 = \cdot 964T_1 + 3\cdot 16w_2$ | $S_2 = \begin{cases}\cdot 262T_1 - 6\cdot 73w_2 \\ \cdot 262T_1 + T_4 - 6\cdot 73w_2\end{cases}$ |
| 3 | $B_3 = 3\cdot 72T_1 + 3\cdot 55T_4 - (20\ 18w_1 + 62\cdot 95w_2)$ | $P_3 = \cdot 995T_1 + \cdot 174T_4 + 1\cdot 02w_2$ | $S_3 = \cdot 094T_1 + \cdot 987T_4 - 3\cdot 69w_2$ |
| 4 | $B_4 = 3\cdot 77T_1 + 7\cdot 00T_4 - (20\cdot 18w_1 + 69\cdot 93w_2)$ | $P_4 = \cdot 998T_1 + \cdot 340T_4 + \cdot 02w_2$ | $S_4 = -\cdot 075T_1 + \cdot 942T_4 - \cdot 22w_2$ |

**4. Calculation of the Tensions $T_1$ and $T_4$.**—Let A and I respectively be the area and moment of inertia of the arch section, $A_1$, $A_2$, $A_3$, $A_4$ the areas, and $l_1 = 7\cdot 37$, $l_2 = 2 \times 5\cdot 66$, $l_3 = 7\cdot 42$, $l_4 = 3\cdot 22$ the lengths of the tie-rods; and E the elastic modulus of the iron.

$$\text{Let } \Sigma B^2 = \frac{B_0^2 + 4B_1^2 + 2B_2^2 + 4\,B_3^2 + B_4^2}{3}$$

$$\Sigma P^2 = \frac{P_0^2 + 4P_1^2 + 2P_2^2 + 4P_3^2 + P_4^2}{3}$$

$$\Sigma\frac{lT^2}{A} = \frac{7\cdot 37T_1^2}{A_1} + \frac{5\cdot 66T_2^2}{A_2} + \frac{7\cdot 42T_3^2}{A_3} + \frac{3\cdot 22T_4^2}{A_4}.$$

Remembering that the length of each element 0, 1; etc., measured on the centre line is 3·58 m., we express the internal work of the structure by the following formula, neglecting only the portion due to shear, which has very little effect upon the results :—

$$\text{Internal work} = 2\left[\frac{3\cdot 58}{2EI}\Sigma B^2 + \frac{3\cdot 58}{2EA}\Sigma P^2 + \frac{1}{2E}\Sigma\frac{lT^2}{A}\right].$$

Dividing this by $\frac{2 \times 3\cdot 58}{2EI}$, which will not affect the equations we shall obtain by equating to zero the differential coefficients with regard to $T_1$ and $T_4$, we get

$$\Sigma B^2 + \frac{I}{A}\Sigma P^2 + \frac{I}{3\cdot 58}\Sigma\frac{lT^2}{A} \qquad . \quad . \quad . \quad (1)$$

Taking the section of the arch given by Fig. 6, Pl. VIII, and for the diameters of the tie-rods, $d_1 = d_2 = 42$ ; $d_3 = d_4 = 20$ mm., we get

$$A = \cdot 08 \times \cdot 4 - \cdot 072 \times \cdot 384 - \cdot 008 \times \cdot 274 = \cdot 002160 \text{ sq. m.}$$

$$I = \frac{1}{12}(\cdot 08 \times \cdot 4^3 - \cdot 072 \times \cdot 384^3 - \cdot 008 \times \cdot 274^3) = \cdot 000073 \text{ m.}^4;$$

$$A_1 = A_2 = \cdot 001385 \text{ sq. m.}$$
$$A_3 = A_4 = \cdot 000314 \text{ sq. m.}$$
$$\frac{I}{A} = \cdot 0338$$

and $$\frac{I}{3\cdot 58}\Sigma\frac{lT^2}{A} = \cdot 1085T_1^2 + \cdot 0833T_2^2 + \cdot 4818T_3^2 + \cdot 2091T_4^2 \qquad (2)$$

Now the four tensions must themselves be in equilibrium, so that we have two equations in which $T_2$ and $T_3$ can be expressed in terms of $T_1$ and $T_4$; these two equations may be deduced from the graphical construction in Fig. 5, and are

$$\left.\begin{aligned} T_2 &= \cdot 784T_1 + 1\cdot 498T_4 \\ T_3 &= \cdot 264T_1 - 1\cdot 405T_4 \end{aligned}\right\} \quad . \quad . \quad . \quad (3)$$

we can then eliminate $T_2$ and $T_3$ from formula (2) and thus obtain

$$\frac{I}{3\cdot 58}\Sigma\frac{lT^2}{A} = \cdot 19T_1^2 - \cdot 08 \times 2T_1T_4 + 1\cdot 35T_4^2.$$

Substituting in $\Sigma B^2$ and $\Sigma P^2$ for $B_0$, $B_1$, and $P_0$, $P_1$, etc., their values in terms of $T_1$ and $T_4$ given in the previous table, we obtain

$$\Sigma B^2 = 34\cdot 12T_1^2 + 26\cdot 4 \times 2T_1T_4 + 33\cdot 14T_4^2 - (214\cdot 06w_1 + 546\cdot 11w_2)2T_1 - (149\cdot 74w_1 + 462\cdot 99w_2)2T_4.$$

$$\frac{I}{A}\Sigma P^2 = \cdot 12T_1^2 + \cdot 01 \times 2T_1T_4 + (\cdot 13w_1 + \cdot 34w_2)2T_1 + \cdot 01w_1 \times 2T_4.$$

Thus expression (1) becomes

$$34\cdot 43T_1^2 + 26\cdot 33 \times 2T_1T_4 + 34\cdot 49T_4^2 - (213\cdot 93w_1 + 545\cdot 77w_2) \times 2T_1 - (149\cdot 74w_1 + 462\cdot 98w_2) \times 2T_4$$

Equating to zero the differential coefficients of this expression with regard to $T_1$ and $T_4$, we obtain the following two equations:—

$$34\cdot 43T_1 + 26\cdot 33T_4 = 204\cdot 41w_1 + 544\cdot 33w_2$$
$$26\cdot 33T_1 + 34\cdot 49T_4 = 142\cdot 61w_1 + 461\cdot 12w_2$$

from which we deduce

$$\left.\begin{aligned} T_1 &= \phantom{-}6\cdot 659w_1 + 13\cdot 421w_2 \\ T_4 &= -\cdot 955w_1 + 3\cdot 124w_2 \end{aligned}\right\} \quad . \quad . \quad . \quad (4)$$

**5. Stress per Sq. Metre in the Tie-rods.**—Putting in formulæ (4) $w_1 = 379$, $w_2 = 428$, and substituting in (3) we have:—

$$T_1 = 8268 \text{ kg.}, \quad t_1 = \frac{8268}{\cdot 001385} = 5{,}969{,}000 \text{ kg. per sq. metre.}$$

$$T_2 = 7937 \text{ ,,} \quad t_2 = \frac{7937}{\cdot 001385} = 5{,}730{,}000 \text{ ,, ,,}$$

$$T_3 = 811 \text{ ,,} \quad t_3 = \frac{811}{\cdot 000314} = 2{,}582{,}000 \text{ ,, ,,}$$

$$T_4 = 975 \text{ ,,} \quad t_4 = \frac{975}{\cdot 000314} = 3{,}105{,}000 \text{ ,, ,,}$$

**6. Line of Pressure and Maximum Stresses for the Arch.**— If in the table in § 3 we put $w_1 = 379$, $w_2 = 428$, and give to $T_1$ and $T_4$ their values found above, we get the results set out in the following table. The eccentricity of load is found by dividing the bending moment by the normal pressure, and the stresses are found by means of the formula

$$c = \frac{P}{A} \pm \frac{Bn}{I}.$$

| Section. | Bending Moment, kg. m. | Normal Pressure, kg. | Shear, kg. | Eccentricity of Load, m. | Stresses in kg. per Sq. Metre. At Intrados. | At Extrados. | In Lattice Bracing. |
|---|---|---|---|---|---|---|---|
| 0 | − 386 | 11,048 | + 650 | − ·035 | 4,057,250 | 6,172,350 | 1,642,550 |
| 1 | + 390 | 10,195 | − 163 | + ·038 | 5,793,400 | 3,646,600 | 411,901 |
| 2 | − 1152 | 9,430 | { − 685<br>+ 309 | − ·123 | 1,205,800 | 7,518,200 | { 1,730,995<br>528,143 |
| 3 | − 382 | 8,947 | + 189 | − ·043 | 3,093,200 | 5,191,200 | 477,603 |
| 4 | + 462 | 8,709 | + 214 | + ·053 | 5,300,700 | 2,763,300 | 540,778 |

Finally if we note that the lattice bracing is composed of flat bars 50 mm. × 6 mm.; and if we call

$A_D$ = area of diagonal bracing bars = ·000300 sq. m.,

$I$ = ·000073, the moment of inertia of the arch section,

$M_1 = \frac{1}{8}(\cdot 08 \times \cdot 4^2 - \cdot 072 \times \cdot 384^2 - \cdot 008 \times \cdot 274^2) = \cdot 000198$, the first moment of the half-section about the neutral axis,

$d$ = ·44 m., the diagonal parallel to the arch of the diamond-shaped figure formed by the bracing,

$d_1$ = ·345 m., the other diagonal of the diamond-shaped figure formed by the bracing,

$S$ = the shear at any section,

then the stress in diagonal bracing $= \dfrac{M_1 S \sqrt{d^2 + d_1^2}}{2IA_D} = 2527S$, which gives the figures tabulated in the last column.

**7. Dimensions to be Adopted.**—As the tensile stress in the tie-rods is less than $6 \times 10^6$ kg. per sq. metre, except for the tie-rod $T_1$, where this value is exceeded by a negligible amount, and as in the arch the stress is everywhere less than $8 \times 10^6$ kg. per sq. metre, the dimensions assumed in the calculation may be safely adopted.

The stresses in the lattice bracing should not be regarded as too small because the flat bars are very thin compared with their length.

**8. Examination of the case in which the superload is on the left-hand half of the arch only, the dead load not being taken into account.**

Let $w$ be the load per metre run uniformly distributed on the left-hand half of the arch. Then the total load on the arch will be

$$W = (7\cdot 17 + 7\cdot 39)w = 14\cdot 56w,$$

and the reactions will be

$$\text{At left-hand, } R_L = 11\cdot 11w.$$
$$\text{At right-hand, } R_R = 3\cdot 45w.$$

Let $T_{L1}$, $T_{L2}$, etc., be the tensions in the left-hand tie-rods, and $T_{R1}$, $T_{R2}$, etc., those in the right-hand tie-rods.

We observe that $T_{L2} = T_{R2}$ since they occur in the same rod. It is a simple matter to compile the following table similar to that in § 3 :—

| Section. | Bending Moment. | Normal Pressure. | Shear. |
|---|---|---|---|
| | | *Left-hand Half-arch.* | |
| 0 | $B_0 = \cdot 06T_{L1} - 1\cdot 67w$ | $P_0 = \cdot 818T_{L1} + 7\cdot 27w$ | $S_0 = \cdot 576T_{L1} - 7\cdot 86w$ |
| 1 | $B_1 = 1\cdot 85T_{L1} - 26\cdot 98w$ | $P_1 = \cdot 904T_{L1} + 4\cdot 34w$ | $S_1 = \cdot 425T_{L1} - 6\cdot 16w$ |
| 2 | $B_2 = 3\cdot 09T_{L1} - 44\cdot 63w$ | $P_2 = \cdot 964T_{L1} + 1\cdot 67w$ | $S_2 = \begin{cases} \cdot 262T_{L1} - 3\cdot 55w \\ \cdot 262T_{L1} + T_{L4} - 3\cdot 55w \end{cases}$ |
| 3 | $B_3 = 3\cdot 72T_{L1} + 3\cdot 55T_{L4} - 51\cdot 63w$ | $P_3 = \cdot 995T_{L1} + \cdot 174T_{L4} + \cdot 08w$ | $S_3 = \cdot 094T_{L1} + \cdot 987T_{L4} - \cdot 30w$ |
| 4 | $B_4 = 3\cdot 77T_{L1} + 7T_{L4} - 46\cdot 35w$ | $P_4 = \cdot 998T_{L1} + \cdot 34T_{L4} - \cdot 33w$ | $S_4 = - \cdot 075T_{L1} + \cdot 942T_{L4} + 3\cdot 28w$ |
| | | *Right-hand Half-arch.* | |
| 0 | $B_0' = \cdot 06T_{R1} - \cdot 53w$ | $P_0' = \cdot 818T_{R1} + 2\cdot 49w$ | $S_0' = \cdot 576T_{R1} - 2\cdot 48w$ |
| 1 | $B_1' = 1\cdot 85T_{R1} - 10\cdot 12w$ | $P_1' = \cdot 904T_{R1} + 2\cdot 02w$ | $S_1' = \cdot 425T_{R1} - 2\cdot 88w$ |
| 2 | $B_2' = 3\cdot 09T_{R1} - 20\cdot 98w$ | $P_2' = \cdot 964T_{R1} + 1\cdot 49w$ | $S_2' = \begin{cases} \cdot 262T_{R1} - 3\cdot 18w \\ \cdot 262T_{R1} + T_{R4} - 3\cdot 18w \end{cases}$ |
| 3 | $B_3' = 3\cdot 72T_{R1} + 3\cdot 55T_{R4} - 32\cdot 76w$ | $P_3' = \cdot 995T_{R1} + \cdot 174T_{R4} + \cdot 94w$ | $S_3' = \cdot 094T_{R1} + \cdot 987T_{R4} - 3\cdot 39w$ |
| 4 | $B_4' = 3\cdot 77T_{R1} + 7T_{R4} - 45\cdot 06w$ | $P_4' = \cdot 998T_{R1} + \cdot 34T_{R4} + \cdot 35w$ | $S_4' = - \cdot 075T_{R1} + \cdot 942T_{R4} - 3\cdot 5w$ |

Now, the internal work of the whole truss divided by the factor $\frac{3{\cdot}58}{2EI}$ is given by

$$\frac{1}{3}\Big[(B_0^2 + 4B_1^2 + \ldots) + (B_0'^2 + 4B_1'^2 + \ldots) + \frac{I}{A}\Big\{(P_0^2 + 4P_1^2 + \ldots) + (P_0'^2 + 4P_1'^2 + \ldots)\Big\}\Big]$$
$$+ \frac{I}{3{\cdot}58}\Big\{\frac{7{\cdot}37}{A_1}(T_{L1}^2 + T_{R1}^2) + \frac{5{\cdot}66}{A_2}(T_{L2}^2 + T_{R2}^2) + \frac{7{\cdot}42}{A_3}(T_{L3}^2 + T_{R3}^2) + \frac{3{\cdot}22}{A_4}(T_{L4}^2 + T_{R4}^2)\Big\}.$$

Substituting in this formula for B, B′, P and P′ and putting

$$\left.\begin{aligned} T_{L2} &= {\cdot}784T_{L1} + 1{\cdot}498T_{L4} \\ T_{L3} &= {\cdot}264T_{L1} - 1{\cdot}405T_{L4} \end{aligned}\right\} \quad . \quad . \quad . \quad (5)$$
$$\left.\begin{aligned} T_{R2} &= {\cdot}784T_{R1} + 1{\cdot}498T_{R4} \\ T_{R3} &= {\cdot}264T_{R1} - 1{\cdot}405T_{R4} \end{aligned}\right\} \quad . \quad . \quad . \quad (6)$$

we obtain the following expression:—

$$\left.\begin{aligned} &34{\cdot}43(T_{L1}^2 + T_{R1}^2) + 26{\cdot}33(2T_{L1}T_{L4} + 2T_{R1}T_{R4}) + 34{\cdot}49(T_{L4}^2 + T_{R4}^2) \\ &- (472{\cdot}59w \times 2T_{L1} + 287{\cdot}11w \times 2T_{R1}) - (352{\cdot}56w \times 2T_{L4} + 260{\cdot}16w \times 2T_{R4}) \end{aligned}\right\} \quad (7)$$

Now since $T_{L2} = T_{R2}$, we have from equations (5) and (6)

$${\cdot}784T_{L1} + 1{\cdot}498T_{L4} = {\cdot}784T_{R1} + 1{\cdot}498T_{R4},$$

and therefore $T_{L4} - T_{R4} = -{\cdot}523(T_{L1} - T_{R1})$ . . . (8)

By means of this relation we must eliminate one of the unknowns from expression (7) and equate to zero the differential coefficients of the resulting expression with regard to the three others; if we put

$$T_{L1} + T_{R1} = X_1, \quad T_{L4} + T_{R4} = X_4$$
$$T_{L1} - T_{R1} = Y_1, \quad T_{L4} - T_{R4} = Y_4$$

we deduce

$$\left.\begin{aligned} T_{L1} &= \frac{X_1 + Y_1}{2}, \quad T_{L4} = \frac{X_4 + Y_4}{2} \\ T_{R1} &= \frac{X_1 - Y_1}{2}, \quad T_{R4} = \frac{X_4 - Y_4}{2} \end{aligned}\right\} \quad . \quad . \quad . \quad (9)$$

Substituting these values in (7) and (8), we get

$$\frac{1}{2}\{34{\cdot}43(X_1^2 + Y_1^2) + 26{\cdot}33 \times 2(X_1X_4 + Y_1Y_4) + 34{\cdot}49(X_4^2 + Y_4^2) - 2w(759{\cdot}7X_1 + 612{\cdot}72X_4 + 185{\cdot}48Y_1 + 92{\cdot}40Y_4)\} \quad (10)$$
$$Y_4 = -{\cdot}523Y_1 \quad . \quad . \quad . \quad . \quad (11)$$

We must now eliminate $Y_4$ from (10) and then equate to zero the differential coefficients of the resulting expression with regard to $X_1$, $X_4$, $Y_1$. But we can easily see that the part of expression (10) containing $X_1$ and $X_4$ is exactly the same as that obtained in § 4 by putting $w_1 = w_2 = w$, and therefore the values of $X_1$ and $X_4$ obtained by equating to zero the differential coefficients of (10) with regard to $X_1$ and $X_4$ will be those obtained by putting $w_1 = w_2 = w$ in equations (4); this gives

$$X_1 = 20{\cdot}372w\,; \quad X_4 = 2{\cdot}212w.$$

The other part of expression (10) containing $Y_1$ and $Y_4$ reduces to

$$\frac{1}{2}(16{\cdot}32Y_1^2 - 137{\cdot}15w \times 2Y_1)$$

from which we get by differentiating

$$16{\cdot}32Y_1 - 137{\cdot}15w = 0,$$

i.e.

$$Y_1 = 8{\cdot}402w,$$

and therefore, by equation (11),

$$Y_4 = -\,4{\cdot}394w.$$

From the above values of $X_1$, $X_4$, $Y_1$, and $Y_4$ we deduce by equations (9)

$$T_{L1} = 14{\cdot}387w\,; \quad T_{R1} = 5{\cdot}985w$$
$$T_{L4} = -\,1{\cdot}087w\,; \quad T_{R4} = 3{\cdot}306w.$$

Putting $w = 225$ and then substituting in equations (5) and (6) and in the previous table the results thus obtained, we have the following values in kg. :—

$$T_{L1} = 3237, \quad T_{R1} = 1347$$
$$T_{L2} = 2171, \quad T_{R2} = 2171$$
$$T_{L3} = 1199, \quad T_{R3} = -\,690$$
$$T_{L4} = -\,245, \quad T_{R4} = 744$$

| Section. | Left-hand Side. | | | Right-hand Side. | | |
|---|---|---|---|---|---|---|
| | Bending Moment, kg. m. | Normal Pressure, kg. | Shear, kg. | Bending Moment, kg.-m. | Normal Pressure, kg. | Shear, kg. |
| 0 | − 182 | 4284 | + 96 | − 38 | 1662 | + 218 |
| 1 | − 82 | 3903 | − 10 | + 221 | 1672 | − 76 |
| 2 | − 39 | 3496 | { + 49<br>− 196 | − 558 | 1634 | { − 363<br>+ 381 |
| 3 | − 445 | 3196 | − 5 | + 244 | 1681 | + 98 |
| 4 | + 60 | 3073 | + 264 | + 148 | 1676 | − 188 |

**9. Examination of the case in which the dead load is taken into account and the superload is taken as acting on the left-hand side only.**

We know that this case is deduced from the two previous ones by addition and subtraction, in accordance with the principle of superposition.

The following results are easily obtained :—

*Forces in Tie-rods.*

| Left. | Right. |
|---|---|
| $T_1 - T_{R1} = 7032,$ | $T_1 - T_{L1} = 5142$ |
| $T_2 - T_{R2} = 5887,$ | $T_2 - T_{L2} = 5887$ |
| $T_3 - T_{R3} = 1505,$ | $T_3 - T_{L3} = -384$ * |
| $T_4 - T_{R4} = 250,$ | $T_4 - T_{L4} = 1239$ |

| Section. | Left-hand Side. | | | | Right-hand Side. | | | |
|---|---|---|---|---|---|---|---|---|
| | Bending Moment, kg. m. | Normal Pressure, kg. | Shear, kg. | Eccentricity of Load, m. | Bending Moment, kg. m. | Normal Pressure, kg. | Shear, kg. | Eccentricity of Load, m. |
| 0 | − 348 | 9386 | + 432 | − ·037 | − 204 | 6764 | + 554 | − ·030 |
| 1 | + 169 | 8523 | − 87 | + ·020 | + 472 | 6292 | − 153 | + ·075 |
| 2 | − 594 | 7796 | { − 322<br>− 72 } | − ·076 | − 1113 | 5934 | { − 734<br>+ 505 } | − ·187 |
| 3 | − 626 | 7266 | + 91 | − ·086 | + 63 | 5751 | + 194 | + ·010 |
| 4 | + 314 | 7033 | + 402 | + ·045 | + 402 | 5636 | − 50 | + ·071 |

Comparing these results with those obtained in §§ 5 and 6, we see that the maximum bending moment and normal pressure occur for the case in which the superload is over the whole truss.

On the other hand, the maximum shear occurs for the section 3 of the unloaded side when the other side is loaded.

The maximum stress in the lattice bracing in this case is $2527 \times 734 = 1,854,818$, or less than 2,000,000 kg. per sq. metre.

**10. Temperature Effects.**—The remarks made in the case of the Polonceau truss apply to this truss also.

* It will be noticed that this tension becomes negative (i.e. compression). The force is very small, and it will not be necessary to alter the design of the truss, because the case of full load on one side and no load on the other will hardly ever occur in practice. It is easy to prove by means of the principle of superposition that a small superload on the right-hand side of the truss will be sufficient to bring the rod in question into tension.

## CHAPTER XIX.

### STUDY OF AN ARCHED ROOF-TRUSS WITH SEVERAL TIE-BARS.

(Plate IX.)

**1. General Data:—**

| | |
|---|---|
| Clear span between bearings . . | = 26·7 m. |
| Width of bearing on the walls . | = ·4 m. |
| Rise of intrados . . . . . | = 7·2 m. |
| Distance between principals . . | = 8 m. |
| Distance between purlins measured on the centre line of the arch . | = 1·5 m. |

The arches and purlins are of iron, the latter being formed of two oppositely-facing T-bars, the lower one bent to a circular arc and the upper one straight; ordinary lattice bracing is provided between them.

The covering is of corrugated iron from the bearings up to a certain point, and of glass in the upper part constituting the skylight.

**2. Loads.**—As the covering of ordinary corrugated iron weighs less than the glass, we must make allowance for this difference, and will take the corrugated iron as covering the lower half of each side of the truss and the glass as covering the remainder. In this way we shall assume conditions which are more serious than those which actually obtain in practice, because the glass will not actually cover such a wide area.

*Loads per sq. metre of covering for the lower half of each side.*

| | | |
|---|---|---|
| Superload of snow . . . . . | 60 | kg. |
| Corrugated iron . . . . . . | 11 | ,, |
| Purlins . . . . . . . . | 12 | ,, |
| Total for calculation of purlins . | 83 | kg. |
| Weight of truss itself . . . . . | 10·5 | ,, |
| Transverse bracing . . . . . | 1·5 | ,, |
| Total for calculation of truss . | 95 | kg. |

*Load per sq. metre on the upper half of each side.*

| | |
|---|---|
| Superload of snow . . . . . | 60 kg. |
| Glass, 6 mm. thick . . . . . | 15·6 ,, |
| Iron frames for glass and their supports | 8·4 ,, |
| Purlins . . . . . . . . | 12 ,, |
| Total for calculation of purlins . | 96 kg. |
| Weight of truss itself . . . . . | 10·5 ,, |
| Transverse bracing . . . . . | 1·5 ,, |
| Total for calculation of truss . | 108 kg. |

*Load per metre run of arch* (distance between principals = 8 m.).

1. Lower half—

Dead load = (95 − 60) × 8 = 280 kg.
Superload = 60 × 8 = 480 ,,

Total . . . . 760 kg.

2. Upper half—

Dead load = (108 − 60) × 8 = 384 kg.
Superload = 60 × 8 = 480 ,,

Total . . . . . 864 kg.

As these loads are composed of the superload, acting above the extrados, and the dead load, part of which acts on the extrados and part below, we will assume that the loads carried by the elements 0, 1; 1, 2; 2, 3; 3, 4, act through the mid-points of the extrados of these elements.

**3. Notation and General Formulæ.**—Let $T_1$, $T_2$, $T_3$, $T_4$ be the tensions in the tie-bars, and let us divide into four equal elements the lattice-braced portion of the arch comprised between the web-plates at crown and support. Draw normals to the centre line of the arch at the points of division, one of these normals coinciding with the direction of the force $T_4$ in one of the tie-bars. The lattice-braced portion of the arch is thus divided into four elements on each side, the length being 3·71 m. measured on the centre line and 3·76 m. measured on the extrados.

The reactions will be :—

$$R = 3·76 \times 2 \times 760 + (2 \times 3·76 + ·28)\, 864 = 12{,}454·4 \text{ kg.}$$

We will now calculate the bending moment, normal pressure, and shear for each of the sections 0, 1 . . . 4; for the section 1, for instance, we shall have the following formulæ :—

$$B_1 = 1{\cdot}98T_1 - 3{\cdot}03R + 3{\cdot}76 \times 760 \times 1{\cdot}68,$$
$$P_1 = {\cdot}905T_1 + {\cdot}578(R - 3{\cdot}76 \times 760),$$
$$S_1 = {\cdot}424T_1 - {\cdot}815(R - 3{\cdot}76 \times 760),$$

where ·578 and ·815 are the values of the sine and cosine of the angle which the section 1 makes with the vertical (Fig. 2, Pl. IX); and ·905 and ·424 are the values of the sine and cosine of the angle which the same section makes with the direction of the tie-rod $T_1$ (Fig. 3).

For the other sections we shall have similar formulæ, except that for the sections 2, 3, and 4 we must also take the tie-rod $T_4$ into account. All these formulæ with the necessary numerical simplifications are tabulated below, $T_1$ and $T_4$ being the only unknowns.

| Section. | Bending Moment. | Normal Pressure. | Shear. |
|---|---|---|---|
| 0 | $B_0 = {\cdot}13T_1 - 2488$ | $P_0 = {\cdot}824T_1 + 8780$ | $S_0 = {\cdot}570T_1 - 8892$ |
| 1 | $B_1 = 1{\cdot}98T_1 - 32936$ | $P_1 = {\cdot}905T_1 + 5547$ | $S_1 = {\cdot}424T_1 - 7821$ |
| 2 | $B_2 = 3{\cdot}27T_1 - 58064$ | $P_2 = {\cdot}965T_1 + 2932$ | $S_2 = \begin{cases} {\cdot}266T_1 - 6065 \\ {\cdot}266T_1 + T_4 - 6065 \end{cases}$ |
| 3 | $B_3 = 3{\cdot}98T_1 + 3{\cdot}68T_4 - 75440$ | $P_3 = {\cdot}995T_1 + {\cdot}165T_4 + 970$ | $S_3 = T_1 + {\cdot}990T_4 - 3352$ |
| 4 | $B_4 = 4{\cdot}07T_1 + 7{\cdot}26T_4 - 82000$ | $P_4 = {\cdot}997T_1 + {\cdot}326T_4 + 28$ | $S_4 = -{\cdot}060T_1 + {\cdot}950T_4 - 240$ |

**4. Calculation of the Tensions $T_1$ and $T_4$.**—We will neglect the internal work due to shear, which will have a very small influence upon the results of our calculation, and we will call

A and I the area and moment of inertia of the arch section.

$A_1$, $A_2$, $A_3$, $A_4$ the areas and $l_1 = 7{\cdot}8$, $l_2 = 2 \times 5{\cdot}85$, $l_3 = 7{\cdot}8$, $l_4 = 3{\cdot}1$ the lengths of the tie-bars.

E the elastic modulus of the iron.

Remembering that the length of each element is 3·68 m. and taking as an approximation

$$\Sigma B^2 = \frac{B_0^2 + 4B_1^2 + 2B_2^2 + \ldots B_4^2}{3}$$

$$\Sigma P^2 = \frac{P_0^2 + 4P_1^2 + 2P_2^2 + \ldots P_4^2}{3}$$

$$\Sigma\frac{lT^2}{A} = \frac{7{\cdot}8T_1^2}{A_1} + \frac{5{\cdot}85T_2^2}{A_2} + \frac{7{\cdot}8T_3^2}{A_3} + \frac{3{\cdot}1T_4^2}{A_4},$$

we have the following expression for the internal work of the whole structure,

$$2\left[\frac{3\cdot68}{2EI}\Sigma B^2 + \frac{3\cdot68}{2EA}\Sigma P^2 + \frac{1}{2E}\Sigma\frac{lT^2}{A}\right];$$

dividing by $\frac{2 \times 3\cdot68}{2EI}$, which will have no effect upon the equations obtained by equating to zero the differential coefficients with regard to $T_1$, $T_4$, we get

$$\Sigma B^2 + \frac{I}{A}\Sigma P^2 + \frac{I}{3\cdot68}\Sigma\frac{lT^2}{A} \quad . \quad . \quad . \quad (1)$$

Taking the section of arch shown in Fig. 6, Pl. IX, and for the diameters of the tie-bars

$$d_1 = 60, \quad d_2 = 56, \quad d_3 = 25, \quad d_4 = 22 \text{ mm.},$$

we obtain the following results :—

$$A = \cdot13 \times \cdot6 - \cdot1205 \times \cdot579 - \cdot0095 \times \cdot456 = \cdot003898 \text{ sq. m.};$$

$$I = \frac{1}{12}(\cdot13 \times \cdot6^3 - \cdot1205 \times \cdot579^3 - \cdot0095 \times \cdot456^3) = \cdot000315 \text{ m.}^4;$$

$A_1 = \cdot002827$; $A_2 = \cdot002463$; $A_3 = \cdot000491$; $A_4 = \cdot000380$ sq. m. and therefore

$$\frac{I}{A} = \cdot0811; \quad \frac{I}{A_1} = \cdot1114; \quad \frac{I}{A_2} = \cdot1279; \quad \frac{I}{A_3} = \cdot6415; \quad \frac{I}{A_4} = \cdot8289.$$

$$\frac{I}{3\cdot68}\Sigma\frac{lT^2}{A} = \cdot2361T_1^2 + \cdot2033T_2^2 + 1\cdot3597T_3^2 + \cdot6982T_4^2 \quad . \quad (2)$$

Now from Fig. 5, Pl. IX, we can deduce the following two equations between the tensions in the tie-bars :—

$$\left.\begin{aligned} T_2 &= \cdot786T_1 + 1\cdot466T_4 \\ T_3 &= \cdot264T_1 - 1\cdot368T_4 \end{aligned}\right\} \quad . \quad . \quad . \quad (3)$$

We can thus eliminate $T_2$ and $T_3$ from formula (2), thus obtaining

$$\frac{I}{3\cdot68}\Sigma\frac{lT^2}{A} = \cdot46T_1^2 - \cdot26 \times 2T_1T_4 + 3\cdot68T_4^2.$$

Substituting in $\Sigma B^2$, $\Sigma P^2$ for $B_0$, $B_1$, etc., $P_0$, $P_1$, etc., their values given in tabular form in § 3, we get

$$\Sigma B^2 = 39T_1^2 + 29\cdot37 \times 2T_1T_4 + 35\cdot62T_4^2 - 725216 \times 2T_1 - 568592 \times 2T_4$$

$$\frac{I}{A}\Sigma P^2 = \cdot29T_1^2 + \cdot03 \times 2T_1T_4 + \cdot01T_4^2 + 1009 \times 2T_1 + 13 \times 2T_4$$

Expression (1) therefore becomes

$$39\cdot75T_1^2 + 29\cdot14 \times 2T_1T_4 + 39\cdot31T_4^2 - 724207 \times 2T_1 - 568579 \times 2T_4.$$

Equating to zero the differential coefficients of this formula with regard to $T_1$, $T_4$, we obtain the following two equations :—

$$39{\cdot}75\mathrm{T}_1 + 29{\cdot}14\mathrm{T}_4 = 724207,$$
$$29{\cdot}14\mathrm{T}_1 + 39{\cdot}31\mathrm{T}_4 = 568579,$$

from which we deduce

$$\mathrm{T}_1 = 16689{\cdot}68 \text{ kg.}$$
$$\mathrm{T}_4 = 2093{\cdot}04 \text{ ,,}$$

**5. Maximum Stresses per Sq. Metre in Tie-bars.**—With these values substituted in equations (3), we get

$$\mathrm{T}_2 = 16186{\cdot}48 \text{ kg.}$$
$$\mathrm{T}_3 = 1542{\cdot}80 \text{ ,,}$$

These values give us the following stresses in kg. per sq. metre :—

$$t_1 = \frac{16689{\cdot}68}{{\cdot}002827} = 5{,}903{,}000\,; \quad t_2 = \frac{16186{\cdot}48}{{\cdot}002463} = 6{,}776{,}000\,;$$

$$t_3 = \frac{1542{\cdot}80}{{\cdot}000491} = 3{,}142{,}000\,; \quad t_4 = \frac{2093{\cdot}04}{{\cdot}000380} = 5{,}508{,}000.$$

**6. Line of pressure of the arch and maximum stresses per sq. metre.**

If in the table of § 3 we give the above values to $\mathrm{T}_1$, $\mathrm{T}_4$ we obtain the values of the bending moments, normal pressures, and shears tabulated below.

Dividing each bending moment by the normal pressure, we find the eccentricity of load, and by the previous formula $c = \frac{\mathrm{P}}{\mathrm{A}} \pm \frac{\mathrm{B}n}{\mathrm{I}}$ the stresses at intrados and extrados are obtained.

Finally if we note that the lattice bracing consists of flat bars having a sectional area $\mathrm{A}_d = {\cdot}05 \times {\cdot}006 = {\cdot}0003$ sq. m., and forming quadrilateral figures whose diagonals are $d_1 = {\cdot}66$ and $d_2 = {\cdot}53$, and that for the arch section the first moment of the half-section about the neutral axis

$$\mathrm{M}_1 = \frac{1}{8}({\cdot}13 \times {\cdot}6^2 - {\cdot}1205 \times {\cdot}579^2 - {\cdot}0095 \times {\cdot}456^2) = {\cdot}000803,$$

we have the maximum compressive or tensile stress in the bracing

$$= \frac{\mathrm{M}_1\mathrm{S}\sqrt{d_1^2 + d_2^2}}{2\mathrm{IA}_d} = 3494{\cdot}66\mathrm{S},$$

where S is the shearing force at the section under consideration. The results of all the calculations are collected in the following table :—

| Section. | Bending Moment, kg.-m. | Normal Pressure, kg. | Shear, kg. | Eccentricity of Load, m.* | Stresses in kg. per Square Metre. | | |
|---|---|---|---|---|---|---|---|
| | | | | | Intrados. | Extrados. | Lattice Bracing. |
| 0 | − 318 | 22,532 | + 621 | − ·014 | 5,477,600 | 6,083,200 | 2,170,180 |
| 1 | + 110 | 26,651 | − 745 | + ·004 | 6,905,500 | 6,696,500 | 2,603,480 |
| 2 | − 3489 | 19,038 | { − 1626<br>+ 476 | − ·183 | 1,696,000 | 8,072,000 | { 4,280,400<br>1,663,470 |
| 3 | − 1313 | 17,922 | + 390 | − ·073 | 3,345,600 | 5,850,000 | 1,362,900 |
| 4 | + 1122 | 17,350 | + 747 | + ·065 | 5,521,700 | 3,384,500 | 2,610,460 |

We thus see that the maximum stress in the iron occurs at the section 2, and is equal to 8,072,000 kg. per sq. metre.

The maximum stress in the lattice bracing occurs at section 2, and is equal to 4,280,000 kg. per sq. metre.

We can thus conclude that the truss is satisfactory from the combined standpoints of economy and stability.†

* The line of pressure is shown in Fig. 7, Pl. IX.

† We have not regarded it as necessary to examine also the case in which the superload is on one side only, because we saw from the previous study of a truss of this type that the maximum stresses for full load are greater than those which occur when only one side is loaded, except for the lattice bracing, in which the stresses may be greater in the latter case, the difference being, however, not worth considering.

With regard to temperature variations, we have already pointed out in the previous chapter that in these structures they have no effect if they are uniform.

## CHAPTER XX.

### STUDY OF AN ARCHED ROOF-TRUSS WITH SEVERAL TIE-BARS (CRESCENT TRUSS).

(Plates X and XI.)

**1. General Data** (Fig. 1, Pl. X).—The trusses rest at one end on the wall and at the other end on cast-iron columns.

| | |
|---|---|
| Distance from wall to centre-line of columns . . . . | = 30 m. |
| Rise of intrados . . . . | = 5·5 m. |
| Radius of intrados . . . | = 23·068 m. |
| Radius of circle of which tie-bars are chords . . . . | = 70·645 m. |
| Distance between principals . . | = 3·8 m. |
| Distance between purlins measured on the centre-line of arch . | = 1·356 m. |

**2. Loads.**—We will suppose that the roof is covered with galvanised corrugated iron 1·25 mm. thick, fixed directly to the purlins, but that towards the top there are glass lights. The weight per sq. metre of the glass is a little greater than that of the corrugated iron, but we will assume it to be the same, and will adopt a superload slightly in excess of that which strictly should be taken.

The total load per sq. metre is obtained as follows:—

| | |
|---|---|
| Superload due to wind or snow . | 70 kg. per sq. m. |
| Galvanised corrugated iron, 1·25 mm. thick . . . . . . | 14 ,, ,, |
| Purlins . . . . . . . . | 18 ,, ,, |
| Total for calculation of purlins . | 102 ,, ,, |
| Weight of truss per sq. metre of covering . . . . . . | 16 ,, ,, |
| Transverse bracing . . . . . | 1 ,, ,, |
| Total for calculation of truss . | 119 ,, ,, |

As the distance between principals is 3·8 m., the load per metre run will be

$$119 \times 3{\cdot}8 = 452{\cdot}2 \text{ kg.}$$

**3. Calculations for the superload on both sides of the truss. Notation and general formulæ.**

The half-arch (Fig. 2, Pl. XI) is divided into six equal parts having a length of 2·713 m. along the centre line. Each part will carry a load of 452·2 × 2·713 = 1226·82 kg., which we will assume to act at the mid-point of each part as shown.

The vertical reaction at each end will be equal to half the total load, i.e.

$$R = 1226{\cdot}82 \times 6 = 7360{\cdot}92 \text{ kg.}$$

We will call $T_1$, $T_2$, . . . $T_9$ the tensions in the tie-bars. Knowing the reactions and the loading, we can express in terms of these unknown tensions the bending moment, normal pressure, and shear at each section. Fig. 2, Pl. XI, gives the lever-arms of the reactions, loads, and tensions; Fig. 3 gives the sines and cosines of the angles which the various sections make with the vertical; Fig. 4 gives the sines and cosines of the angles which the two tensions $T_2$, $T_5$ make with the sections 0, 1, 2; Fig. 5 gives the sines and cosines of the angles which the tensions $T_3$, $T_7$ make with the sections 2, 3, 4; and Fig. 6 gives the sines and cosines of the angles which the tensions $T_4$, $T_9$ make with the sections 4, 5, 6.

With these data we can without difficulty compose the following table :—

| Section. | Bending Moment. | Normal Pressure. | Shear. |
|---|---|---|---|
| 0 | $B_0 = -736$ | $P_0 = {\cdot}848T_2 + {\cdot}982T_5 + 4{,}748$ | $S_0 = -{\cdot}530T_2 + {\cdot}180T_5 + 5{,}638$ |
| 1 | $B_1 = 1{\cdot}15T_2 + {\cdot}95T_5 - 15{,}372$ | $P_1 = {\cdot}904T_2 + {\cdot}955T_5 + 3{,}386$ | $S_1 = -{\cdot}430T_2 + {\cdot}295T_5 + 5{,}134$ |
| 2 | $B_2 = 2{\cdot}17T_2 - 28{,}168$ | $P_2 = {\cdot}914T_2 + {\cdot}950T_5 + 2{,}208$ | $S_2 = -{\cdot}320T_2 + {\cdot}402T_5 + 4{,}407$ |
| | | $P_2' = {\cdot}922T_3 + {\cdot}842T_7 + 2{,}208$ | $S_2' = -{\cdot}385T_3 + {\cdot}540T_7 + 4{,}407$ |
| 3 | $B_3 = 2{\cdot}88T_3 + 1{\cdot}84T_7 - 38{,}804$ | $P_3 = {\cdot}960T_3 + {\cdot}770T_7 + 1.259$ | $S_3 = -{\cdot}272T_3 + {\cdot}635T_7 + 3{,}467$ |
| 4 | $B_4 = 3{\cdot}47T_3 - 46{,}815$ | $P_4 = {\cdot}988T_3 + {\cdot}698T_7 + 574$ | $S_4 = -{\cdot}165T_3 + {\cdot}715T_7 + 2{,}390$ |
| | | $P_4' = {\cdot}970T_4 + {\cdot}720T_9 + 574$ | $S_4' = -{\cdot}225T_4 + {\cdot}690T_9 + 2{,}390$ |
| 5 | $D_5 = 3{\cdot}75T_4 + 2{\cdot}20T_9 - 51{,}686$ | $P_5 = {\cdot}990T_4 + {\cdot}630T_9 + 141$ | $S_5 = -{\cdot}115T_4 + {\cdot}775T_9 + 1{,}217$ |
| 6 | $B_6 = 3{\cdot}92T_4 - 53{,}354$ | $P_6 = T_4 + {\cdot}540T_9$ | $S_6 = {\cdot}840T_9$ |

It will be noted that two values of P and S are given for the sections 2 and 4. This is due to the fact that there is a sudden change in these quantities at the points 2, 4, 6, due to the tie-rods meeting at these points. This cannot cause a sudden change in the bending moment, because the tie-rods are attached to the arch on the centre line, so that, for example, the tensions $T_5$ and $T_6$ cause no bending moment at the section 2.

Now the three tensions meeting at B, Fig. 2, and the four tensions meeting at C and D must be in equilibrium among themselves. The vector polygons given in Figs. 7, 8, and 9 enable us to obtain the following equations between the tensions:—

$$\left.\begin{aligned} T_1 &= \quad 1{\cdot}072T_2 \\ T_5 &= \quad {\cdot}090T_2 \\ T_6 &= -\ {\cdot}780T_2 + {\cdot}825T_3 \\ T_7 &= \quad {\cdot}635T_2 - {\cdot}575T_3 \\ T_8 &= -\ {\cdot}835T_3 + {\cdot}880T_4 \\ T_9 &= \quad {\cdot}834T_3 - {\cdot}785T_4 \end{aligned}\right\} \quad . \quad . \quad . \quad (1)$$

We can now substitute in the previous table for $T_5$, $T_7$, $T_9$ their expressions in terms of $T_2$, $T_3$, $T_4$, thus obtaining the following table, in which $T_2$, $T_3$, and $T_4$ are the only unknowns:—

| Section. | Bending Moment. | Normal Pressure. | Shear. |
|---|---|---|---|
| 0 | $B_0 = -736$ | $P_0 = {\cdot}936T_2 + 4{,}778$ | $S_0 = -{\cdot}514T_2 + 5{,}638$ |
| 1 | $B_1 = 1{\cdot}235T_2 - 15{,}372$ | $P_1 = {\cdot}990T_2 + 3{,}386$ | $S_1 = -{\cdot}403T_2 + 5{,}134$ |
| 2 | $B_2 = 2{\cdot}170T_3 - 28{,}168$ | $P_2 = {\cdot}999T_2 + 2{,}208$ | $S_2 = -{\cdot}284T_2 + 4{,}407$ |
| | | $P_2' = {\cdot}535T_2 + {\cdot}438T_3 + 2{,}208$ | $S_2' = +{\cdot}343T_2 - {\cdot}695T_3 + 4{,}407$ |
| 3 | $B_3 = 1{\cdot}168T_2 + 1{\cdot}822T_3 - 38{,}804$ | $P_3 = {\cdot}489T_2 + {\cdot}517T_3 + 1{,}259$ | $S_3 = +{\cdot}403T_2 - {\cdot}637T_3 + 3{,}467$ |
| 4 | $B_4 = 3{\cdot}470T_3 - 46{,}815$ | $P_4 = {\cdot}443T_2 + {\cdot}587T_3 + 574$ | $S_4 = +{\cdot}454T_2 - {\cdot}576T_3 + 2{,}390$ |
| | | $P_4' = {\cdot}600T_3 + {\cdot}405T_4 + 574$ | $S_4' = +{\cdot}575T_3 - {\cdot}767T_4 + 2{,}390$ |
| 5 | $B_5 = 1{\cdot}835T_3 + 2{\cdot}023T_4 - 51{,}686$ | $P_5 = {\cdot}525T_3 + {\cdot}495T_4 + 141$ | $S_5 = +{\cdot}646T_3 - {\cdot}723T_4 + 1{,}217$ |
| 6 | $B_6 = 3{\cdot}920T_4 - 53{,}354$ | $P_6 = {\cdot}450T_3 + {\cdot}576T_4$ | $S_6 = +{\cdot}701T_3 - {\cdot}659T_4$ |

**4. Determination of the Unknowns $T_2$, $T_3$, $T_4$.**—We will adopt the same notation as in the previous chapter; then remembering that the length of each element is 2·713 m., and that the length of each tie-bar is as shown on the drawing, we see that, neglecting the portion due to shear (which will have no appreciable effect upon the results), the internal work of the whole structure is given by

$$2\left(\frac{2{\cdot}713}{2EI}\Sigma B^2 + \frac{2{\cdot}713}{2EA}\Sigma P^2 + \frac{1}{2E}\Sigma\frac{lT^2}{A}\right) \quad . \quad . \quad (2)$$

We will take

$$\left.\begin{aligned} \Sigma B^2 &= \frac{1}{3}(B_0^2 + 4B_1^2 + 2B_2^2 + \ .\ .\ .\ B_6^2) \\ \Sigma P^2 &= \frac{1}{3}(P_0^2 + 4P_1^2 + P_2^2 + P_2'^2 + 4P_3^2 + P_4^2 + P_4'^2 + 4P_5^2 + P_6^2) \\ \Sigma\frac{lT^2}{A} &= \frac{2{\cdot}510T_1^2}{A_1} + \frac{5{\cdot}020T_2^2}{A_2} + \frac{5{\cdot}020T_3^2}{A_3} + \frac{2{\cdot}510T_4^2}{A_4} + \frac{3{\cdot}210T_5^2}{A_5} \\ &\quad + \frac{3{\cdot}425T_6^2}{A_6} + \frac{4{\cdot}240T_7^2}{A_7} + \frac{4{\cdot}370T_8^2}{A_8} + \frac{4{\cdot}630T_9^2}{A_9} \end{aligned}\right\} (3)$$

To determine the values of the unknowns $T_2$, $T_3$, $T_4$, we must equate to zero the differential coefficients of formula (2) with regard to these quantities; but we shall obtain the same results if we divide through by $\frac{2 \times 2{\cdot}713}{2EI}$, thus obtaining

$$\Sigma B^2 + \frac{I}{A}\Sigma P^2 + \frac{I}{2{\cdot}713}\Sigma\frac{lT^2}{A} \quad . \quad . \quad . \quad (4)$$

We will now substitute for $B_0$, $B_1$ . . . $P_0$, $P_1$ . . . their expressions in terms of $T_2$, $T_3$, $T_4$ given in § 3, and for $T_1$, $T_5$, $T_6$, $T_7$, $T_8$, and $T_9$ their expressions as functions of these same unknowns.

We will take the section of the arch shown in Fig. 10, Pl. XI, and the diameters of the tie-bars as

$$d_1 = d_2 = d_3 = d_4 = {\cdot}06 \text{ m.}$$
$$d_5 = {\cdot}024 \text{ m.}$$
$$d_6 = d_7 = d_8 = d_9 = {\cdot}02 \text{ m.}$$

We thus obtain the following values:—

$$A = {\cdot}12 \times {\cdot}175 - {\cdot}104 \times {\cdot}155 = {\cdot}00488 \text{ sq. m.}$$
$$I = \frac{1}{12}({\cdot}12 \times {\cdot}175^3 - {\cdot}104 \times {\cdot}155^3) = {\cdot}00002132 \text{ m.}^4$$
$$A_1 = A_2 = A_3 = A_4 = {\cdot}002826 \text{ sq. m.}$$
$$A_5 = {\cdot}000452 \quad ,,$$
$$A_6 = A_7 = A_8 = A_9 = {\cdot}000314 \quad ,,$$

We then find

$$\Sigma B^2 = 6{\cdot}9918T_2^2 + 16{\cdot}9431T_3^2 + 10{\cdot}5788T_4^2 + 2{\cdot}8375 \times 2T_2T_3 + 4{\cdot}9496 \times 2T_3T_4 - 126493 \times 2T_2 - 329025 \times 2T_3 - 209130 \times 2T_4.$$

$$\frac{I}{A}\Sigma P^2 = {\cdot}0107T_2^2 + {\cdot}0048T_3^2 + {\cdot}0022T_4^2 + {\cdot}0022 \times 2T_2T_3 + {\cdot}0023 \times 2T_3T_4 + 35{\cdot}14 \times 2T_2 + 6{\cdot}67 \times 2T_3 + {\cdot}75 \times 2T_4.$$

$$\frac{I}{2{\cdot}713}\Sigma\frac{lT^2}{A} = {\cdot}1169T_2^2 + {\cdot}2638T_3^2 + {\cdot}1625T_4^2 - {\cdot}0938 \times 2T_2T_3 - {\cdot}1559 \times 2T_3T_4$$

Thus expression (4) becomes

$$7{\cdot}1194T_2^2 + 17{\cdot}2117T_3^2 + 10{\cdot}7435T_4^2 + 2{\cdot}7459 \times 2T_2T_3 + 4{\cdot}7960 \times 2T_3T_4 - 126458 \times 2T_2 - 329018 \times 2T_3 - 209129 \times 2T_4,$$

and on equating to zero its differential coefficients with regard to $T_2$, $T_3$, $T_4$ we obtain the equations

$$7{\cdot}1194T_2 + 2{\cdot}7459T_3 - 126458 = 0,$$
$$2{\cdot}7459T_2 + 17{\cdot}2117T_3 + 4{\cdot}7960T_4 - 329018 = 0,$$
$$4{\cdot}7960T_3 + 10{\cdot}7435T_4 - 209129 = 0;$$

from which we deduce

$$T_2 = 12{,}619 \text{ kg.} \quad T_3 = 13{,}338 \text{ kg.} \quad T_4 = 13{,}511 \text{ kg.} \quad . \quad (5)$$

**5. Stresses per Sq. Metre in the Tie-bars and in the Arch.**—Substituting these values of $T_2$, $T_3$, and $T_4$ in formulæ (1) we get:—

$$T_1 = 13{,}528 \text{ kg.} \quad T_5 = 1136 \text{ kg.} \quad T_6 = 1161 \text{ kg.}$$
$$T_7 = 344 \text{ kg.} \quad T_8 = 752 \text{ kg.} \quad T_9 = 518 \text{ kg.}$$

Dividing these by the corresponding values of $A_1$, $A_2$, etc., we have the following stresses in the tie-bars in kg. per sq. metre:—

$$t_1 = 4{,}787{,}000, \quad t_2 = 4{,}465{,}000, \quad t_3 = 4{,}719{,}000,$$
$$t_4 = 4{,}781{,}000, \quad t_5 = 2{,}512{,}000, \quad t_6 = 3{,}697{,}000,$$
$$t_7 = 1{,}095{,}000, \quad t_8 = 2{,}396{,}000, \quad t_9 = 1{,}648{,}000.$$

We now substitute the above values of $T_2$, $T_3$, $T_4$ in the table given at the end of § 3, and obtain the bending moment, normal pressure, and shear for each section, the eccentricity of load and stresses at intrados and extrados being found in the same manner as in the previous example.

Finally, to obtain the maximum shear stress, we have

$$M_1 = \frac{1}{8}({\cdot}12 \times {\cdot}175^2 - {\cdot}104 \times {\cdot}155^2) = {\cdot}000147,$$

and $b = {\cdot}008 \times 2 = {\cdot}016$, the breadth of the arch section at the neutral axis,

and hence maximum shear stress $= s = \dfrac{M_1 S}{bI} = 431S$,

which gives the results in the last column of the following table:—

| Section. | Bending Moment, kg.-m. | Normal Pressure, kg. | Shear, kg. | Eccentricity of Load,* m. | Stresses in kg. per Sq. Metre at Intrados. | Extrados. | Neutral Axis (Shear). |
|---|---|---|---|---|---|---|---|
| 0 | −736 | +16,560 | −848 | −·044 | + 372,500 | +6,414,500 | −365,488 |
| 1 | +213 | +15,879 | + 48 | +·013 | +4,128,100 | +2,379,500 | + 20,688 |
| 2 | −784 | +14,815 | +823 | −·053 | − 181,600 | +6,253,400 | +354,713 |
| | | +14,801 | −535 | −·052 | − 184,500 | +6,250,500 | −230,585 |
| 3 | +238 | +14,325 | + 56 | +·016 | +3,905,500 | +1,951,900 | + 10,775 |
| 4 | −531 | +13,994 | +436 | −·037 | + 688,500 | +5,046,900 | +187,916 |
| | | +14,049 | −304 | −·037 | + 699,700 | +5,058,100 | −131,024 |
| 5 | +123 | +13,832 | + 65 | +·009 | +3,339,300 | +2,329,700 | + 28,015 |
| 6 | −390 | +13,785 | +446 | −·028 | +1,224,200 | +4,425,400 | +192,226 |

* The line of pressure for the truss fully loaded is shown by the continuous line in Fig. 11, Pl. X.

**6. Dimensions to be Adopted.**—As the calculations show that the maximum stress is 6,414,500 kg. per sq. metre, we can safely adopt the arch section that has been assumed in the calculations.

As for the tie-bars, their tensile stress is seen to be less than 6 kg. per sq. mm., so that their diameter may be somewhat diminished. We will only make this reduction for tie-bars 1-4, because the diameter of the others is already quite small.

We will therefore take $d_1 = d_2 = d_3 = d_4 = \cdot 054$ m.,

$$\text{i.e. } A_1 = A_2 = A_3 = A_4 = \cdot 002290 \text{ sq. m.},$$

thus giving the following stresses in kg. per sq. metre :—

$$t_1 = 5{,}908{,}000, \quad t_2 = 5{,}510{,}000, \quad t_3 = 5{,}824{,}000, \quad t_4 = 5{,}900{,}000.$$

**7. Examination of the case in which the superload acts only on the left-hand half of the arch, and the dead load is not taken into account.**

Since the superload is 70 kg. per sq. metre and the principals are 3·8 m. apart, the superload per metre run will be 70 × 3·8 = 266 kg.

To determine the reactions we observe that if we assume the truss to be freely supported, the sum of the reactions $R_L$ and $R_R$ must be equal to the total load, and the sum of the moments of the reactions about any point—say section 6—must be equal to the sum of the moments of the load about the same point,

$$\text{i.e. } R_L + R_R = 266 \times 2{\cdot}713 \times 6 = 4330, \text{ and}$$

$$(R_L - R_R)15{\cdot}05 = (1{\cdot}36 + 4{\cdot}05 + 6{\cdot}67 + 9{\cdot}23 + 11{\cdot}63 + 13{\cdot}87)266 \times 2{\cdot}713,$$

whence $R_L = 3288$ kg. ; $R_R = 1042$ kg.

Now let $T_{L1}$, $T_{L2}$, etc., be the tensions in the tie-bars on the left-hand side and $T_{R1}$, $T_{R2}$, etc., those for the right-hand side.

We note that $T_{L4} = T_{R4}$, since they occur in the same rod. Then, knowing the reactions, we can easily calculate the bending moment, normal pressure, and shear for each section on the right and left-hand sides.

We thus compile a table corresponding to the first table in § 3, which will differ only in the terms independent of the tensions, the terms involving the tensions being the same with $T_L$ and $T_R$ substituted for T.

As before we shall clearly have the following equations between the tensions :—

$$\left.\begin{aligned} T_{L_1} &= 1{\cdot}072T_{L_2} \\ T_{L_5} &= {\cdot}090T_{L_2} \\ T_{L_6} &= -{\cdot}780T_{L_2} + {\cdot}825T_{L_3} \\ T_{L_7} &= {\cdot}635T_{L_2} - {\cdot}575T_{L_3} \\ T_{L_8} &= -{\cdot}835T_{L_3} + {\cdot}880T_{L_4} \\ T_{L_9} &= {\cdot}834T_{L_3} - {\cdot}785T_{L_4} \end{aligned}\right\} \quad . \quad . \quad . \quad (5\text{A})$$

$$\left.\begin{aligned} T_{R_1} &= 1{\cdot}072T_{R_2} \\ T_{R_5} &= {\cdot}090T_{R_2} \\ T_{R_6} &= -{\cdot}780T_{R_2} + {\cdot}825T_{R_3} \\ T_{R_7} &= {\cdot}635T_{R_2} - {\cdot}575T_{R_3} \\ T_{R_8} &= -{\cdot}835T_{R_3} + {\cdot}880T_{R_4} \\ T_{R_9} &= {\cdot}834T_{R_3} - {\cdot}785T_{R_4} \end{aligned}\right\} \quad . \quad . \quad . \quad (6)$$

We can eliminate $T_{L_5}$, $T_{L_7}$, $T_{L_9}$ and the similar tensions for the other side, and thus obtain the following table (on opposite page), of which the two portions differ from that given at the end of § 3 by the purely numerical terms, the terms involving the tensions being the same with the substitution of $T_L$ or $T_R$ for T.

The internal work of the truss is given by the formula

$$\frac{2{\cdot}713}{2EI}(\Sigma B^2 + \Sigma B'^2) + \frac{2{\cdot}713}{2EA}(\Sigma P^2 + \Sigma P'^2) + \frac{1}{2E}\Sigma\frac{l(T_L^2 + T_R^2)}{A} \quad (7)$$

in which we have

$$\left.\begin{aligned} \Sigma B^2 &= \tfrac{1}{3}(B_0^2 + 4B_1^2 + 2B_2^2 + \ . \ . \ . \ + B_6^2) \\ \Sigma B'^2 &= \tfrac{1}{3}(B_0'^2 + 4B_1'^2 + 2B_2'^2 + \ . \ . \ . \ + B_6'^2) \\ \Sigma P^2 &= \tfrac{1}{3}(P_0^2 + 4P_1^2 + P_2^2 + P_{2'}^2 + 4P_3^2 + P_4^2 + P_{4'}^2 + 4P_5^2 + P_6^2) \\ \Sigma P'^2 &= \tfrac{1}{3}(P_0'^2 + 4P_1'^2 + P_2'^2 + P_{2'}'^2 + 4P_3'^2 + P_4'^2 + P_{4'}'^2 + 4P_5'^2 + P_6'^2) \end{aligned}\right\} (8)$$

$$\left.\begin{aligned} &\sum\frac{l(T_L^2 + T_R^2)}{A} \\ &= \frac{2{\cdot}51(T_{L_1}^2 + T_{R_1}^2)}{A_1} + \frac{5{\cdot}02(T_{L_2}^2 + T_{R_2}^2)}{A_2} + \frac{5{\cdot}02(T_{L_3}^2 + T_{R_3}^2)}{A_3} \\ &+ \frac{2{\cdot}51(T_{L_4}^2 + T_{R_4}^2)}{A_4} + \frac{3{\cdot}21(T_{L_5}^2 + T_{R_5}^2)}{A_5} + \frac{3{\cdot}425(T_{L_6}^2 + T_{R_6}^2)}{A_6} \\ &+ \frac{4{\cdot}24(T_{L_7}^2 + T_{R_7}^2)}{A_7} + \frac{4{\cdot}37(T_{L_8}^2 + T_{R_8}^2)}{A_8} + \frac{4{\cdot}63(T_{L_9}^2 + T_{R_9}^2)}{A_9} \end{aligned}\right\} (8\text{A})$$

To determine the unknown tensions we can divide formula (7) through by $\frac{2{\cdot}713}{2EI}$ and obtain

$$\Sigma B^2 + \Sigma B'^2 + \frac{I}{A}(\Sigma P^2 + \Sigma P'^2) + \frac{I}{2{\cdot}713}\Sigma\frac{l(T_L^2 + T_R^2)}{A} \quad (9)$$

| Section. | Bending Moment. | Normal Pressure. | Shear. |
|---|---|---|---|
| | | *Left-hand Half-arch.* | |
| 0 | $B_0 = -329$ | $P_0 = \cdot936T_{L_2} + 2{,}121$ | $S_0 = -\cdot514T_{L_2} + 2{,}519$ |
| 1 | $B_1 = 1\cdot235T_{L_2} - 6{,}677$ | $P_1 = \cdot990T_{L_2} + 1{,}417$ | $S_1 = -\cdot403T_{L_2} + 2{,}148$ |
| 2 | $B_2 = 2\cdot170T_{L_2} - 11{,}787$ | $P_2 = \cdot999T_{L_2} + 830$ | $S_2 = -\cdot284T_{L_2} + 1{,}657$ |
| | | $P_2' = \cdot535T_{L_2} + \cdot438T_{L_3} + 830$ | $S_2' = +\cdot343T_{L_2} - \cdot695T_{L_3} + 1{,}657$ |
| 3 | $B_3 = 1\cdot168T_{L_2} - 1\cdot822T_{L_3} - 15{,}459$ | $P_3 = \cdot489T_{L_2} + \cdot517T_{L_3} + 384$ | $S_3 = +\cdot403T_{L} - \cdot637T_{L_3} + 1{,}058$ |
| 4 | $B_4 = 3\cdot470T_{L_3} - 17{,}452$ | $P_4 = \cdot443T_{L_2} + \cdot587T_{L_3} + 94$ | $S_4 = +\cdot454T_{L_2} - \cdot576T_{L_3} + 391$ |
| | | $P_4' = \cdot600T_{L_3} + \cdot405T_{L_4} + 94$ | $S_4' = +\cdot575T_{L_3} - \cdot767T_{L_4} + 391$ |
| 5 | $B_5 = 1\cdot835T_{L_3} + 2\cdot023T_{L_4} - 17{,}556$ | $P_5 = \cdot525T_{L_3} + \cdot495T_{L_4} - 37$ | $S_5 = +\cdot646T_{L_3} - \cdot723T_{L_4} - 318$ |
| 6 | $B_6 = 3\cdot920T_{L_4} - 15{,}692$ | $P_6 = \cdot450T_{L_3} + \cdot576T_{L_4}$ | $S_6 = +\cdot701T_{L_3} - \cdot659T_{L_4} - 1{,}042$ |
| | | *Right-hand Half-arch.* | |
| 0 | $B_0' = -104$ | $P_0' = \cdot936T_{R_2} + 672$ | $S_0' = -\cdot514T_{R_2} + 798$ |
| 1 | $B_1' = 1\cdot235T_{R_2} - 2{,}365$ | $P_1' = \cdot990T_{R_2} + 575$ | $S_1' = -\cdot403T_{R_2} + 872$ |
| 2 | $B_2' = 2\cdot170T_{R_2} - 4{,}783$ | $P_2' = \cdot999T_{R_2} + 469$ | $S_2' = -\cdot284T_{R_2} + 936$ |
| | | $P_2'' = \cdot535T_{R_2} + \cdot438T_{R_3} + 469$ | $S_2'' = +\cdot343T_{R_2} - \cdot695T_{R_3} + 936$ |
| 3 | $B_3' = 1\cdot168T_{R_2} - 1\cdot822T_{R_3} - 7{,}367$ | $P_3' = \cdot489T_{R_2} + \cdot517T_{R_3} + 356$ | $S_3' = +\cdot403T_{R_2} - \cdot637T_{R_3} + 982$ |
| 4 | $B_4' = 3\cdot470T_{R_3} - 10{,}087$ | $P_4' = \cdot443T_{R_2} + \cdot587T_{R_3} + 243$ | $S_4' = +\cdot454T_{R_2} - \cdot576T_{R_3} + 1{,}015$ |
| | | $P_4'' = \cdot600T_{R_3} + \cdot405T_{R_4} + 243$ | $S_4'' = +\cdot575T_{R_3} - \cdot767T_{R_4} + 1{,}015$ |
| 5 | $B_5' = 1\cdot835T_{R_3} + 2\cdot023T_{R_4} - 12{,}848$ | $P_5' = \cdot525T_{R_3} + \cdot495T_{R_4} + 120$ | $S_5' = +\cdot646T_{R_3} - \cdot723T_{R_4} + 1{,}034$ |
| 6 | $B_6' = 3\cdot920T_{R_4} - 15{,}692$ | $P_6' = \cdot450T_{R_3} + \cdot576T_{R_4}$ | $S_6' = +\cdot701T_{R_3} - \cdot659T_{R_4} + 1{,}042$ |

If now in formulæ (8) we substitute for B, P, etc., their expressions given in the above table, and if, further, we eliminate $T_{L_1}, T_{L_5}, T_{L_6} \ldots, T_{R_1}, T_{R_5}, T_{R_6} \ldots$ from formula (8A) by means of equations (5) and (6), then (9) becomes

$$7{\cdot}1194(T_{L_2}^2 + T_{R_2}^2) + 17{\cdot}2117(T_{L_3}^2 + T_{R_3}^2) + 10{\cdot}7435(T_{L_4}^2 + T_{R_4}^2)$$
$$+ 2{\cdot}7459 \times 2(T_{L_2}T_{L_3} + T_{R_2}T_{R_3}) + 4{\cdot}796 \times 2(T_{L_3}T_{L_4} + T_{R_3}T_{R_4})$$
$$- (52107 \times 2T_{L_2} + 22280 \times 2T_{R_2}) - (120879 \times 2T_{L_3} + 72665 \times 2T_{R_3})$$
$$- (67873 \times 2T_{L_4} + 55147 \times 2T_{R_4}) \quad . \quad . \quad (10)$$

It is useful to note that the coefficients of the terms of the second degree in the tensions are the same in the above formula as in that obtained in § 4; this shows that to evaluate the above we have only to calculate the terms of the first degree —an operation which is both short and simple.

We have now to equate to zero the differential coefficients of expression (10) with regard to the unknowns. But for simplification we will take

$$T_{L_2} + T_{R_2} = X_2, \quad T_{L_3} + T_{R_3} = X_3, \quad T_{L_4} = T_{R_4} = \frac{X_4}{2}$$
$$T_{L_2} - T_{R_2} = Y_2, \quad T_{L_3} - T_{R_3} = Y_3$$

which give

$$\left.\begin{aligned} T_{L_2} &= \frac{X_2 + Y_2}{2}, \quad T_{R_2} = \frac{X_2 - Y_2}{2} \\ T_{L_3} &= \frac{X_3 + Y_3}{2}, \quad T_{R_3} = \frac{X_3 - Y_3}{2} \\ T_{L_4} &= \frac{X_4}{2}, \quad T_{R_4} = \frac{X_4}{2} \end{aligned}\right\} \quad . \quad . \quad (11)$$

and further

$$T_{L_2}^2 + T_{R_2}^2 = \frac{X_2^2 + Y_2^2}{2}$$
$$T_{L_3}^2 + T_{R_3}^2 = \frac{X_3^2 + Y_3^2}{2}$$
$$T_{L_4}^2 + T_{R_4}^2 = \frac{X_4^2}{2}$$
$$2(T_{L_2}T_{L_3} + T_{R_2}T_{R_3}) = X_2X_3 + Y_2Y_3$$
$$2(T_{L_3}T_{L_4} + T_{R_3}T_{R_4}) = X_3X_4.$$

Thus, substituting in expression (10), we get

$$\frac{1}{2}(7{\cdot}1194X_2^2 + 17{\cdot}2117X_3^2 + 10{\cdot}7435X_4^2 + 2{\cdot}7459 \times 2X_2X_3$$
$$+ 4{\cdot}7960 \times 2X_3X_4 - 74387 \times 2X_2 - 193544 \times 2X_3 - 123020 \times 2X_4)$$
$$+ \frac{1}{2}(7{\cdot}1194Y_2^2 + 17{\cdot}2117Y_3^2 + 2{\cdot}7459 \times 2Y_2Y_3 - 29827 \times 2Y_2$$
$$- 48214 \times 2Y_3).$$

We then have to equate to zero the differential coefficients of this expression with regard to $X_2$, $X_3$, $X_4$, $Y_2$, $Y_3$, thus obtaining five equations, from the first three of which $X_2$, $X_3$, and $X_4$ can be determined, $Y_2$, $Y_3$ being found from the last two.

Now from the principle of superposition it follows that if we assume the superload to be on both sides—neglecting the dead load—the tensions in the bars BC, CD, and DE will be

$$T_{L2} + T_{R2} = X_2, \quad T_{L3} + T_{R3} = X_3, \quad T_{L4} + T_{R4} = X_4.$$

It follows, moreover, that the respective tensions in a tie-rod for the superload on the whole arch and for the superload and dead load will bear the ratio 70 : 119, so that we can get $X_2$, $X_3$, and $X_4$ by multiplying by $\frac{70}{119}$ the values of $T_2$, $T_3$, and $T_4$ obtained in § 4, which will give

$$X_2 = 7423, \quad X_3 = 7846, \quad X_4 = 7947 \quad . \quad . \quad (12)$$

Now the two equations for $Y_2$, $Y_3$ obtained from the above expression are as follows :—

$$7{\cdot}1194Y_2 + 2{\cdot}7459Y_3 - 29827 = 0$$
$$2{\cdot}7459Y_2 + 17{\cdot}2117Y_3 - 48214 = 0$$

from which we deduce

$$Y_2 = 3313, \quad Y_3 = 2273.$$

Knowing thus $X_2$, $X_3$, $X_4$, $Y_2$ and $Y_3$, formulæ (11) give the following values in kg. :—

$$T_{L2} = 5368, \quad T_{R2} = 2055,$$
$$T_{L3} = 5059, \quad T_{R3} = 2786,$$
$$T_{L4} = T_{R4} = 3974.$$

Substituting these results in formulæ (5A) and (6), and in the succeeding tables, we obtain the following results :—

$$\begin{aligned} T_{L1} &= 5755, & T_{R1} &= 2203, \\ T_{L5} &= 483, & T_{R5} &= 185, \\ T_{L6} &= -13, & T_{R6} &= 696, \\ T_{L7} &= 500, & T_{R7} &= -297, \\ T_{L8} &= -728, & T_{R8} &= 1170, \\ T_{L9} &= 1100. & T_{R9} &= -795. \end{aligned}$$

| Section, | Left-hand Half-arch. | | | Right-hand Half-arch. | | |
|---|---|---|---|---|---|---|
| | Bending Moment, kg.-m. | Normal Pressure, kg. | Shear, kg. | Bending Moment, kg.-m. | Normal Pressure, kg. | Shear, kg. |
| 0 | − 329 | 7145 | − 240 | − 104 | 2595 | − 258 |
| 1 | + 436 | 6731 | − 15 | + 357 | 2609 | + 44 |
| 2 | − 138 | {6193<br>5918 | {+ 132<br>− 18 | − 324 | {2549<br>2816 | {+ 352<br>− 295 |
| 3 | + 29 | 5625 | − 1 | + 110 | 2801 | + 36 |
| 4 | + 103 | {5422<br>4739 | {− 85<br>+ 252 | − 418 | {2789<br>3524 | {+ 343<br>− 431 |
| 5 | − 234 | 4586 | + 76 | + 304 | 3550 | − 39 |
| 6 | − 115 | 4565 | − 115 | − 115 | 2543 | + 376 |

The case considered in the present paragraph has in itself no practical value because we have left the dead load out of account; but it serves to reduce to simple additions and multiplications the examination of different cases which may arise.

**8. Determination of stresses in the case in which the dead load is taken into account, and in which the superload on the left-hand side is 70 kg. per sq. metre, and on the right-hand side 35 kg. per sq. metre.**

It follows from the principle of superposition that this case can be deduced from the two cases considered in the previous paragraphs, i.e. the truss fully loaded, and superload on the left-hand side only, the dead load being neglected.

We may imagine that the superload of 70 kg. per sq. metre is spread over the whole truss, and that 35 kg. per sq. metre is then taken off from the right-hand side. It is then clear that the tensions, as well as the bending moments, etc., will be diminished by the values which these quantities would have for a superload of 35 kg. per sq. metre on the right-hand side (the dead load being neglected). We thus obtain the following values for the tensions in kg. :—

| *Left-hand side.* | *Right-hand side.* |
|---|---|
| $T_1 - \frac{1}{2}T_{R1} = 12427$, | $T_1 - \frac{1}{2}T_{L1} = 10651$, |
| $T_2 - \frac{1}{2}T_{R2} = 11592$, | $T_2 - \frac{1}{2}T_{L2} = 9035$, |
| $T_3 - \frac{1}{2}T_{R3} = 11945$, | $T_3 - \frac{1}{2}T_{L3} = 10808$, |
| $T_4 - \frac{1}{2}T_{R4} = 11524$, | $T_4 - \frac{1}{2}T_{L4} = 11524$, |
| $T_5 - \frac{1}{2}T_{R5} = 1043$, | $T_5 - \frac{1}{2}T_{L5} = 894$, |
| $T_6 - \frac{1}{2}T_{R6} = 813$, | $T_6 - \frac{1}{2}T_{L6} = 1168$, |
| $T_7 - \frac{1}{2}T_{R7} = 492$, | $T_7 - \frac{1}{2}T_{L7} = 94$, |
| $T_8 - \frac{1}{2}T_{R8} = 167$, | $T_8 - \frac{1}{2}T_{L8} = 1116$, |
| $T_9 - \frac{1}{2}T_{R9} = 915$. | $T_9 - \frac{1}{2}T_{L9} = -32$. |

The values for bending moment, etc., are given in the table on the opposite page.

If we compare these results with those obtained for the superload of 70 kg. per sq. metre over the whole span, we see that

1. In the tie-rods forming the main tension members the stresses are greatest for the full load.
2. In the other tie-rods (the diagonal members) the maximum stress occurs when the superload on one side is greater than that on the other.

| Section. | Left-hand Half-arch. | | | | Right-hand Half-arch. | | | |
|---|---|---|---|---|---|---|---|---|
| | Bending Moment, kg.-m. | Normal Pressure, kg. | Shear, kg. | Eccentricity of Load, m.* | Bending Moment, kg.-m. | Normal Pressure, kg. | Shear, kg. | Eccentricity of Load, m. |
| 0 | − 684 | 15,263 | − 719 | − ·045 | − 572 | 12,988 | − 728 | − ·044 |
| 1 | + 35 | 14,575 | + 26 | + ·002 | − 5 | 12,514 | + 56 | − ·000 |
| 2 | − 622 | 13,541 | + 647 | − ·046 | − 715 | 11,769 | + 762 | − ·060 |
| 3 | | 13,393 | − 387 | − ·046 | | 11,842 | − 526 | − ·060 |
| 3 | + 183 | 12,925 | + 38 | + ·014 | + 224 | 11,513 | + 57 | + ·019 |
| 4 | − 322 | 12,600 | + 264 | − ·025 | − 582 | 11,273 | + 479 | − ·051 |
| | | 12,787 | − 88 | − ·025 | | 11,680 | − 430 | − ·049 |
| 5 | − 29 | 12,057 | + 85 | − ·002 | + 240 | 11,539 | + 27 | + ·021 |
| 6 | − 332 | 12,014 | + 258 | − ·028 | − 332 | 11,503 | + 504 | − ·028 |

3. In the second case the bar $T_9$ is in compression, but this compression is very small and will change again to a tension as soon as the smaller superload becomes slightly greater.
4. The maximum bending moment and normal pressure and therefore the maximum stress occurs for the case in which the full superload is over the whole span.

**9. Temperature effects.**—As we assume that one of the ends of the truss can slide freely upon its abutment, and as, moreover, all the members are of iron and therefore have the same coefficient of expansion, it follows that a uniform variation of temperature will induce no stress in the structure.

* The line of pressure for this case is shown in broken lines in Fig. 11, Pl. X.

# CHAPTER XXI.

## STUDY OF AN IRON ARCH BRIDGE WITH FLAT SPRINGINGS.

(Plates XII and XIII.)

**1. General Data.**—

| | |
|---|---|
| Clear span between springings . . | = 45·00 m. |
| Chord of intrados . . . . . . | = 45·00 ,, |
| Rise ,, ,, . . . . . . . | = 5·50 ,, |
| Radius of ,, . . . . . . . | = 48·773 ,, |
| Angle subtended at centre by intrados . | = 54° 56′ 42″ |
| Depth of arch at crown . . . . . | = 1·50 m. |
| ,, ,, springings . . . | = 2·00 ,, |
| Chord of extrados . . . . . . | = 46·846 ,, |
| Rise ,, ,, . . . . . . . | = 5·226 ,, |
| Radius of ,, . . . . . . . | = 55·102 ,, |
| Angle subtended at centre by extrados . | = 50° 18′ 42″ |
| Distance between cross-girders . . | = 1·760 m. |
| Breadth of bridge . . . . . . | = 4·50 ,, |

**2. Loads.**—The dead load per metre run of the bridge is made up as follows:—

| | |
|---|---|
| Rails and fittings (39 × 2) . . . . | 78 kg. |
| Sleepers (40 × 2) . . . . . . | 80 ,, |
| Timber for foot-ways (60 × 2) . . | 120 ,, |
| Metal plates between rails . . . | 30 ,, |
| Hand-rail . . . . . . . | 32 ,, |
| Cross-girders and stringers . . . | 500 ,, |
| Arch segments, posts, and beams . . | 1000 ,, |
| Total dead load . . . . | 1840 ,, |
| For superload we will take the same as for a girder beam of the same span, viz. . . . . . . . . | 5200 ,, |
| Thus the total load per metre run will be | 7040 ,, |

As there are two arch-ribs, it is clear that for each we shall have

| | |
|---|---|
| Dead load per metre run . . . . | 920 kg. |
| Superload ,, ,, ,, . . . . | 2600 ,, |
| Total . . . | 3520 ,, |

**3. Notation and general formulæ in the case in which the superload is taken as on the left-hand half of the bridge only, the dead load being left out of account.**

As is seen in Fig. 1, Pl. XIII, the centre line of the left-hand half of the arch has been divided into six equal parts, the length of each being 3·964 m. At the points of division we have drawn normals to the centre line, i.e. radii; and from the intersections of these normals with the extrados we have drawn verticals. We assume that the dead load and superload are uniformly distributed over the horizontal projection of the extrados, i.e. that for any part or element the load acts through the centre of the horizontal projection of that element of the extrados.

The iron arch rib has at its extremities flat ends normal to the intrados which abut against cast-iron bearings fixed to the springings.

In the calculations we must regard the arch as fixed at the ends, because owing to their being flat they cannot turn, but if tension results near the extrados in the portions regarded as fixed, it will follow that this assumption does not obtain.

We will take as the unknowns of the problem :—

B the bending moment at the crown,
H the horizontal thrust ,, ,,
S the shear at the crown.

We can then readily express in terms of these quantities the bending moment, normal pressure, and shear at any section to the right or left of the crown.

If we consider the arch as cut through at the crown, we shall not alter the equilibrium of the two halves provided that we apply at the crown of each part forces H and S at the centre of the section and a couple equal to the bending moment B. Thus for the section 4, for example, of the left-hand half-arch, all the forces to the right of it will be: the two forces H and S at the centre of the crown, the bending moment B, and the load of 2600 kg. per metre run on the horizontal projection of the extrados of the two elements 4, 5 and 5, 6. Therefore to express the bending moment $B_4$, pressure $P_4$, and shear $S_4$ on the section under consideration, we shall have the following formulæ :—

$$B_4 = B - \cdot 60H - 7{\cdot}86S + 2600 \times 7{\cdot}99 \times \left(7{\cdot}86 - \frac{7{\cdot}99}{2}\right)^{*}$$
$$P_4 = \cdot 990H - \cdot 160S + 2600 \times 7{\cdot}99 \times \cdot 160$$
$$S_4 = \cdot 160H + \cdot 990S - 2600 \times 7{\cdot}99 \times \cdot 990$$

For section 4 of the right-hand half-arch, the bending moment, normal pressure and shear will be given by omitting from the above formulæ the numerical terms due to the load of 2600 kg. per metre run, and by changing the signs of the terms in S. We thus have

$$B_4' = B - \cdot 60H + 7{\cdot}86S$$
$$P_4' = \cdot 990H + \cdot 160S$$
$$S_4' = \cdot 160H - \cdot 990S$$

We thus compile the table given on the opposite page.

**4. Determination of the Unknowns.**—The two half-arches, considered separately, are strained under the action of the forces applied to them; but the strains must be such that the crown section of one always remains in contact with the crown section of the other. We shall thus have the following three geometrical conditions :—

1 and 2. The horizontal and vertical displacements of the centre of the crown for the left-hand half-arch must be equal and opposite to the corresponding displacements for the right-hand half-arch.

3. The angular deflection of the crown section for the left-hand half-arch must be equal and opposite to that for the right-hand half-arch.

We find the values of the unknowns B, H, and S which satisfy these three conditions by expressing the internal work of the whole arch in terms of these unknowns, and equating to zero its differential coefficients with regard to each.

We will call

E the elastic modulus of the iron,

$A_0, A_1, \ldots A_6$ the areas of the sections 0, 1, . . . 6 of the arch,

$I_0, I_1, \ldots I_6$ their moments of inertia.

* It will be noticed that Fig. 1 of Pl. XIII gives the lever arms of the forces which enter into the expression for the bending moments, while Fig. 3 gives the sines and cosines of the angles which the loads and the horizontal thrust make with the sections. These sines and cosines are used in the expressions for the normal thrusts and shears.

| Section. | Bending Moment. | Normal Pressure. | Shear. |
|---|---|---|---|
| | | **1. *Left-hand Half-arch.*** | |
| 0 | $B_0 = B - 5·36H - 22·96S + 685,035$ | $P_0 = ·894H - ·460S + 28,010$ | $S_0 = ·460H + ·894S - 54,437$ |
| 1 | $B_1 = B - 3·72H - 19·33S + 485,831$ | $P_1 = 926H - ·388S + 19,863$ | $S_1 = ·388H + ·920S - 47,406$ |
| 2 | $B_2 = B - 2·38H - 15·59S + 315,868$ | $P_2 = ·954H - ·315S + 12,989$ | $S_2 = ·315H + ·954S - 39,339$ |
| 3 | $B_3 = B - 1·35H - 11·76S + 179,735$ | $P_3 = ·976H - ·237S + 7,370$ | $S_3 = ·237H + ·976S - 30,350$ |
| 4 | $B_4 = B - ·60H - 7·86S + 80,395$ | $P_4 = ·990H - ·160S + 3,324$ | $S_4 = ·160H + ·990S - 20,566$ |
| 5 | $B_5 = B - ·15H - 3·94S + 20,176$ | $P_5 = ·999H - ·082S + 853$ | $S_5 = ·082H + ·999S - 10,390$ |
| 6 | $B_6 = B$ | $P_6 = 1·000H$ | $S_6 = 1·000S$ |
| | | **2. *Right-hand Half-arch.*** | |
| 0 | $B_0' = B - 5·36H + 22·96S$ | $P_0' = ·894H + ·460S$ | $S_0' = ·460H - ·894S$ |
| 1 | $B_1' = B - 3·72H + 19·33S$ | $P_1' = ·926H + ·388S$ | $S_1' = ·388H - ·926S$ |
| 2 | $B_2' = B - 2·38H + 15·59S$ | $P_2' = ·954H + ·315S$ | $S_2' = ·315H - ·954S$ |
| 3 | $B_3' = B - 1·35H + 11·76S$ | $P_3' = ·976H + ·237S$ | $S_3' = ·237H - ·976S$ |
| 4 | $B_4' = B - ·60H + 7·86S$ | $P_4' = ·990H + ·160S$ | $S_4' = ·160H - ·990S$ |
| 5 | $B_5' = B - ·15H + 3·94S$ | $P_5' = ·999H + ·082S$ | $S_5' = ·082H - ·999S$ |
| 6 | $B_6' = B$ | $P_6' = 1·000H$ | $S_6' = -1·000S$ |

We will take

$$\Sigma \frac{B^2 + B'^2}{I} = \frac{1}{3}\left[\frac{B_0^2 + B_0'^2}{I_0} + \frac{4(B_1^2 + B_1'^2)}{I_1} + \frac{2(B_2^2 + B_2'^2)}{I_2} + \ldots + \frac{B_6^2 + B_6'^2}{I_6}\right]$$

$$\Sigma \frac{P^2 + P'^2}{A} = \frac{1}{3}\left[\frac{P_0^2 + P_0'^2}{A_0} + \frac{4(P_1^2 + P_1'^2)}{A_1} + \frac{2(P_2^2 + P_2'^2)}{A_2} + \ldots + \frac{P_6^2 + P_6'^2}{A_6}\right]$$

Remembering that the length of each element along the centre line is 3·964 m., and neglecting the internal work due to shear, we shall have the following equation for expressing the internal work of the whole arch :—

$$\text{Internal work} = \frac{3·964}{2E}\left[\Sigma\frac{B^2 + B'^2}{I} + \Sigma\frac{P^2 + P'^2}{A}\right].$$

Dividing by the term $\frac{3·964}{2E}$, which will make no difference to the formulæ found by equating to zero the differential coefficients with regard to B, H, S, we get

$$\Sigma\frac{B^2 + B'^2}{I} + \Sigma\frac{P^2 + P'^2}{A} \quad . \quad . \quad . \quad (1)$$

The depth of the sections of the arch decreases from the springings, but except for the web, the depths of which at the sections 0, 1, 2, . . . 6 are written in Fig. 1, Pl. XIII, the other members composing the arch are of the same section throughout. The general cross-section of the arch is shown in Fig. 4.

We reproduce in the following table the areas and moments of inertia of the sections 0, 1, 2, . . . 6, together with the logarithms of their reciprocals, which are employed for the calculation of formula (1) :—

| Section. | Area (A). | Moment of Inertia (I). | Log $\frac{1}{A}$. | Log $\frac{1}{I}$. |
|---|---|---|---|---|
| 0 | $A_0$ = ·039024 | $I_0$ = ·022582 | 1·4086682 | 1·6462376 |
| 1 | $A_1$ = ·037224 | $I_1$ = ·018779 | 1·4291770 | 1·7263275 |
| 2 | $A_2$ = ·035666 | $I_2$ = ·015822 | 1·4477456 | 1·8007386 |
| 3 | $A_3$ = ·034464 | $I_3$ = ·013754 | 1·4626343 | 1·8615710 |
| 4 | $A_4$ = ·033624 | $I_4$ = ·012409 | 1·4733506 | 1·9062632 |
| 5 | $A_5$ = ·033144 | $I_5$ = ·011675 | 1·4795951 | 1·9327431 |
| 6 | $A_6$ = ·033024 | $I_6$ = ·011498 | 1·4811703 | 1·9393777 |

If now we substitute in formula (1) for $B_0$, $B_0'$, $B_1$, $B_1'$, etc., $P_0$, $P_0'$, $P_1$, $P_1'$, etc., their values in terms of B, H, S tabulated in § 3, and for $I_0$, $I_1$, . . ., $A_0$, $A_1$, . . . their numerical values given above, and make the numerical calculations, which can be done simply and systematically by the aid of the logarithms, we get—

$$\Sigma\frac{B^2 + B'^2}{I} = 843·51B^2 - 1247·51 \times 2BH + 81961733 \times 2B$$
$$+ 3686·7H^2 - 240653093 \times 2H + 126101S^2 \quad 1354307066 \times 2S\,;$$
$$\Sigma\frac{P^2 + P'^2}{A} = 322·5H^2 + 1494668 \times 2H + 24S^2 - 555444 \times 2S\,;$$

and formula (1) becomes

$$843{\cdot}51B^2 - 1247{\cdot}51 \times 2BH + 81961733 \times 2B + 4009{\cdot}2H^2 - 239158425 \times 2H + 126125S^2 - 1354922510 \times 2S \quad (2)$$

Equating to zero the differential coefficients of this expression with regard to B, H, and S, we get the following three equations:—

$$\begin{aligned} 843{\cdot}51B - 1247{\cdot}51H + 81961733 &= 0, \\ -1247{\cdot}51B + 4009{\cdot}2H - 239158425 &= 0, \\ 126125S - 1354922510 &= 0. \end{aligned}$$

In these equations it should be noted that the first two contain only B and H and the third contains only S, and from them we deduce

$$\begin{aligned} B &= -16{,}527 \\ H &= 54{,}523 \\ S &= 10{,}743. \end{aligned}$$

**5. Line of pressure; maximum stresses per sq. metre and deflection at the centre, account being taken of the dead load and superload on both sides of the crown.**

In this case, the loading being symmetrical, we have $S = 0$; moreover, if we leave out of account the dead load and consider only the superload over the whole bridge we shall have at the crown

$$\begin{aligned} B &= -16527 \times 2 = -33{,}054 \\ H &= 54523 \times 2 = 109{,}046. \end{aligned}$$

Now the superload is 2600 kg. per m. run and the total load is 3520 kg. per m. run, so that we have, for full load

$$B = -33054 \times \frac{3520}{2600} = -44{,}749$$

$$H = 109046 \times \frac{3520}{2600} = 147{,}626.$$

Thus, knowing the bending moment and thrust at the crown for the total load, we can easily calculate the bending moment, normal pressure, and shear for each of the sections 0, 1, 2 . . . 6. Dividing the bending moments by the corresponding normal pressures, we obtain the eccentricity of loading. Finally, by means of the formula

$$c = \frac{P}{A} \pm \frac{Bn}{I},$$

in which P and B represent the normal pressure and bending moment for any section, and A, I, $n$ the area, moment of inertia, and half-depth of that section, we find the maximum stresses per sq. metre at the intrados and extrados of the sections 0, 1, 2, . . . 6.

All these results are collected in the following table :—

| Section. | Bending Moment. | Normal Pressure. | Shear. | Eccentricity of Load. | Stresses per sq. m. at | |
|---|---|---|---|---|---|---|
| | | | | | Intrados. | Extrados. |
| 0 | + 91,387 | 169,924 | − 5788 | + ·538 | 8,401,300 | 307,500 |
| 1 | + 63,807 | 163,657 | − 6901 | + ·390 | 7,539,600 | 1,253,600 |
| 2 | + 31,527 | 158,444 | − 6755 | + ·199 | 6,155,900 | 2,728,700 |
| 3 | − 719 | 154,082 | − 6101 | − ·005 | 4,428,467 | 4,513,133 |
| 4 | − 24,626 | 150,673 | − 4223 | − ·163 | 2,943,300 | 6,018,900 |
| 5 | − 39,584 | 148,656 | − 1961 | − ·266 | 1,925,300 | 7,044,900 |
| 6 | − 44,755 | 147,648 | 0 | − ·303 | 1,551,700 | 7,390,300 |

If we wish to calculate the deflection of the arch at the crown, we employ the formula *

$$d_n = \frac{3{\cdot}964}{\mathrm{E}} \cdot \frac{1}{3}\left\{\left(\frac{\mathrm{B}_0 x_0}{\mathrm{I}_0} + 4\frac{\mathrm{B}_1 x_1}{\mathrm{I}_1} + 2\frac{\mathrm{B}_2 x_2}{\mathrm{I}_2} + \ldots + 4\frac{\mathrm{B}_5 x_5}{\mathrm{I}_5}\right) + \left(\frac{\mathrm{P}_0 \sin\phi_0}{\mathrm{A}_0} + \frac{4\mathrm{P}_1 \sin\phi_1}{\mathrm{A}_1} + \frac{2\mathrm{P}_2 \sin\phi_2}{\mathrm{A}_2} + \ldots + \frac{4\mathrm{P}_5 \sin\phi_5}{\mathrm{A}_5}\right)\right\},$$

in which $x_0, x_1, x_2, \ldots x_5$ are the distances from the centres of the sections 0, 1, 2, . . . 5 to the vertical plane through the crown, and $\phi_0, \phi_1, \phi_2, \ldots \phi_5$ the angles which these sections make with that plane.

Substituting in this formula for the bending moments $\mathrm{B}_0, \mathrm{B}_1, \mathrm{B}_2, \ldots$ and the normal pressures $\mathrm{P}_0, \mathrm{P}_1, \mathrm{P}_2, \ldots$, their values given in the foregoing table, for $x_0, x_1, x_2, \ldots, \sin\phi_0, \sin\phi_1, \sin\phi_2, \ldots$, their values resulting from the figures, and finally for $\mathrm{A}_0, \mathrm{A}_1, \mathrm{A}_2, \ldots \mathrm{I}_0, \mathrm{I}_1, \mathrm{I}_2, \ldots$, their values given in § 4, we get

$$\frac{1}{3}\left(\frac{\mathrm{B}_0 x_0}{\mathrm{I}_0} + 4\frac{\mathrm{B}_1 x_1}{\mathrm{I}_1} + 2\frac{\mathrm{B}_2 x_2}{\mathrm{I}_2} + \ldots + 4\frac{\mathrm{B}_5 x_5}{\mathrm{I}_5}\right) = 110{,}212{,}000$$

$$\frac{1}{3}\left(\frac{\mathrm{P}_0 \sin\phi_0}{\mathrm{A}_0} + \frac{4\mathrm{P}_1 \sin\phi_1}{\mathrm{A}_1} + \frac{2\mathrm{P}_2 \sin\phi_2}{\mathrm{A}_2} + \ldots + \frac{4\mathrm{P}_5 \sin\phi_5}{\mathrm{A}_5}\right) = 6{,}220{,}000,$$

so that, taking $\mathrm{E} = 18 \times 10^9$, we have

$$d_n = {\cdot}026 \text{ m.}$$

* This formula is a general form of that adopted in Chapter XV, § 5, and is deduced in the same manner.

**6. Line of pressure and maximum stresses per sq. metre, taking into account the dead load, and supposing the superload to be on the *left*-hand half of the bridge only.**

In this case, the values of B, H, and S are obtained by adding to the values which these quantities have for the dead load only, the values which they have for the superload on the left-hand half of the bridge.

Now for the dead load alone we shall have

$$B = -\ 33054 \times \frac{920}{2600} = -\ 11{,}696;$$

$$H = \ 109046 \times \frac{920}{2600} = \ 38{,}585;$$

$$S = \ 0,$$

and for the superload on the left-hand half only, we shall have the values of B, H, and S obtained in § 4.

In the present case we shall thus have

$$B = -\ 11{,}696 - 16{,}527 = -\ 28{,}223;$$

$$H = \ 38{,}585 + 54{,}523 = \ 93{,}108;$$

$$S = \ 10{,}743.$$

Now as the bending moment, normal pressure, and shear at the crown are known, we can calculate the three corresponding quantities for the sections 0, 1, 2, . . . ; and dividing the bending moments by the corresponding normal pressures, we obtain the distances from the line of pressure to the centre line of the arch (i.e. the eccentricities of the load). We thus compile the following tables :—

| Left-hand Half-arch. | | | | | Right-hand Half-arch. | | | | |
|---|---|---|---|---|---|---|---|---|---|
| Section. | Bending Moment. | Normal Pressure. | Shear. | Eccentricity of Load. | Section. | Bending Moment. | Normal Pressure. | Shear. | Eccentricity of Load. |
| 0 | +153,498 | 116,223 | – 21,257 | +1·321 | 0 | – 38,218 | 98,112 | +13,964 | – ·389 |
| 1 | + 75,498 | 109,001 | – 18,172 | + ·693 | 1 | + 4,991 | 97,444 | + 9,467 | + ·051 |
| 2 | + 10,336 | 103,045 | – 13,681 | + ·100 | 2 | +29,434 | 96,824 | + 5,160 | + ·304 |
| 3 | – 36,924 | 98,322 | 8,530 | – ·375 | 3 | +36,017 | 96,044 | + 842 | + ·375 |
| 4 | – 59,725 | 94,976 | – 2,311 | – ·629 | 4 | +28,660 | 95,090 | – 3,016 | + ·301 |
| 5 | – 57,206 | 93,307 | + 4,300 | – ·612 | 5 | + 7,273 | 94,215 | – 6,774 | + ·077 |
| 6 | – 28,228 | 93,125 | +10,743 | – ·303 | 6 | – 28,228 | 93,125 | – 10,743 | – ·303 |

If we compare these tables with that which was given in the previous paragraph, we see that the maximum bending moments are much less in the case in which the superload is over the whole span than when it is on one half only. The maximum stress per sq. metre in the iron in the latter case is

$$\frac{116223}{\cdot 039024} + \frac{153498 \times 1\cdot 00}{\cdot 022582} = 9{,}775{,}000,$$

a value very much in excess of that for the case of the superload over the whole span.

The maximum shear stress occurs at the left-hand section 0, where the shearing force is a maximum. For this section we have

$$M_1 = \frac{1}{8}(\cdot 250 \times 2\cdot 024^2 - \cdot 038 \times 2^2 - \cdot 176 \times 1\cdot 976^2 - \cdot 024 \times 1\cdot 8^2) = \cdot 013397,$$

$$I_0 = \cdot 022582;$$

the maximum shear stress, per sq. metre, will be

$$\frac{21257 \times \cdot 013397}{\cdot 022582 \times \cdot 012} = 1{,}050{,}900.$$

**7. Temperature Effects.**—Suppose that the arch was erected at a temperature $t°$, at which it had the exact dimensions to rest upon the springings without inducing any other stresses than those due to its own weight and to the loads carried.

If the arch were free, it is clear that at a temperature $t_1°$ it would have a form geometrically similar to its initial form, so that the angle of the extreme sections would be exactly the same in the two cases; but the distance between the centres of these sections, instead of being $\frac{45\cdot 000 + 46\cdot 846}{2} = 45\cdot 923$ m. (i.e. equal to the distance between the centres of the springings) would have increased by an amount $\frac{45\cdot 923}{82500}(t_1° - t°)$, taking $\frac{1}{82500}$ as the coefficient of expansion of the iron.

If, therefore, the arch had to be erected at the temperature $t_1°$, the work could not be carried out without inducing stresses due only to the increase of temperature $(t_1° - t°)$ and independent of the weight of the arch and its load.

Now the internal work of the arch, due to the bending moment B, and forces H, S applied at the crown, is given by formula (2) of § 4, after having multiplied by $\frac{3\cdot 964}{2E}$ and having omitted the terms of the first degree in B, H, and S, which depend upon the loads; this internal work is then expressed by the formula

$$\frac{3\cdot 964}{2E}(843\cdot 51B^2 - 1247\cdot 51 \times 2BH + 4009\cdot 2H^2 + 126125S^2).$$

To find the values of B, H, S due only to the change $(t_1° - t°)$ in temperature, we must add to this expression for the internal work the term $-\frac{45·923}{82500}(t_1° - t°)H$, and then equate to zero the differential coefficients with regard to B, H, and S. Multiplying the equations thus obtained by the numerical factor $\frac{E}{3·964}$, and observing that if we take $E = 18 \times 10^9$, we have

$$\frac{E}{3·964} \times \frac{45·923}{82500} = 2,534,000,$$

the three aforesaid equations become

$$843·51B - 1247·51H = 0,$$
$$-1247·51B + 4009·2H = 2534000(t_1° - t°),$$
$$126125S = 0.$$

The last equation gives us at once $S = 0$; this is what we should expect, because variations in temperature have the same effect in the two halves of the arch and so cannot cause any shearing force in the section at the crown.

The other two equations give us

$$B = 1743(t_1° - t°),$$
$$H = 1171(t_1° - t°).$$

Having obtained the values of B, H, S, we substitute them in the second half of the table of § 3, which relates to the half-arch without load, and calculate the bending moment, normal pressure, and shear for each of the sections 0, 1, 2, . . . 6. We thus compile the following table:—

| Section. | Bending Moment. | Normal Pressure. | Shear. |
|---|---|---|---|
| 0 | $B_0' = -4533(t_1° - t°)$ | $P_0' = 1047(t_1° - t°)$ | $S_0' = 539(t_1° - t°)$ |
| 1 | $B_1' = -2613(t_1° - t°)$ | $P_1' = 1084(t_1° - t°)$ | $S_1' = 454(t_1° - t°)$ |
| 2 | $B_2' = -1044(t_1° - t°)$ | $P_2' = 1117(t_1° - t°)$ | $S_2' = 369(t_1° - t°)$ |
| 3 | $B_3' = +163(t_1° - t°)$ | $P_3' = 1143(t_1° - t°)$ | $S_3' = 277(t_1° - t°)$ |
| 4 | $B_4' = +1040(t_1° - t°)$ | $P_4' = 1159(t_1° - t°)$ | $S_4' = 187(t_1° - t°)$ |
| 5 | $B_5' = +1567(t_1° - t°)$ | $P_5' = 1170(t_1° - t°)$ | $S_5' = 96(t_1° - t°)$ |
| 6 | $B_6' = +1743(t_1° - t°)$ | $P_6' = H' = 1171(t_1° - t°)$ | $S_6' = 0$ |

If the mean temperature at erection of the arch is 15° C., and if the range of temperature is from 40° C. to − 10° C., it is clear that to consider the two extreme cases we must put

For the temperature of 40° C., $t_1° - t° = 40 - 15 = 25°$ C.

" " − 10° C., $t_1° - t° = -10 - 15 = -25°$ C.

We will examine only the latter case, because then the bending moments are positive near to the springings and negative near to the crown, such as we found for each half of the arch in the case of the superload over the whole span, and for the left-hand half of the arch, when that half alone carries the superload.

Thus, putting in the last table $t_1^\circ - t^\circ = -25$, and adding the results to those which we found in § 6 for the left-hand half of the arch, we obtain the following table:—

*Left-hand Half of the Arch.*

| Section. | Bending Moment. | Normal Pressure. | Shear. | Eccentricity of Load. |
|---|---|---|---|---|
| 0 | + 267,923 | 90,048 | − 34,732 | + 2·975 |
| 1 | + 140,823 | 81,901 | − 29,522 | + 1·719 |
| 2 | + 36,436 | 75,120 | − 22,906 | + ·484 |
| 3 | − 40,999 | 69,747 | − 15,463 | − ·587 |
| 4 | − 85,725 | 66,001 | − 6,986 | − 1·298 |
| 5 | − 96,331 | 64,057 | + 1,900 | − 1·503 |
| 6 | − 72,803 | 63,850 | + 10,743 | − 1·124 |

We see from these results what a large influence temperature variation has in iron arches with flat ends. The bending moment at the springings, which is 153,498 kg. m. at the mean temperature, increases to 267,923 kg. m. for the same load when the temperature falls to − 10° C. The maximum compressive stress per sq. metre occurs at the intrados of the springing section, and its value is

$$c = \frac{90048}{\cdot 039024} + \frac{266823 \times 1\cdot 00}{\cdot 022582} = 14{,}122{,}000\,;$$

from which it follows that the arch should be strengthened near to the springings.

The stress at the extrados at the same section is

$$\frac{90{,}048}{\cdot 039024} - \frac{266{,}823 \times 1\cdot 00}{\cdot 022582} = -\ 9{,}508{,}000\,;$$

the negative value indicates that the stress is a tension.

**8. Conclusion.**—This last result shows that the arch tends to open up at the springings near to the extrados, and that all the thrust tends to be carrried on the bearing towards the intrados.

The arch cannot, therefore, behave as though fixed at the ends, and it would not be easy to secure fixity, having regard to the high value of the bending moment at the springings.

It is important to point out that arches with flat ends and with small rise compared with the span, which, according to the intention of the designers, should behave as fixed at the ends, do not generally fulfil this condition, but lift from the bearings near to the extrados, and the whole pressure has to be carried near to the intrados, where it may happen that the masonry will be cracked in consequence.

This explains why British engineers, who more than others have employed this form of arch, have had finally to abandon it on account of the rapid deterioration of the masonry near to the springings.

It is for this reason that at the present time iron arches are generally constructed with rounded ends.

# CHAPTER XXII.

## STUDY OF AN IRON ARCH BRIDGE WITH ROUNDED ENDS.

(Plates XII and XIII.)

**1. General Data.**—These are exactly the same as for the bridge previously considered.

**2. Loads.**—The loads also will be taken as the same.

**3. Principles and General Formulæ for the Case in which the Whole Span is Loaded.**—As will be seen from Fig. 2, Pl. XIII, the centre line of the half-arch has been divided into six equal voussoirs, the length of each being 3·964 m. From the points of division we have drawn normals to the centre line and erected verticals from the points in which these cut the extrados. We assume that each element has to carry, in addition to its own weight, the weight of the portion of the superstructure and of the superload comprised between the two verticals erected from the ends of its extrados; we will also take these loads to be uniformly distributed along the horizontal projection of the extrados, so that their centre of gravity is midway between these two verticals.

The arch has cylindrical surfaces at its ends, which bear upon cylindrical plates having a smaller curvature.

If the ends of the arch and the bearing plates were not compressible, the contact at each end would only be along the generating line of the cylinder which is intersected by the centre line of the arch; but on account of the compressibility of the material, contact will be over a small surface (sufficient nevertheless to prevent cracking of the end of the arch or of the bearing plate); the centre of compression may still be taken, however, with a very close approximation, as the point in which the centre line of the arch intersects the bearing plate.

The advantage of cylindrical bearings lies in the fact that owing to the thrust being applied at the centre of the bearing, it is transmitted uniformly over the masonry whatever the distribution of the load on the arch. This advantago is so great that for iron arches cylindrical bearings should always be adopted in preference to plane ones.

The bending moment at the ends of the arch is zero, and therefore for all distributions of load the horizontal thrust is the only unknown.

To determine this quantity we have the following geometrical condition to be satisfied :—

*In the deformation, the distance between the ends of the centre line (i.e. the span) must remain constant.*

The load on each of the elements 0, 1 ; 1, 2, etc., is obtained by multiplying the horizontal projection of its extrados by the load of 3520 kg. previously obtained; and the vertical reaction at each end, which is half the total load on the arch, has a value

$$R = 3520 \times 23{\cdot}423 = 82{,}449 \text{ kg.}$$

It is now easy to express in terms of the horizontal thrust H and of the known loads, the bending moment, normal pressure, and shear for each of the sections 0, 1, 2, . . . 6. For example, for section 2 we have

$$B_2 = 2{\cdot}98H \quad - 737R + 3520 \times 7{\cdot}56\left(7{\cdot}83 - \frac{7{\cdot}56}{2}\right),$$
$$P_2 = \quad {\cdot}954H + {\cdot}315(R - 3520 \times 7{\cdot}56),$$
$$S_2 = \quad {\cdot}315H - {\cdot}954(R - 3520 \times 7{\cdot}56).$$

Substituting for R its numerical value and simplifying, we compile the following tablo in which H is the only unknown :—

| Section. | Bending Moment. | Normal Pressure. | Shear. |
|---|---|---|---|
| 0 | $B_0 = 0$ | $P_0 = {\cdot}894H + 37{,}922$ | $S_0 = {\cdot}460H - 73{,}700$ |
| 1 | $B_1 = 1{\cdot}64H - 269{,}972$ | $P_1 = {\cdot}926H + 26{,}892$ | $S_1 = {\cdot}388H - 64{,}180$ |
| 2 | $B_2 = 2{\cdot}98H - 499{,}528$ | $P_2 = {\cdot}954H + 17{,}586$ | $S_2 = {\cdot}315H - 53{,}259$ |
| 3 | $B_3 = 4{\cdot}01H - 683{,}698$ | $P_3 = {\cdot}976H + 9{,}978$ | $S_3 = {\cdot}237H - 41{,}089$ |
| 4 | $B_4 = 4{\cdot}76H - 818{,}394$ | $P_4 = {\cdot}990H + 4{,}500$ | $S_4 = {\cdot}160H - 27{,}844$ |
| 5 | $B_5 = 5{\cdot}21H - 899{,}852$ | $P_5 = {\cdot}999H + 1{,}155$ | $S_5 = {\cdot}082H - 14{,}066$ |
| 6 | $B_6 = 5{\cdot}36H - 927{,}167$ | $P_6 = 1{\cdot}000$ | $S_6 = 0$ |

4. **Determination of the Value of the Unknown.**—We will call

E the elastic modulus of the iron,
$A_0$, $A_1$, etc., the areas of the arch sections,
$I_0$, $I_1$, etc., the moments of inertia of these sections,

and take

$$\Sigma\frac{B^2}{I} = \frac{1}{3}\left(\frac{B_0^2}{I_0} + \frac{4B_1^2}{I_1} + \frac{2B_2^2}{I_2} + \ldots \frac{B_6^2}{I_6}\right),$$
$$\Sigma\frac{P^2}{A} = \frac{1}{3}\left(\frac{P_0^2}{A_0} + \frac{4P_1^2}{A_1} + \frac{2P_2^2}{A_2} + \ldots \frac{P_6^2}{A_6}\right).$$

If we remember that the length of each of the elements measured along the centre line is 3·964 m., we have the following formula for the internal work of the whole arch (neglecting as very small that portion due to shear), viz.:—

$$2 \times \frac{3\cdot964}{2E}\left(\Sigma\frac{B^2}{I} + \Sigma\frac{P^2}{A}\right),$$

or the following, neglecting the numerical factor, which will have no effect upon the equation obtained by equating to zero the differential coefficient with regard to H, viz.:—

$$\Sigma\frac{B^2}{I} + \Sigma\frac{P^2}{A} \quad . \quad . \quad . \quad . \quad (1)$$

The section of the arch is shown in Fig. 4, Pl. XIII, the depth being omitted because it varies from one section to another, as shown in Fig. 2.

The areas and moments of inertia of the sections are given in the following table, together with the logarithms of their reciprocals:—

| Section. | Area, A. | Moment of Inertia, I. | $\text{Log}\frac{1}{A}$. | $\text{Log}\frac{1}{I}$. |
|---|---|---|---|---|
| 0 | ·039024 | ·022582 | 1·4086682 | 1·6462376 |
| 1 | ·037224 | ·018779 | 1·4291770 | 1·7263275 |
| 2 | ·035666 | ·015822 | 1·4477456 | 1·8007386 |
| 3 | ·034464 | ·013754 | 1·4626343 | 1·8615710 |
| 4 | ·033624 | ·012409 | 1·4733506 | 1·9062632 |
| 5 | ·033144 | ·011675 | 1·4795951 | 1·9327431 |
| 6 | ·033024 | ·011498 | 1·4811703 | 1·9393777 |

Squaring the expressions given in the previous paragraph for B and P and then dividing by the moments of inertia and areas, we obtain the following results:—

$$\Sigma\frac{B^2}{I} = 7275H^2 - 1248781000 \times 2H,$$

$$\Sigma\frac{P^2}{A} = 161H^2 + 2024000 \times 2H,$$

so that formula (1) becomes

$$7436H^2 - 1246757000 \times 2H \quad . \quad . \quad . \quad (2)$$

To express the geometrical condition of § 3 we must equate to zero the differential coefficient of this expression with regard to H, thus getting

$$7436H - 1246757000 = 0$$

or

$$H = 167670.$$

**5. Line of pressure, maximum stresses per sq. metre in the iron, and deflection of the arch at the crown.**

If in the table given in § 3 we substitute for B and H their values found above, we obtain the values of the bending moment, normal pressure, and shear in the sections 0, 1, 2, . . . 6 of the arch. These values are given in the 2nd, 3rd, and 4th columns of the following table. Dividing the bending moments by the corresponding normal pressures, we obtain the eccentricity of the load.*

By means of the formula

$$c = \frac{P}{A} \pm \frac{Bn}{I},$$

in which B and P represent the bending moment and normal pressure at a section, and A, I, $n$ the area, moment of inertia, and neutral axis depth of the same section, we obtain the stresses per sq. metre at intrados and extrados for the various sections.

| Section. | Bending Moment. | Normal Pressure. | Shear. | Eccentricity of Load. | Stress per Sq. Metre | |
|---|---|---|---|---|---|---|
| | | | | | at Intrados. | at Extrados. |
| 0 | 0 | 187,819 | + 3428 | 0 | 4,800,100 | 4,800,100 |
| 1 | + 5,007 | 182,154 | + 876 | + ·027 | 5,140,130 | 4,646,870 |
| 2 | + 129 | 177,543 | − 443 | + ·001 | 4,984,320 | 4,971,277 |
| 3 | − 11,336 | 173,624 | − 1351 | − ·065 | 4,370,110 | 6,705,290 |
| 4 | − 20,285 | 170,493 | − 1017 | − ·119 | 3,795,500 | 6,345,500 |
| 5 | − 26,291 | 168,657 | − 317 | − ·156 | 3,389,500 | 6,788,800 |
| 6 | − 27,953 | 167,670 | 0 | − ·167 | 3,247,900 | 6,902,900 |

With regard to the shear stress per sq. metre, we know that in solid web girders it is usually very small. If we wish to determine its maximum value at section 0, where the shearing force is a maximum, we calculate for this section the quantity

$$M_{1,0} = \frac{1}{8}(\cdot 250 \times 2\cdot 024^2 - \cdot 038 \times 2^2 - \cdot 176 \times 1\cdot 976^2 - \cdot 024 \times 1\cdot 8^2)$$
$$= \cdot 013397.$$

Then, noting that $I_0 = \cdot 022582$ and that the thickness of the solid web is $b_0 = \cdot 012$, we have

* In Fig. 2, Pl. XIII, the line of pressure for the case of the whole arch loaded has been represented by a continuous curve. The broken curved line represents the line of pressure for the case considered in § 6, in which the superload is on the left-hand half of the arch only.

$$\text{maximum shear stress} = \frac{S_0 M_{1,0}}{b_0 I_0} = \frac{3428 \times \cdot 013397}{\cdot 012 \times \cdot 022582} = 169{,}100,$$

a result which is really very small.

To find the deflection of the arch at the crown we employ the following formula :—*

$$d_n = -\frac{3\cdot 964}{3E}\left\{\left(\frac{4B_1x_1}{I_1} + \frac{2B_2x_2}{I_2} + \ldots + \frac{B_6x_6}{I_6}\right) - \left(\frac{P_0 \sin\phi_0}{A_0} + \frac{4P_1 \sin\phi_1}{A_1} + \ldots + \frac{4P_5 \sin\phi_5}{A_5}\right)\right\}$$

in which $x_1$, $x_2$, . . . $x_6$ are the horizontal distances from the centre of the section 0, i.e. from the point of contact at the springing, to the centres of the sections 1, 2, . . . 6; and $\phi_0$, $\phi_1$, etc., are the angles which these sections make with the vertical.

Substituting in this formula for the bending moments $B_1$, $B_2$ . . . and the normal pressures $P_0$, $P_1$ . . ., their values given in the previous table, for $x_0$, $x_1$ . . ., $\sin\phi_0$, $\sin\phi_1$ . . ., their values obtained from Figs. 2 and 3 of Plate XIII, and finally for $I_0$, $I_1$ . . ., $A_0$, $A_1$ . . ., their values tabulated above, we have

$$d_n = \frac{3\cdot 964}{E}(103148000 + 6996000),$$

and if we then take $E = 18{,}000{,}000{,}000$, we get

$$d_n = \cdot 0244 \text{ m}.$$

* We can easily prove this formula. If in addition to the uniformly distributed load upon the arch we assume a weight $W'$ to be applied to section 6, it is clear that the vertical reaction of the springings will increase by $\frac{W'}{2}$, so that the bending moments upon the sections 0, 1, 2, . . . 6, which were originally 0, $B_1$, . . . $B_6$, will then become 0; $B_1 - \frac{W'x_1}{2}$; $B_2 - \frac{W'x_2}{2}$; . . . $B_6 - \frac{W'x_6}{2}$; and the normal pressures, which were formerly $P_0$, $P_1$, $P_2$, . . . $P_6$, will then become

$$P_0 + \frac{W'\sin\phi_0}{2};\ P_1 + \frac{W'\sin\phi_1}{2};\ P_2 + \frac{W'\sin\phi_2}{2};\ \ldots\ P_6.$$

Then the internal work for the whole arch, allowing for the load $W'$ as well as the uniformly distributed load, will be :—

$$2 \times \frac{3\cdot 964}{2E} \times \frac{1}{3}\left[\left\{\frac{4\left(B_1 - \frac{W'x_1}{2}\right)^2}{I_1} + \frac{2\left(B_2 - \frac{W'x_2}{2}\right)^2}{I_2} + \ldots + \frac{\left(B_6 - \frac{W'x_6}{2}\right)^2}{I_6}\right\} + \left\{\frac{\left(P_0 + \frac{W'\sin\phi_0}{2}\right)^2}{A_0} + \frac{4\left(P_1 + \frac{W'\sin\phi_1}{2}\right)^2}{A_1} + \ldots + \frac{P_6^2}{A_6}\right\}\right].$$

The differential coefficient of this expression with regard to the weight $W'$ gives the deflection of the section 6 in terms of this weight. If the arch is loaded with the distributed load only, we take the differential coefficient with regard to $W'$ and then put $W' = 0$; this will lead to the value given in the text.

**6. Examination of the case in which the superload is taken over the left-hand half of the arch only, the dead load being taken into account.**

In this case, on the left-hand half of the arch we have a load of 3520 kg. per metre run, and therefore a total load of $3520 \times 23{\cdot}423 = 82{,}449$ kg.; whilst on the right-hand half we have a load of 920 kg. per metre run, and therefore a total load of $920 \times 23{\cdot}423 = 21{,}549$ kg.

The vertical reactions at the supports will be as follows:—

At the left-hand support

$$R_L = \frac{\left(22{\cdot}962 + \frac{23{\cdot}423}{2}\right)}{45{\cdot}924} \times 82449 + \frac{\left(22{\cdot}962 - \frac{23{\cdot}423}{2}\right)}{45{\cdot}924} \times 21549 = 67{,}519 \text{ kg.}$$

At the right-hand support

$$R_R = \frac{\left(22{\cdot}962 - \frac{23{\cdot}423}{2}\right)}{45{\cdot}924} \times 82449 + \frac{\left(22{\cdot}962 + \frac{23{\cdot}423}{2}\right)}{45{\cdot}924} \times 21549 = 36{,}467 \text{ kg.}$$

Tho horizontal thrust can be found at once, because it is made up of two parts, of which the one, due to the dead load, will bear to that for full load the ratio of the intensities of the loads in the two cases, i.e. it will be $\frac{920}{3520}$ of the previous value; the other part is due to the superload only on the left-hand half of the arch, and is half that for the superload only over the whole arch. In the case under consideration, therefore, the horizontal thrust will be

$$H = 167670\left(\frac{920}{3520} + \frac{1}{2} \times \frac{2600}{3520}\right) = 105{,}750.$$

Knowing thus the two forces applied at each end of the arch, i.e. the vertical reaction and the horizontal thrust, it is easy to calculate the bending moment, normal pressure, and shear at each of the sections 0, 1, 2. . . . We can thus compile the following tables:—

| Section. | Left-hand Half-arch. | | | | Right-hand Half-arch. | | | |
|---|---|---|---|---|---|---|---|---|
| | Bending Moment. | Normal Pressure. | Shear. | Eccentricity of Load. | Bending Moment. | Normal Pressure. | Shear. | Eccentricity of Load. |
| 0 | 0 | 125,599 | − 11,717 | 0 | 0 | 111,315 | + 16,044 | 0 |
| 1 | − 42,385 | 119,028 | − 9,334 | − ·356 | + 48,707 | 85,106 | + 10,441 | + ·572 |
| 2 | − 74,437 | 113,771 | − 5,715 | − ·654 | + 74,611 | 91,589 | + 5,157 | + ·815 |
| 3 | − 92,538 | 109,654 | − 1,465 | − ·844 | + 78,254 | 97,068 | − 239 | + ·806 |
| 4 | − 89,742 | 106,805 | + 3,847 | − ·840 | + 64,168 | 101,128 | − 5,129 | + ·634 |
| 5 | − 65,127 | 105,575 | + 9,510 | − ·617 | + 31,977 | 104,119 | − 9,911 | + ·307 |
| 6 | − 17,730 | 105,750 | + 14,920 | − ·168 | − 17,730 | 105,750 | − 14,920 | − ·168 |

A comparison of these results with those which we obtained in the previous paragraph shows that the greatest bending moments occur in the case in which the superload is on one half of the arch only.

The maximum stress per sq. metre in the iron, on the left-hand half of the arch, occurs at the extrados of section 3, and its value is

$$c = \frac{109654}{\cdot 034464} + \frac{92538 \times \cdot 822}{\cdot 013754} = 8{,}711{,}000\,;$$

on the right-hand half, the maximum stress occurs at the intrados of section 3, its value being

$$c = \frac{97068}{\cdot 034464} + \frac{78254 \times \cdot 822}{\cdot 013754} = 7{,}492{,}000.$$

In the case of the superload on the whole arch, the maximum stress is 6,705,290 kg. per sq. metre, i.e. it is considerably less than that for the superload on one side only.

As to the shear stress per sq. metre, it is easily seen that it has a maximum value at the crown, where the shearing force has nearly its greatest value, and where the section is the smallest. We have for this section

$$M_{1,6} = \frac{1}{8}(\cdot 25 \times 1\cdot 524^2 - \cdot 038 \times 1\cdot 5^2 - \cdot 176 \times 1\cdot 476^2 - \cdot 024 \times 1\cdot 3^2)$$

$$= \cdot 008894,$$

$$I_6 = \cdot 011498,$$

$$b_6 = \cdot 012,$$

and therefore shear stress $= \dfrac{S_6 M_{1,6}}{b_6 I_6} = \dfrac{14920 \times \cdot 008894}{\cdot 012 \times \cdot 011498}$

$= 1{,}686{,}000$ kg. per sq. metre.

**7. Effect of Temperature Variation.**—Suppose that the arch was erected at a temperature $t^\circ$, at which it had the exact dimensions necessary for it to be put in place without inducing any stresses other than those due to the loading.

If the arch were free to expand, it is clear that at a higher temperature $t_1^\circ$ the centre-line span, instead of being

$$\frac{45 + 46{\cdot}846}{2} = 45{,}923 \text{ m.},$$

would be increased by an amount

$$\frac{45{\cdot}923}{82500}(t_1^\circ - t^\circ),$$

$\frac{1}{82500}$ being the coefficient of expansion of the iron.

At the temperature $t_1^\circ$, therefore, it is clear that the horizontal thrust will be equal to that due to the load at the initial temperature plus an amount $H_T$.

To determine this quantity, we note that the internal work of the arch, considered as subjected only to horizontal forces $H_T$ applied at the supports, is given by formula (2), § 4, by multiplying by $2 \times \frac{3{\cdot}964}{2E}$ and neglecting the terms of the first order in $H_T$ which depend upon the loading. This internal work is therefore given by the expression

$$2 \times \frac{3{\cdot}964}{2E} \times 7436 H_T^2.$$

We must now add to this expression the term

$$-\frac{45{\cdot}923}{82500}(t_1^\circ - t^\circ) H_T,$$

and then equate to zero the differential coefficient with regard to $H_T$. We thus obtain the equation

$$2 \times \frac{3{\cdot}964}{E} \times 7436 H_T - \frac{45{\cdot}923}{82500}(t_1^\circ - t^\circ) = 0,$$

and taking $E = 18 \times 10^9$, we obtain

$$H_T = 170{\cdot}38(t_1^\circ - t^\circ).$$

To obtain the bending moment, normal pressure, and shear in the sections 0, 1, 2, . . . the value of $H_T$ should be substituted in the table of § 3, first neglecting the purely numerical terms, which depend upon the loads. We thus compile the following table :—

| Section. | Bending Moment. | Normal Pressure. | Shear. |
|---|---|---|---|
| 0 | $B_0' = 0$ | $P_0' = 152(t_1^\circ - t^\circ)$ | $S_0' = 78(t_1^\circ - t^\circ)$ |
| 1 | $B_1' = 279(t_1^\circ - t^\circ)$ | $P_1' = 158(t_1^\circ - t^\circ)$ | $S_1' = 66(t_1^\circ - t^\circ)$ |
| 2 | $B_2' = 408(t_1^\circ - t^\circ)$ | $P_2' = 163(t_1^\circ - t^\circ)$ | $S_2' = 54(t_1^\circ - t^\circ)$ |
| 3 | $B_3' = 683(t_1^\circ - t^\circ)$ | $P_3' = 166(t_1^\circ - t^\circ)$ | $S_3' = 40(t_1^\circ - t^\circ)$ |
| 4 | $B_4' = 811(t_1^\circ - t^\circ)$ | $P_4' = 169(t_1^\circ - t^\circ)$ | $S_4' = 27(t_1^\circ - t^\circ)$ |
| 5 | $B_5' = 888(t_1^\circ - t^\circ)$ | $P_5' = 170(t_1^\circ - t^\circ)$ | $S_5' = 14(t_1^\circ - t^\circ)$ |
| 6 | $B_6' = 914(t_1^\circ - t^\circ)$ | $P_6' = 170(t_1^\circ - t^\circ)$ | $S_6' = 0$ |

If the mean temperature at erection is 15° C. and the limits of variation are + 40° C. and − 10° C., it is clear that to consider these two limiting cases we must first take

$$(t_1^\circ - t^\circ) = 40 - 15 = 25;$$

and in the second case

$$(t_1^\circ - t^\circ) = -10 - 15 = -25.$$

We will add the results thus obtained to those obtained in § 6 for the superload on the left-hand half of the arch only. It should be noted that the maximum bending moments on the left-hand half of the arch will occur at the temperature − 10° C. and on the right at 40° C. We thus obtain the following results:—

| Section. | Left-hand Half-arch. Temperature − 10° C. | | | | Right-hand Half-arch. Temperature 40° C. | | | |
|---|---|---|---|---|---|---|---|---|
| | Bending Moment. | Normal Pressure. | Shear. | Eccentricity of Load. | Bending Moment. | Normal Pressure. | Shear. | Eccentricity of Load. |
| 0 | 0 | 121,799 | − 13,667 | 0 | 0 | 115,115 | + 14,094 | 0 |
| 1 | − 49,350 | 115,078 | − 10,984 | − ·428 | + 55,682 | 89,056 | + 8,791 | + ·625 |
| 2 | − 84,637 | 109,696 | − 7,065 | − ·771 | + 84,811 | 95,664 | + 3,807 | + ·886 |
| 3 | − 109,613 | 105,504 | − 2,465 | − 1·038 | + 95,329 | 101,218 | − 1,239 | + ·941 |
| 4 | − 100,017 | 102,580 | + 3,172 | − ·975 | + 84,443 | 105,353 | − 5,804 | + ·802 |
| 5 | − 87,327 | 101,325 | + 9,160 | − ·861 | + 54,177 | 108,369 | − 10,261 | + ·499 |
| 6 | − 40,580 | 101,500 | + 14,920 | − ·399 | + 5,120 | 110,000 | − 14,920 | + ·046 |

The maximum stress in the iron in the left-hand half-arch occurs at the extrados of section 3, and is equal to

$$c = \frac{105504}{\cdot 034464} + \frac{109613 \times \cdot 822}{\cdot 013754} = 9{,}516{,}000 \text{ kg. per sq. metre,}$$

whilst that on the right-hand side occurs at the intrados of section 3, and is equal to

$$c = \frac{101218}{\cdot 034464} + \frac{95329 \times \cdot 822}{\cdot 013754} = 8{,}550{,}000 \text{ kg. per sq. metre.}$$

**8. Conclusion.**—These last results show that with the sections adopted for the arch ribs, the bridge would not have a sufficient factor of safety.

It will therefore be necessary to strengthen the ribs by adding to the extrados and intrados a second plate ·25 m. broad and ·012 m. thick, extending from the middle of the element 1, 2 up to the middle of the element 4, 5, i.e. over a length of about 12 m. Then the area and moment of inertia of the section 3 will have the following values :—

$$A_3 = \cdot 034464 + 2 \times \cdot 25 \times \cdot 012 = \cdot 040464,$$

$$I_3 = \cdot 013754 + \frac{\cdot 25}{12}(1\cdot 668^3 - 1\cdot 644^3) = \cdot 017035,$$

and the maximum stress in the iron will be

$$c = \frac{105504}{\cdot 040464} + \frac{109613 \times \cdot 834}{\cdot 017035} = 7{,}970{,}000 \text{ kg. per sq. metre.}$$

This stress should not be regarded as excessive, since we have already allowed for temperature.*

* Comparing the maximum stress in the arch with flat ends, i.e. ends regarded as fixed, with that in the arch with cylindrical ends, i.e. hinged, we might think that the fixing of the ends is uneconomical of material, which is contrary to elementary principles. But if a full comparison were made between the results obtained in the two cases, we should recognise that if we could restrict the amount of metal employed to that which is strictly necessary, the arch with flat ends would be considerably more economical than that with cylindrical ends.

In practice the reverse will often hold true :—

1. Because the fixing of the ends can often only be obtained by adopting special construction involving the use of a certain quantity of metal which does not enter into these comparisons.

2. Because it is not possible to build up the arch with the exact distribution of metal necessitated by the theory.

# CHAPTER XXIII.

## STUDY OF A BRICK MASONRY ARCH BRIDGE CONSTRUCTED OVER THE OGLIO FOR THE MAIN LINE OF THE MILAN TO VENICE RAILWAY.

### *Investigation of the Stability of the Central Main Arch.* (Plate XIV.)

**1. General Data.**—

| | | |
|---|---|---|
| Clear span * . . . . . . | = | 42·00 m. |
| Rise . . . . . . . | = | 11·90 ,, |
| Radius of intrados . . . . . | = | 24·479 ,, |
| Angle subtended at centre by intrados | = | 118° 5′ 42″ |
| Breadth of bridge . . . . . | = | 7·50 m. |
| Thickness of arch-ring at crown . | = | 1·40 ,, |
| ,, ,, ,, ,, springings | = | 2·38 ,, |

The extrados of the arch is stepped as shown in Plate XIV.

**2. Loads.**—In the following calculations we will substitute for the actual stepped extrados a cylindrical surface whose section is formed of circular arcs, each of which passes through the bottoms of three successive steps. We must not forget, therefore, that the arch to which the results of the following calculations will strictly apply is the theoretical arch having for extrados the above-mentioned cylindrical surface. It is clear, however, that these results will apply with sufficient accuracy to the actual arch.

We will follow the usual practice of considering a breadth of arch of 1 m. only, although that does not simplify the calculation or possess any other advantage; nor is it at all necessary.

The arch ring is constructed of specially selected bricks, the mortar being made with Palazzolo hydraulic lime.

* The clear span of 42 m. was prescribed by the Chief Board of Public Works, following the opposition which had been aroused by other schemes involving the construction of piles in the river bed. The particular design was the work of M. Cesare Bermini, Engineer to the Italian State Railways and Chief of the Second Division of Maintenance and Works, who also directed its execution.

The weight of the masonry is 2000 kg. per cub. metre.

The sides of the arch are also constructed of brickwork, and the arch ring is covered with a layer of cement mortar to prevent infiltration of water. On this bed of mortar, gravel, weighing 1600 kg. per cub. metre, is laid up to the level of the rails.

The greatest superload, due to the trains, which the bridge is designed to carry is 8000 kg. per metre length of the bridge, i.e.

$$\frac{8000}{7{\cdot}50} = 1067 \text{ kg. per sq. metre of area.}$$

To facilitate the calculations, we will reduce the gravel above the bed of mortar and the superload to equivalent depths of masonry; thus we will reduce the vertical heights of the gravel in the ratio $\frac{1600}{2000}$, and will represent the superload by a layer of masonry of depth $\frac{1067}{2000} = {\cdot}533$ m.

If we now adopt as unit weight the weight of a cubic metre of masonry, viz. 2000 kg., it is clear that the weight of any portion of the arch and its load will be numerically represented by the corresponding area between the intrados and the line representing the equivalent depth.

If, therefore, we divide the centre line of the half-arch into six equal parts or voussoirs, and draw normals to this centre line at these points, and verticals at the points where these normals cut the extrados, and if, moreover, we *assume* that each portion of the arch ring carries the load between the corresponding verticals, we can obtain from the dimensions given on the figure the area of each voussoir and of the corresponding portion of the load diagram. Moreover, if we determine graphically the centroid of each of these areas, and measure the horizontal distance of each from the crown, we can then find the moment about the crown of each of these areas, and therefore the combined moments of arch ring weight and load of the elements 0, 1; 0, 2, etc.

After having obtained these moments about the crown, we can easily deduce the moments of the same loads about the centres of the sections 0, 1, 2, . . . 5, i.e. the moment about the centre of section 0 of the whole load of the element 6, 0; and similarly the moments of the elements 6, 1; 6, 2; 6, 3; . . . 6, 5, about the centres of the sections 1, 2, 3, . . . 5.

In carrying out these calculations, we find the following results. It should be noted that the moments given in columns

3, 5, 7 are not those about the crown, but are those about the centres of the sections 0, 1, 2, . . . 5; for the element 6, 3, for example, the moments given are taken about the centre of the section 3.

| Element. | Arch Alone. | | Load. | | Arch and Load. | |
|---|---|---|---|---|---|---|
| 1 | Area. 2 | Moment. 3 | Area. 4 | Moment. 5 | Area. 6 | Moment. 7 |
| 6, 0 | 51·35 | 458·49 | 94·67 | 556·19 | 146·02 | 1014·68 |
| 6, 1 | 41·17 | 339·55 | 66·14 | 378·42 | 107·31 | 717·97 |
| 6, 2 | 31·42 | 224·96 | 42·19 | 227·85 | 73·61 | 452·81 |
| 6, 3 | 22·11 | 127·66 | 24·81 | 118·63 | 46·92 | 246·29 |
| 6, 4 | 13·59 | 56·28 | 13·12 | 48·80 | 26·71 | 105·08 |
| 6, 5 | 6·34 | 14·52 | 5·52 | 11·46 | 11·86 | 25·98 |

**3. Principles and General Formula for the Condition when the Centering is Struck, i.e. when the arch supports only its own weight.**—In this case, the load being symmetrical about the crown, the shear at this point will be zero; we will therefore take as unknowns the bending moment B and the horizontal thrust H at the crown, and it will be easy to express as functions of these quantities the bending moment, normal pressure, and shear at any other point. If we consider the arch as cut through at the crown, it is clear that each half will be in equilibrium so long as we apply at the crown a *uniformly varying* pressure such that its resultant is H and its resultant moment about the centre of the crown is B.

Therefore to express the bending moment, normal pressure, and shear at any point, for example at section 4, it will be necessary to consider the pressure applied at the crown and the weight of the element 6, 4, whose moment about section 4 is given by column 3 of the previous table.

We thus obtain

$$B_4 = B - 1{\cdot}28H + 56{\cdot}28,$$
$$P_4 = {\cdot}943H + 13{\cdot}59 \times {\cdot}335,$$
$$S_4 = {\cdot}335H - 13{\cdot}59 \times {\cdot}943.$$

By deriving similar formulæ for all the sections 0, 1, 2, . 5, and collecting the results, we obtain the following table, in which the only unknowns are B and H:—

| Section. | Bending Moment. | Normal Pressure. | Shear. |
|---|---|---|---|
| 0 | $B_0 = B - 11{\cdot}96H + 458{\cdot}49$ | $P_0 = {\cdot}517H + 44{\cdot}01$ | $S_0 = {\cdot}857H - 26{\cdot}55$ |
| 1 | $B_1 = B - 8{\cdot}47H + 339{\cdot}55$ | $P_1 = {\cdot}655H + 31{\cdot}12$ | $S_1 = {\cdot}756H - 26{\cdot}97$ |
| 2 | $B_2 = B - 5{\cdot}45H + 224{\cdot}96$ | $P_2 = {\cdot}775H + 19{\cdot}95$ | $S_2 = {\cdot}635H - 24{\cdot}35$ |
| 3 | $B_3 = B - 2{\cdot}98H + 127{\cdot}66$ | $P_3 = {\cdot}870H + 10{\cdot}88$ | $S_3 = {\cdot}492H - 19{\cdot}24$ |
| 4 | $B_4 = B - 1{\cdot}28H + 56{\cdot}28$ | $P_4 = {\cdot}943H + 4{\cdot}55$ | $S_4 = {\cdot}335H - 12{\cdot}82$ |
| 5 | $B_5 = B - {\cdot}38H + 14{\cdot}52$ | $P_5 = {\cdot}987H + 1{\cdot}08$ | $S_5 = {\cdot}170H - 6{\cdot}26$ |
| 6 | $B_6 = B$ | $P_6 = 1{\cdot}00H$ | $S_6 = 0$ |

**4. Determination of the Unknowns.**—Although the arch under consideration is not homogeneous, it may be so regarded, because the distribution of the bricks and mortar of which it is composed is the same throughout. When the centering is struck the arch becomes deformed, but it is clear from considerations of symmetry that the section at the crown, which was vertical before the centering was struck, will remain vertical during the process, and not move horizontally either to the right or to the left. If then we assume one half of the arch to be removed, and that we apply to the crown of the other half a pressure H and a bending moment B to maintain equilibrium, it is clear that we should be able to determine H and B from the condition that in the deformation of the half-arch under the action of the pressure H, the bending moment B, and its own weight, the crown section must remain vertical, i.e.,

1. That its horizontal displacement must be zero.
2. That its angular displacement must be zero.

We can find the values of B and H which satisfy these conditions by expressing the internal work of the half-arch as a function of the two unknowns and equating to zero the differential coefficients with regard to each.

We will call

E the mean elastic modulus in compression of the arch, $A_0$, $A_1$, $A_2$, . . . $A_6$ the areas of the sections 0, 1, 2, . . . 6; $I_0$, $I_1$, $I_2$, . . . $I_6$ the moments of inertia of these same sections.

We will put for brevity

$$\Sigma\frac{B^2}{I} = \frac{1}{3}\left(\frac{B_0^2}{I_0} + \frac{4B_1^2}{I_1} + \frac{2B_2^2}{I_2} + \frac{4B_3^2}{I_3} + \frac{2B_4^2}{I_4} + \frac{4B_5^2}{I_5} + \frac{B_6^2}{I_6}\right)$$

$$\Sigma\frac{P^2}{A} = \frac{1}{3}\left(\frac{P_0^2}{A_0} + \frac{4P_1^2}{A_1} + \frac{2P_2^2}{A_2} + \frac{4P_3^2}{A_3} + \frac{2P_4^2}{A_4} + \frac{4P_5^2}{A_5} + \frac{P_6^2}{A_6}\right).$$

The expression for the internal work of the half-arch (neglecting the portion due to shear as having a very small effect upon the result) is

$$\frac{4{\cdot}37}{2E}\left(\Sigma\frac{B^2}{I}+\Sigma\frac{P^2}{A}\right).$$

We may neglect the numerical coefficient $\frac{4{\cdot}37}{2E}$, since it will have no effect upon the equations which we shall obtain later by equating to zero the differential coefficients with regard to B and H, and employ the expression

$$\Sigma\frac{B^2}{I}+\Sigma\frac{P^2}{A} \quad . \quad . \quad . \quad . \quad . \quad (1)$$

We have arranged in the following table the values of the areas $A_0$, $A_1$, etc., the moments of inertia $I_0$, $I_1$, etc., and the logarithms of the reciprocals of these quantities:—

| Sections.<br>1 | Areas.<br>2 | Moments of Inertia.<br>3 | Logarithms of Reciprocals of Areas.<br>4 | Logarithms of Reciprocals of Moments of Inertia.<br>5 |
|---|---|---|---|---|
| 0 | 2·38 | 1·1234 | $\bar{1}$·62342 | $\bar{1}$·94947 |
| 1 | 2·28 | ·9877 | $\bar{1}$·64207 | ·00537 |
| 2 | 2·18 | ·8633 | $\bar{1}$·66154 | ·06384 |
| 3 | 2·08 | ·7499 | $\bar{1}$·68194 | ·12500 |
| 4 | 1·82 | ·5024 | $\bar{1}$·73993 | ·29895 |
| 5 | 1·50 | ·2812 | $\bar{1}$·82391 | ·55098 |
| 6 | 1·40 | ·2287 | $\bar{1}$·85387 | ·64073 |

Substituting in formula (1) for the moments $B_0$, $B_1$, etc., and the pressures $P_0$, $P_1$, etc., their values tabulated in § 3, and for $A_0$, $A_1$, etc., $I_0$, $I_1$, etc., their numerical values given above, we have

$$\Sigma\frac{B^2}{I}=11{\cdot}72B^2-27{\cdot}90\,.\,2BH+180{\cdot}80H^2+1138{\cdot}68\,.\,2B-7254{\cdot}40\,.\,2H$$

$$\Sigma\frac{P^2}{A}=2{\cdot}38H^2+28{\cdot}42\,.\,2H.$$

Formula (1) then becomes

$$11{\cdot}72B^2-27{\cdot}90\,.\,2BH+183{\cdot}18H^2+1138{\cdot}68\,.\,2B-7225{\cdot}98\,.\,2H.$$

Equating to zero the differential coefficients of this expression with regard to B and H, we obtain the following equations:—

$$11{\cdot}72B-27{\cdot}90H=-1138{\cdot}68,$$
$$-27{\cdot}90B+183{\cdot}18H=7225{\cdot}98,$$

from which we obtain

$$B=-5{\cdot}096,$$
$$H=38{\cdot}671.$$

**5. Line of Pressure and Stress per Sq. Metre at Intrados and Extrados.**—Substituting in the table given in § 3 the values of B and H obtained above, we find the values of the bending moment, normal pressure, and shear at the sections 0, 1, 2, . . . 6. Dividing the bending moments by the corresponding normal pressures, we obtain the distances between the centre line and the load points of the sections (the eccentricities of the load). Finally, from the values of the bending moments and normal pressures, we find the stresses at intrados and extrados of the arch by the formula

$$c = \frac{\mathrm{P}}{\mathrm{A}} \pm \frac{\mathrm{B}n}{\mathrm{I}},$$

in which P and B represent the normal pressure and bending moment at a section, and A, I, $n$, the area, moment of inertia, and half-depth of the same section.

The results of all these calculations are collected in the following table:—

| Section. | Bending Moment. | Normal Pressure. | Shear. | Eccentricity of Load. | Compression Stress in kg. per sq. metre. | |
|---|---|---|---|---|---|---|
| | | | | | Intrados. | Extrados. |
| 1 | 2 | 3 | 4 | 5 | 6 | 7 |
| 0 | − 9·11 | 64·00 | + 6·59 | − 0·142 | 34,504 | 73,040 |
| 1 | + 6·91 | 56·45 | + 2·27 | + 0·122 | 65,478 | 33,558 |
| 2 | + 9·11 | 49·92 | + 0·21 | + 0·182 | 69,858 | 21,738 |
| 3 | + 7·32 | 44·52 | − 0·21 | + 0·164 | 63,110 | 22,506 |
| 4 | + 1·69 | 41·02 | + 0·13 | + 0·041 | 51,200 | 38,956 |
| 5 | − 5·27 | 39·25 | + 0·31 | − 0·135 | 24,822 | 79,846 |
| 6 | − 5·10 | 38·67 | 0·00 | − 0·132 | 24,022 | 86,462 |

**6. Deflection of the Arch at the Crown due to Striking of the Centering.**—The following is the formula for giving the deflection due to striking of the centering:—

$$d_n = \frac{4{\cdot}37}{\mathrm{E}} \cdot \frac{1}{3}\left\{\left(\frac{\mathrm{B}_0 x_0}{\mathrm{I}_0} + \frac{4\mathrm{B}_1 x_1}{\mathrm{I}_1} + \frac{2\mathrm{B}_2 x_2}{\mathrm{I}_2} + \ldots + \frac{4\mathrm{B}_5 x_5}{\mathrm{I}_5}\right) + \left(\frac{\mathrm{P}_0 \sin\phi_0}{\mathrm{A}_0} + \frac{4\mathrm{P}_1 \sin\phi_1}{\mathrm{A}_1} + \ldots + \frac{4\mathrm{P}_5 \sin\phi_5}{\mathrm{A}_5}\right)\right\}$$

in which $\mathrm{B}_0$, $\mathrm{B}_1$, . . ., $\mathrm{P}_0$, $\mathrm{P}_1$, . . . are the bending moments and normal pressures, the numerical values of which are given in columns 2 and 3 of the above table; $\mathrm{I}_0$, $\mathrm{I}_1$, . . ., $\mathrm{A}_0$, $\mathrm{A}_1$, . . . the moments of inertia and areas of the sections; $x_0$, $x_1$, . . . the distances between the centres of these sections and the vertical plane through the crown; and $\phi_0$, $\phi_1$, . . . the angles which the sections make with the vertical. It will be noted that $x_6 = 0$ and $\sin \phi_6 = 0$.

Substituting for the symbols their numerical values from the previous table or from Plate XIV, we find

$$\frac{1}{3}\left(\frac{B_0 x_0}{I_0} + \frac{4B_1 x_1}{I_1} + \frac{2B_2 x_2}{I_2} + \ldots + \frac{4B_5 x_5}{I_5}\right) = 310{\cdot}20,$$

$$\frac{1}{3}\left(\frac{P_0 \sin \phi_0}{A_0} + \frac{4P_1 \sin \phi_1}{A_1} + \frac{2P_2 \sin \phi_2}{A_2} + \ldots + \frac{4P_5 \sin \phi_5}{A_5}\right) = 65{\cdot}01.$$

These results must be multiplied by the weight per cub. metre of the masonry, i.e. by 2000, so that the deflection of the arch at the crown is given by the formula

$$d_n = \frac{3280000}{E}.$$

If we knew the value of E, i.e. the mean compression elastic modulus of the masonry of the arch at the moment of striking of the centering, we could determine the deflection of the arch at the crown by the above formula.

Now, this deflection during the striking of the centering was found to be 0·045 m., so that we will deduce the value of E from the equation

$$0{\cdot}045 = \frac{3280000}{E},$$

i.e.

$$E = 72{,}889{,}000.$$

**7. Examination of the case in which, in addition to the weight of the arch itself, allowance is made for the permanent load which it has to support and also for the superload.**

If we preserve the same notation as in §§ 3 and 4, and note that for the case of the complete load we have in columns 6 and 7 of the table given in § 2 the weights of the elements 6, 0; 6, 1; 6, 2; . . . 6, 5; and their moments about the centres of the sections 0, 1, 2, . . . 5, we can easily compile the following table:—

| Section. 1 | Bending Moment. 2 | Normal Pressure. 3 | Shearing Force. 4 |
|---|---|---|---|
| 0 | $B_0 = B - 11{\cdot}96H + 1014{\cdot}68$ | $P_0 = 0{\cdot}517H + 125{\cdot}14$ | $S_0 = 0{\cdot}857H - 75{\cdot}49$ |
| 1 | $B_1 = B - 8{\cdot}47H + 717{\cdot}97$ | $P_1 = 0{\cdot}655H + 81{\cdot}13$ | $S_1 = 0{\cdot}756H - 70{\cdot}29$ |
| 2 | $B_2 = B - 5{\cdot}45H + 452{\cdot}81$ | $P_2 = 0{\cdot}755H + 46{\cdot}74$ | $S_2 = 0{\cdot}635H - 57{\cdot}05$ |
| 3 | $B_3 = B - 2{\cdot}98H + 246{\cdot}29$ | $P_3 = 0{\cdot}870H + 23{\cdot}08$ | $S_3 = 0{\cdot}492H - 40{\cdot}82$ |
| 4 | $B_4 = B - 1{\cdot}28H + 105{\cdot}08$ | $P_4 = 0{\cdot}943H + 8{\cdot}95$ | $S_4 = 0{\cdot}335H - 25{\cdot}19$ |
| 5 | $B_5 = B - 0{\cdot}38H + 25{\cdot}98$ | $P_5 = 0{\cdot}987H + 2{\cdot}02$ | $S_5 = 0{\cdot}170H - 11{\cdot}71$ |
| 6 | $B_6 = B$ | $P_6 = 1{\cdot}000H$ | $S_6 = 0$ |

In this table there are no unknowns except B and H, which we determine exactly as in § 4, i.e. by substituting in formula (1) for the bending moments $B_0$, $B_1$, etc., and the normal pressures $P_0$, $P_1$, etc., their values in terms of B and H, and then differentiating and equating to zero, thus obtaining the two equations required.

In this way we find in the first place

$$\Sigma\frac{B^2}{I} = 11{\cdot}72B^2 - 27{\cdot}90\,.\,2BH + 180{\cdot}80H^2 + 2320{\cdot}33\,.\,2B$$
$$- 15243{\cdot}21\,.\,2H,$$
$$\Sigma\frac{P^2}{A} = 2{\cdot}38H^2 + 68{\cdot}67\,.\,2H\,;$$

and formula (1) becomes

$$11{\cdot}72B^2 - 27{\cdot}90\,.\,2BH + 183{\cdot}18H^2 + 2320{\cdot}33\,.\,2B - 15174{\cdot}54\,.\,2H,$$

from which we get, by equating to zero the partial differential coefficients, the two equations

$$11{\cdot}72B - 27{\cdot}90H + 2320{\cdot}33 = 0,$$
$$-27{\cdot}90B + 183{\cdot}18H - 15174{\cdot}54 = 0,$$

giving

$$B = -1{\cdot}220,$$
$$H = 82{\cdot}653.$$

The value of B being negative, it follows that the line of pressure at the crown is above the centre line.

The values of B and H being now known, we will substitute them in the expressions tabulated above for the bending moment, normal pressure, and shear; and then by carrying out the same operations as in § 5, we compile the following table, in which will be found for each of the sections 0, 1, 2, . . . 6, the bending moment, normal pressure, shear, eccentricity, and, finally, the stress per sq. metre at intrados and extrados:—

| Section. | Bending Moment. | Normal Pressure. | Shear. | Eccentricity. | Stress per Sq. Metre. | |
|---|---|---|---|---|---|---|
| | | | | | At Intrados. | At Extrados. |
| 0 | + 24·93 | 167·87 | − 4·66 | + 0·152 | 183,880 | 88,048 |
| 1 | + 16·68 | 135·27 | − 7·80 | + 0·121 | 157,162 | 80,154 |
| 2 | + 1·13 | 109·14 | − 4·57 | + 0·010 | 102,980 | 97,272 |
| 3 | − 1·24 | 94·99 | + 0·33 | − 0·013 | 87,898 | 94,778 |
| 4 | − 1·94 | 86·89 | + 2·50 | − 0·022 | 88,180 | 103,108 |
| 5 | − 4·99 | 83·60 | + 2·34 | − 0·060 | 84,850 | 138,086 |
| 6 | − 1·22 | 82·65 | 0·00 | − 0·015 | 110,602 | 125,538 |

To determine the deflection at the crown of the arch in the case of full loading, we employ the formula given in § 6, substituting for $B_0$, $B_1$, $B_2$, . . ., $P_0$, $P_1$, $P_2$, . . . their values given in columns 2 and 3 of the preceding table. We thus get

$$\frac{1}{3}\left(\frac{B_0x_0}{I_0}+\frac{4B_1x_1}{I_1}+\frac{2B_2x_2}{I_2}+\ldots+\frac{4B_5x_5}{I_5}\right)=462{\cdot}07,$$

$$\frac{1}{3}\left(\frac{P_0\sin\phi_0}{A_0}+\frac{4P_1\sin\phi_1}{A_1}+\frac{2P_2\sin\phi_2}{A_2}+\ldots+\frac{4P_5\sin\phi_5}{A_5}\right)=153{\cdot}56,$$

and consequently (after multiplying these results by 2000, i.e. the weight per cub. metre of the masonry)

$$d_n=\frac{5380000}{E}.$$

If we assume that after the construction is completed, the elastic modulus E has the same value as when the centering was struck, i.e. 72,889,000, the total deflection of the arch at the crown should be

$$d_n=\frac{5380000}{72889000}=0{\cdot}074 \text{ m}.$$

As the deflection of the arch at the crown under its own weight during the striking of the centering was equal to 0·045 m., we see that the load should cause an additional deflection equal to 0·074 − 0·045 = 0·029.

This result shows that the weight of the arch itself has a larger effect upon its deflection at the crown than that of the load which it has to carry. That is explained by the different manner in which the weights are distributed; for the weight of the arch itself is distributed almost uniformly along its length, whilst the load is greater at the springings than at the crown.

We should, however, point out that the deflection of the arch at the crown due to the load must be less than that which we have calculated, viz. less than 0·029 m., for the following reasons :—

1. During the building of the masonry of the spandrels the mortar becomes compacted and its compressibility diminishes; moreover, the deflection of 0·029 m. is caused for the greater part by the ballast and superload, which are uniformly distributed. Now when the ballast is placed in position and the superload is applied the mortar has already become compacted, and its compressibility is therefore less.

2. The spandrel masonry does not constitute a load without cohesion, as we have assumed in the calculations, but becomes monolithic with the abutments. It forms as it were two solid

bodies built in at the springings, which not only prevent their own load from coming on to the arch-ring, but also take part of the weight of the ballast and superload. This last consideration does not have much effect upon very flat arches, in which the spandrel masonry has only a small depth even at the springings.

# CHAPTER XXIV.

## STUDY OF A STONE BRIDGE ERECTED OVER THE DOIRE AT TURIN BY CHARLES MOSCA.

(Plate XV.)

**1. Data.**—

| | |
|---|---|
| Span of intrados . . . . . | = 45·00 m. |
| Rise ,, ,, . . . . . | = 5·50 ,, |
| Radius of ,, . . . . . | = 48·773 ,, |
| Angle subtended by intrados . . | = 54° 56′ 42″ |
| Thickness of ring at crown . . | = 1·50 m. |
| ,, ,, ,, ,, springings . | = 2·00 ,, |
| Span of extrados . . . . . | = 46·846 ,, |
| Rise ,, ,, . . . . . | = 5·226 ,, |
| Radius of ,, . . . . . | = 55·102 ,, |
| Angle subtended by extrados . . | = 50° 18′ 42″ |
| Breadth of bridge, over parapets . | = 12·60 m. |

**2. Loading.**—The arch-ring is of granite obtained from Manalaggio, near Pignerol (Italy), the weight of which is 2750 kg. per cub. metre. The filling is of ashlar masonry weighing 2300 kg. per cub. metre. Above the masonry there is a layer of mortar ·15 m. thick, upon which is earth (placed in layers and beaten down) up to the level of the macadam. The weight of the mortar and earth is 1600 kg. per cub. metre, and that of the macadam is 1800 kg. per cub. metre.

To simplify, we reduce the ashlar filling, mortar, earth, and macadam to equivalent heights of the granite of which the arch-ring is composed. Thus we reduce the height of the ashlar masonry in the ratio $\frac{2300}{2750}$, the heights of the mortar and earth, and macadam, being reduced by $\frac{1600}{2750}$ and $\frac{1800}{2750}$ respectively. We can thus draw above the extrados a load line of equivalent weight of granite.

We will allow a superload of 600 kg. per sq. metre, as some authorities specify, although it may be pointed out that the densest crowd will not weigh more than about 400 kg. per sq. metre. The equivalent height of granite of this superload is $\frac{600}{2750} = \cdot 218$ m., so that if we draw above our equivalent dead load line a parallel line ·218 m. vertically above it, we shall have the total load line of the bridge.

Then, taking as a unit load the weight of a cubic metre of granite (2750 kg.), and considering a metre width of the bridge, it is clear that the weight of an element of the arch-ring and the load above it will be represented by the corresponding area of the load curve.

If, therefore, we divide the centre line of the half arch-ring into six equal elements, and draw normals to the centre line at the points of division, and verticals at the points where these normals intersect the extrados, and if, further, we assume that each of the six elements has to carry the load between the corresponding verticals, we can find the area of each element of the arch-ring and of the corresponding portion of the load curve. We can then find graphically the centroids of these areas, and by measuring their horizontal distances from the vertical line through the crown, we can obtain the moment of each of the areas about this line, and thus the sum of all the moments.

Knowing therefore the moment about the crown section of any element of the arch-ring and of its corresponding load, for instance the element 6, 3, we can deduce the moment about the centre of the section 3, which is that required for the calculation, by subtracting the moment already found from the product of the weight of the element 6, 3, and the horizontal distance between the centroids of the sections 3 and 6 (i.e. 11·81). In general, if we call

$w_3$ the weight carried by the element 6, 3,

$d_3$ and $d_3'$ the horizontal distances of its centroid from the centroids of the sections 6, 3,

$D_3$ the horizontal distance between the centroids of these sections,

$m_3$ and $m_3'$ the moments of the weight $p_3$ about the crown vertical and the centroid of the section 3,

we shall then have

$$m_3 = w_3 d_3, \quad m_3' = w_3 d_3',$$

but

$$d_3' = D_3 - d_3,$$

so that

$$m_3' = w_3 D_3 - w_3 d_3 = w_3 D_3 - m_3.$$

Working in this way we obtain the following results:—

| Element. | Arch-ring. | | Load Curve. | | Arch-ring and Load Curve. | | Arch-ring and Load Curve in Kilogrammes. | |
|---|---|---|---|---|---|---|---|---|
| 1 | Area. 2 | Moment. 3 | Area. 4 | Moment. 5 | Area. 6 | Moment. 7 | Area. 8 | Moment. 9 |
| 6-0 | 39·400 | 422·70 | 43·756 | 312·55 | 83·156 | 735·25 | 228,679 | 2,021,937 |
| 6-1 | 31·816 | 292·89 | 28·876 | 186·49 | 60·692 | 479·38 | 166,903 | 1,318,295 |
| 6-2 | 24·783 | 187·93 | 18·046 | 103·15 | 42·829 | 291·08 | 117,780 | 800,470 |
| 6-3 | 18·203 | 105·67 | 10·539 | 50·67 | 28·739 | 156·34 | 79,033 | 429,935 |
| 6-4 | 11·958 | 47·16 | 5·700 | 20·46 | 17·658 | 67·62 | 48,559 | 185,955 |
| 6-5 | 5·930 | 11·92 | 2·500 | 4·95 | 8·430 | 16·37 | 23,182 | 46,392 |

It should be noted that the moments in columns 3, 5, 7, and 9 are not those about the crown vertical but those about the centres of the sections 0, 1, 2, . . . 6 respectively; for the element 6-3, for instance, the tabulated moments are those about the centre of section 3.

The values in columns 8, 9 were obtained from those in columns 6, 7 by multiplying by 2750.

**3. Principles and General Formulæ.**—A masonry arch should always be regarded as fixed at the ends, because if it is properly designed—as should always be assumed in preliminary calculations—the pressure on the abutments should be distributed over the whole extent of the sections at the springings.

If, as a result of the calculation on the assumption of fixed ends, there is found to be tension across the section at the springings, and if this tension cannot exist, owing either to no mortar being employed or to the material not possessing any tensile strength, the calculation should be re-made on the assumption that the sections are reduced to those which have been found to be under compression. These sections may have to be corrected by a third calculation, and so on.

We will suppose then at the outset that the arch-ring of the Mosca bridge is fixed to the abutments and fully loaded, so that for the elements 6-0, 6-1, 6-2 . . . we have the weights and moments given in the last two columns of the above table.

Since the load is symmetrical about the crown, the shearing force at this section must be zero; we will therefore take as the unknowns of our problem the bending moment B and the horizontal thrust H at the crown, in terms of which we can easily express the bending moment, thrust and shear at any other section. If we imagine the two halves of the arch-ring to be separated, it is clear that they will remain in equilibrium provided that there is applied at the crown of each half a pressure, varying in some symmetrical manner, and such that its total value is H and its moment about the centre line is B.

We thus obtain the bending moment, thrust, and shear for any section, for example section 4, by considering the thrust applied at the crown and the weight of the element 6-4, the moment of which about the centre of the section 4 is given in the previous table; we thus have:—

$$B_4 = B - \cdot 60H + 185{,}955,$$
$$P_4 = \cdot 988H + \cdot 154 \times 48{,}559,$$
$$S_4 = \cdot 154H - \cdot 988 \times 48{,}559.$$

By deriving similar equations for all the sections 0, 1, . . . 6, and simplifying, we obtain the results tabulated below:—

| Sect. | Bending Moment. | Normal Thrust. | Shear. |
|---|---|---|---|
| 1 | 2 | 3 | 4 |
| 0 | $B_0 = B - 5{\cdot}36H + 2{,}021{,}937$ | $P_0 = \cdot 895H + 102{,}217$ | $S_0 = \cdot 447H - 204{,}655$ |
| 1 | $B_1 = B - 3{\cdot}76H + 1{,}318{,}295$ | $P_1 = \cdot 925H + 63{,}085$ | $S_1 = \cdot 378H - 154{,}385$ |
| 2 | $B_2 = B - 2{\cdot}40H + 800{,}470$ | $P_2 = \cdot 950H + 35{,}915$ | $S_2 = \cdot 305H - 111{,}897$ |
| 3 | $B_3 = B - 1{\cdot}35H + 429{,}935$ | $P_3 = \cdot 973H + 18{,}177$ | $S_3 = \cdot 230H - 76{,}890$ |
| 4 | $B_4 = B - \cdot 60H + 185{,}955$ | $P_4 = \cdot 988H + 7{,}480$ | $S_4 = \cdot 154H - 47{,}987$ |
| 5 | $B_5 = B - \cdot 15H + 46{,}392$ | $P_5 = \cdot 996H + 1{,}815$ | $S_5 = \cdot 078H - 22{,}962$ |
| 6 | $B_6 = B$ | $P_6 = 1{\cdot}000H$ | $S_6 = 0$ |

**4. Determination of the Unknown Quantities.**—It is clear that the values of H and B vary according to whether or not the arch-ring is homogeneous and according to whether the mortar is of constant thickness or varies in thickness from intrados to extrados.

It is therefore desirable to give a detailed description of the arch.* It was formed of 93 voussoirs of granite, which were shaped to form the arch-ring with an intrados span of 45 m. and a rise of 5·50 m. without mortar. On the other hand, the centering was constructed to a circular arc of 45 m. chord and 5·75 m. rise, the length of arc exceeding that of the designed intrados by ·162 m.

This arrangement was adopted so as to leave between the voussoirs at the springings an appreciable thickness of mortar at the intrados and very little at the extrados. These thicknesses of mortar were a maximum at the springings and diminished gradually to zero at the 11th voussoir in the following manner:—

* The particulars here given are from a memoir by M. Carlo Mosca, Engineer to the Italian Railways (nephew of the celebrated builder of the bridge).

| Joints. | Thickness at Intrados. | Thickness at Extrados. |
|---|---|---|
| 1 | ·009 | ·0037 |
| 2 | ·008 | ·0027 |
| 3 | ·008 | ·0034 |
| 4 | ·007 | ·0033 |
| 5 | ·007 | ·0032 |
| 6 | ·006 | ·0030 |
| 7 | ·005 | ·0024 |
| 8 | ·005 | ·0023 |
| 9 | ·004 | ·0020 |
| 10 | ·003 | ·0014 |
| 11 | ·002 | ·0009 |
| | ·064 | ·0283 |

In this way there was an increase of ·064 near each springing, making ·128 for the whole arch. It was expected that the remaining ·034 would be accounted for by the slight imperfections in the voussoirs, and this proved to be the case.

From the 11th to the 36th joint there was no mortar. However, on account of the arrangement of the voussoirs near the springings the angle which the 36th joint made with the vertical plane through the crown was greater than the sum of the angles of the voussoirs between the 36th joint and the crown; this difference was taken up between the 36th and 47th voussoirs by disposing them so that they were in contact at the intrados, the gaps at the extrados being filled by mortar. Thus near the crown there were cuneiform joints, similar to those at the springings but reversed. The thickness of these joints is given by the following table:—

| Joints. | Thickness at Intrados. | Thickness at Extrados. |
|---|---|---|
| 37 | 0 | ·001 |
| 38 | 0 | ·001 |
| 39 | 0 | ·002 |
| 40 | 0 | ·002 |
| 41 | 0 | ·003 |
| 42 | 0 | ·003 |
| 43 | 0 | ·004 |
| 44 | 0 | ·006 |
| 45 | 0 | ·007 |
| 46 | 0 | ·008 |
| 47 | 0 | ·008 |
| | 0 | ·045 |

It should be noted that the sizes of the joints near the springings were first of all maintained by iron wedges, which were not replaced by mortar until the construction of the arch had reached the 39th voussoir.

This method of construction was adopted in order to obtain a good distribution of the pressures, and was successful, as the following calculations show. We will assume, to simplify the work, that all the joints were without mortar, except the two joints at the springings, which we will consider as made up of all the layers of mortar in the first eleven joints, and the joint at the crown, at which we will consider to be concentrated all the mortar from the 37th joint on the left to the 37th joint on the right.

We will thus consider at each springing a bed of mortar ·064 m. thick at the intrados and ·0283 m. thick at the extrados; at the crown we have another layer of mortar varying in thickness from zero at the intrados to ·045 × 2 = ·09 m. at the extrados.

We will use the following notation :—

E, E′, the elastic modulus in compression for granite and mortar respectively.

$d_0, d_1, d_2, \ldots d_6$, the depth of the arch-ring at the sections 0, 1, 2, . . . 6.

$A_0, A_1, A_2, \ldots A_6$, the areas of these sections, which are numerically equal to the depths, because a unit breadth has been taken.

$I_0 = \frac{d_0^3}{12}, I_1 = \frac{d_1^3}{12} \ldots$, the moments of inertia of these sections.

The values of $A_0$, $A_1$, etc., and $I_0$, $I_1$, etc., together with the logarithms of their reciprocals are given in the following table :—

| Section. | Area, A. | Moment of Inertia, I. | Logarithm of Reciprocal of A. | Logarithm of Reciprocal of I. |
|---|---|---|---|---|
| 0 | 2·01 | ·67672 | $\bar{1}$·69680 | ·16959 |
| 1 | 1·85 | ·52764 | $\bar{1}$·73283 | ·27766 |
| 2 | 1·72 | ·42404 | $\bar{1}$·76447 | ·37259 |
| 3 | 1·62 | ·35429 | $\bar{1}$·79049 | ·45064 |
| 4 | 1·55 | ·31032 | $\bar{1}$·80967 | ·50819 |
| 5 | 1·51 | ·28691 | $\bar{1}$·82102 | ·54225 |
| 6 | 1·50 | ·28125 | $\bar{1}$·82391 | ·55091 |

If the entire arch were of granite, i.e. if the three theoretical joints at springings and crown were also of granite, the internal work for the whole arch, neglecting that due to shear, would be given by the expression

$$W_i = 2 \cdot \frac{4 \cdot 00}{2E}\left[\Sigma\frac{B^2}{I} + \Sigma\frac{P^2}{A}\right], \quad . \quad . \quad . \quad (1)$$

in which the expressions $\Sigma\frac{B^2}{I}$ and $\Sigma\frac{P^2}{A}$ connote the same quantities as in previous cases.

To make this expression exact, we must subtract from it the internal work of the three granite wedges which are absent, and add to it that of the mortar wedges which take their place.

Now for a granite voussoir or wedge subjected to a normal pressure P and a bending moment B, the breadth at the intrados being $b_i$ and at the extrados $b_e$, the internal work is given by the expression

$$W_i = \frac{1}{2E} \cdot \frac{(b_i + b_e)}{2}\left\{\frac{P^2}{A} + \frac{B^2}{I} + \frac{(b_i - b_e)}{(b_i + b_e)} \cdot \frac{4BP}{Ad}\right\}.^*$$

* This may be proved as follows: Let ABCDD′C′B′A′ (Fig. 4, Pl. XV) be a granite wedge, the edges AA′, BB′, CC′, and DD′ of which are parallel to each other and perpendicular to the common plane through their centres. The wedge is symmetrical about this plane, which we will call the *Plane of Symmetry*. We will suppose that all the faces except ABCD and A′B′C′D′ are free, and that the latter are subjected to forces which are perpendicular to the plane of symmetry. To maintain the equilibrium of the wedge under the action of the external forces, the resultant forces upon the two faces must be equal and opposite.

Let P be the resultant pressure on the face ABCD, B the resultant moment about the centroid axis FG of this face, and I the moment of inertia and A the area of the section parallel to the plane of symmetry, i.e. perpendicular to the edge AA′.

If the distribution of stress on the face ABCD follows a linear law, we know that the intensity of compressive stress at any point in the middle section is given by the formula

$$c = \frac{P}{A} + \frac{Bz}{I},$$

$z$ being the distance of the point from the centroid axis.

Next let us consider an element HKK′H′ of the wedge, the faces of which are parallel to the faces AB′ and AD′, and of which the section is a very small rectangle of sides $dy$ and $dz$.

If the edge HH′ is at a distance $z$ from the plane FGG′F′ containing the centroid axes of the sections, it is clear that this element will be subjected to a pressure

$$dc = \left(\frac{P}{A} + \frac{Bz}{I}\right)dydz$$

parallel to the edge HH′: then the internal work of this element of volume will be equal to

$$\begin{aligned} dW_i &= \frac{1}{2E}HH'\left(\frac{P}{A} + \frac{Bz}{I}\right)^2 dydz \\ &= \frac{1}{2E}NN'\left(\frac{P}{A} + \frac{Bz}{I}\right)^2 dydz \\ &= \frac{1}{2E}\left(\frac{b_i + b_e}{2} - \frac{b_e - b_i}{d} \cdot z\right)\left(\frac{P}{A} + \frac{Bz}{I}\right)^2 dydz. \end{aligned}$$

For a mortar wedge equal in size to the granite one and subjected to the same forces, the internal work will be the same with E changed to E′, so that the difference between the values of internal work becomes

$$\frac{1}{2}\left(\frac{1}{\mathrm{E}'}-\frac{1}{\mathrm{E}}\right)\left(\frac{b_e+b_i}{2}\right)\left\{\frac{\mathrm{P}^2}{\mathrm{A}}+\frac{\mathrm{B}^2}{\mathrm{I}}-4\left(\frac{b_e-b_i}{b_e+b_i}\right)\frac{\mathrm{PB}}{d\mathrm{A}}\right\}.$$

Now for each of the joints at the springings we have

$$\mathrm{P}=\mathrm{P}_0;\ \mathrm{B}=\mathrm{B}_0;\ \mathrm{A}=\mathrm{A}_0;\ \mathrm{I}=\mathrm{I}_0;\ d=d_0.$$
$$b_e=\cdot 0283;\ b_i=\cdot 064,$$

and for the joint at the crown

$$\mathrm{P}=\mathrm{P}_6;\ \mathrm{B}=\mathrm{B}_6;\ \mathrm{A}=\mathrm{A}_6;\ \mathrm{I}=\mathrm{I}_6;\ d=d_6.$$
$$b_e=\cdot 09\ ;\ b_i=0.$$

Therefore the formula to be added to (1) to get the total internal work of the arch will be

$$\mathrm{W}_i'=\frac{1}{2}\left(\frac{1}{\mathrm{E}'}-\frac{1}{\mathrm{E}}\right)\left[\cdot 0923\left(\frac{\mathrm{P}_0^2}{\mathrm{A}_0}+\frac{\mathrm{B}_0^2}{\mathrm{I}_0}+1\cdot 547\frac{\mathrm{B}_0\mathrm{P}_0}{d_0\mathrm{A}_0}\right)\right.$$
$$\left.+\cdot 045\left(\frac{\mathrm{P}_6^2}{\mathrm{A}_6}+\frac{\mathrm{B}_6^2}{\mathrm{I}_6}-\frac{4\mathrm{P}_6\mathrm{B}_6}{d_6\mathrm{A}_6}\right)\right] \quad . \quad (2)$$

---

To obtain the internal work for the whole wedge, we must integrate the formula, first with reference to $y$, considering $z$ constant, and then with reference to $z$.

The first integration should be carried between the limits 0 and the width AB of the wedge, which we take as equal to 1 m.; then as the function does not contain the variable $y$ and as $\int_0^1 dy=1$, the result of the first integration will be

$$\frac{1}{2\mathrm{E}}\left(\frac{b_i+b_e}{2}-\frac{b_e-b_i}{d}\cdot z\right)\left(\frac{\mathrm{P}}{\mathrm{A}}+\frac{\mathrm{B}z}{\mathrm{I}}\right)^2 dz$$
$$\text{or } \frac{1}{2\mathrm{E}}\left(\frac{b_i+b_e}{2}\right)\left\{1-2\left(\frac{b_e-b_i}{b_e+b_i}\right)\frac{z}{d}\right\}\left(\frac{\mathrm{P}}{\mathrm{A}}+\frac{\mathrm{B}z}{\mathrm{I}}\right)^2 dz.$$

The second integration must be taken between the limits $-\frac{d}{2}$ and $+\frac{d}{2}$, and if we separate out the terms we have

$$\frac{1}{2\mathrm{E}}\left(\frac{b_e+b_i}{2}\right)\left[\frac{\mathrm{P}^2}{\mathrm{A}^2}+2\left\{\frac{\mathrm{PB}}{\mathrm{AI}}-\left(\frac{b_e-b_i}{b_e+b_i}\right)\cdot\frac{\mathrm{P}^2}{d\mathrm{A}^2}\right\}z\right.$$
$$\left.+\left\{\frac{\mathrm{B}^2}{\mathrm{I}^2}-4\left(\frac{b_e-b_i}{b_e+b_i}\right)\frac{\mathrm{PB}}{d\mathrm{AI}}\right\}z^2-\frac{2}{d}\left(\frac{b_e-b_i}{b_e+b_i}\right)\frac{\mathrm{B}^2}{\mathrm{I}^2}z^3\right]dz\ ;$$

and since we have

$$\int_{-\frac{d}{2}}^{+\frac{d}{2}}dz=d=\mathrm{A};\quad \int_{-\frac{d}{2}}^{+\frac{d}{2}}z\,dz=0;\quad \int_{-\frac{d}{2}}^{+\frac{d}{2}}z^2dz=\frac{d^3}{12}=\mathrm{I};\quad \int_{-\frac{d}{2}}^{+\frac{d}{2}}z^3dz=0$$

the result of the integration thus becomes

$$\mathrm{W}_i=\frac{1}{2\mathrm{E}}\left(\frac{b_e+b_i}{2}\right)\left\{\frac{\mathrm{P}^2}{\mathrm{A}}+\frac{\mathrm{B}^2}{\mathrm{I}}-4\left(\frac{b_e-b_i}{b_e+b_i}\right)\frac{\mathrm{PB}}{d\mathrm{A}}\right\}.$$

Substituting for $B_0$, $B_1$, $B_2$, . . . $P_0$, $P_1$, $P_2$, . . . their values in terms of B and H given in the table in § 3, and giving the quantities $I_0$, $I_1$, $I_2$, . . ., $A_0$, $A_1$, $A_2$ . . ., their numerical values, we get

$$\Sigma\frac{B^2}{I} = 16{\cdot}34B^2 - 22{\cdot}98 \times 2BH + 66{\cdot}67H^2 + 7818892 \times 2B - 23340905 \times 2H.$$

$$\Sigma\frac{P^2}{A} = 3{\cdot}39H^2 + 91020 \times 2H.$$

$$\frac{P_0^2}{A_0} + \frac{B_0^2}{I_0} + 1{\cdot}547\frac{B_0P_0}{d_0A_0} = 1{\cdot}50B^2 - 7{\cdot}87 \times 2BH + 41{\cdot}64H^2 + 3052775 \times 2B - 15966225 \times 2H.$$

$$\frac{P_6^2}{A_6} + \frac{B_6^2}{I_6} + \frac{4B_6P_6}{d_6A_6} = 3{\cdot}56B^2 - {\cdot}889 \times 2BH + {\cdot}67H^2.$$

Thus formula (1) becomes

$$W_i = 2\,.\,\frac{4{\cdot}00}{2E}(16{\cdot}34B^2 - 22{\cdot}98\,.\,2BH + 70{\cdot}06H^2 + 7818893\,.\,2B - 23249885\,.\,2H) \quad . \quad (3)$$

and formula (2) becomes

$$W'_i = \frac{1}{2}\left(\frac{1}{E'} - \frac{1}{E}\right)({\cdot}2984B^2 - {\cdot}766\,.\,2BH + 3{\cdot}876H^2 + 281765\,.\,2B - 1466630\,.\,2H) \quad . \quad (4)$$

We must now add these two expressions (3) and (4); but we must first know the value of the two moduli E, E′, or, at least, their relative values. Now, we have no reliable information upon the elastic modulus in compression for granite and mortar; but seeing that while mortar is fresh it must be easily compressible, as was proved to be the case when the centering of the Mosca bridge was taken down, while granite must compress very little, we will assume that the compressibility of the fresh mortar used in the bridge is 100 times as great as that of granite; i.e. we will suppose that

$$\frac{E'}{E} = \frac{1}{100}.$$

With this relation formula (4) becomes

$$W'_i = 2\,.\,\frac{4{\cdot}00}{2E}(3{\cdot}69B^2 - 9{\cdot}48\,.\,2BH + 47{\cdot}94H^2 + 3486835\,.\,2B - 18148625\,.\,2H);$$

and adding this to formula (3) we get for the total internal work of the arch

$$W_i + W_i' = 2 \cdot \frac{4{\cdot}00}{2E}(20{\cdot}03B^2 - 32{\cdot}46 \cdot 2BH + 118{\cdot}00H^2$$
$$+ 11305728 \cdot 2B - 41398510 \cdot 2H).$$

Now the arch in straining must satisfy the following conditions :—

1. *The section at the crown must remain verticle under strain, i.e. its angular movement must be zero.*
2. *This section must descend vertically ; i.e. its horizontal movement must be zero.*

The values of B and H which satisfy these two conditions are those obtained by equating to zero the differential coefficients of the internal work with regard to B and H.

We thus obtain the following equations :—

$$20{\cdot}03B - 32{\cdot}46H + 11305728 = 0,$$
$$-32{\cdot}46B + 118{\cdot}00H - 41398510 = 0,$$

from which we obtain

$$B = 7690$$
$$H = 352990.$$

**5. Line of Pressure and Maximum Stresses per Sq. Metre.**— If in the table of § 3 we substitute the above values for B and H, we obtain the values of the bending moment, normal pressure, and shear at each of the sections 0, 1, 2, . . . of the bridge, as given in the following table.

Dividing each bending moment by the corresponding normal pressure, we obtain the values given in column 5 of the eccentricity of the load, i.e. of the distance between the centre line of the arch and the line of pressure.

| Section. | Bending Moment. | Normal Pressure. | Shear. | Eccentricity of Load. | Compressive Stress per Sq. Metre | |
|---|---|---|---|---|---|---|
| | | | | | Intrados. | Extrados. |
| 1 | 2 | 3 | 4 | 5 | 6 | 7 |
| 0 | + 137,627 | 418,137 | – 46,865 | + ·329 | 412,425 | 3,635 |
| 1 | – 1,315 | 389,595 | – 20,955 | – ·003 | 212,905 | 208,305 |
| 2 | – 38,990 | 371,245 | – 4,237 | – ·105 | 294,915 | 136,765 |
| 3 | – 38,895 | 361,627 | + 4,296 | – ·108 | 312,155 | 134,306 |
| 4 | – 18,145 | 356,230 | + 6,372 | – ·051 | 275,146 | 184,514 |
| 5 | + 1,135 | 353,385 | + 4,571 | + ·003 | 237,011 | 231,045 |
| 6 | + 7,690 | 352,990 | 0 | + ·022 | 255,997 | 214,993 |

By means of the formula

$$c = \frac{P}{A} \pm \frac{Bn}{I}$$

we obtain the stress per sq. metre at intrados and extrados of the sections under consideration.

It will be seen that the distance between the centre line and the line of pressure is everywhere less than one-sixth of the depth of the section, i.e. the line of pressure keeps within the middle-third of the section, and consequently there will be no tension stress at any point of the arch.

The greatest compressive stress occurs at the intrados near to the springings and is equal to 41·24 kg. per sq. cm.

This stress is not too high for the granite employed in the Mosca bridge, but it would be too high for the mortar, which usually fails at a stress of less than 40 kg. per sq. cm.

An important point should, however, be noted here. If the centering of an arch is struck when the mortar has already well set, the latter is likely to crush under the compression of 41·24 kg. per sq. cm. which we have calculated above, and to drop gradually from the joints, thus endangering the stability of the arch; if on the other hand the centering is struck before the mortar has set, the latter will be subjected to a heavy compression, but the grains of sand of which it is composed cannot fall out because they are held by the lime which has not yet set. In this case the mortar sets under the pressure to which it will be subjected, and so will no longer be in danger of crushing.

It appears to us from these considerations that for large arches in which the compressive stress attains a fairly high value, the centering ought to be struck a short time after the keystone has been placed in position. In small arches in which the compressive stress on the mortar will only be a small fraction of the crushing stress, it does not matter whether the centering is struck before or after the mortar has set.

**6. Consideration of the internal forces which would occur in the arch if there were no mortar in any joint.**—It is clear that it is impossible for tension to be induced across any joint, because the granite voussoirs are simply placed in contact with each other. We must, therefore, first investigate whether tension tends to be induced at any point, and if this is the case it follows that at some sections, the true resisting section will be less than the apparent section; it will then be necessary to determine the actual dimensions of the resisting sections, by successive approximations.

To determine whether tension tends to arise, we must first consider the arch as a monolithic structure fixed at the springings, instead of being simply supported there.

In this case the internal work of the arch is given by formula (1) of paragraph (4) or by formula (3).

By equating to zero the differential coefficients of the latter with regard to B and H, and then dividing by the numerical coefficient $2 \cdot \frac{4}{E}$, we obtain the following equations:—

$$16{\cdot}34B - 22{\cdot}98H - 7818893 = 0$$
$$-22{\cdot}98B + 70{\cdot}06H - 23249885 = 0,$$

from which we obtain

$$B = -21800^{*}$$
$$H = 324710.$$

Substituting these values in the table given at the end of paragraph 3, we see that the line of pressure keeps within the middle-third for the whole arch, except for a portion about 2·8 m. long near to each springing. In these portions it passes outside on the lower side, so that the arch tends to open out at the extrados. This might have been expected in view of the small rise of the arch for its span.

For the section 0 at the springings we obtain

$$B_0 = 259737,$$
$$P_0 = 392827;$$

the resulting eccentricity being $\frac{259737}{392827} = {\cdot}66$ m.; this is more than one-sixth of the depth of the section, i.e. more than $\frac{2{\cdot}01}{6} = {\cdot}335$.

If the arch were monolithic and fixed at the ends so that tension could be resisted, we could now find on which portion of the section there was compression, and on which portion there was tension.

For this it would be necessary to determine the neutral axis, along which the stress is zero: if we call $y$ the distance from

* The negative value of the bending moment at the crown shows that the line of pressure is above the centre line of the arch at that point, at a distance equal to

$$e = \frac{21800}{324710} = {\cdot}067 \text{ m.}$$

Now we have seen that when we take account of the mortar in the joints, there is compressive stress over every section, and the arch behaves as a monolithic one; but the line of pressure passes below the centre line of the arch at the crown. We see, therefore, how great is the effect of the mortar (which is much more compressible than the granite) upon the distribution of the stresses.

this axis, which will be in the upper half, to the centre line of the arch, we obtain the formula

$$\frac{P_0}{A_0} - \frac{B_0 y}{I_0} = 0,$$

which gives $y = \cdot 508$, on substituting their numerical values for $A_0$, $B_0$, $I_0$, and $P_0$.

The depth of the compressed portion of the section will therefore be

$$\frac{2\cdot 01}{2} + \cdot 508 = 1\cdot 513 \text{ m.},$$

and the depth under tension will be

$$\frac{2\cdot 01}{2} - \cdot 508 = \cdot 497 \text{ m.}$$

These results do not hold for the case under consideration, because the arch can open up. To determine to what depth from the extrados the opening up will extend at different sections, we will now suppose that the true resisting section at the springings is very nearly equal to the depth found to be in compression in the hypothetical cases considered above, i.e. 1·50 m. up from the intrados.

For the other sections 1, 2, 3, . . . 6, the full depth is effective, because, as we have seen above, there is compressive stress over their whole depth.

The effective depth of the sections at the springings being changed, the centre line of the arch may be considered also changed, i.e. to be now ·75 m. from the intrados; the bending moment, thrust (normal pressure), and shear will also be changed, because the centre line of the arch for the element 0, 1 will no longer be the same, and thus we shall no longer have the same leverage for the thrust H, nor for the loads, nor will the angles which the forces make with the last element of the centre line of the arch be the same.

The tables given in paragraphs 3 and 4, therefore, will hold for the present case, except the expressions for $B_0$, $P_0$, $S_0$, which will become *

$$\begin{aligned} B_0 &= B - 5\cdot 58H + 1{,}996{,}782, \\ P_0 &= \cdot 860H + 116{,}626, \\ S_0 &= \cdot 510H - 196{,}664; \end{aligned}$$

and the values of $A_0$, $I_0$ will now be 1·50 and ·28125 respectively.

* The centre line of the arch is drawn so that it passes ·75 m. above the intrados at the springings. To avoid confusion, this has not been shown on the illustration, but the reader can do so without difficulty. To make the matter

With these new values of $B_0$, $P_0$, $I_0$, $A_0$, the internal work for the arch given by formula (1), § 4 becomes (omitting the factor $2 \cdot \frac{4{\cdot}00}{E}$)

$$\frac{1}{2}(17{\cdot}03B^2 - 26{\cdot}95 \cdot 2BH + 92{\cdot}86H^2 + 9189493 \cdot 2B - 31309780 \cdot 2H).$$

Equating to zero the differential coefficients of this expression with regard to B and H, we get the following two equations:—

$$17{\cdot}03B - 26{\cdot}95H + 9189493 = 0$$
$$-26{\cdot}95B + 92{\cdot}86H - 31309780 = 0.$$

from which we get

$$H = 333960$$
$$B = -11120.$$

Substituting these values in the above expressions for $B_0$ and $P_0$ we get

$$B_0 = 122166,$$
$$P_0 = 403832,$$

therefore the eccentricity $e$ at the springings is equal to

$$\frac{122166}{403832} = {\cdot}30.$$

---

quite clear, we will give here the quantities necessary to make up the expressions for $B_0$, $P_0$, $S_0$, for the two cases.

| | Depth of Section 0. | |
|---|---|---|
| | 2·01 m. | 1·50 m. |
| Vertical distance from the centre of section 0 to the horizontal line through the centre of the crown. | 5·36 | 5·58 |
| Horizontal distance from the centre of section 0 to the vertical line through crown . . . . | 23·05 | 23·05 − ·11 |
| Sine of angle which section 0 makes with the vertical | ·447 | ·510 |
| Cosine of angle which section 0 makes with the vertical . . . . . . . . . . . | ·895 | ·860 |

Remembering that the weight of the portion 0, 6 is 228,679 kg., and that its moment about the centre of the original section 0 is 2,021,937 kg. m. (see table at end of § 2, col. 9, line 1), we obtain as follows the values given above for the new expressions for $B_0$, $P_0$, $S_0$:—

$$1{,}996{,}782 = 2{,}021{,}937 - 228{,}679 \times {\cdot}11,$$
$$116{,}626 = 228{,}679 \times {\cdot}510,$$
$$196{,}664 = 228{,}679 \times {\cdot}860.$$

As this is still greater than $\frac{1{\cdot}50}{6} = 0{\cdot}25$, i.e. greater than one-sixth of the assumed depth of the section, it follows that the line of pressure still passes outside the middle-third near to the springings.

We will also investigate in the present case the extent of the compressed portion of the section, assuming that the other portion is capable of resisting tension; if we call $y$ the distance from the neutral axis to the centre line, we shall have the equation

$$\frac{P_0}{A_0} - \frac{B_0 y}{I_0} = 0,$$

i.e.

$$269221 - 434350y = 0,$$

or

$$y = {\cdot}62 \text{ m.}$$

Thus the compressed portion of the section will be

$$\frac{1{\cdot}50}{2} + {\cdot}62 = 1{\cdot}37 \text{ m.};$$

and the part under tension

$$\frac{1{\cdot}50}{2} - {\cdot}62 = {\cdot}13 \text{ m.}$$

We could now make a third approximate calculation, taking 1·37 m. for the depth of the section at the springings; but in practice a second approximation is usually sufficient, and we may conclude with sufficient accuracy that if the arch under consideration had been constructed without mortar, the sections at the springings would have been compressed only over a depth of 1·37 m from the intrados, and that consequently the arch would have opened up for a depth of $2{\cdot}01 - 1{\cdot}37 = {\cdot}64$ m. from the extrados.

If we wished to obtain a closer approximation without making a third calculation, we might proceed as follows:—

Call $1{\cdot}37 - x$ the true compressed section at the springings; then in the first calculation in which we took the depth at springings equal to 2·01 m., i.e. deeper than the true compressed section by an amount ${\cdot}64 + x$, we found the compressed depth to be 1·513 m., the error being ${\cdot}143 + x$; in the second calculation in which we took the section as 1·50 m., i.e. ${\cdot}13 + x$ greater than the true section, the compressed depth was found to be 1·37 m., the error being $x$. If then we assume that the initial errors are approximately proportional to the final errors, we shall have the proportion

$${\cdot}64 + x : {\cdot}143 + x = {\cdot}13 + x : x,$$

which reduces to a linear equation giving

$$x = \cdot 042.$$

The true compressed section, therefore, to a close degree of approximation will be

$$1 \cdot 37 - \cdot 042 = 1 \cdot 328 \text{ m.},$$

and therefore the arch will open up for a depth of $2 \cdot 01 - 1 \cdot 328 = \cdot 682$ m. from the extrados.

The normal pressure at the springings may be taken as equal to that which we obtained in the second approximation; and as now the line of pressure passes, at the springings, to the limit of the middle-third of the effective compressed section, it follows that the maximum compressive stress per sq. metre will be double the mean stress, and that its value will be $\frac{403832}{1328} \times 2 = 608{,}200$ kg. per sq. m., or the maximum compressive stress will be $60 \cdot 82$ kg. per sq. cm.

**7. Conclusion.**—As this stress will not be excessive for the granite of which the arch is constructed, we see that it could have been built of sufficient strength without adopting the particular arrangement designed by the celebrated engineer, Carlo Mosca.

But a comparison of the results obtained for the two cases which we have investigated shows to what extent the design has contributed to the stability of the arch; because the maximum compression stress has been reduced to $41 \cdot 24$ kg. per sq. cm., whereas for an arch without mortar it would have been $60 \cdot 82$ kg. per sq. cm.

# INDEX.

# PLATES

PLATE I

Fig. 1.

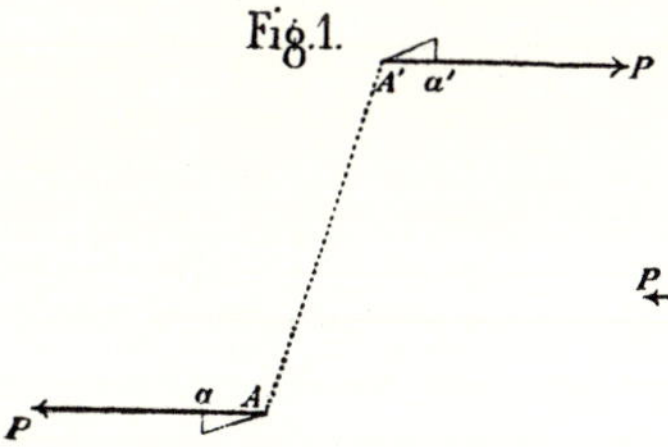


Fig. 2.

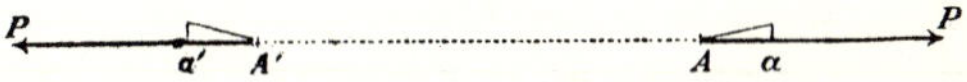


Fig. 3.

Fig. 4.

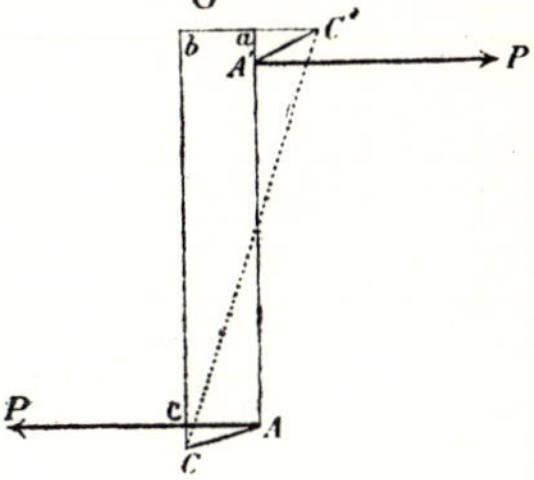


Fig. 5.

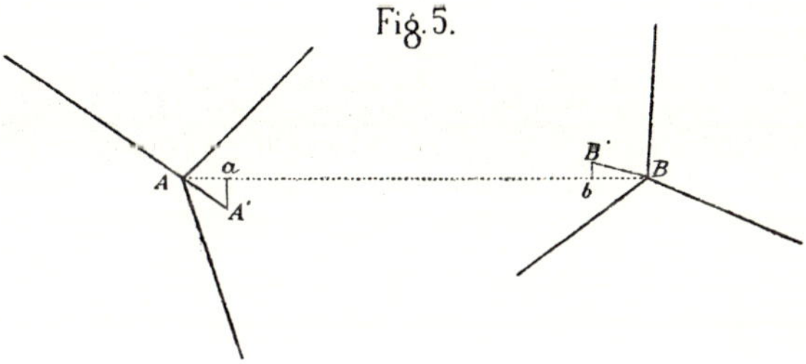


Fig. 7.

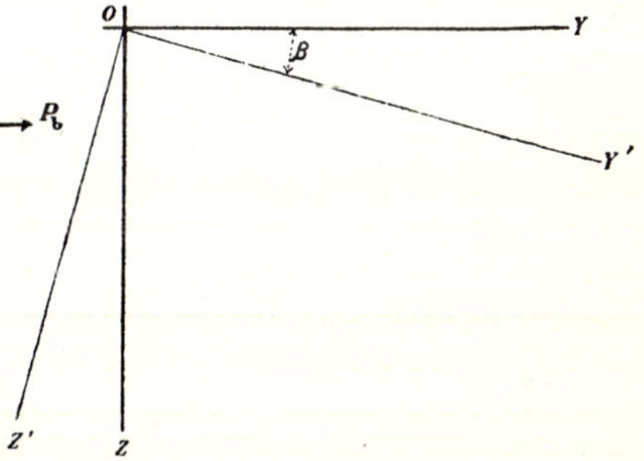


Fig. 6.

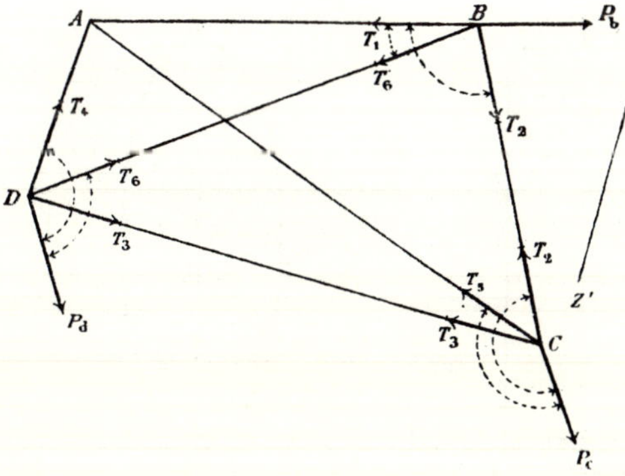

PLATE I *(continued)*

Fig. 8.

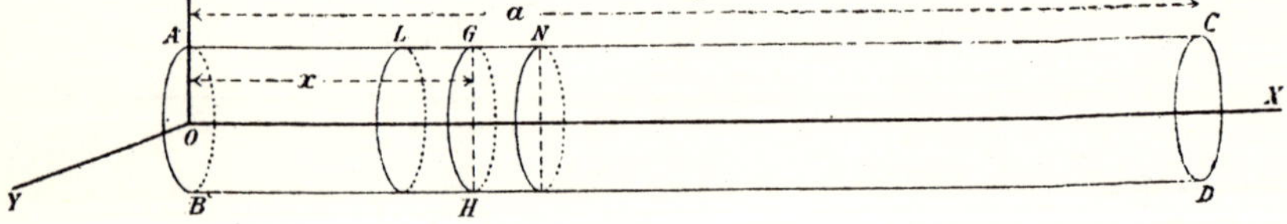


Fig. 9.

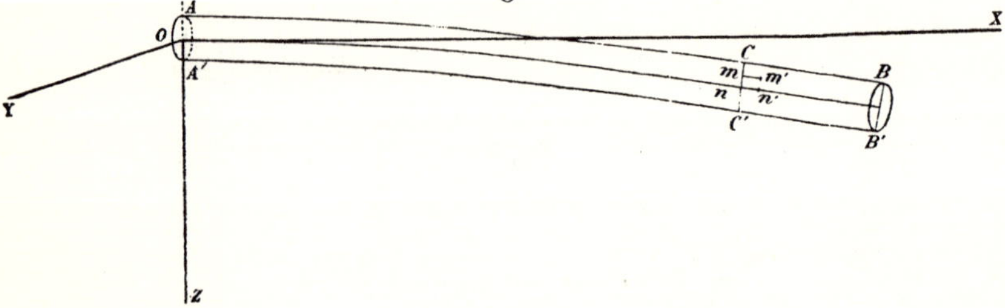


Fig. 10.

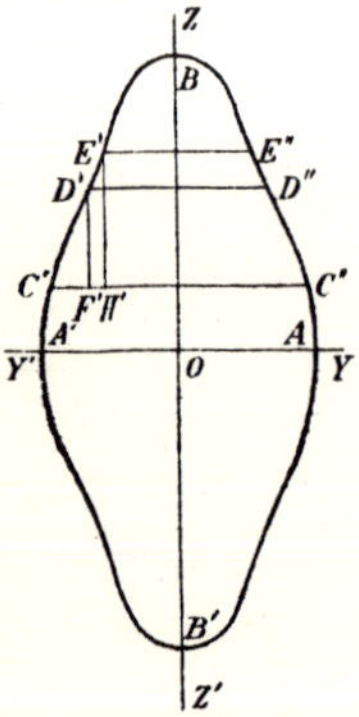


Fig. 11

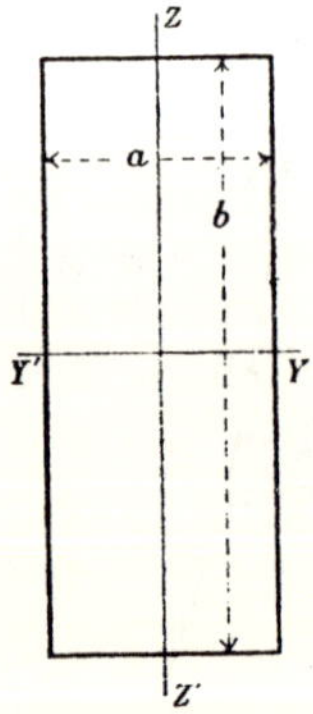


Fig. 12.

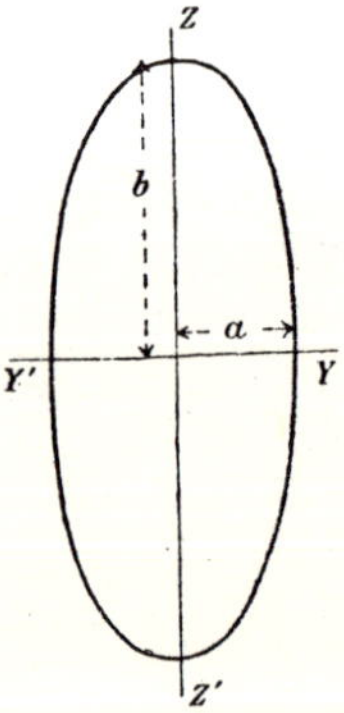

PLATE I *(concluded)*

Fig. 13.

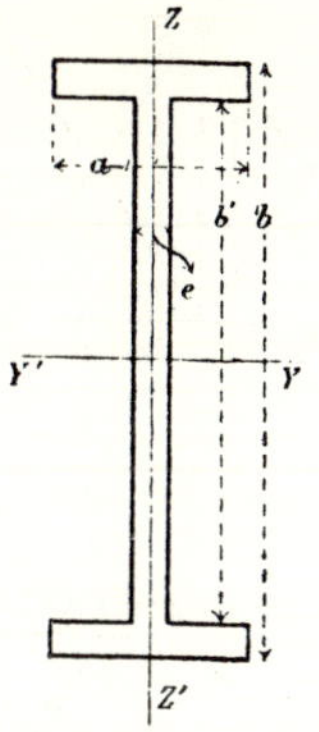


Fig. 14.

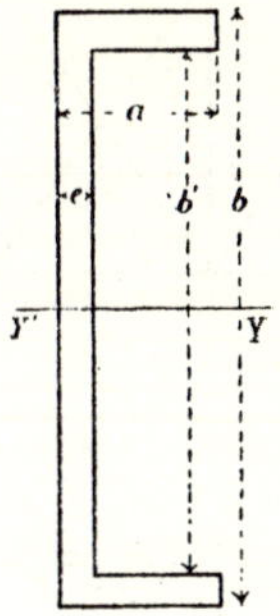


Fig. 15

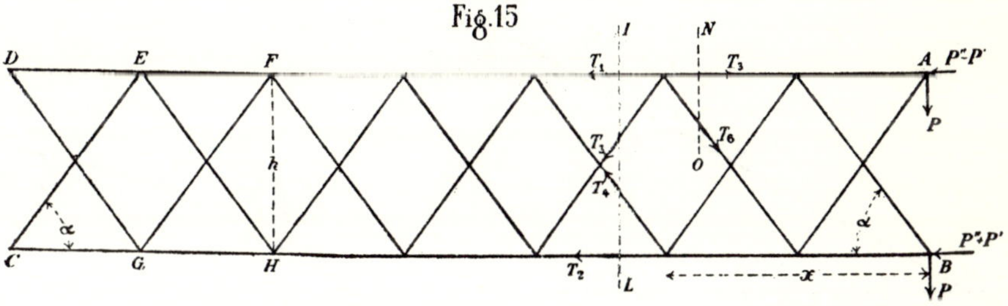


Fig. 16.

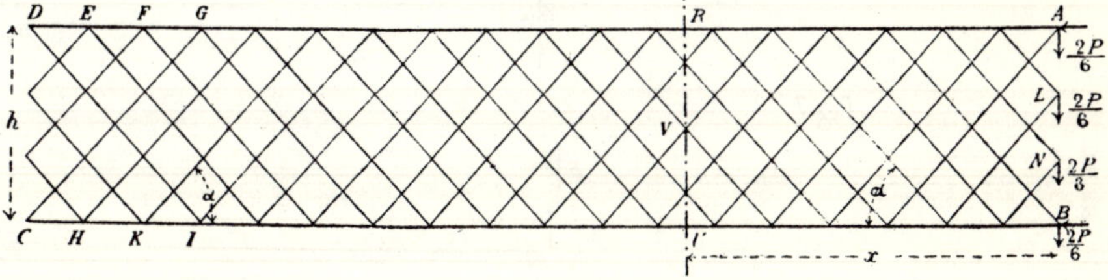

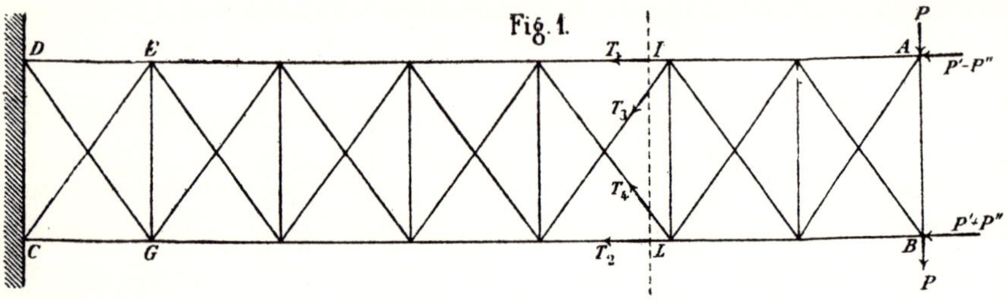

Fig. 1.

Fig. 2.

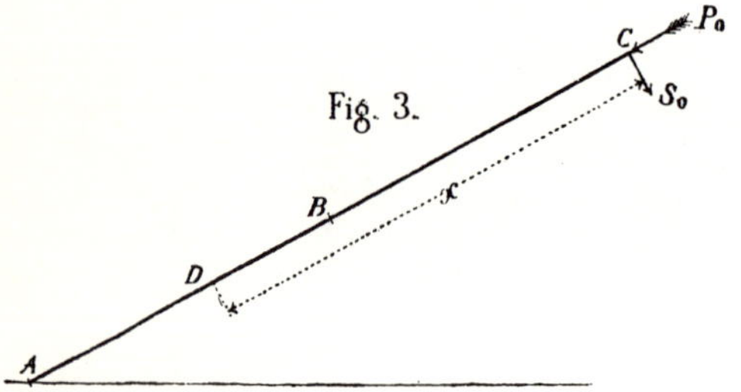

Fig. 3.

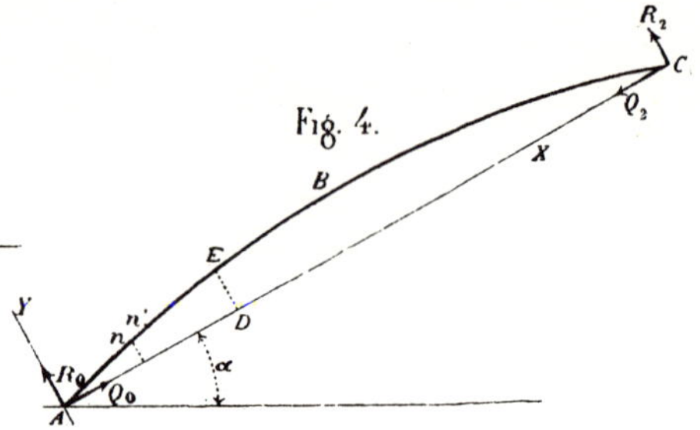

Fig. 4.

Fig. 5.
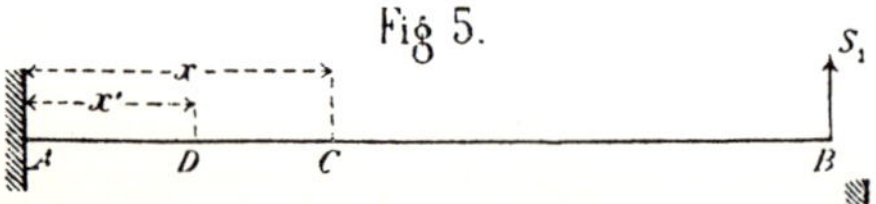


Fig. 6.

Fig. 7

Fig. 8
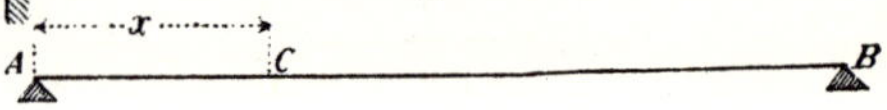

PLATE II *(continued)*

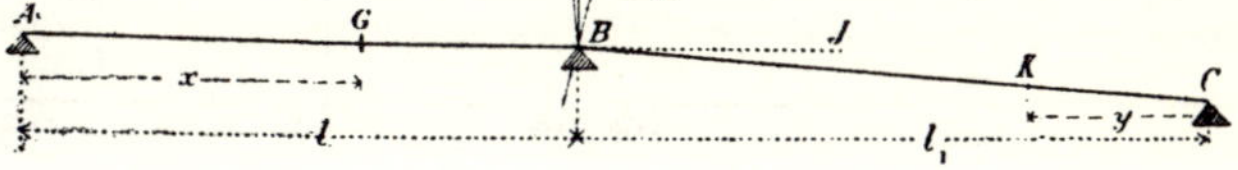


Fig. 9.

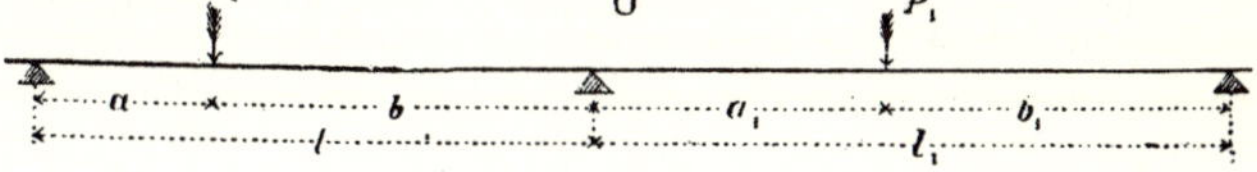


Fig. 10

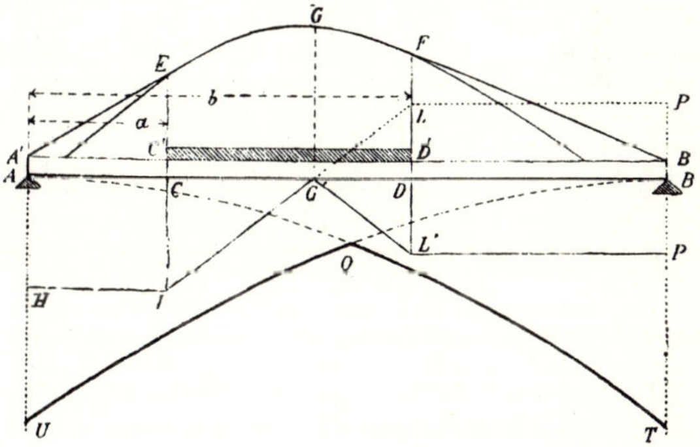


Fig. 11

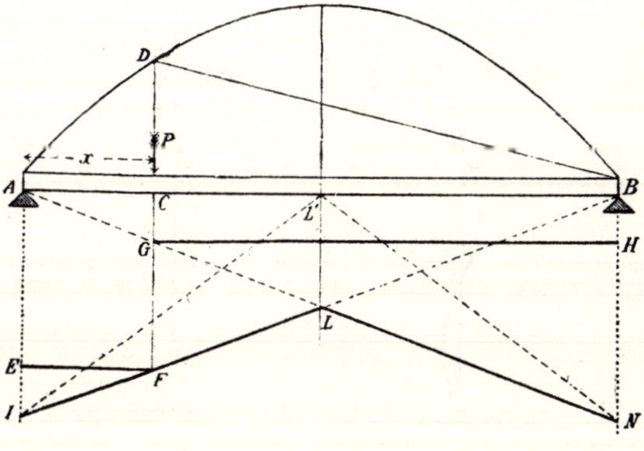


Fig. 12

Fig. 13.

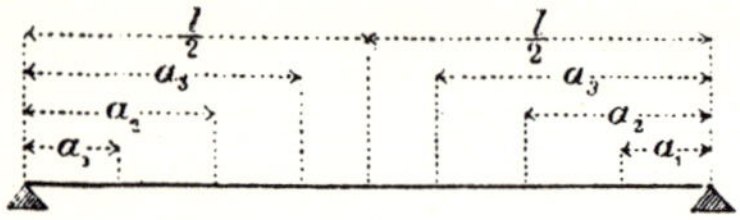


Fig. 14

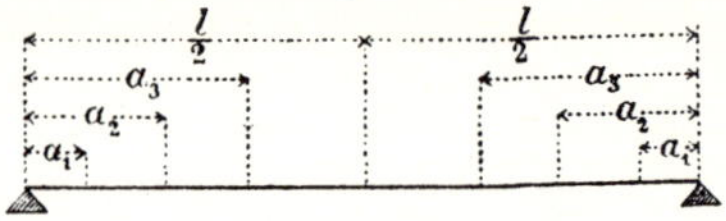


Fig. 15.

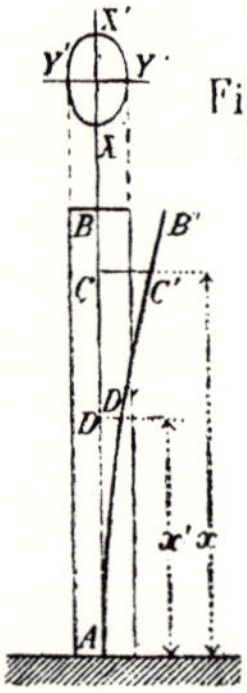


Fig. 16.

Fig. 17.

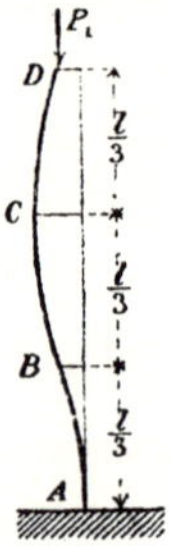


Fig. 18.

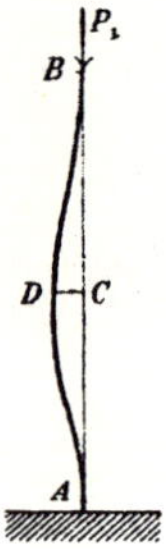


Fig. 19.

Fig. 20.

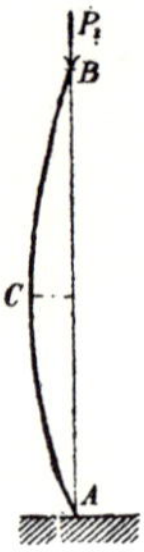


Fig. 21.

Fig. 22.

PLATE III

Fig. 1.

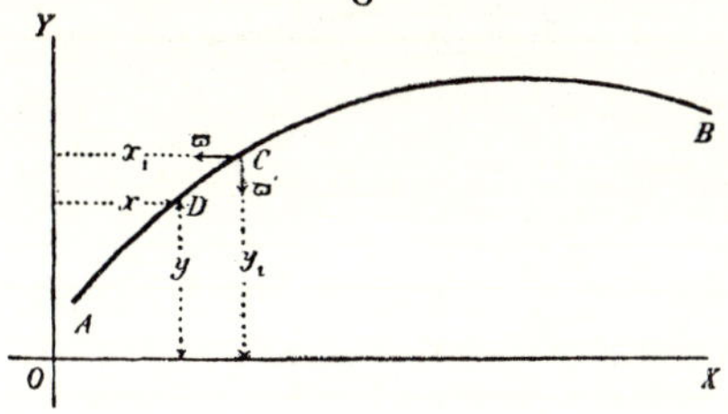


Fig. 2.

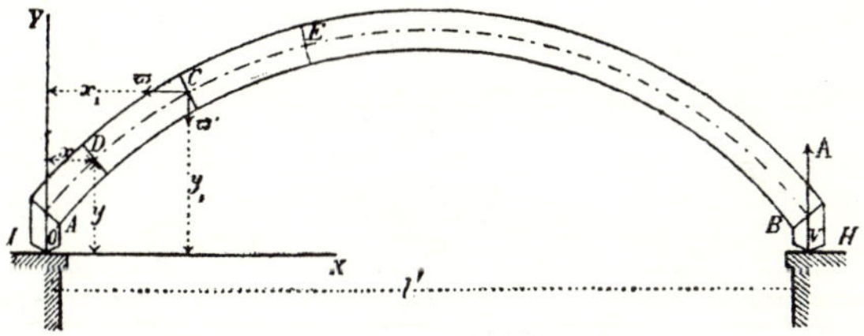


Fig. 3.

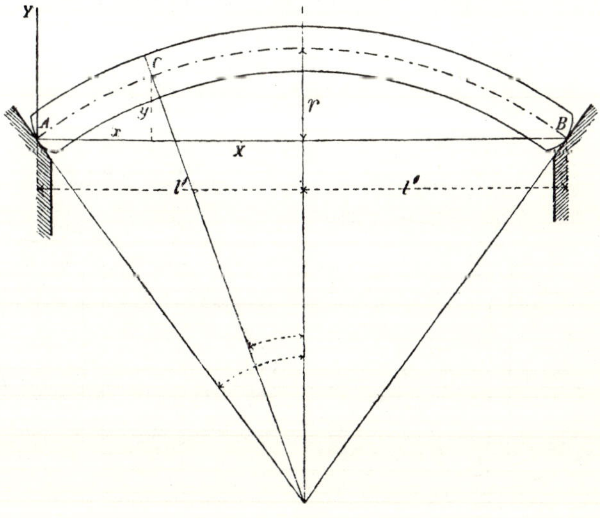

Fig. 4

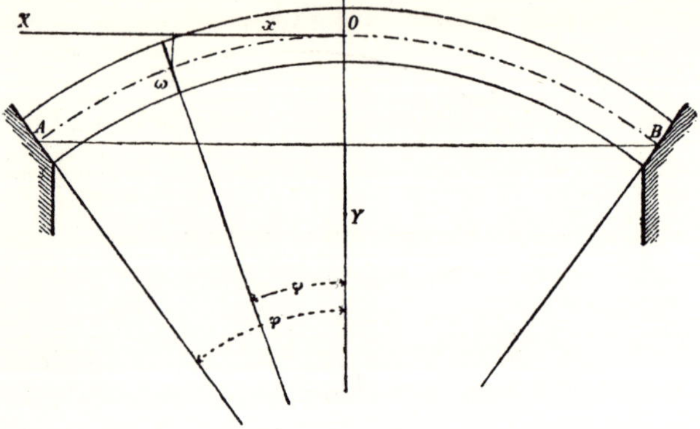


Fig. 5.

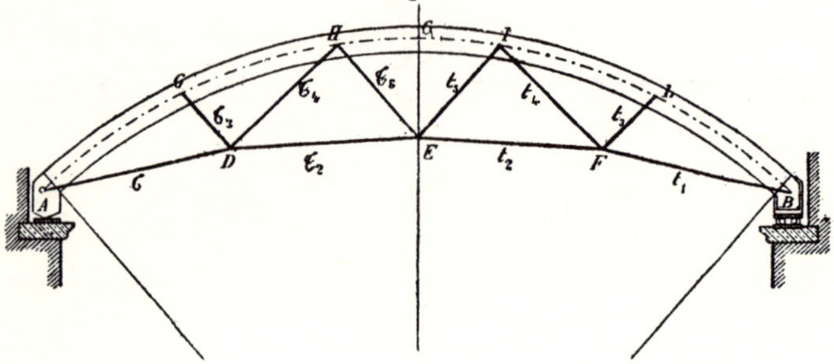


Fig. 6.

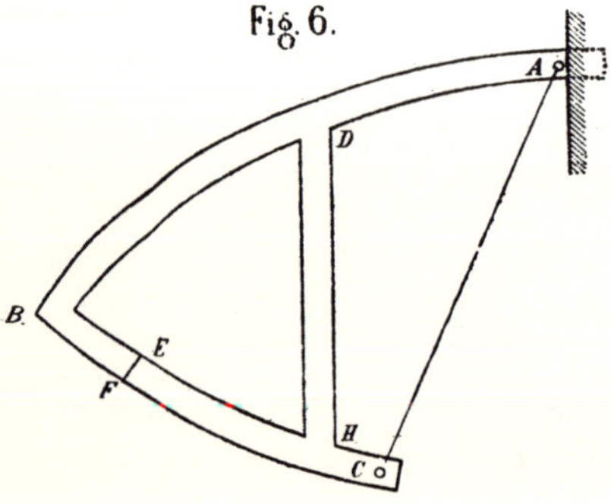


Fig. 7.

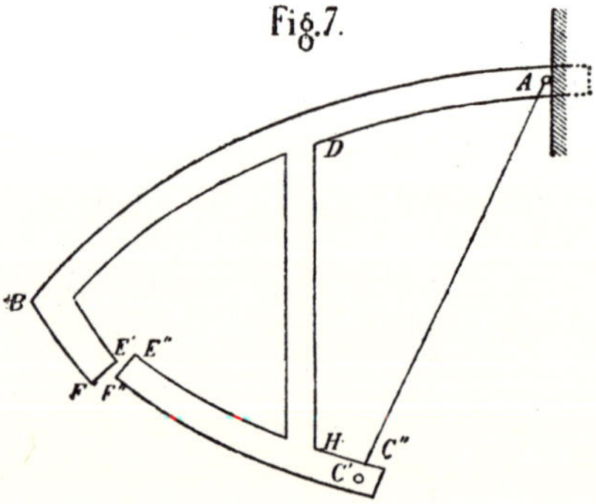

PLATE III *(concluded)*

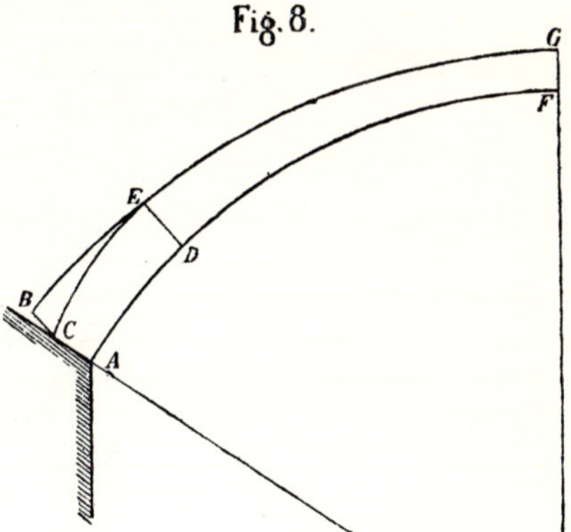


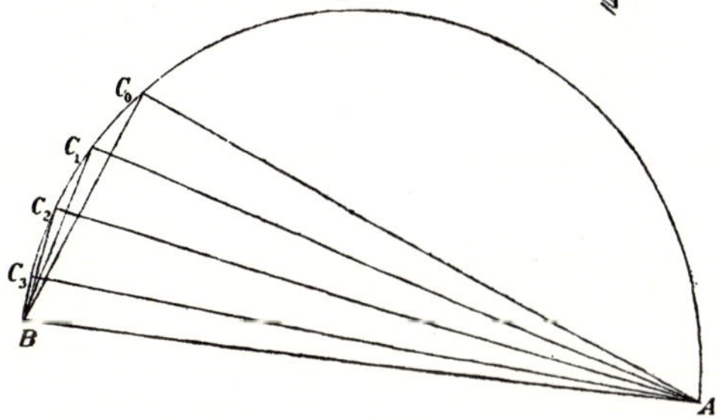


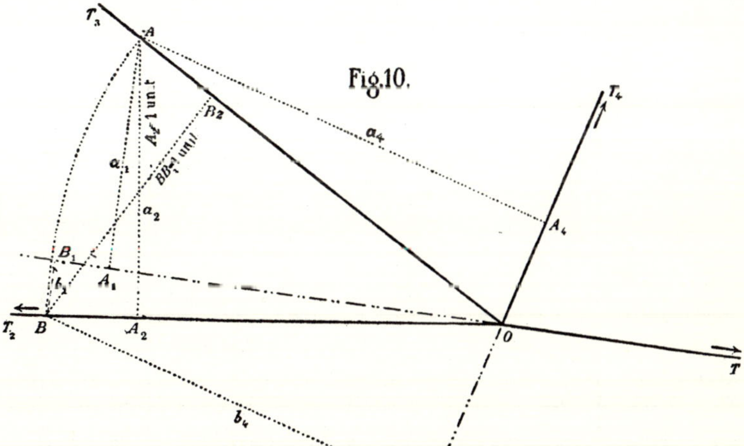

PLATE IV

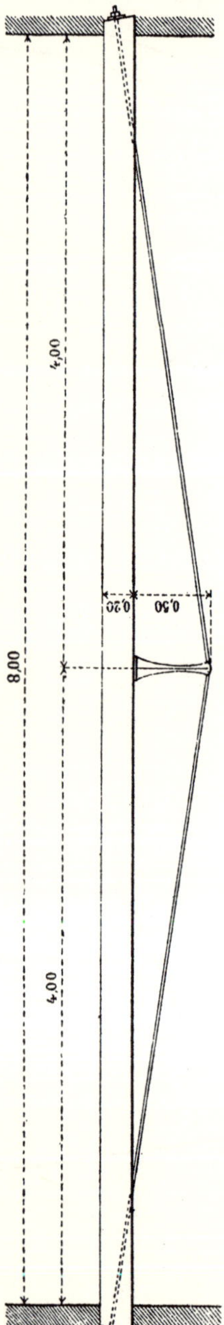


Fig. 1.

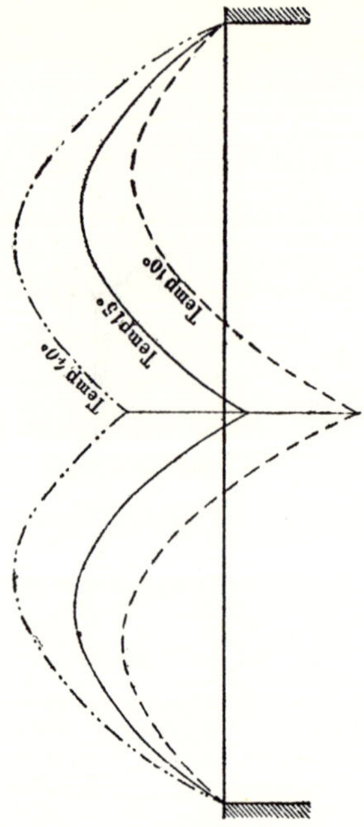


Fig. 2.

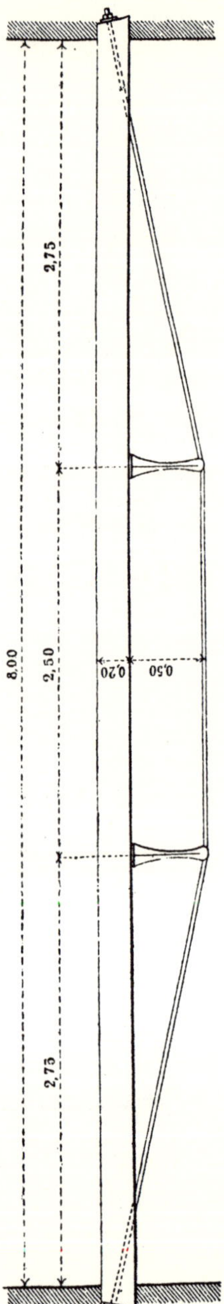


Fig. 3

## PLATE IV *(concluded)*

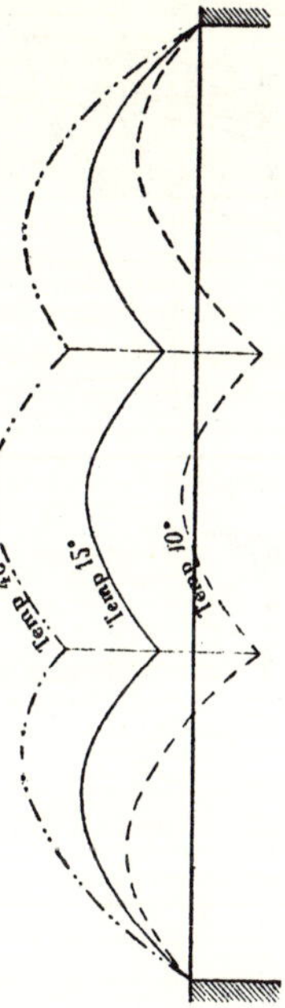


Fig. 4.

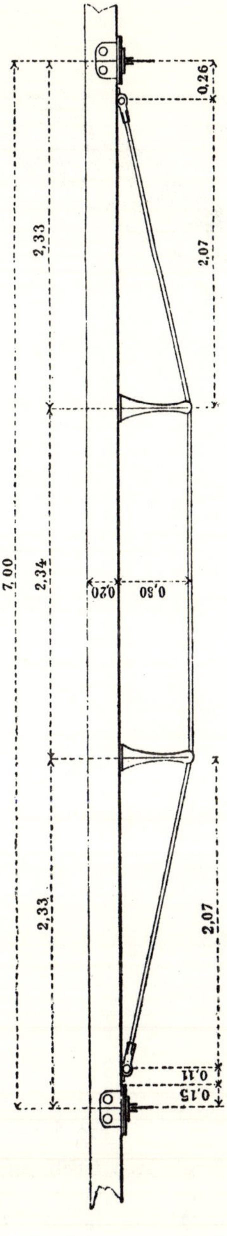


Fig. 5

SCALE OF FIGS. 1, 3, 5.

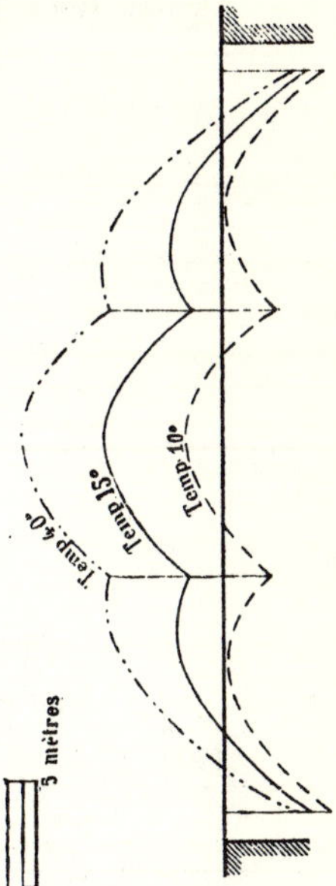


Fig. 6.

BENDING MOMENT DIAGRAMS.

SCALE—1 cm. = 100 kg. metres.

SCALE OF FIGS. 2, 4, 6.

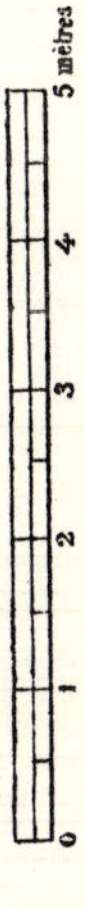

Arched Roof Truss without Tie-Rods.
(*See* p. 254.)

Fig. 1.

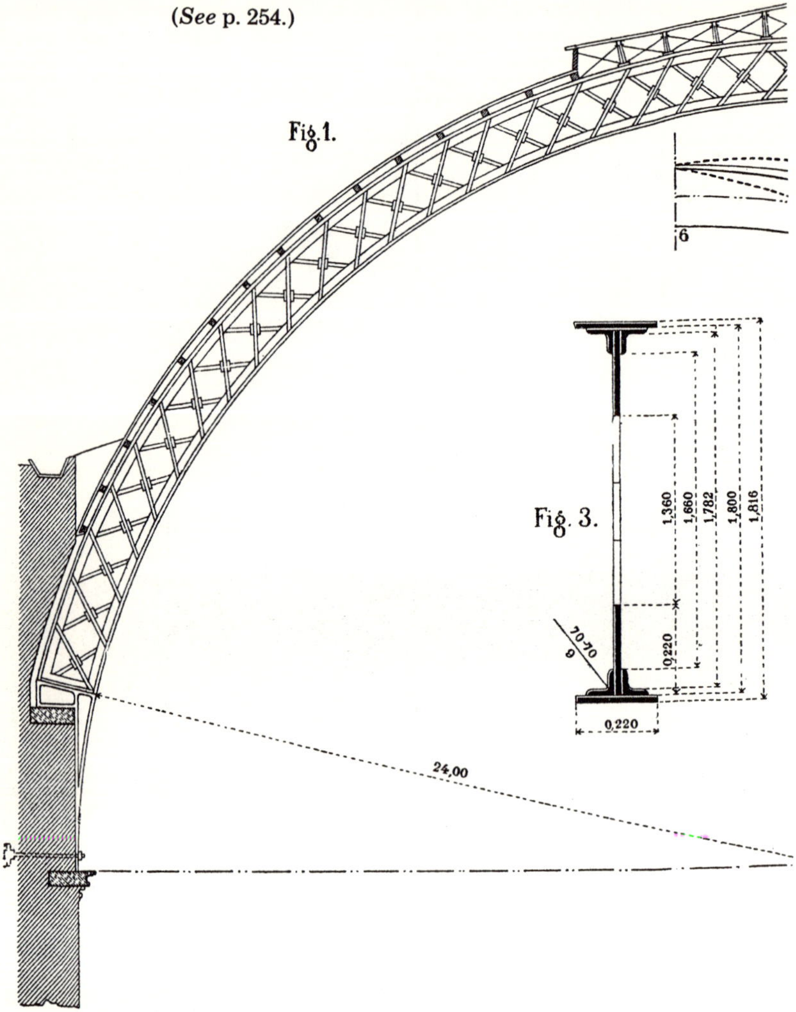


Fig. 4.

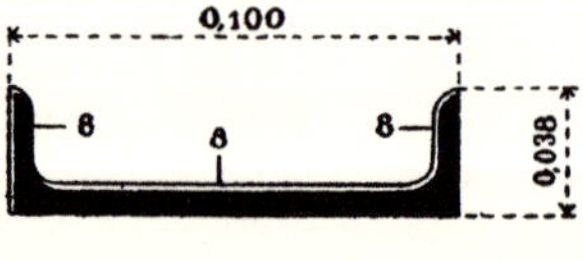

PLATE V *(concluded)*

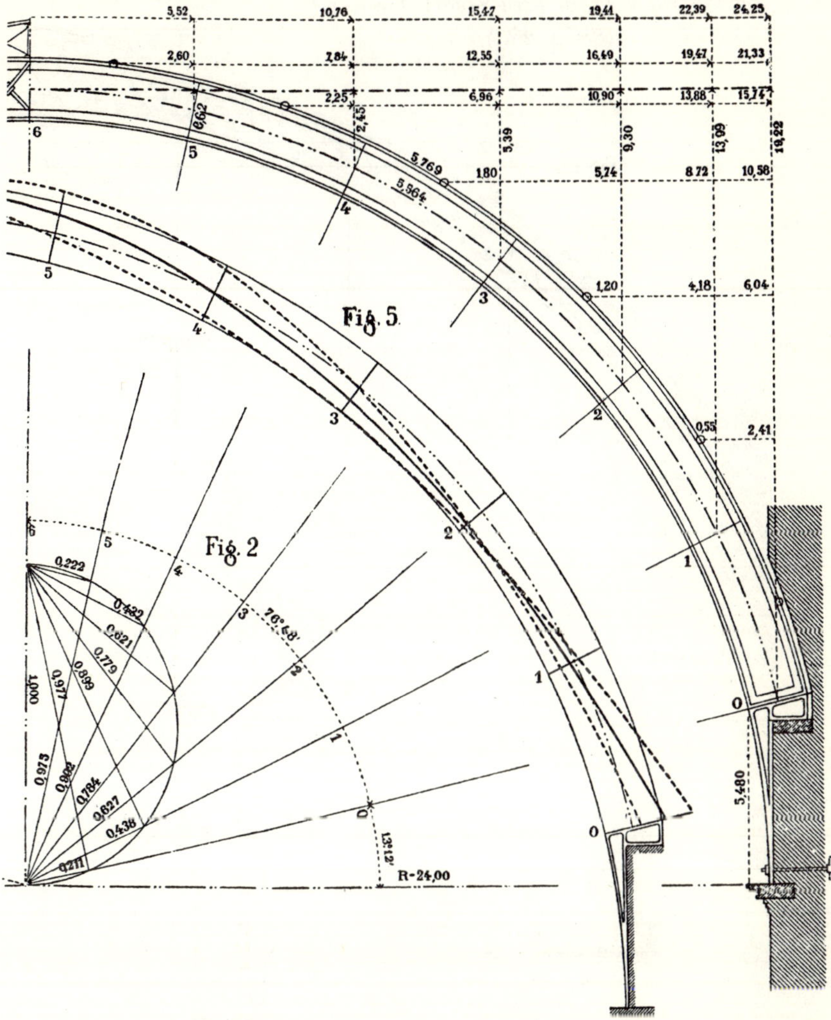


SCALE OF FIG. 2.

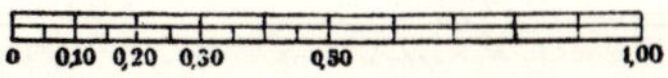


SCALE OF FIGS. 1 AND 5.

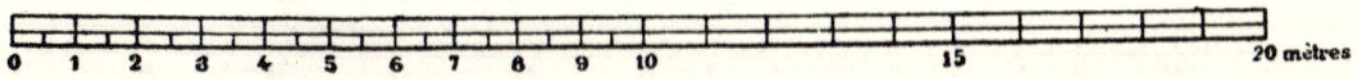

PLATE VI

ARCHED ROOF TRUSS WITH SINGLE TIE-ROD.

(*See* p. 267.)

Fig. 1.

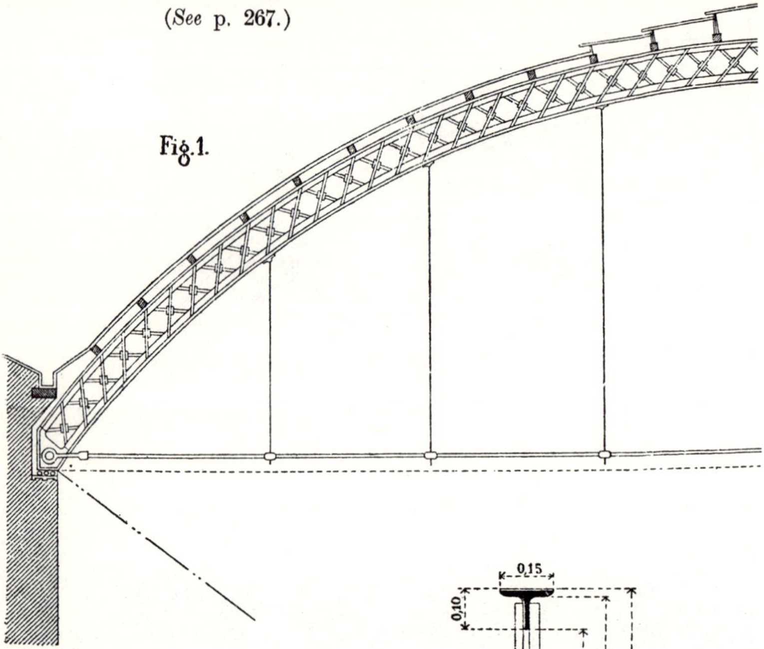

Fig. 3.

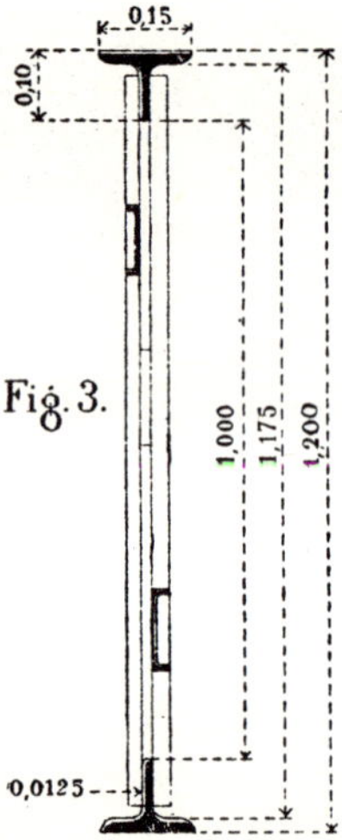


Fig. 4.

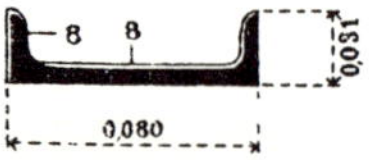


SCALE OF FIGS. 1 AND 5.

PLATE VI *(concluded)*

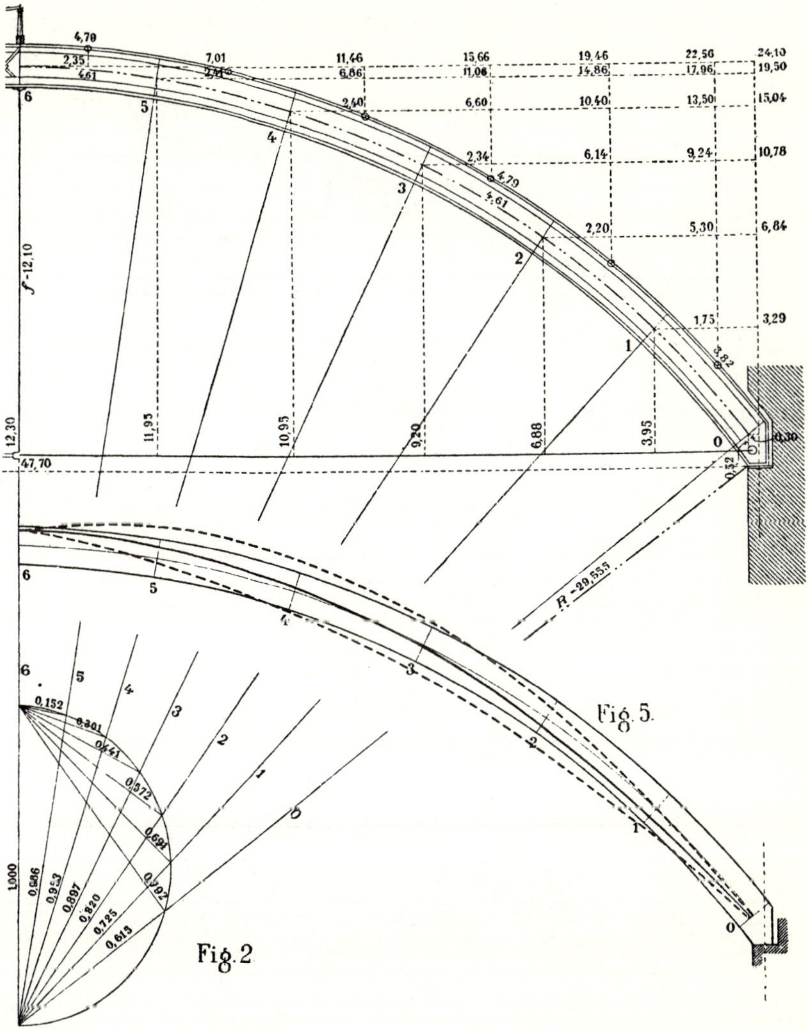


Fig. 5

Fig. 2

Scale of Fig. 2.

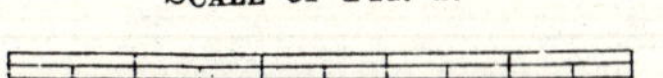

PLATE VII

ROOF TRUSS OF POLONCEAU TYPE.

(*See* p. 274.)

Fig. 1.

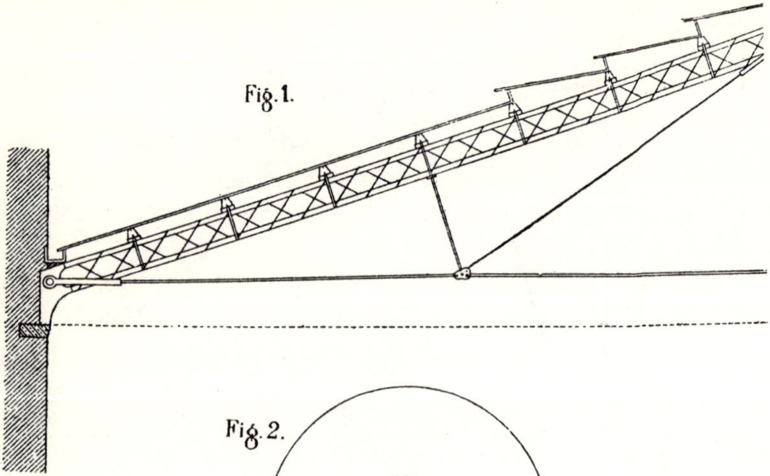

Fig. 2.

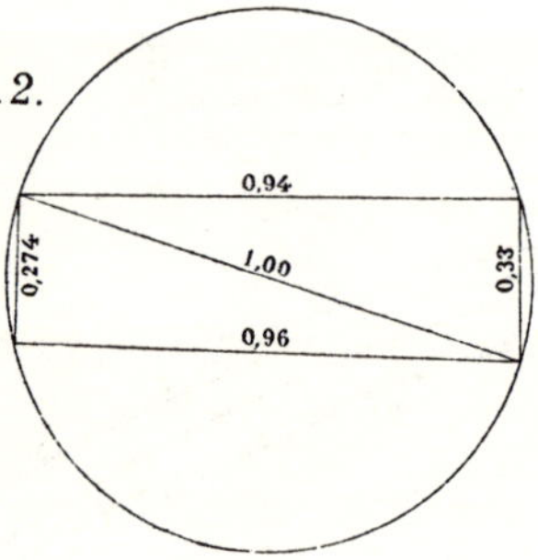


Fig. 3

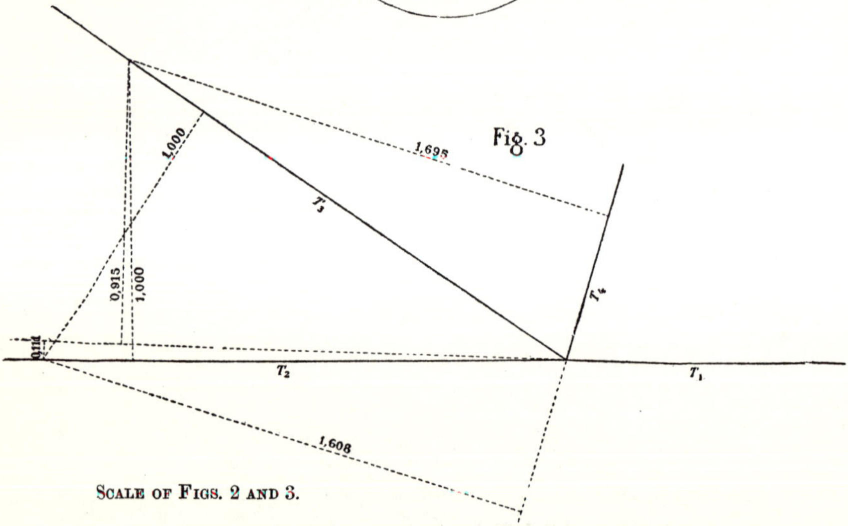


SCALE OF FIGS. 2 AND 3.

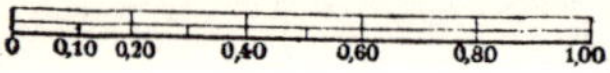

PLATE VII *(concluded)*

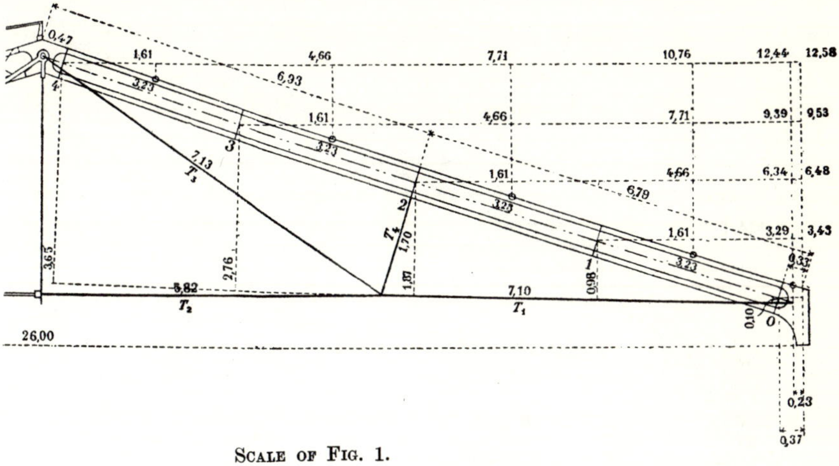


Scale of Fig. 1.

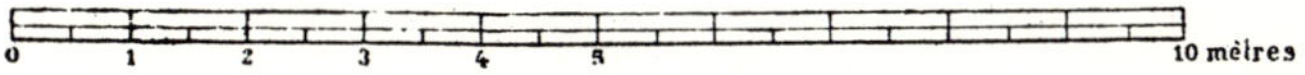


Fig. 4.

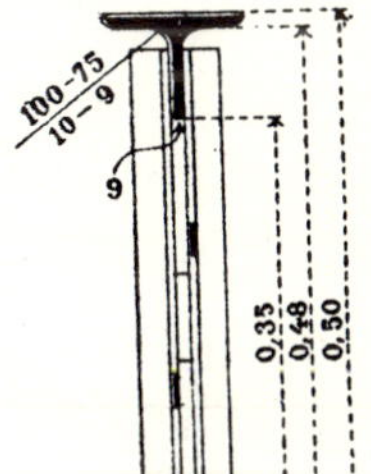


Fig. 5.

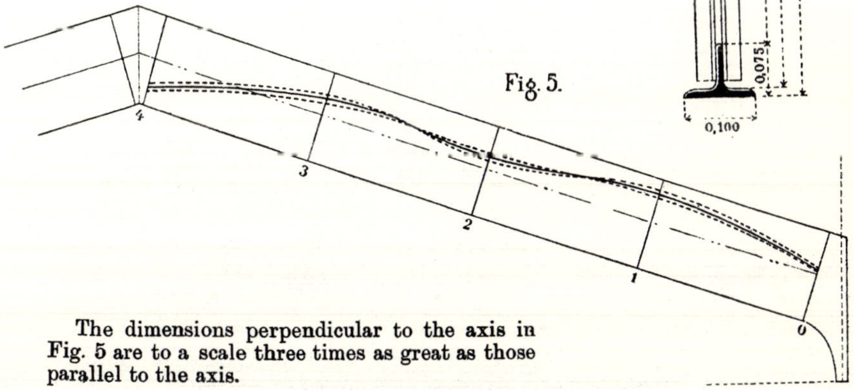

The dimensions perpendicular to the axis in Fig. 5 are to a scale three times as great as those parallel to the axis.

**PLATE VIII**

Arched Roof Truss with Several Tie-Rods.

(*See* p. 284.)

Fig. 1.

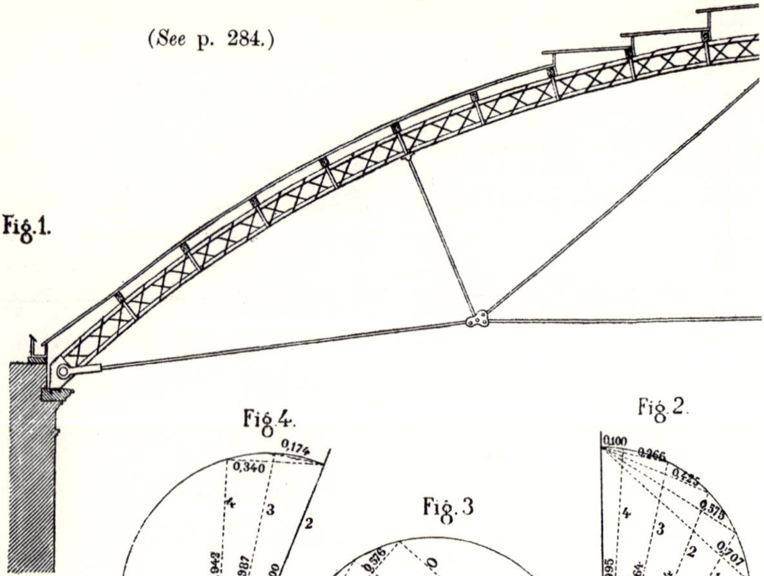

Fig. 4.

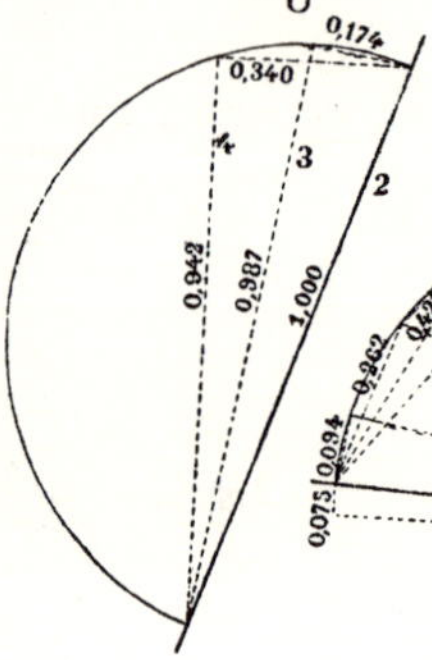


Fig. 3

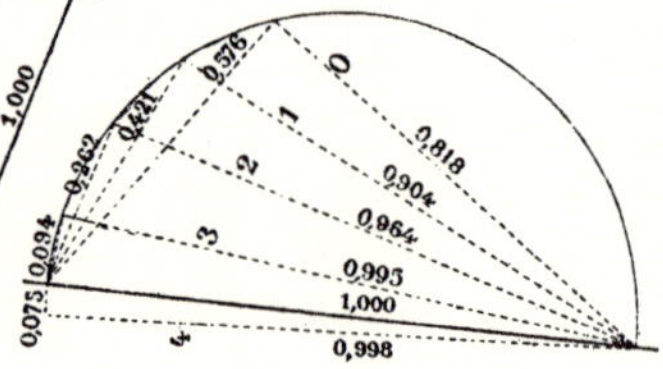


Fig. 2.

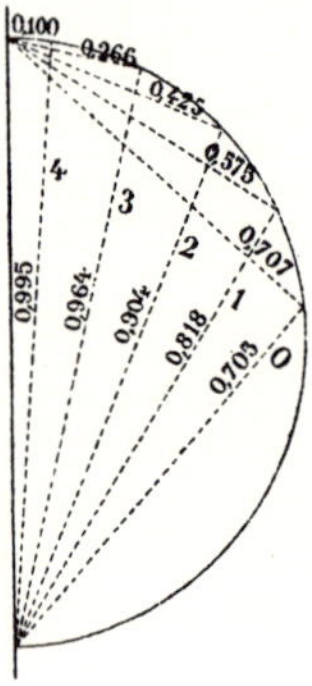


Fig. 6

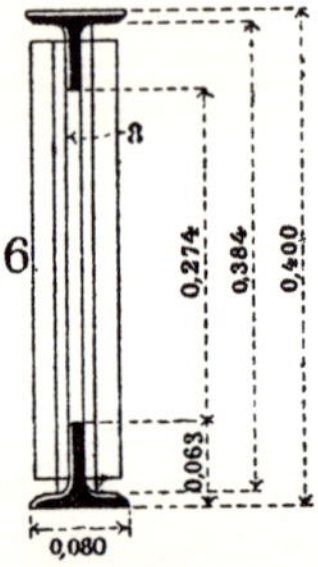


Scale of Fig. 1.

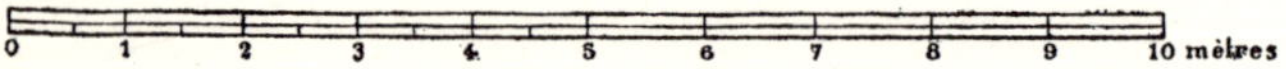


Scale of Figs. 2-5.

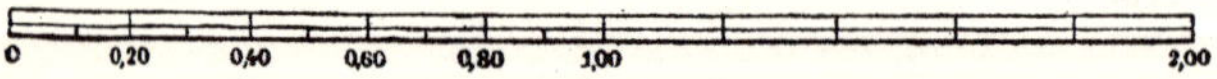

PLATE VIII *(concluded)*

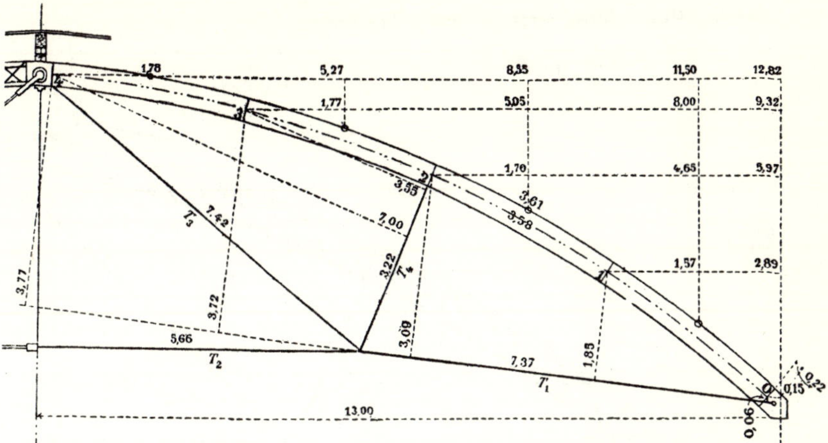


The dimensions perpendicular to the axis in Fig. 7 are to a scale 2·5 times as great as those parallel to the axis.

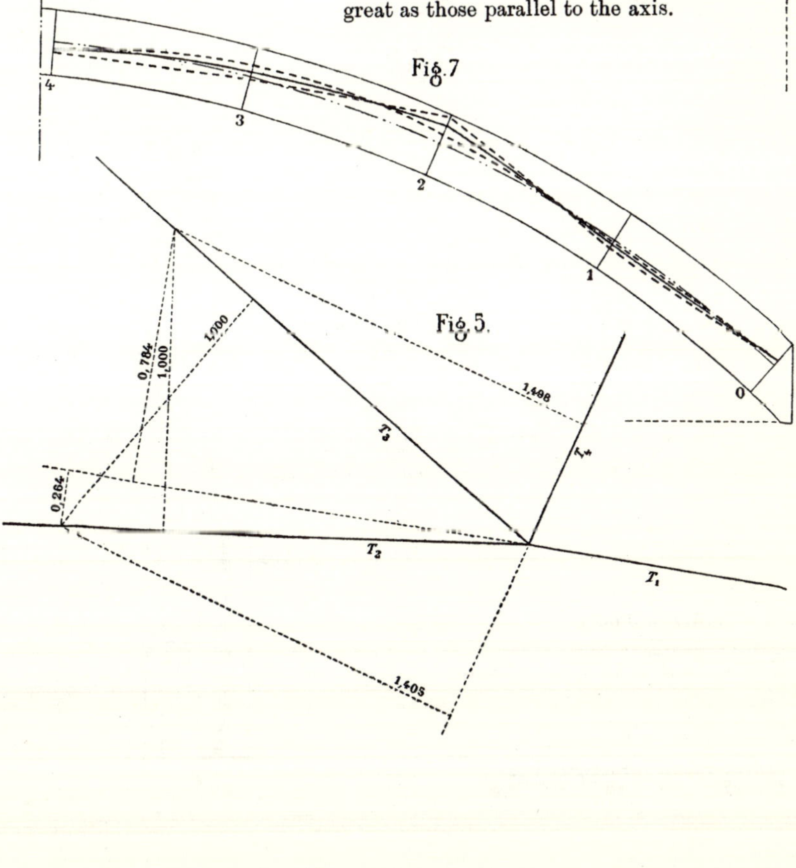

## Arched Roof Truss with Several Tie-Rods.

(*See* p. 292.)

Fig. 1.

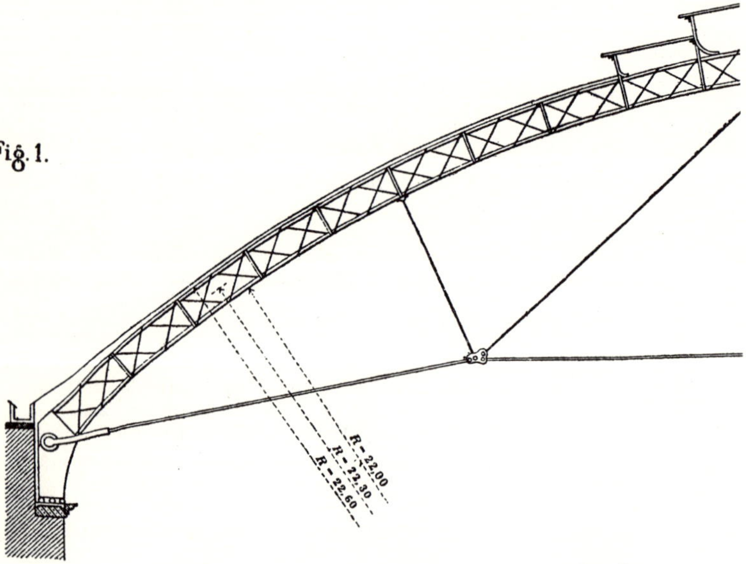


Fig. 4.

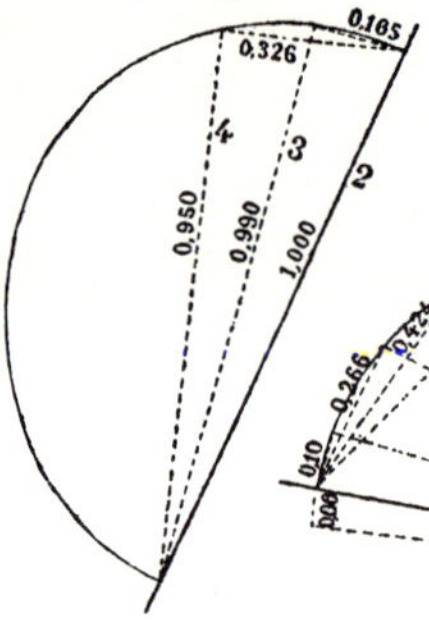


Fig. 3.

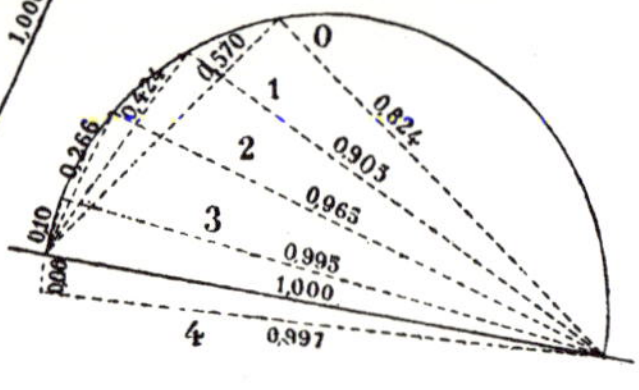


Fig. 2.

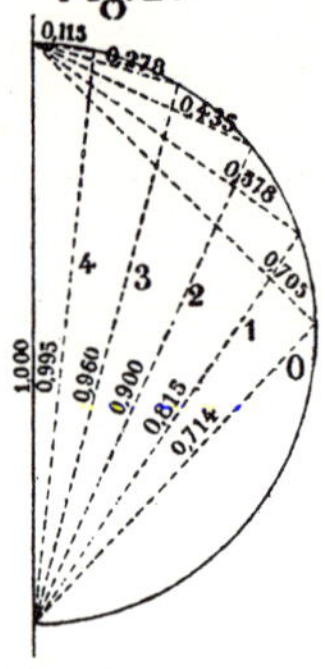


Fig. 6.

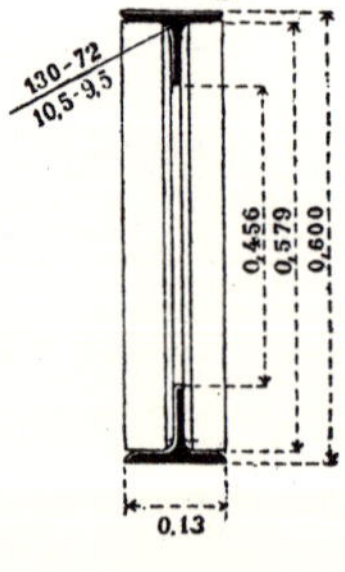


Scale of Fig. 1.

Scale of Figs. 2-5.

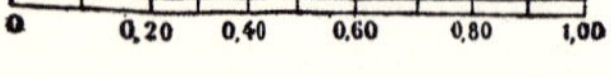

PLATE IX *(concluded)*

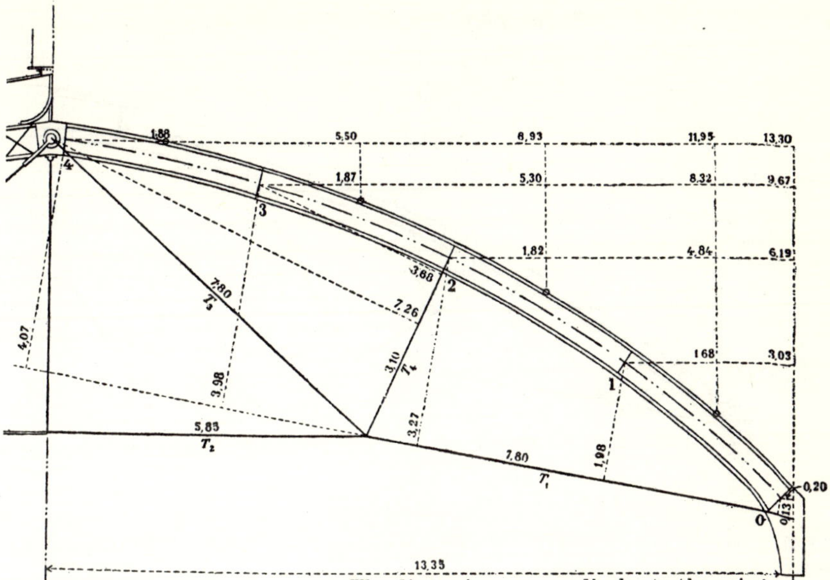


The dimensions perpendicular to the axis in Fig. 7 are to a scale twice as great as those parallel to the axis.

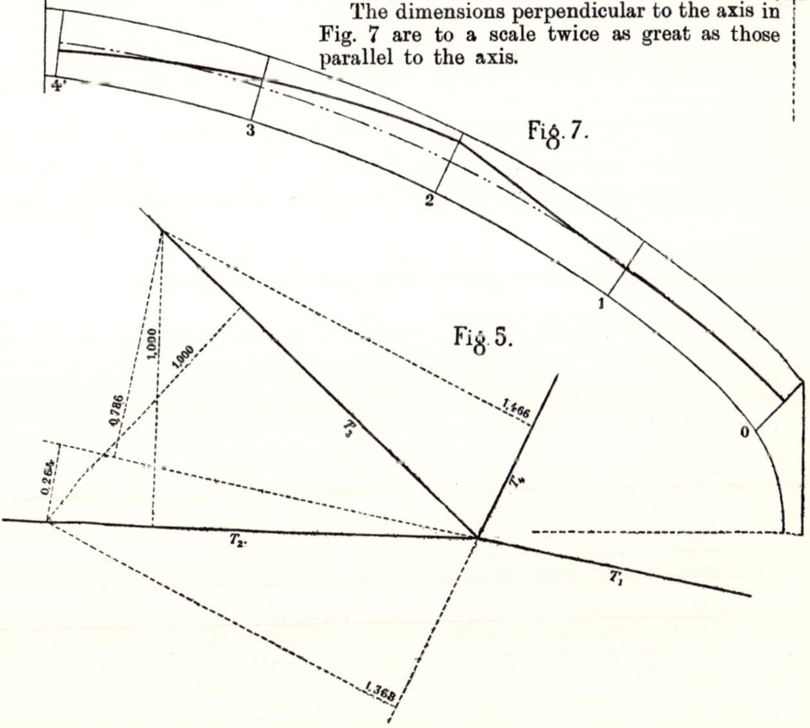


Fig. 7.

Fig. 5.

Arched Roof Truss with Several Tie-Rods.

(*See* p. 298.)

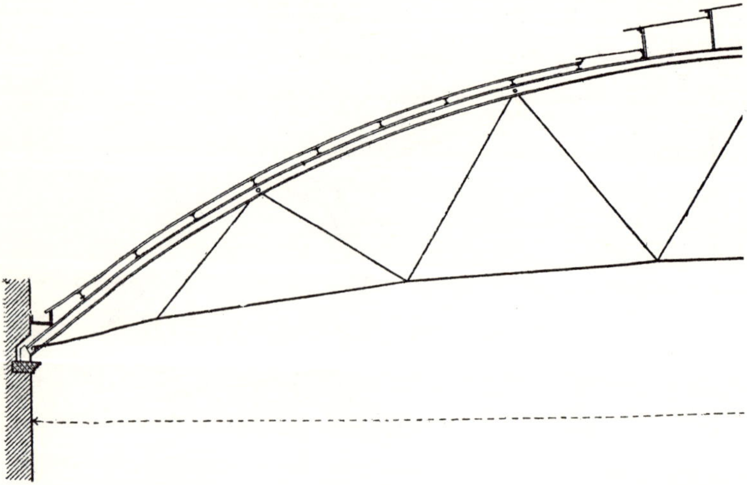

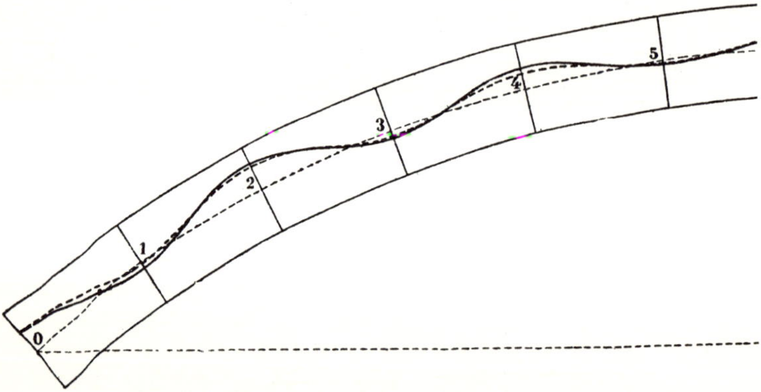


Scale of Figs. 1 and 11.

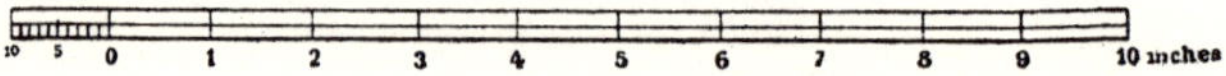

PLATE X *(concluded)*

Fig. 1.

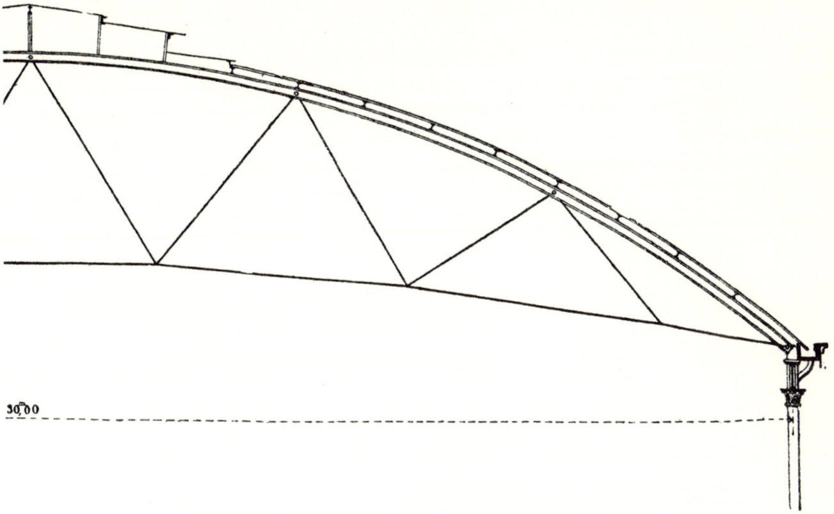


For Figs. 2-10, *see* Plate XI.

Fig. 11.

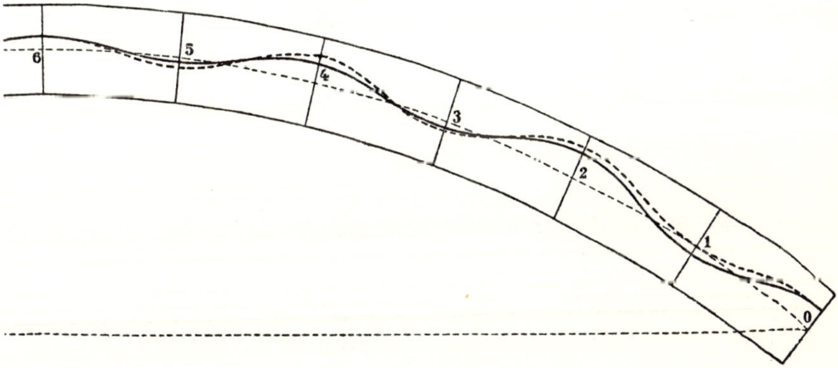


The dimensions perpendicular to the axis in Fig. 11 are to a scale ten times as great as those parallel to the axis.

PLATE XI

Fig. 2.

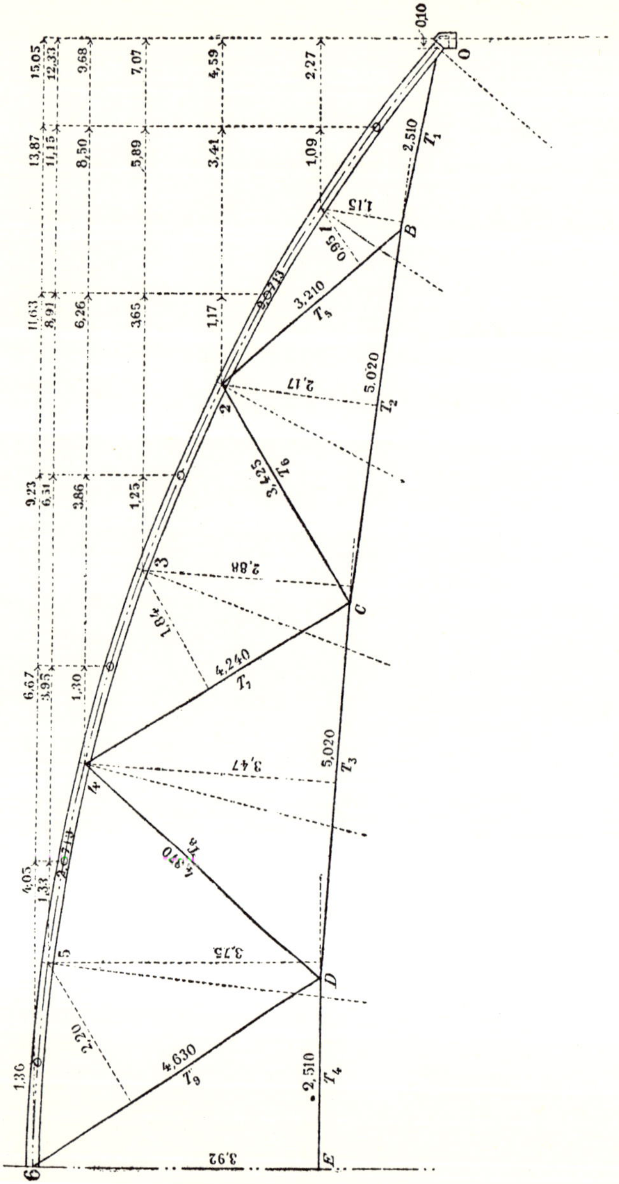


Scale of Fig. 2.

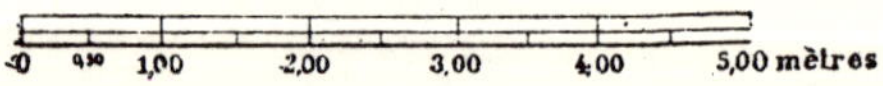

PLATE XI *(continued)*

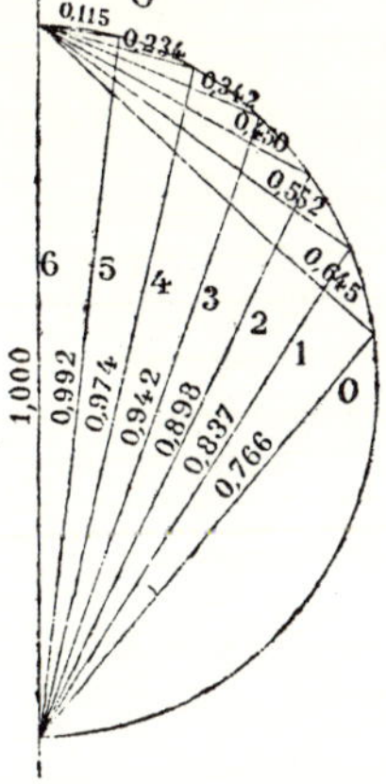


Fig. 3.

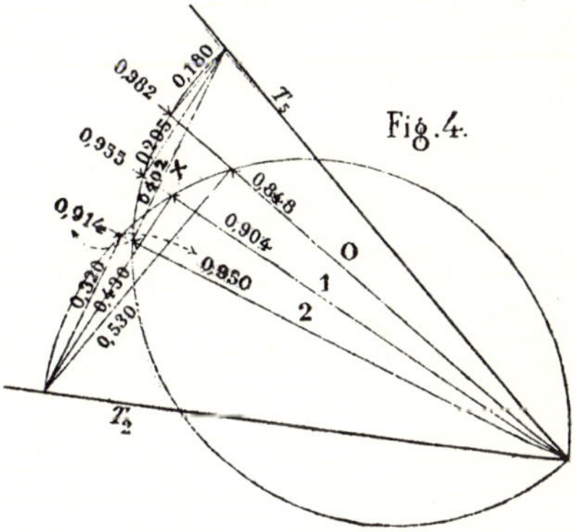


Fig. 4.

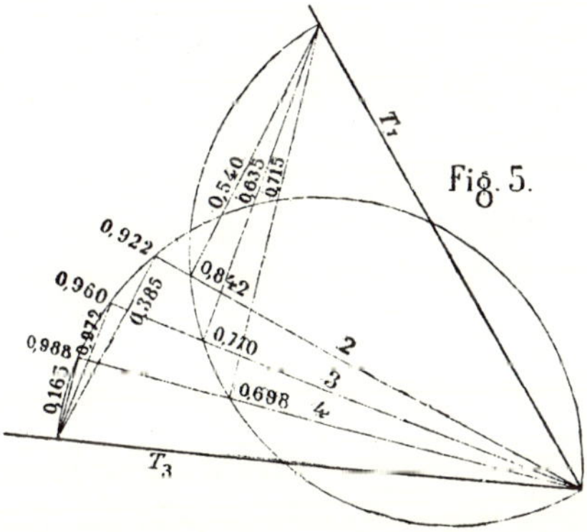


Fig. 5.

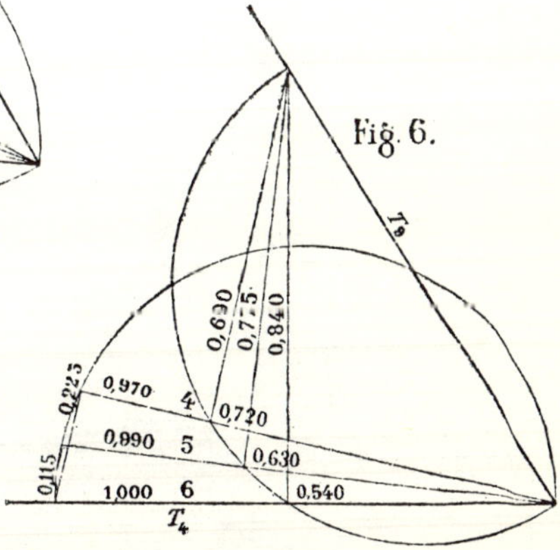


Fig. 6.

SCALE OF FIGS. 3-9.

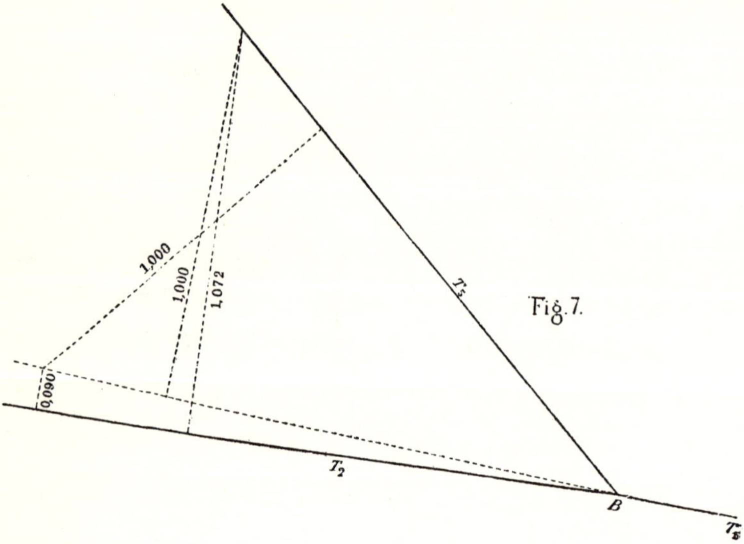
1,000
1,000
1,072
0,090
$T_5$
$T_2$
B
$T_5$

Fig. 7.

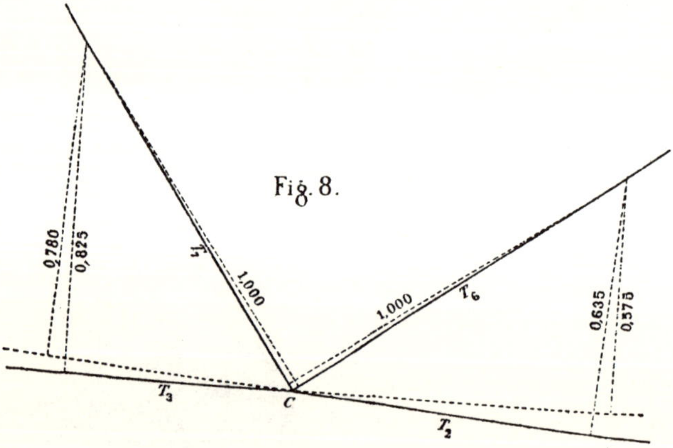
0,780
0,825
$T_1$
1,000
1,000
$T_6$
0,635
0,575
$T_3$
C
$T_2$

Fig. 8.

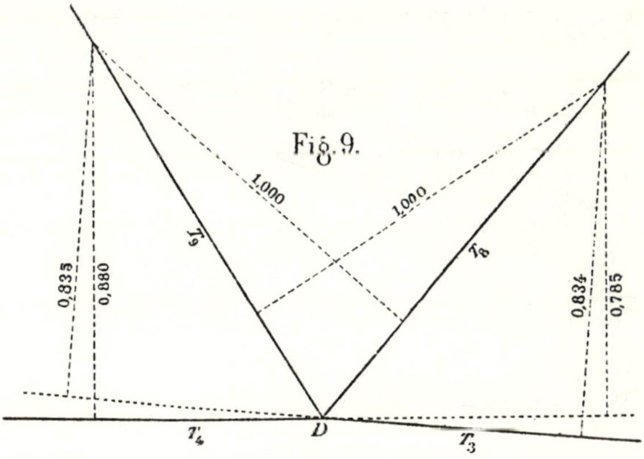
1,000
1,000
T9
T8
0,835
0,880
0,834
0,785
T4
D
T3

Fig. 9.

Fig. 10.

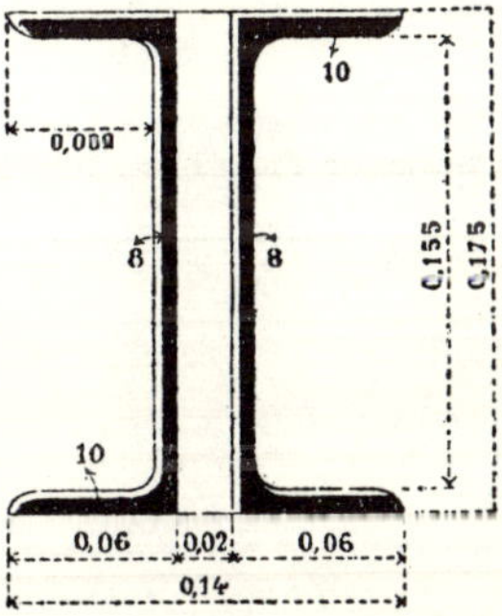
10
0,009
8
8
0,155
0,175
10
0,06
0,02
0,06
0,14

Arch Bridges with Flat and Rounded Ends.
(*See* p. 310.)

Fig 1. Elevation.

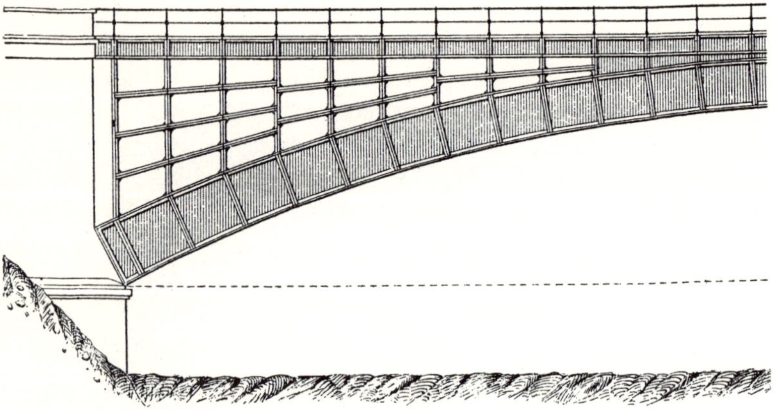

Scale of Figs. 1 and 2.

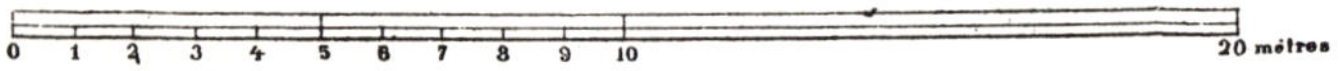

PLATE XII *(continued)*

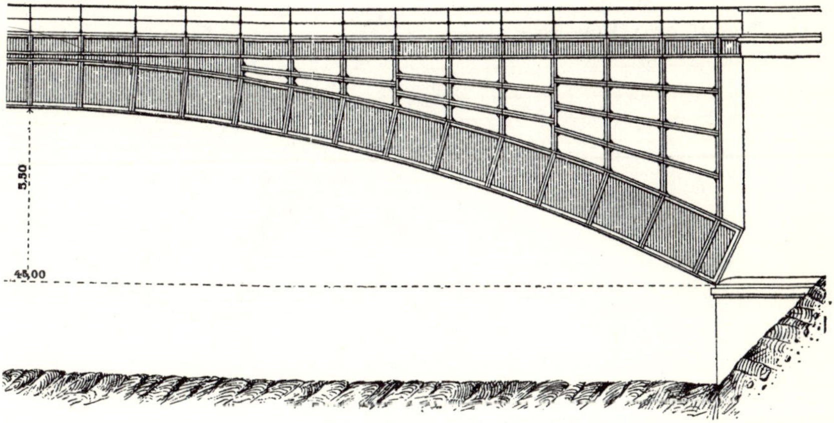


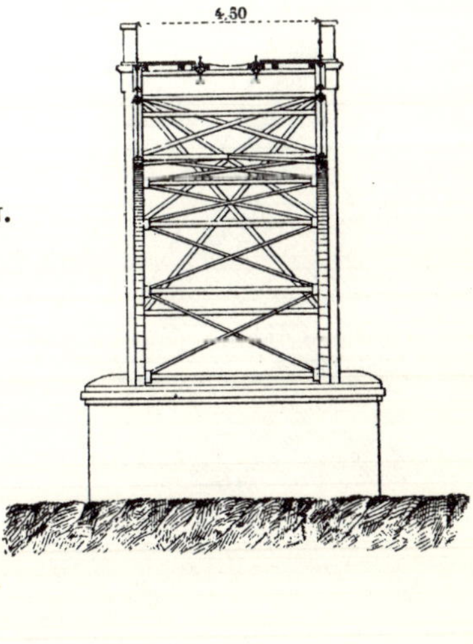


FIG. 2.
CENTRAL SECTION.

FIG. 3.
DETAIL OF A FLAT END.

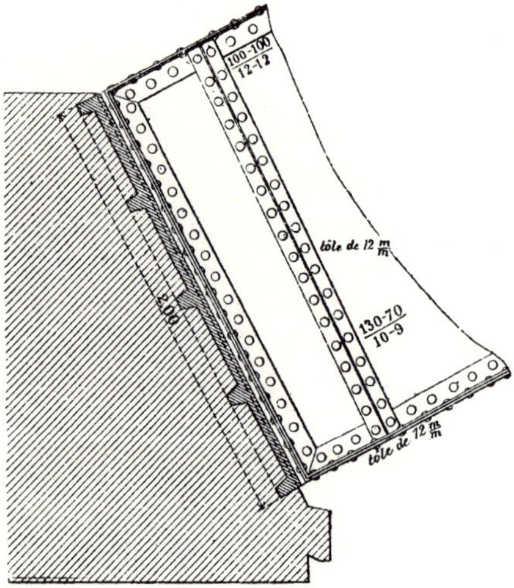


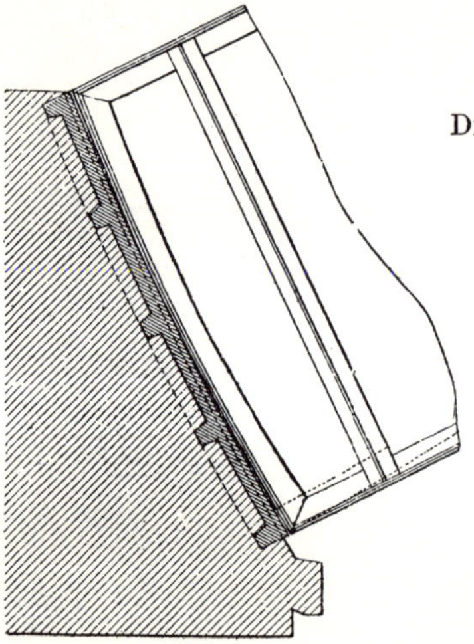

FIG. 4.
DETAIL OF A ROUNDED END

PLATE XII *(concluded)*

FIG. 5.
END DETAIL OF TOP MEMBER.

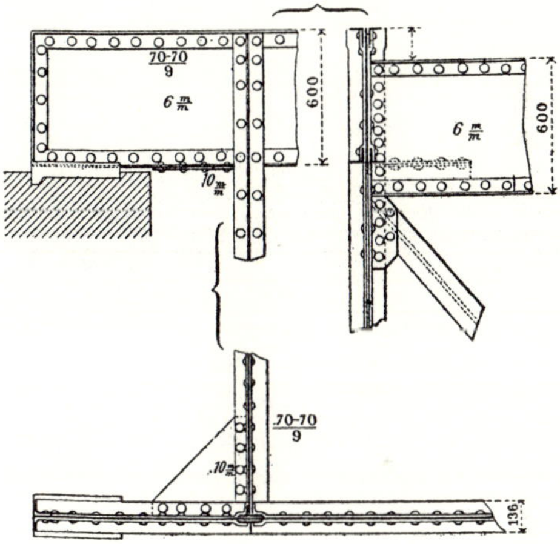


SCALE OF FIGS. 3-5.

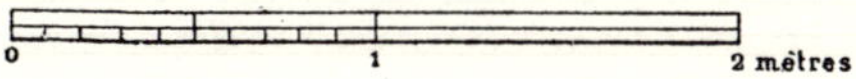

# PLATE XIII

## Arch Bridge with Flat Ends.

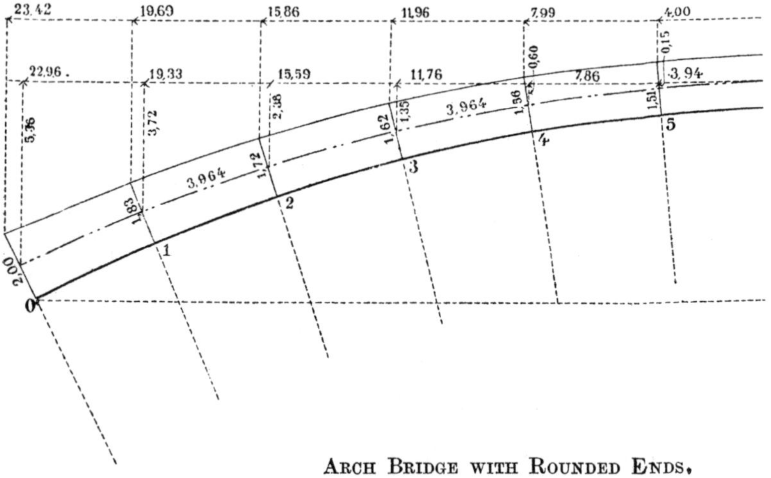


## Arch Bridge with Rounded Ends.

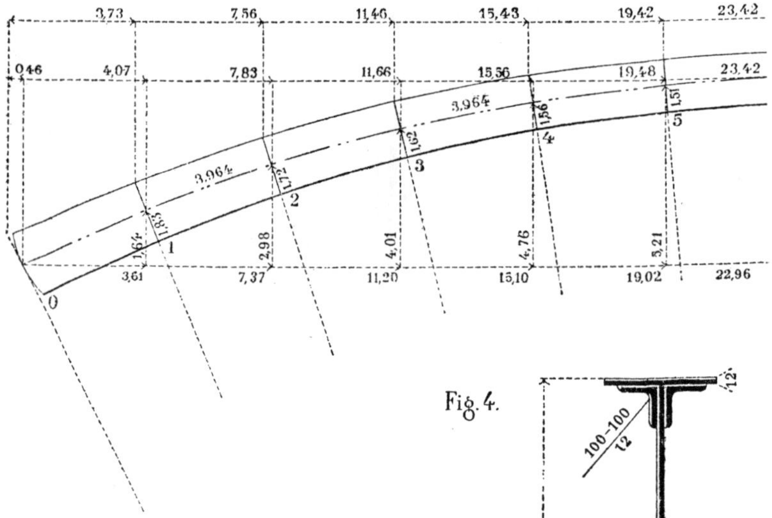


Fig. 4.

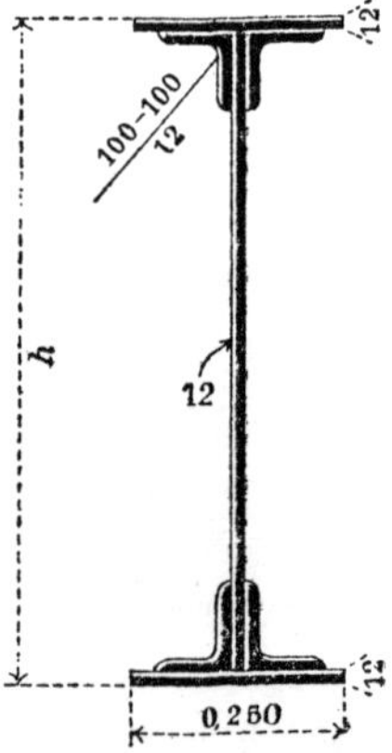


Scale of Figs. 1 and 2.

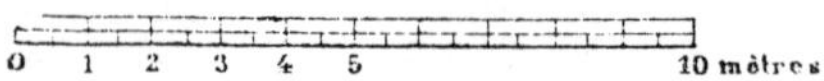

## PLATE XIII *(concluded)*

Fig. 1.

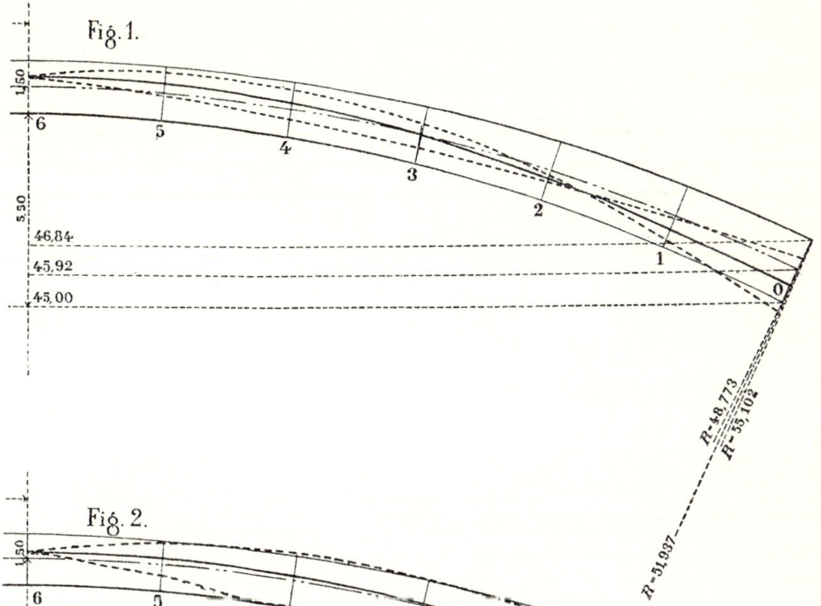


Fig. 2.

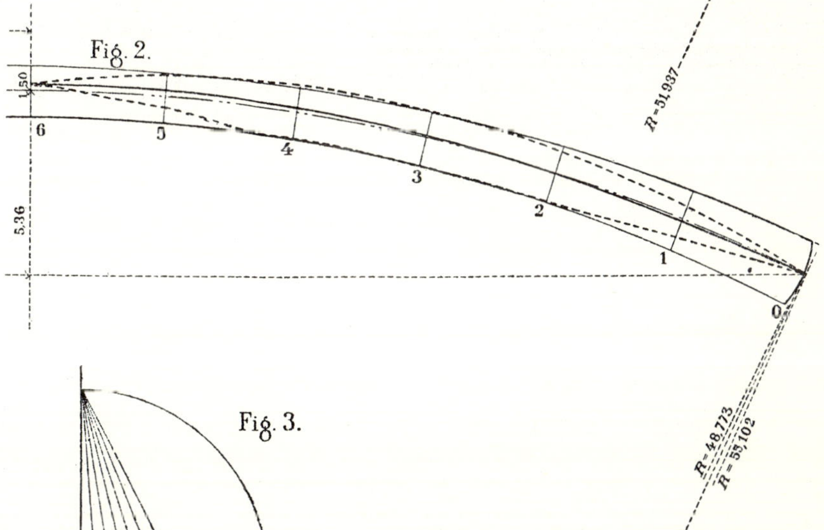


Fig. 3.

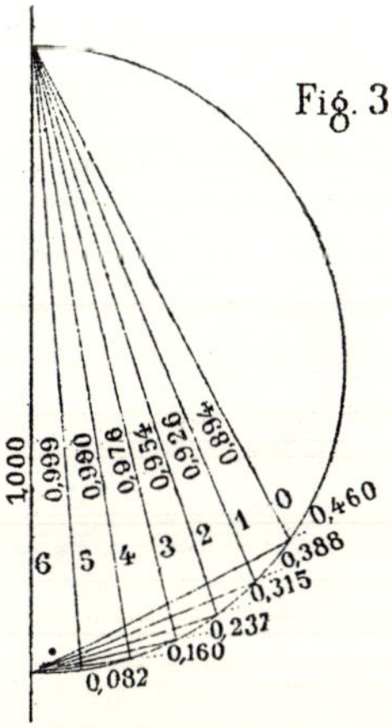


SCALE OF FIG. 3.

MASONRY ARCH BRIDGE OVER RIVER OGLIO.

(*See* p. 332.)

Fig. 1.

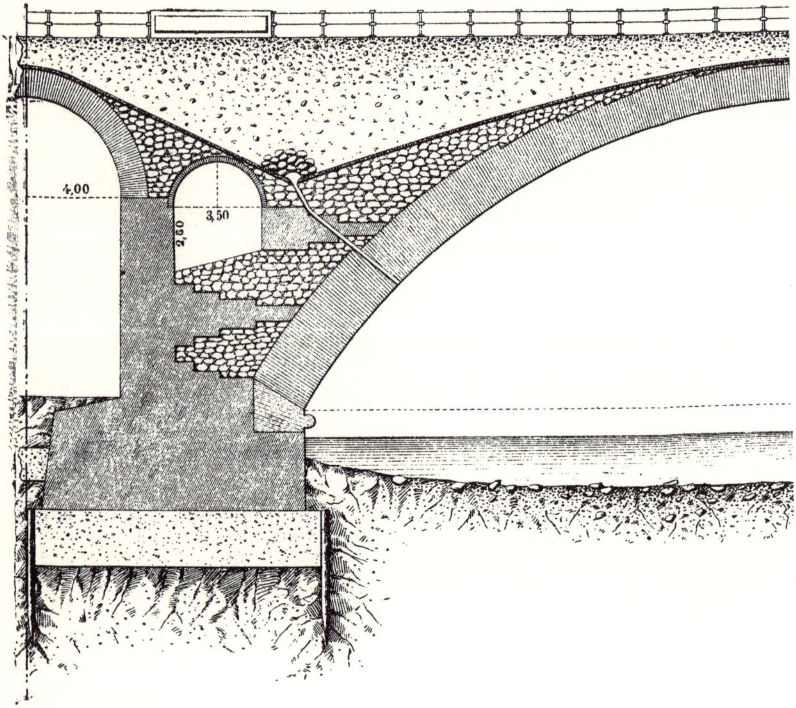


SCALE OF FIG. 2.

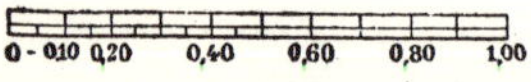


SCALE OF FIGS. 1 AND 3.

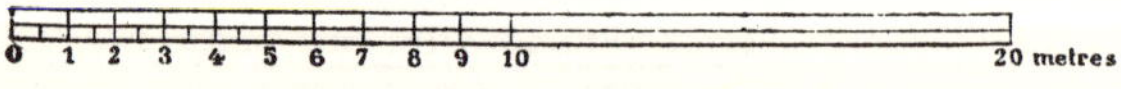

PLATE XIV *(concluded)*

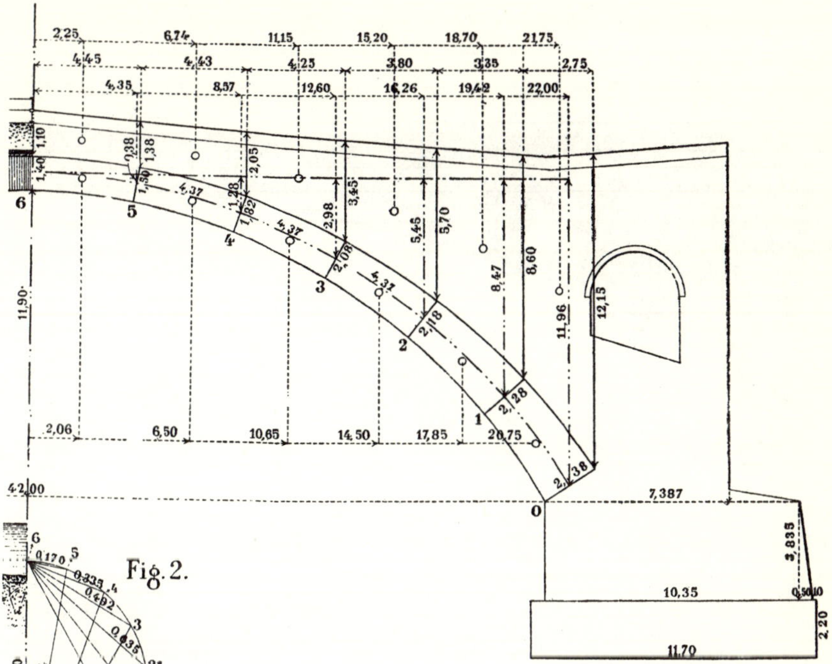


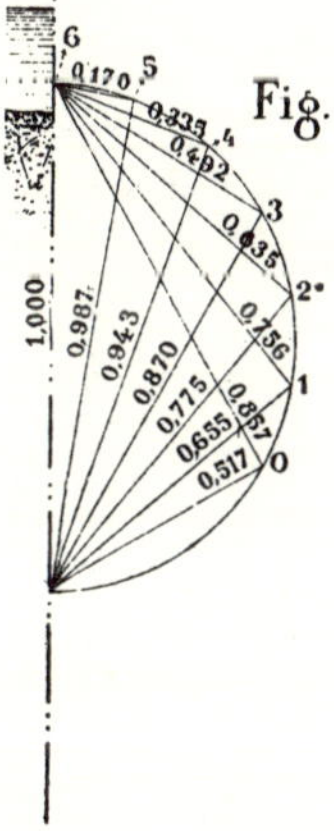


Fig. 2.

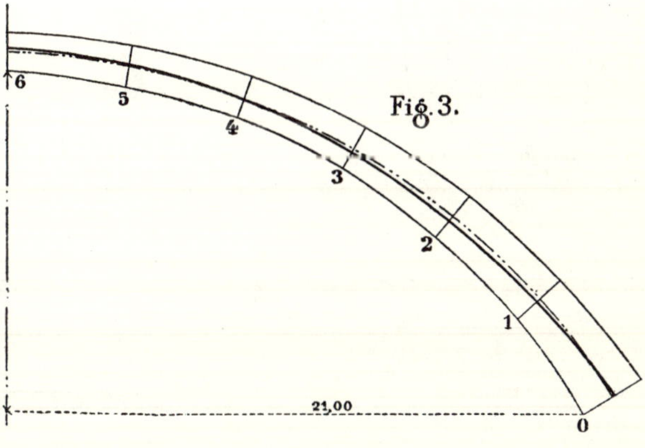


Fig. 3.

PLATE XV

Stone Bridge over River Doire.

(*See* p. 342.)

Fig. 1.

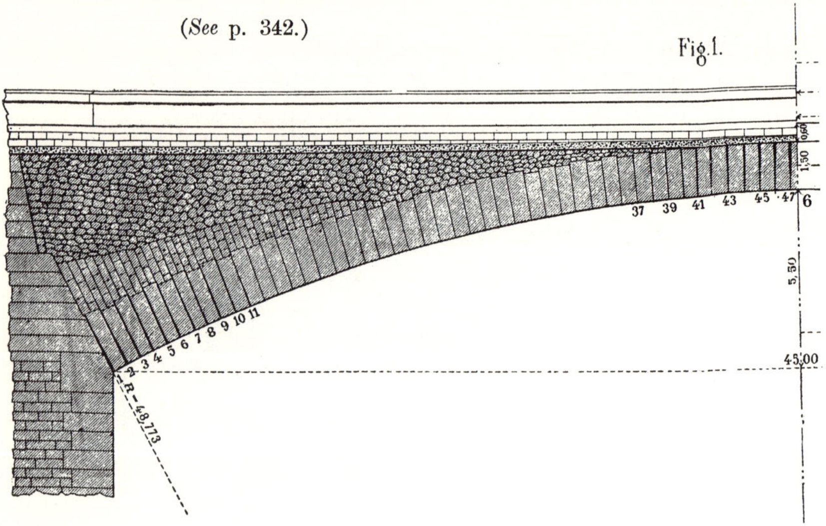


Fig. 2.

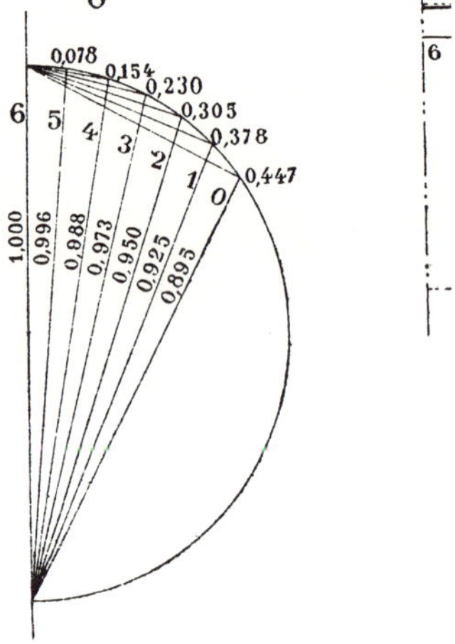


Fig. 4.

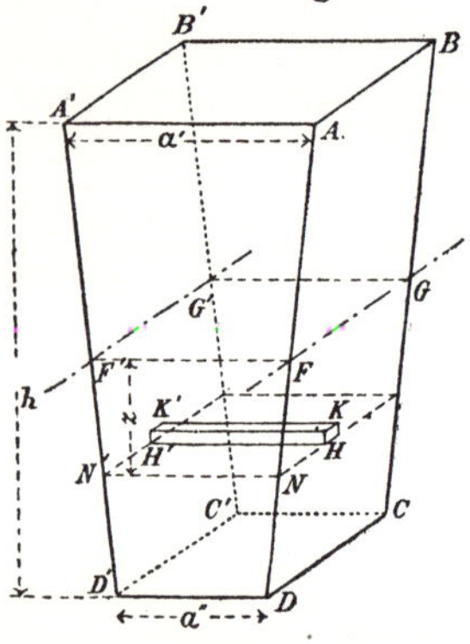


Scale of Fig. 2.

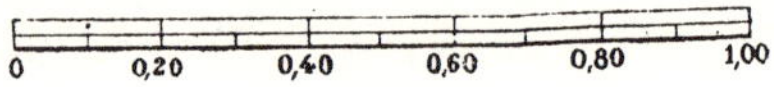

PLATE XV *(concluded)*

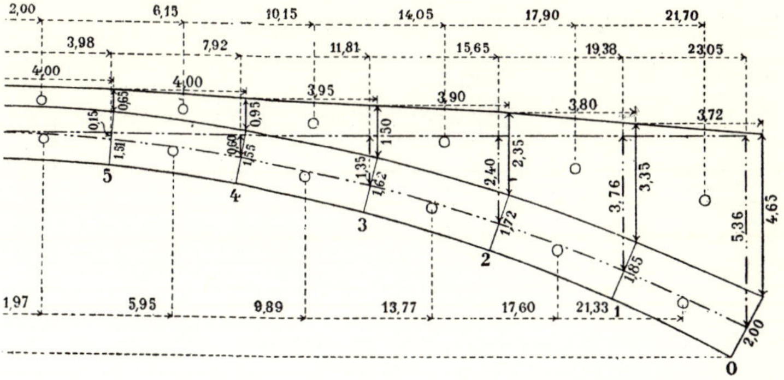


Fig. 3.

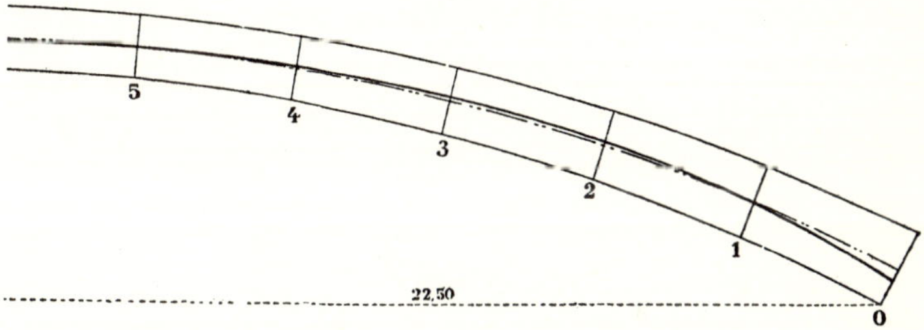


Scale of Figs. 1 and 3.

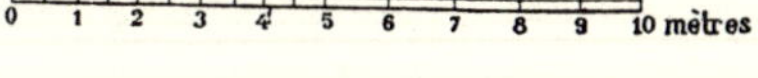

# CATALOGUE OF DOVER BOOKS

# MATHEMATICS—INTERMEDIATE TO ADVANCED

## General

**INTRODUCTION TO APPLIED MATHEMATICS, Francis D. Murnaghan.** A practical and thoroughly sound introduction to a number of advanced branches of higher mathematics. Among the selected topics covered in detail are: vector and matrix analysis, partial and differential equations, integral equations, calculus of variations, Laplace transform theory, the vector triple product, linear vector functions, quadratic and bilinear forms, Fourier series, spherical harmonics, Bessel functions, the Heaviside expansion formula, and many others. Extremely useful book for graduate students in physics, engineering, chemistry, and mathematics. Index. 111 study exercises with answers. 41 illustrations. ix + 389pp. 5⅜ x 8½.
S1042 Paperbound **$2.00**

**OPERATIONAL METHODS IN APPLIED MATHEMATICS, H. S. Carslaw and J. C. Jaeger.** Explanation of the application of the Laplace Transformation to differential equations, a simple and effective substitute for more difficult and obscure operational methods. Of great practical value to engineers and to all workers in applied mathematics. Chapters on: Ordinary Linear Differential Equations with Constant Coefficients;; Electric Circuit Theory; Dynamical Applications; The Inversion Theorem for the Laplace Transformation; Conduction of Heat; Vibrations of Continuous Mechanical Systems; Hydrodynamics; Impulsive Functions; Chains of Differential Equations; and other related matters. 3 appendices. 153 problems, many with answers. 22 figures. xvi + 359pp. 5⅜ x 8½. S1011 Paperbound **$2.25**

**APPLIED MATHEMATICS FOR RADIO AND COMMUNICATIONS ENGINEERS, C. E. Smith.** No extraneous material here!—only the theories, equations, and operations essential and immediately useful for radio work. Can be used as refresher, as handbook of applications and tables, or as full home-study course. Ranges from simplest arithmetic through calculus, series, and wave forms, hyperbolic trigonometry, simultaneous equations in mesh circuits, etc. Supplies applications right along with each math topic discussed. 22 useful tables of functions, formulas, logs, etc. Index. 166 exercises, 140 examples, all with answers. 95 diagrams. Bibliography. x + 336pp. 5⅜ x 8. S141 Paperbound **$1.75**

## Algebra, group theory, determinants, sets, matrix theory

**ALGEBRAS AND THEIR ARITHMETICS, L. E. Dickson.** Provides the foundation and background necessary to any advanced undergraduate or graduate student studying abstract algebra. Begins with elementary introduction to linear transformations, matrices, field of complex numbers; proceeds to order, basal units, modulus, quaternions, etc.; develops calculus of linears sets, describes various examples of algebras including invariant, difference, nilpotent, semi-simple. "Makes the reader marvel at his genius for clear and profound analysis," Amer. Mathematical Monthly. Index. xii + 241pp. 5⅜ x 8. S616 Paperbound **$1.50**

**THE THEORY OF EQUATIONS WITH AN INTRODUCTION TO THE THEORY OF BINARY ALGEBRAIC FORMS, W. S. Burnside and A. W. Panton.** Extremely thorough and concrete discussion of the theory of equations, with extensive detailed treatment of many topics curtailed in later texts. Covers theory of algebraic equations, properties of polynomials, symmetric functions, derived functions, Horner's process, complex numbers and the complex variable, determinants and methods of elimination, invariant theory (nearly 100 pages), transformations, introduction to Galois theory, Abelian equations, and much more. Invaluable supplementary work for modern students and teachers. 759 examples and exercises. Index in each volume. Two volume set. Total of xxiv + 604pp. 5⅜ x 8. S714 Vol I Paperbound **$1.85**
S715 Vol II Paperbound **$1.85**
The set **$3.70**

**COMPUTATIONAL METHODS OF LINEAR ALGEBRA, V. N. Faddeeva,** translated by **C. D. Benster.** First English translation of a unique and valuable work, the only work in English presenting a systematic exposition of the most important methods of linear algebra—classical and contemporary. Shows in detail how to derive numerical solutions of problems in mathematical physics which are frequently connected with those of linear algebra. Theory as well as individual practice. Part I surveys the mathematical background that is indispensable to what follows. Parts II and III, the conclusion, set forth the most important methods of solution, for both exact and iterative groups. One of the most outstanding and valuable features of this work is the 23 tables, double and triple checked for accuracy. These tables will not be found elsewhere. Author's preface. Translator's note. New bibliography and index. x + 252pp. 5⅜ x 8. S424 Paperbound **$1.95**

**ALGEBRAIC EQUATIONS, E. Dehn.** Careful and complete presentation of Galois' theory of algebraic equations; theories of Lagrange and Galois developed in logical rather than historical form, with a more thorough exposition than in most modern books. Many concrete applications and fully-worked-out examples. Discusses basic theory (very clear exposition of the symmetric group); isomorphic, transitive, and Abelian groups; applications of Lagrange's and Galois' theories; and much more. Newly revised by the author. Index. List of Theorems. xi + 208pp. 5⅜ x 8. S697 Paperbound **$1.45**

**ALGEBRAIC THEORIES, L. E. Dickson.** Best thorough introduction to classical topics in higher algebra develops theories centering around matrices, invariants, groups. Higher algebra, Galois theory, finite linear groups, Klein's icosahedron, algebraic invariants, linear transformations, elementary divisors, invariant factors; quadratic, bi-linear, Hermitian forms, singly and in pairs. Proofs rigorous, detailed; topics developed lucidly, in close connection with their most frequent mathematical applications. Formerly "Modern Algebraic Theories." 155 problems. Bibliography. 2 indexes. 285pp. 5⅜ x 8. S547 Paperbound **$1.50**

**LECTURES ON THE ICOSAHEDRON AND THE SOLUTION OF EQUATIONS OF THE FIFTH DEGREE, Felix Klein.** The solution of quintics in terms of rotation of a regular icosahedron around its axes of symmetry. A classic & indispensable source for those interested in higher algebra, geometry, crystallography. Considerable explanatory material included. 230 footnotes, mostly bibliographic. 2nd edition, xvi + 289pp. 5⅜ x 8. S314 Paperbound **$2.25**

**LINEAR GROUPS, WITH AN EXPOSITION OF THE GALOIS FIELD THEORY, L. E. Dickson.** The classic exposition of the theory of groups, well within the range of the graduate student. Part I contains the most extensive and thorough presentation of the theory of Galois Fields available, with a wealth of examples and theorems. Part II is a full discussion of linear groups of finite order. Much material in this work is based on Dickson's own contributions. Also includes expositions of Jordan, Lie, Abel, Betti-Mathieu, Hermite, etc. "A milestone in the development of modern algebra," W. Magnus, in his historical introduction to this edition. Index. xv + 312pp. 5⅜ x 8. S482 Paperbound **$1.95**

**INTRODUCTION TO THE THEORY OF GROUPS OF FINITE ORDER, R. Carmichael.** Examines fundamental theorems and their application. Beginning with sets, systems, permutations, etc., it progresses in easy stages through important types of groups: Abelian, prime power, permutation, etc. Except 1 chapter where matrices are desirable, no higher math needed. 783 exercises, problems. Index. xvi + 447pp. 5⅜ x 8. S300 Paperbound **$2.25**

**THEORY OF GROUPS OF FINITE ORDER, W. Burnside.** First published some 40 years ago, this is still one of the clearest introductory texts. Partial contents: permutations, groups independent of representation, composition series of a group, isomorphism of a group with itself, Abelian groups, prime power groups, permutation groups, invariants of groups of linear substitution, graphical representation, etc. 45pp. of notes. Indexes. xxiv + 512pp. 5⅜ x 8. S38 Paperbound **$2.75**

**CONTINUOUS GROUPS OF TRANSFORMATIONS, L. P. Eisenhart.** Intensive study of the theory and geometrical applications of continuous groups of transformations; a standard work on the subject, called forth by the revolution in physics in the 1920's. Covers tensor analysis, Riemannian geometry, canonical parameters, transitivity, imprimitivity, differential invariants, the algebra of constants of structure, differential geometry, contact transformations, etc. "Likely to remain one of the standard works on the subject for many years . . . principal theorems are proved clearly and concisely, and the arrangement of the whole is coherent," MATHEMATICAL GAZETTE. Index. 72-item bibliography. 185 exercises. ix + 301pp. 5⅜ x 8. S781 Paperbound **$2.00**

**THE THEORY OF GROUPS AND QUANTUM MECHANICS, H. Weyl.** Discussions of Schroedinger's wave equation, de Broglie's waves of a particle, Jordan-Hoelder theorem, Lie's continuous groups of transformations, Pauli exclusion principle, quantization of Maxwell-Dirac field equations, etc. Unitary geometry, quantum theory, groups, application of groups to quantum mechanics, symmetry permutation group, algebra of symmetric transformation, etc. 2nd revised edition. Bibliography. Index. xxii + 422pp. 5⅜ x 8. S269 Paperbound **$2.35**

**APPLIED GROUP-THEORETIC AND MATRIX METHODS, Bryan Higman.** The first systematic treatment of group and matrix theory for the physical scientist. Contains a comprehensive, easily-followed exposition of the basic ideas of group theory (realized through matrices) and its applications in the various areas of physics and chemistry: tensor analysis, relativity, quantum theory, molecular structure and spectra, and Eddington's quantum relativity. Includes rigorous proofs available only in works of a far more advanced character. 34 figures, numerous tables. Bibliography. Index. xiii + 454pp. 5⅜ x 8⅜. S1147 Paperbound **$2.50**

**THE THEORY OF GROUP REPRESENTATIONS, Francis D. Murnaghan.** A comprehensive introduction to the theory of group representations. Particular attention is devoted to those groups—mainly the symmetric and rotation groups—which have proved to be of fundamental significance for quantum mechanics (esp. nuclear physics). Also a valuable contribution to the literature on matrices, since the usual representations of groups are groups of matrices. Covers the theory of group integration (as developed by Schur and Weyl), the theory of 2-valued or spin representations, the representations of the symmetric group, the crystallographic groups, the Lorentz group, reducibility (Schur's lemma, Burnside's Theorem, etc.), the alternating group, linear groups, the orthogonal group, etc. Index. List of references. xi + 369pp. 5⅜ x 8½. S1112 Paperbound **$2.35**

**THEORY OF SETS, E. Kamke.** Clearest, amplest introduction in English, well suited for independent study. Subdivision of main theory, such as theory of sets of points, are discussed, but emphasis is on general theory. Partial contents: rudiments of set theory, arbitrary sets and their cardinal numbers, ordered sets and their order types, well-ordered sets and their cardinal numbers. Bibliography. Key to symbols. Index. vii + 144pp. 5⅜ x 8. S141 Paperbound **$1.35**

**THEORY AND APPLICATIONS OF FINITE GROUPS, G. A. Miller, H. F. Blichfeldt, L. E. Dickson.** Unusually accurate and authoritative work, each section prepared by a leading specialist: Miller on substitution and abstract groups, Blichfeldt on finite groups of linear homogeneous transformations, Dickson on applications of finite groups. Unlike more modern works, this gives the concrete basis from which abstract group theory arose. Includes Abelian groups, prime-power groups, isomorphisms, matrix forms of linear transformations, Sylow groups, Galois' theory of algebraic equations, duplication of a cube, trisection of an angle, etc. 2 Indexes. 267 problems. xvii + 390pp. 5⅜ x 8. S216 Paperbound **$2.00**

**THE THEORY OF DETERMINANTS, MATRICES, AND INVARIANTS, H. W. Turnbull.** Important study includes all salient features and major theories. 7 chapters on determinants and matrices cover fundamental properties, Laplace identities, multiplication, linear equations, rank and differentiation, etc. Sections on invariants gives general properties, symbolic and direct methods of reduction, binary and polar forms, general linear transformation, first fundamental theorem, multilinear forms. Following chapters study development and proof of Hilbert's Basis Theorem, Gordan-Hilbert Finiteness Theorem, Clebsch's Theorem, and include discussions of apolarity, canonical forms, geometrical interpretations of algebraic forms, complete system of the general quadric, etc. New preface and appendix. Bibliography. xviii + 374pp. 5⅜ x 8. S699 Paperbound **$2.25**

**AN INTRODUCTION TO THE THEORY OF CANONICAL MATRICES, H. W. Turnbull and A. C. Aitken.** All principal aspects of the theory of canonical matrices, from definitions and fundamental properties of matrices to the practical applications of their reduction to canonical form. Beginning with matrix multiplications, reciprocals, and partitioned matrices, the authors go on to elementary transformations and bilinear and quadratic forms. Also covers such topics as a rational canonical form for the collineatory group, congruent and conjunctive transformation for quadratic and hermitian forms, unitary and orthogonal transformations, canonical reduction of pencils of matrices, etc. Index. Appendix. Historical notes at chapter ends. Bibliographies. 275 problems. xiv + 200pp. 5⅜ x 8. S177 Paperbound **$1.55**

**A TREATISE ON THE THEORY OF DETERMINANTS, T. Muir.** Unequalled as an exhaustive compilation of nearly all the known facts about determinants up to the early 1930's. Covers notation and general properties, row and column transformation, symmetry, compound determinants, adjugates, rectangular arrays and matrices, linear dependence, gradients, Jacobians, Hessians, Wronskians, and much more. Invaluable for libraries of industrial and research organizations as well as for student, teacher, and mathematician; very useful in the field of computing machines. Revised and enlarged by W. H. Metzler. Index. 485 problems and scores of numerical examples. iv + 766pp. 5⅜ x 8. S670 Paperbound **$3.00**

**THEORY OF DETERMINANTS IN THE HISTORICAL ORDER OF DEVELOPMENT, Sir Thomas Muir.** Unabridged reprinting of this complete study of 1,859 papers on determinant theory written between 1693 and 1900. Most important and original sections reproduced, valuable commentary on each. No other work is necessary for determinant research: all types are covered—each subdivision of the theory treated separately; all papers dealing with each type are covered; you are told exactly what each paper is about and how important its contribution is. Each result, theory, extension, or modification is assigned its own identifying numeral so that the full history may be more easily followed. Includes papers on determinants in general, determinants and linear equations, symmetric determinants, alternants, recurrents, determinants having invariant factors, and all other major types. "A model of what such histories ought to be," NATURE. "Mathematicians must ever be grateful to Sir Thomas for his monumental work," AMERICAN MATH MONTHLY. Four volumes bound as two. Indices. Bibliographies. Total of lxxxiv + 1977pp. 5⅜ x 8. S672-3 The set, Clothbound **$12.50**

## Calculus and function theory, Fourier theory, infinite series, calculus of variations, real and complex functions

### FIVE VOLUME "THEORY OF FUNCTIONS' SET BY KONRAD KNOPP

This five-volume set, prepared by Konrad Knopp, provides a complete and readily followed account of theory of functions. Proofs are given concisely, yet without sacrifice of completeness or rigor. These volumes are used as texts by such universities as M.I.T., University of Chicago, N. Y. City College, and many others. "Excellent introduction . . . remarkably readable, concise, clear, rigorous," JOURNAL OF THE AMERICAN STATISTICAL ASSOCIATION.

**ELEMENTS OF THE THEORY OF FUNCTIONS, Konrad Knopp.** This book provides the student with background for further volumes in this set, or texts on a similar level. Partial contents: foundations, system of complex numbers and the Gaussian plane of numbers, Riemann sphere of numbers, mapping by linear functions, normal forms, the logarithm, the cyclometric functions and binomial series. "Not only for the young student, but also for the student who knows all about what is in it," MATHEMATICAL JOURNAL. Bibliography. Index. 140pp. 5⅜ x 8. S154 Paperbound **$1.35**

**THEORY OF FUNCTIONS, PART I, Konrad Knopp.** With volume II, this book provides coverage of basic concepts and theorems. Partial contents: numbers and points, functions of a complex variable, integral of a continuous function, Cauchy's integral theorem, Cauchy's integral formulae, series with variable terms, expansion of analytic functions in power series, analytic continuation and complete definition of analytic functions, entire transcendental functions, Laurent expansion, types of singularities. Bibliography. Index. vii + 146pp. 5⅜ x 8. S156 Paperbound **$1.35**

**THEORY OF FUNCTIONS, PART II, Konrad Knopp.** Application and further development of general theory, special topics. Single valued functions, entire, Weierstrass, Meromorphic functions. Riemann surfaces. Algebraic functions. Analytical configuration, Riemann surface. Bibliography. Index. x + 150pp. 5⅜ x 8. S157 Paperbound **$1.35**

**PROBLEM BOOK IN THE THEORY OF FUNCTIONS, VOLUME 1, Konrad Knopp.** Problems in elementary theory, for use with Knopp's THEORY OF FUNCTIONS, or any other text, arranged according to increasing difficulty. Fundamental concepts, sequences of numbers and infinite series, complex variable, integral theorems, development in series, conformal mapping. 182 problems. Answers. viii + 126pp. 5⅜ x 8. S158 Paperbound **$1.35**

**PROBLEM BOOK IN THE THEORY OF FUNCTIONS, VOLUME 2, Konrad Knopp.** Advanced theory of functions, to be used either with Knopp's THEORY OF FUNCTIONS, or any other comparable text. Singularities, entire & meromorphic functions, periodic, analytic, continuation, multiple-valued functions, Riemann surfaces, conformal mapping. Includes a section of additional elementary problems. "The difficult task of selecting from the immense material of the modern theory of functions the problems just within the reach of the beginner is here masterfully accomplished," AM. MATH. SOC. Answers. 138pp. 5⅜ x 8. S159 Paperbound **$1.35**

**A COURSE IN MATHEMATICAL ANALYSIS, Edouard Goursat.** Trans. by E. R. Hedrick, O. Dunkel. Classic study of fundamental material thoroughly treated. Exceptionally lucid exposition of wide range of subject matter for student with 1 year of calculus. Vol. 1: Derivatives and Differentials, Definite Integrals, Expansion in Series, Applications to Geometry. Problems. Index. 52 illus. 556pp. Vol. 2, Part I: Functions of a Complex Variable, Conformal Representations, Doubly Periodic Functions, Natural Boundaries, etc. Problems. Index. 38 illus. 269pp. Vol. 2, Part 2: Differential Equations, Cauchy-Lipschitz Method, Non-linear Differential Equations, Simultaneous Equations, etc. Problems. Index. 308pp. 5⅜ x 8.
Vol. 1 S554 Paperbound **$2.50**
Vol. 2 part 1 S555 Paperbound **$1.85**
Vol. 2 part 2 S556 Paperbound **$1.85**
3 vol. set **$6.20**

**MODERN THEORIES OF INTEGRATION, H. Kestelman.** Connected and concrete coverage, with fully-worked-out proofs for every step. Ranges from elementary definitions through theory of aggregates, sets of points, Riemann and Lebesgue integration, and much more. This new revised and enlarged edition contains a new chapter on Riemann-Stieltjes integration, as well as a supplementary section of 186 exercises. Ideal for the mathematician, student, teacher, or self-studier. Index of Definitions and Symbols. General Index. Bibliography. x + 310pp. 5⅝ x 8⅜. S572 Paperbound **$2.25**

**THEORY OF MAXIMA AND MINIMA, H. Hancock.** Fullest treatment ever written; only work in English with extended discussion of maxima and minima for functions of 1, 2, or n variables, problems with subsidiary constraints, and relevant quadratic forms. Detailed proof of each important theorem. Covers the Scheeffer and von Dantscher theories, homogeneous quadratic forms, reversion of series, fallacious establishment of maxima and minima, etc. Unsurpassed treatise for advanced students of calculus, mathematicians, economists, statisticians. Index. 24 diagrams. 39 problems, many examples. 193pp. 5⅜ x 8. S665 Paperbound **$1.50**

**AN ELEMENTARY TREATISE ON ELLIPTIC FUNCTIONS, A. Cayley.** Still the fullest and clearest text on the theories of Jacobi and Legendre for the advanced student (and an excellent supplement for the beginner). A masterpiece of exposition by the great 19th century British mathematician (creator of the theory of matrices and abstract geometry), it covers the addition-theory, Landen's theorem, the 3 kinds of elliptic integrals, transformations, the q-functions, reduction of a differential expression, and much more. Index. xii + 386pp. 5⅜ x 8. S728 Paperbound **$2.00**

**THE APPLICATIONS OF ELLIPTIC FUNCTIONS, A. G. Greenhill.** Modern books forego detail for sake of brevity—this book offers complete exposition necessary for proper understanding, use of elliptic integrals. Formulas developed from definite physical, geometric problems; examples representative enough to offer basic information in widely useable form. Elliptic integrals, addition theorem, algebraical form of addition theorem, elliptic integrals of 2nd, 3rd kind, double periodicity, resolution into factors, series, transformation, etc. Introduction. Index. 25 illus. xi + 357pp. 5⅜ x 8. S603 Paperbound **$1.75**

**THE THEORY OF FUNCTIONS OF REAL VARIABLES, James Pierpont.** A 2-volume authoritative exposition, by one of the foremost mathematicians of his time. Each theorem stated with all conditions, then followed by proof. No need to go through complicated reasoning to discover conditions added without specific mention. Includes a particularly complete, rigorous presentation of theory of measure; and Pierpont's own work on a theory of Lebesgue integrals, and treatment of area of a curved surface. Partial contents, Vol. 1: rational numbers, exponentials, logarithms, point aggregates, maxima, minima, proper integrals, improper integrals, multiple proper integrals, continuity, discontinuity, indeterminate forms. Vol. 2: point sets, proper integrals, series, power series, aggregates, ordinal numbers, discontinuous functions, sub-, infra-uniform convergence, much more. Index. 95 illustrations. 1229pp. 5⅜ x 8. S558-9, 2 volume set, paperbound **$5.20**

**FUNCTIONS OF A COMPLEX VARIABLE, James Pierpont.** Long one of best in the field. A thorough treatment of fundamental elements, concepts, theorems. A complete study, rigorous, detailed, with carefully selected problems worked out to illustrate each topic. Partial contents: arithmetical operations, real term series, positive term series, exponential functions, integration, analytic functions, asymptotic expansions, functions of Weierstrass, Legendre, etc. Index. List of symbols. 122 illus. 597pp. 5⅜ x 8. S560 Paperbound **$2.45**

**MODERN OPERATIONAL CALCULUS: WITH APPLICATIONS IN TECHNICAL MATHEMATICS, N. W. McLachlan.** An introduction to modern operational calculus based upon the Laplace transform, applying it to the solution of ordinary and partial differential equations. For physicists, engineers, and applied mathematicians. Partial contents: Laplace transform, theorems or rules of the operational calculus, solution of ordinary and partial linear differential equations with constant coefficients, evaluation of integrals and establishment of mathematical relationships, derivation of Laplace transforms of various functions, etc. Six appendices deal with Heaviside's unit function, etc. Revised edition. Index. Bibliography. xiv + 218pp. 5⅜ x 8½. S192 Paperbound **$1.75**

**ADVANCED CALCULUS, E. B. Wilson.** An unabridged reprinting of the work which continues to be recognized as one of the most comprehensive and useful texts in the field. It contains an immense amount of well-presented, fundamental material, including chapters on vector functions, ordinary differential equations, special functions, calculus of variations, etc., which are excellent introductions to these areas. For students with only one year of calculus, more than 1300 exercises cover both pure math and applications to engineering and physical problems. For engineers, physicists, etc., this work, with its 54 page introductory review, is the ideal reference and refresher. Index. ix + 566pp. 5⅜ x 8. S504 Paperbound **$2.45**

**ASYMPTOTIC EXPANSIONS, A. Erdélyi.** The only modern work available in English, this is an unabridged reproduction of a monograph prepared for the Office of Naval Research. It discusses various procedures for asymptotic evaluation of integrals containing a large parameter and solutions of ordinary linear differential equations. Bibliography of 71 items. vi + 108pp. 5⅜ x 8. S318 Paperbound **$1.35**

**INTRODUCTION TO ELLIPTIC FUNCTIONS: with applications, F. Bowman.** Concise, practical introduction to elliptic integrals and functions. Beginning with the familiar trigonometric functions, it requires nothing more from the reader than a knowledge of basic principles of differentiation and integration. Discussion confined to the Jacobian functions. Enlarged bibliography. Index. 173 problems and examples. 56 figures, 4 tables. 115pp. 5⅜ x 8. S922 Paperbound **$1.25**

**ON RIEMANN'S THEORY OF ALGEBRAIC FUNCTIONS AND THEIR INTEGRALS: A SUPPLEMENT TO THE USUAL TREATISES, Felix Klein.** Klein demonstrates how the mathematical ideas in Riemann's work on Abelian integrals can be arrived at by thinking in terms of the flow of electric current on surfaces. Intuitive explanations, not detailed proofs given in an extremely clear exposition, concentrating on the kinds of functions which can be defined on Riemann surfaces. Also useful as an introduction to the origins of topological problems. Complete and unabridged. Approved translation by Frances Hardcastle. New introduction. 43 figures. Glossary. xii + 76pp. 5⅜ x 8½. S1072 Paperbound **$1.25**

**COLLECTED WORKS OF BERNHARD RIEMANN.** This important source book is the first to contain the complete text of both 1892 Werke and the 1902 supplement, unabridged. It contains 31 monographs, 3 complete lecture courses, 15 miscellaneous papers, which have been of enormous importance in relativity, topology, theory of complex variables, and other areas of mathematics. Edited by R. Dedekind, H. Weber, M. Noether, W. Wirtinger. German text. English introduction by Hans Lewy. 690pp. 5⅜ x 8. S226 Paperbound **$3.75**

**THE TAYLOR SERIES, AN INTRODUCTION TO THE THEORY OF FUNCTIONS OF A COMPLEX VARIABLE, P. Dienes.** This book investigates the entire realm of analytic functions. Only ordinary calculus is needed, except in the last two chapters. Starting with an introduction to real variables and complex algebra, the properties of infinite series, elementary functions, complex differentiation and integration are carefully derived. Also biuniform mapping, a thorough two part discussion of representation and singularities of analytic functions, overconvergence and gap theorems, divergent series, Taylor series on its circle of convergence, divergence and singularities, etc. Unabridged, corrected reissue of first edition. Preface and index. 186 examples, many fully worked out. 67 figures. xii + 555pp. 5⅜ x 8. S391 Paperbound **$2.75**

**INTRODUCTION TO BESSEL FUNCTIONS, Frank Bowman.** A rigorous self-contained exposition providing all necessary material during the development, which requires only some knowledge of calculus and acquaintance with differential equations. A balanced presentation including applications and practical use. Discusses Bessel Functions of Zero Order, of Any Real Order; Modified Bessel Functions of Zero Order; Definite Integrals; Asymptotic Expansions; Bessel's Solution to Kepler's Problem; Circular Membranes; much more. "Clear and straightforward . . . useful not only to students of physics and engineering, but to mathematical students in general," Nature. 226 problems. Short tables of Bessel functions. 27 figures. Index. x + 135pp. 5⅜ x 8. S462 Paperbound **$1.35**

**ELEMENTS OF THE THEORY OF REAL FUNCTIONS, J. E. Littlewood.** Based on lectures given at Trinity College, Cambridge, this book has proved to be extremely successful in introducing graduate students to the modern theory of functions. It offers a full and concise coverage of classes and cardinal numbers, well-ordered series, other types of series, and elements of the theory of sets of points. 3rd revised edition. vii + 71pp. 5⅜ x 8.
S171 Clothbound **$2.85**
S172 Paperbound **$1.25**

**TRANSCENDENTAL AND ALGEBRAIC NUMBERS, A. O. Gelfond.** First English translation of work by leading Soviet mathematician. Thue-Siegel theorem, its p-adic analogue, on approximation of algebraic numbers by numbers in fixed algebraic field; Hermite-Lindemann theorem on transcendency of Bessel functions, solutions of other differential equations; Gelfond-Schneider theorem on transcendency of alpha to power beta; Schneider's work on elliptic functions, with method developed by Gelfond. Translated by L. F. Boron. Index. Bibliography. 200pp. 5⅜ x 8. S615 Paperbound **$1.75**

**ELLIPTIC INTEGRALS, H. Hancock.** Invaluable in work involving differential equations containing cubics or quartics under the root sign, where elementary calculus methods are inadequate. Practical solutions to problems that occur in mathematics, engineering, physics: differential equations requiring integration of Lamé's, Briot's, or Bouquet's equations; determination of arc of ellipse, hyperbola, lemniscate; solutions of problems in elastica; motion of a projectile under resistance varying as the cube of the velocity; pendulums; many others. Exposition is in accordance with Legendre-Jacobi theory and includes rigorous discussion of Legendre transformations. 20 figures. 5 place table. Index. 104pp. 5⅛ x 8.
S484 Paperbound **$1.25**

**LECTURES ON THE THEORY OF ELLIPTIC FUNCTIONS, H. Hancock.** Reissue of the only book in English with so extensive a coverage, especially of Abel, Jacobi, Legendre, Weierstrasse, Hermite, Liouville, and Riemann. Unusual fullness of treatment, plus applications as well as theory, in discussing elliptic function (the universe of elliptic integrals originating in works of Abel and Jacobi), their existence, and ultimate meaning. Use is made of Riemann to provide the most general theory. 40 page table of formulas. 76 figures. xxiii + 498pp.
S483 Paperbound **$2.55**

**THE THEORY AND FUNCTIONS OF A REAL VARIABLE AND THE THEORY OF FOURIER'S SERIES, E. W. Hobson.** One of the best introductions to set theory and various aspects of functions and Fourier's series. Requires only a good background in calculus. Provides an exhaustive coverage of: metric and descriptive properties of sets of points; transfinite numbers and order types; functions of a real variable; the Riemann and Lebesgue integrals; sequences and series of numbers; power-series; functions representable by series sequences of continuous functions; trigonometrical series; representation of functions by Fourier's series; complete exposition (200pp.) on set theory; and much more. "The best possible guide," Nature. Vol. I: 88 detailed examples, 10 figures. Index. xv + 736pp. Vol. II: 117 detailed examples, 13 figures. Index. x + 780pp. 6⅛ x 9¼. Vol. I: S387 Paperbound **$3.00**
Vol. II: S388 Paperbound **$3.00**

**ALMOST PERIODIC FUNCTIONS, A. S. Besicovitch.** This unique and important summary by a well-known mathematician covers in detail the two stages of development in Bohr's theory of almost periodic functions: (1) as a generalization of pure periodicity, with results and proofs; (2) the work done by Stepanoff, Wiener, Weyl, and Bohr in generalizing the theory. Bibliography. xi + 180pp. 5⅜ x 8. S18 Paperbound **$1.75**

**THE ANALYTICAL THEORY OF HEAT, Joseph Fourier.** This book, which revolutionized mathematical physics, is listed in the Great Books program, and many other listings of great books. It has been used with profit by generations of mathematicians and physicists who are interested in either heat or in the application of the Fourier integral. Covers cause and reflection of rays of heat, radiant heating, heating of closed spaces, use of trigonometric series in the theory of heat, Fourier integral, etc. Translated by Alexander Freeman. 20 figures. xxii + 466pp. 5⅜ x 8. S93 Paperbound **$2.50**

**AN INTRODUCTION TO FOURIER METHODS AND THE LAPLACE TRANSFORMATION, Philip Franklin.** Concentrates upon essentials, enabling the reader with only a working knowledge of calculus to gain an understanding of Fourier methods in a broad sense, suitable for most applications. This work covers complex qualities with methods of computing elementary functions for complex values of the argument and finding approximations by the use of charts; Fourier series and integrals with half-range and complex Fourier series; harmonic analysis; Fourier and Laplace transformations, etc.; partial differential equations with applications to transmission of electricity; etc. The methods developed are related to physical problems of heat flow, vibrations, electrical transmission, electromagnetic radiation, etc. 828 problems with answers. Formerly entitled "Fourier Methods." Bibliography. Index. x + 289pp. 5⅜ x 8.
S452 Paperbound **$2.00**

**THE FOURIER INTEGRAL AND CERTAIN OF ITS APPLICATIONS, Norbert Wiener.** The only book-length study of the Fourier integral as link between pure and applied math. An expansion of lectures given at Cambridge. Partial contents: Plancherel's theorem, general Tauberian theorem, special Tauberian theorems, generalized harmonic analysis. Bibliography. viii + 201pp. 5⅜ x 8. S272 Paperbound **$1.50**

## Differential equations, ordinary and partial; integral equations

**INTRODUCTION TO THE DIFFERENTIAL EQUATIONS OF PHYSICS, L. Hopf.** Especially valuable to the engineer with no math beyond elementary calculus. Emphasizing intuitive rather than formal aspects of concepts, the author covers an extensive territory. Partial contents: Law of causality, energy theorem, damped oscillations, coupling by friction, cylindrical and spherical coordinates, heat source, etc. Index. 48 figures. 160pp. 5⅜ x 8.
S120 Paperbound **$1.25**

**INTRODUCTION TO THE THEORY OF LINEAR DIFFERENTIAL EQUATIONS, E. G. Poole.** Authoritative discussions of important topics, with methods of solution more detailed than usual, for students with background of elementary course in differential equations. Studies existence theorems, linearly independent solutions; equations with constant coefficients; with uniform analytic coefficients; regular singularities; the hypergeometric equation; conformal representation; etc. Exercises. Index. 210pp. 5⅜ x 8. S629 Paperbound **$1.65**

**DIFFERENTIAL EQUATIONS FOR ENGINEERS, P. Franklin.** Outgrowth of a course given 10 years at M. I. T. Makes most useful branch of pure math accessible for practical work. Theoretical basis of D.E.'s; solution of ordinary D.E.'s and partial derivatives arising from heat flow, steady-state temperature of a plate, wave equations; analytic functions; convergence of Fourier Series. 400 problems on electricity, vibratory systems, other topics. Formerly "Differential Equations for Electrical Engineers." Index 41 illus. 307pp. 5⅜ x 8.
S601 Paperbound **$1.65**

**DIFFERENTIAL EQUATIONS, F. R. Moulton.** A detailed, rigorous exposition of all the non-elementary processes of solving ordinary differential equations. Several chapters devoted to the treatment of practical problems, especially those of a physical nature, which are far more advanced than problems usually given as illustrations. Includes analytic differential equations; variations of a parameter; integrals of differential equations; analytic implicit functions; problems of elliptic motion; sine-amplitude functions; deviation of formal bodies; Cauchy-Lipschitz process; linear differential equations with periodic coefficients; differential equations in infinitely many variations; much more. Historical notes. 10 figures. 222 problems. Index. xv + 395pp. 5⅜ x 8. S451 Paperbound **$2.00**

**DIFFERENTIAL AND INTEGRAL EQUATIONS OF MECHANICS AND PHYSICS (DIE DIFFERENTIAL- UND INTEGRALGLEICHUNGEN DER MECHANIK UND PHYSIK), edited by P. Frank and R. von Mises.** Most comprehensive and authoritative work on the mathematics of mathematical physics available today in the United States: the standard, definitive reference for teachers, physicists, engineers, and mathematicians—now published (in the original German) at a relatively inexpensive price for the first time! Every chapter in this 2,000-page set is by an expert in his field: Carathéodory, Courant, Frank, Mises, and a dozen others. Vol I, on mathematics, gives concise but complete coverages of advanced calculus, differential equations, integral equations, and potential, and partial differential equations. Index. xxiii + 916pp. Vol. II (physics): classical mechanics, optics, continuous mechanics, heat conduction and diffusion, the stationary and quasi-stationary electromagnetic field, electromagnetic oscillations, and wave mechanics. Index. xxiv + 1106pp. Two volume set. Each volume available separately. 5⅝ x 8⅜.
S787 Vol I Clothbound **$7.50**
S788 Vol II Clothbound **$7.50**
The set **$15.00**

**LECTURES ON CAUCHY'S PROBLEM, J. Hadamard.** Based on lectures given at Columbia, Rome, this discusses work of Riemann, Kirchhoff, Volterra, and the author's own research on the hyperbolic case in linear partial differential equations. It extends spherical and cylindrical waves to apply to all (normal) hyperbolic equations. Partial contents: Cauchy's problem, fundamental formula, equations with odd number, with even number of independent variables; method of descent. 32 figures. Index. iii + 316pp. 5⅜ x 8. S105 Paperbound **$1.75**

**THEORY OF DIFFERENTIAL EQUATIONS, A. R. Forsyth.** Out of print for over a decade, the complete 6 volumes (now bound as 3) of this monumental work represent the most comprehensive treatment of differential equations ever written. Historical presentation includes in 2500 pages every substantial development. Vol. 1, 2: EXACT EQUATIONS, PFAFF'S PROBLEM; ORDINARY EQUATIONS, NOT LINEAR: methods of Grassmann, Clebsch, Lie, Darboux; Cauchy's theorem; branch points; etc. Vol. 3, 4: ORDINARY EQUATIONS, NOT LINEAR; ORDINARY LINEAR EQUATIONS: Zeta Fuchsian functions, general theorems on algebraic integrals, Brun's theorem, equations with uniform periodic coffiecients, etc. Vol. 4, 5: PARTIAL DIFFERENTIAL EQUATIONS: 2 existence-theorems, equations of theoretical dynamics, Laplace transformations, general transformation of equations of the 2nd order, much more. Indexes. Total of 2766pp. 5⅜ x 8. S576-7-8 Clothbound: the set **$15.00**

**PARTIAL DIFFERENTIAL EQUATIONS OF MATHEMATICAL PHYSICS, A. G. Webster.** A keystone work in the library of every mature physicist, engineer, researcher. Valuable sections on elasticity, compression theory, potential theory, theory of sound, heat conduction, wave propagation, vibration theory. Contents include: deduction of differential equations, vibrations, normal functions, Fourier's series, Cauchy's method, boundary problems, method of Riemann-Volterra. Spherical, cylindrical, ellipsoidal harmonics, applications, etc. 97 figures. vii + 440pp. 5⅜ x 8. S263 Paperbound **$2.00**

**ORDINARY DIFFERENTIAL EQUATIONS, E. L. Ince.** A most compendious analysis in real and complex domains. Existence and nature of solutions, continuous transformation groups, solutions in an infinite form, definite integrals, algebraic theory, Sturmian theory, boundary problems, existence theorems, 1st order, higher order, etc. "Deserves the highest praise, a notable addition to mathematical literature," BULLETIN, AM. MATH. SOC. Historical appendix. Bibliography. 18 figures. viii + 558pp. 5⅜ x 8. S349 Paperbound **$2.75**

**INTRODUCTION TO NONLINEAR DIFFERENTIAL AND INTEGRAL EQUATIONS, Harold T. Davis.** A thorough introduction to this important area, of increasing interest to mathematicians and scientists. First published by the United States Atomic Energy Commission, it includes chapters on the differential equation of the first order, the Riccati equation (as a bridge between linear and nonlinear equations), existence theorems, second order equations, elliptic integrals, elliptic functions, and theta functions, second order differential equations of polynomial class, continuous analytic continuation, the phase plane and its phenomena, nonlinear mechanics, the calculus of variations, etc. Appendices on Painlevé transcendents and Van der Pol and Volterra equations. Bibliography of 350 items. 137 problems. Index. xv + 566pp. 5⅜ x 8½. S971 Paperbound **$2.00**

**THEORY OF FUNCTIONALS AND OF INTEGRAL AND INTEGRO-DIFFERENTIAL EQUATIONS, Vito Volterra.** Unabridged republication of the only English translation. An exposition of the general theory of the functions depending on a continuous set of values of another function, based on the author's fundamental notion of the transition from a finite number of variables to a continually infinite number. Though dealing primarily with integral equations, much material on calculus of variations is included. The work makes no assumption of previous knowledge on the part of the reader. It begins with fundamental material and proceeds to Generalization of Analytic Functions, Integro-Differential Equations, Functional Derivative Equations, Applications, Other Directions of Theory of Functionals, etc. New introduction by G. C. Evans. Bibliography and criticism of Volterra's work by E. Whittaker. Bibliography. Index of authors cited. Index of subjects. xxxx + 226pp. 5⅜ x 8. S502 Paperbound **$1.75**

**LINEAR INTEGRAL EQUATIONS, W. V. Lovitt.** Systematic survey of general theory, with some application to differential equations, calculus of variations, problems of math, physics. Partial contents: integral equation of 2nd kind by successive substitutions; Fredholm's equation as ratio of 2 integral series in lambda, applications of the Fredholm theory, Hilbert-Schmidt theory of symmetric kernels, application, etc. Neumann, Dirichlet, vibratory problems. Index. ix + 253pp. 5⅜ x 8. S176 Paperbound **$1.60**

## Foundations of mathematics

**THE CONTINUUM AND OTHER TYPES OF SERIAL ORDER, E. V. Huntington.** This famous book gives a systematic elementary account of the modern theory of the continuum as a type of serial order. Based on the Cantor-Dedekind ordinal theory, which requires no technical knowledge of higher mathematics, it offers an easily followed analysis of ordered classes, discrete and dense series, continuous series, Cantor's transfinite numbers. 2nd edition. Index. viii + 82pp. 5⅜ x 8. S130 Paperbound **$1.00**

**CONTRIBUTIONS TO THE FOUNDING OF THE THEORY OF TRANSFINITE NUMBERS, Georg Cantor.** These papers founded a new branch of mathematics. The famous articles of 1895-7 are translated, with an 82-page introduction by P. E. B. Jourdain dealing with Cantor, the background of his discoveries, their results, future possibilities. Bibliography. Index. Notes. ix + 211 pp. 5⅜ x 8. S45 Paperbound **$1.35**

**ELEMENTARY MATHEMATICS FROM AN ADVANCED STANDPOINT, Felix Klein.**

This classic text is an outgrowth of Klein's famous integration and survey course at Göttingen. Using one field of mathematics to interpret, adjust, illuminate another, it covers basic topics in each area, illustrating its discussion with extensive analysis. It is especially valuable in considering areas of modern mathematics. "Makes the reader feel the inspiration of . . . a great mathematician, inspiring teacher . . . with deep insight into the foundations and interrelations," BULLETIN, AMERICAN MATHEMATICAL SOCIETY.

**Vol. 1. ARITHMETIC, ALGEBRA, ANALYSIS.** Introducing the concept of function immediately, it enlivens abstract discussion with graphical and geometrically perceptual methods. Partial contents: natural numbers, extension of the notion of number, special properties, complex numbers. Real equations with real unknowns, complex quantities. Logarithmic, exponential functions, goniometric functions, infinitesimal calculus. Transcendence of e and pi, theory of assemblages. Index. 125 figures. ix + 274pp . 5⅜ x 8. S150 Paperbound **$1.85**

**Vol. 2. GEOMETRY.** A comprehensive view which accompanies the space perception inherent in geometry with analytic formulas which facilitate precise formulation. Partial contents: Simplest geometric manifolds: line segment, Grassmann determinant principles, classification of configurations of space, derivative manifolds. Geometric transformations: affine transformations, projective, higher point transformations, theory of the imaginary. Systematic discussion of geometry and its foundations. Indexes. 141 illustrations. ix + 214pp. 5⅜ x 8. S151 Paperbound **$1.75**

**THE ADVANCED GEOMETRY OF PLANE CURVES AND THEIR APPLICATIONS, C. Zwikker.** An unusual study of many important curves, their geometrical properties and their applications, including discussions of many less well-known curves not often treated in textbooks on synthetic and analytic Euclidean geometry. Includes both algebraic and transcendental curves such as the conic sections, kinked curves, spirals, lemniscates, cycloids, etc. and curves generated as involutes, evolutes, anticaustics, pedals, envelopes and orthogonal trajectories. Dr. Zwikker represents the points of the curves by complex numbers instead of two real Cartesian coordinates, allowing direct and even elegant proofs. Formerly: "Advanced Plane Geometry." 273 figures. xii + 299pp. 5⅜ x 8½. S1078 Paperbound **$2.00**

**A TREATISE ON THE DIFFERENTIAL GEOMETRY OF CURVES AND SURFACES, L. P. Eisenhart.** Introductory treatise especially for the graduate student, for years a highly successful textbook. More detailed and concrete in approach than most more recent books. Covers space curves, osculating planes, moving axes, Gauss' method, the moving trihedral, geodesics, conformal representation, etc. Last section deals with deformation of surfaces, rectilinear congruences, cyclic systems, etc. Index. 683 problems. 30 diagrams. xii + 474pp. 5⅜ x 8. S667 Paperbound **$2.75**

**A TREATISE ON ALGEBRAIC PLANE CURVES, J. L. Coolidge.** Unabridged reprinting of one of few full coverages in English, offering detailed introduction to theory of algebraic plane curves and their relations to geometry and analysis. Treats topological properties, Riemann-Roch theorem, all aspects of wide variety of curves including real, covariant, polar, containing series of a given sort, elliptic, polygonal, rational, the pencil, two parameter nets, etc. This volume will enable the reader to appreciate the symbolic notation of Aronhold and Clebsch. Bibliography. Index. 17 illustrations. xxiv + 513pp. 5⅜ x 8. S543 Paperbound **$2.75**

**AN INTRODUCTION TO THE GEOMETRY OF N DIMENSIONS, D. M. Y. Sommerville.** An introduction presupposing no prior knowledge of the field, the only book in English devoted exclusively to higher dimensional geometry. Discusses fundamental ideas of incidence, parallelism, perpendicularity, angles between linear space; enumerative geometry; analytical geometry from projective and metric points of view; polytopes; elementary ideas in analysis situs; content of hyper-spacial figures. Bibliography. Index. 60 diagrams. 196pp. 5⅜ x 8. S494 Paperbound **$1.50**

**GEOMETRY OF FOUR DIMENSIONS, H. P. Manning.** Unique in English as a clear, concise introduction. Treatment is synthetic, and mostly Euclidean, although in hyperplanes and hyperspheres at infinity, non-Euclidean geometry is used. Historical introduction. Foundations of 4-dimensional geometry. Perpendicularity, simple angles. Angles of planes, higher order. Symmetry, order, motion; hyperpyramids, hypercones, hyperspheres; figures with parallel elements; volume, hypervolume in space, regular polyhedroids. Glossary. 78 figures. ix + 348pp. 5⅜ x 8. S182 Paperbound **$2.00**

**CONVEX FIGURES AND POLYHEDRA, L. A. Lyusternik.** An excellent elementary discussion by a leading Russian mathematician. Beginning with the basic concepts of convex figures and bodies and their supporting lines and planes, the author covers such matters as centrally symmetric convex figures, theorems of Euler, Cauchy, Steinitz and Alexandrov on convex polyhedra, linear systems of convex bodies, planar sections of convex bodies, the Brunn-Minkowski inequality and its consequences, and many other related topics. No more than a high school background in mathematics needed for complete understanding. First English translation by T. J. Smith. 182 illustrations. Index. x + 176pp. 5⅜ x 8½. S1021 Paperbound **$1.50**

**NON-EUCLIDEAN GEOMETRY, Roberto Bonola.** The standard coverage of non-Euclidean geometry. It examines from both a historical and mathematical point of view the geometries which have arisen from a study of Euclid's 5th postulate upon parallel lines. Also included are complete texts, translated, of Bolyai's SCIENCE OF ABSOLUTE SPACE. Lobachevsky's THEORY OF PARALLELS. 180 diagrams. 431pp. 5⅜ x 8. S27 Paperbound **$2.00**

**ELEMENTS OF NON-EUCLIDEAN GEOMETRY, D. M. Y. Sommerville.** Unique in proceeding step-by-step, in the manner of traditional geometry. Enables the student with only a good knowledge of high school algebra and geometry to grasp elementary hyperbolic, elliptic, analytic non-Euclidean geometries; space curvature and its philosophical implications; theory of radical axes; homothetic centres and systems of circles; parataxy and parallelism; absolute measure; Gauss' proof of the defect area theorem; geodesic representation; much more, all with exceptional clarity. 126 problems at chapter endings provide progressive practice and familiarity. 133 figures. Index. xvi + 274pp. 5⅜ x 8. S460 Paperbound **$1.50**

**INTRODUCTORY NON-EUCLIDEAN GEOMETRY, H. P. Manning.** Sound elementary introduction to non-Euclidean geometry. The first two thirds (Pangeometry and the Hyperbolic Geometry) require a grasp of plane and solid geometry and trigonometry. The last sections (the Elliptic Geometry and Analytic Non-Euclidean Geometry) necessitate also basic college calculus for understanding the text. The book does not propose to investigate the foundations of geometry, but rather begins with the theorems common to Euclidean and non-Euclidean geometry and then takes up the specific differences between them. A simple and direct account of the bases of this important branch of mathematics for teachers and students. 94 figures. vii + 95pp. 5⅜ x 8. S310 Paperbound **$1.00**

**ELEMENTARY CONCEPTS OF TOPOLOGY, P. Alexandroff.** First English translation of the famous brief introduction to topology for the beginner or for the mathematician not undertaking extensive study. This unusually useful intuitive approach deals primarily with the concepts of complex, cycle, and homology, and is wholly consistent with current investigations. Ranges from basic concepts of set-theoretic topology to the concept of Betti groups. "Glowing example of harmony between intuition and thought," David Hilbert. Translated by A. E. Farley. Introduction by D. Hilbert. Index. 25 figures. 73pp. 5⅜ x 8. S747 Paperbound **$1.00**

## Number theory

**INTRODUCTION TO THE THEORY OF NUMBERS, L. E. Dickson.** Thorough, comprehensive approach with adequate coverage of classical literature, an introductory volume beginners can follow. Chapters on divisibility, congruences, quadratic residues & reciprocity, Diophantine equations, etc. Full treatment of binary quadratic forms without usual restriction to integral coefficients. Covers infinitude of primes, least residues, Fermat's theorem, Euler's phi function, Legendre's symbol, Gauss's lemma, automorphs, reduced forms, recent theorems of Thue & Siegel, many more. Much material not readily available elsewhere. 239 problems. Index. I figure. viii + 183pp. 5⅜ x 8. S342 Paperbound **$1.65**

**ELEMENTS OF NUMBER THEORY, I. M. Vinogradov.** Detailed 1st course for persons without advanced mathematics; 95% of this book can be understood by readers who have gone no farther than high school algebra. Partial contents: divisibility theory, important number theoretical functions, congruences, primitive roots and indices, etc. Solutions to both problems and exercises. Tables of primes, indices, etc. Covers almost every essential formula in elementary number theory! Translated from Russian. 233 problems, 104 exercises. viii + 227pp. 5⅜ x 8. S259 Paperbound **$1.60**

**THEORY OF NUMBERS and DIOPHANTINE ANALYSIS, R. D. Carmichael.** These two complete works in one volume form one of the most lucid introductions to number theory, requiring only a firm foundation in high school mathematics. "Theory of Numbers," partial contents: Eratosthenes' sieve, Euclid's fundamental theorem, G.C.F. and L.C.M. of two or more integers, linear congruences, etc "Diophantine Analysis": rational triangles, Pythagorean triangles, equations of third, fourth, higher degrees, method of functional equations, much more. "Theory of Numbers": 76 problems. Index. 94pp. "Diophantine Analysis": 222 problems. Index. 118pp. 5⅜ x 8. S529 Paperbound **$1.35**

## Numerical analysis, tables

**MATHEMATICAL TABLES AND FORMULAS, Compiled by Robert D. Carmichael and Edwin R. Smith.** Valuable collection for students, etc. Contains all tables necessary in college algebra and trigonometry, such as five-place common logarithms, logarithmic sines and tangents of small angles, logarithmic trigonometric functions, natural trigonometric functions, four-place antilogarithms, tables for changing from sexagesimal to circular and from circular to sexagesimal measure of angles, etc. Also many tables and formulas not ordinarily accessible, including powers, roots, and reciprocals, exponential and hyperbolic functions, ten-place logarithms of prime numbers, and formulas and theorems from analytical and elementary geometry and from calculus. Explanatory introduction. viii + 269pp. 5⅜ x 8½. S111 Paperbound **$1.00**

**MATHEMATICAL TABLES, H. B. Dwight.** Unique for its coverage in one volume of almost every function of importance in applied mathematics, engineering, and the physical sciences. Three extremely fine tables of the three trig functions and their inverse functions to thousandths of radians; natural and common logarithms; squares, cubes; hyperbolic functions and the inverse hyperbolic functions; $(a^2 + b^2)$ exp. ½a; complete elliptic integrals of the 1st and 2nd kind; sine and cosine integrals; exponential integrals Ei(x) and Ei( —x); binomial coefficients; factorials to 250; surface zonal harmonics and first derivatives; Bernoulli and Euler numbers and their logs to base of 10; Gamma function; normal probability integral; over 60 pages of Bessel functions; the Riemann Zeta function. Each table with formulae generally used, sources of more extensive tables, interpolation data, etc. Over half have columns of differences, to facilitate interpolation. Introduction. Index. viii + 231pp. 5⅜ x 8. S445 Paperbound **$1.75**

**TABLES OF FUNCTIONS WITH FORMULAE AND CURVES, E. Jahnke & F. Emde.** The world's most comprehensive 1-volume English-text collection of tables, formulae, curves of transcendent functions. 4th corrected edition, new 76-page section giving tables, formulae for elementary functions—not in other English editions. Partial contents: sine, cosine, logarithmic integral; factorial function; error integral; theta functions; elliptic integrals, functions; Legendre, Bessel, Riemann, Mathieu, hypergeometric functions, etc. Supplementary books. Bibliography. Indexed. "Out of the way functions for which we know no other source," SCIENTIFIC COMPUTING SERVICE, Ltd. 212 figures. 400pp. 5⅜ x 8. S133 Paperbound **$2.00**

**JACOBIAN ELLIPTIC FUNCTION TABLES, L. M. Milne-Thomson.** An easy to follow, practical book which gives not only useful numerical tables, but also a complete elementary sketch of the application of elliptic functions. It covers Jacobian elliptic functions and a description of their principal properties; complete elliptic integrals; Fourier series and power series expansions; periods, zeros, poles, residues, formulas for special values of the argument; transformations, approximations, elliptic integrals, conformal mapping, factorization of cubic and quartic polynomials; application to the pendulum problem; etc. Tables and graphs form the body of the book: Graph, 5 figure table of the elliptic function sn (u m); cn (u m); dn (u m). 8 figure table of complete elliptic integrals K, K′, E, E′, and the nome q. 7 figure table of the Jacobian zeta-function Z(u). 3 figures. xi + 123pp. 5⅜ x 8.
S194 Paperbound **$1.35**

**TABLES OF INDEFINITE INTEGRALS, G. Petit Bois.** Comprehensive and accurate, this orderly grouping of over 2500 of the most useful indefinite integrals will save you hours of laborious mathematical groundwork. After a list of 49 common transformations of integral expressions, with a wide variety of examples, the book takes up algebraic functions, irrational monomials, products and quotients of binomials, transcendental functions, natural logs, etc. You will rarely or never encounter an integral of an algebraic or transcendental function not included here; any more comprehensive set of tables costs at least $12 or $15. Index. 2544 integrals. xii + 154pp. 6⅛ x 9¼. S225 Paperbound **$2.00**

**SUMMATION OF SERIES, Collected by L. B. W. Jolley.** Over 1100 common series collected, summed, and grouped for easy reference—for mathematicians, physicists, computer technicians, engineers, and students. Arranged for convenience into categories, such as arithmetical and geometrical progressions, powers and products of natural numbers, figurate and polygonal numbers, inverse natural numbers, exponential and logarithmic series, binomial expansions, simple inverse products, factorials, and trigonometric and hyperbolic expansions. Also included are series representing various Bessel functions, elliptic integrals; discussions of special series involving Legendre polynomials, the zeta function, Bernoulli's function, and similar expressions. Revised, enlarged second edition. New preface. xii + 251pp. 5⅜ x 8½.
S23 Paperbound **$2.25**

**A TABLE OF THE INCOMPLETE ELLIPTIC INTEGRAL OF THE THIRD KIND, R. G. Selfridge, J. E. Maxfield.** The first complete 6-place tables of values of the incomplete integral of the third kind, prepared under the auspices of the Research Department of the U.S. Naval Ordnance Test Station. Calculated on an IBM type 704 calculator and thoroughly verified by echo-checking and a check integral at the completion of each value of **a.** Of inestimable value in problems where the surface area of geometrical bodies can only be expressed in terms of the incomplete integral of the third and lower kinds; problems in aero-, fluid-, and thermodynamics involving processes where nonsymmetrical repetitive volumes must be determined; various types of seismological problems; problems of magnetic potentials due to circular current; etc. Foreword. Acknowledgment. Introduction. Use of table. xiv + 805pp. 5⅝ x 8⅜.
S501 Clothbound **$7.50**

**PRACTICAL ANALYSIS, GRAPHICAL AND NUMERICAL METHODS, F. A. Willers.** Translated by R. T. Beyer. Immensely practical handbook for engineers, showing how to interpolate, use various methods of numerical differentiation and integration, determine the roots of a single algebraic equation, system of linear equations, use empirical formulas, integrate differential equations, etc. Hundreds of shortcuts for arriving at numerical solutions. Special section on American calculating machines, by T. W. Simpson. 132 illustrations. 422pp. 5⅜ x 8.
S273 Paperbound **$2.00**

**NUMERICAL INTEGRATION OF DIFFERENTIAL EQUATIONS, A. A. Bennett, W. E. Milne, H. Bateman.** Republication of original monograph prepared for National Research Council. New methods of integration of differential equations developed by 3 leading mathematicians: THE INTERPOLATIONAL POLYNOMIAL and SUCCESSIVE APPROXIMATIONS by A. A. Bennett; STEP-BY-STEP METHODS OF INTEGRATION by W. W. Milne; METHODS FOR PARTIAL DIFFERENTIAL EQUATIONS by H. Bateman. Methods for partial differential equations, transition from difference equations to differential equations, solution of differential equations to non-integral values of a parameter will interest mathematicians and physicists. 288 footnotes, mostly bibliographic; 235-item classified bibliography. 108pp. 5⅜ x 8. S305 Paperbound **$1.35**

**INTRODUCTION TO RELAXATION METHODS, F. S. Shaw.** Fluid mechanics, design of electrical networks, forces in structural frameworks, stress distribution, buckling, etc. Solve linear simultaneous equations, linear ordinary differential equations, partial differential equations, Eigen-value problems by relaxation methods. Detailed examples throughout. Special tables for dealing with awkwardly-shaped boundaries. Indexes. 253 diagrams. 72 tables. 400pp. 5⅜ x 8.
S244 Paperbound **$2.45**

**NUMERICAL SOLUTIONS OF DIFFERENTIAL EQUATIONS, H. Levy & E. A. Baggott.** Comprehensive collection of methods for solving ordinary differential equations of first and higher order. All must pass 2 requirements: easy to grasp and practical, more rapid than school methods. Partial contents: graphical integration of differential equations, graphical methods for detailed solution. Numerical solution. Simultaneous equations and equations of 2nd and higher orders. "Should be in the hands of all in research in applied mathematics, teaching," NATURE. 21 figures. viii + 238pp. 5⅜ x 8. S168 Paperbound **$1.75**

# MATHEMATICS, HISTORIES AND CLASSICS

**HISTORY OF MATHEMATICS, D. E. Smith.** Most comprehensive non-technical history of math in English. Discusses lives and works of over a thousand major and minor figures, with footnotes supplying technical information outside the book's scheme, and indicating disputed matters. Vol I: A chronological examination, from primitive concepts through Egypt, Babylonia, Greece, the Orient, Rome, the Middle Ages, the Renaissance, and up to 1900. Vol 2: The development of ideas in specific fields and problems, up through elementary calculus. Two volumes, total of 510 illustrations, 1355pp. 5⅜ x 8. Set boxed in attractive container. T429, 430 Paperbound, the set **$5.00**

**A SHORT ACCOUNT OF THE HISTORY OF MATHEMATICS, W. W. R. Ball.** Most readable non-technical history of mathematics treats lives, discoveries of every important figure from Egyptian, Phoenician mathematicians to late 19th century. Discusses schools of Ionia, Pythagoras, Athens, Cyzicus, Alexandria, Byzantium, systems of numeration; primitive arithmetic; Middle Ages, Renaissance, including Arabs, Bacon, Regiomontanus, Tartaglia, Cardan, Stevinus, Galileo, Kepler; modern mathematics of Descartes, Pascal, Wallis, Huygens, Newton, Leibnitz, d'Alembert, Euler, Lambert, Laplace, Legendre, Gauss, Hermite, Weierstrass, scores more. Index. 25 figures. 546pp. 5⅜ x 8. S630 Paperbound **$2.25**

**A HISTORY OF GEOMETRICAL METHODS, J. L. Coolidge.** Full, authoritative history of the techniques which men have employed in dealing with geometric questions . . . from ancient times to the modern development of projective geometry. Critical analyses of the original works. Contents: Synthetic Geometry—the early beginnings, Greek mathematics, non-Euclidean geometries, projective and descriptive geometry; Algebraic Geometry—extension of the system of linear coordinates, other systems of point coordinates, enumerative and birational geometry, etc.; and Differential Geometry—intrinsic geometry and moving axes, Gauss and the classical theory of surfaces, and projective and absolute differential geometry. The work of scores of geometers analyzed: Pythagoras, Archimedes, Newton, Descartes, Leibniz, Lobachevski, Riemann, Hilbert, Bernoulli, Schubert, Grassman, Klein, Cauchy, and many, many others. Extensive (24-page) bibliography. Index. 13 figures. xviii + 451pp. 5⅜ x 8½. S1006 Paperbound **$2.25**

**THE MATHEMATICS OF GREAT AMATEURS, Julian Lowell Coolidge.** Enlightening, often surprising, accounts of what can result from a non-professional preoccupation with mathematics. Chapters on Plato, Omar Khayyam and his work with cubic equations, Piero della Francesca, Albrecht Dürer, as the true discoverer of descriptive geometry, Leonardo da Vinci and his varied mathematical interests, John Napier, Baron of Merchiston, inventor of logarithms, Pascal, Diderot, l'Hospital, and seven others known primarily for contributions in other fields. Bibliography. 56 figures. viii + 211pp. 5⅜ x 8½. S1009 Paperbound **$1.50**

**ART AND GEOMETRY, Wm. M. Ivins, Jr.** A controversial study which propounds the view that the ideas of Greek philosophy and culture served not to stimulate, but to stifle the development of Western thought. Through an examination of Greek art and geometrical inquiries and Renaissance experiments, this book offers a concise history of the evolution of mathematical perspective and projective geometry. Discusses the work of Alberti, Dürer, Pelerin, Nicholas of Cusa, Kepler, Desargues, etc. in a wholly readable text of interest to the art historian, philosopher, mathematician, historian of science, and others. x + 113pp. 5⅜ x 8⅜. T941 Paperbound **$1.00**

**A SOURCE BOOK IN MATHEMATICS, D. E. Smith.** Great discoveries in math, from Renaissance to end of 19th century, in English translation. Read announcements by Dedekind, Gauss, Delamain, Pascal, Fermat, Newton, Abel, Lobachevsky, Bolyai, Riemann, De Moivre, Legendre, Laplace, others of discoveries about imaginary numbers, number congruence, slide rule, equations, symbolism, cubic algebraic equations, non-Euclidean forms of geometry, calculus, function theory, quaternions, etc. Succinct selections from 125 different treatises, articles, most unavailable elsewhere in English. Each article preceded by biographical, historical introduction. Vol. I: Fields of Number, Algebra. Index. 32 illus. 338pp. 5⅜ x 8. Vol. II: Fields of Geometry, Probability, Calculus, Functions, Quaternions. 83 illus. 432pp. 5⅜ x 8.
Vol. 1: S552 Paperbound **$2.00**
Vol. 2: S553 Paperbound **$2.00**
2 vol. set, **$4.00**

**A COLLECTION OF MODERN MATHEMATICAL CLASSICS, edited by R. Bellman.** 13 classic papers, complete in their original languages, by Hermite, Hardy and Littlewood, Tchebychef, Fejér, Fredholm, Fuchs, Hurwitz, Weyl, van der Pol, Birkhoff, Kellogg, von Neumann, and Hilbert. Each of these papers, collected here for the first time, triggered a burst of mathematical activity, providing useful new generalizations or stimulating fresh investigations. Topics discussed include classical analysis, periodic and almost periodic functions, analysis and number theory, integral equations, theory of approximation, non-linear differential equations, and functional analysis. Brief introductions and bibliographies to each paper. xii + 292pp. 6 x 9. S730 Paperbound **$2.00**

**THE WORKS OF ARCHIMEDES,** edited by **T. L. Heath.** All the known works of the great Greek mathematician are contained in this one volume, including the recently discovered Method of Archimedes. Contains: On Sphere & Cylinder, Measurement of a Circle, Spirals, Conoids, Spheroids, etc. This is the definitive edition of the greatest mathematical intellect of the ancient world. 186-page study by Heath discusses Archimedes and the history of Greek mathematics. Bibliography. 563pp. 5⅜ x 8. S9 Paperbound **$2.45**